Advanced Unsaturated Soil Mechanics

Unsaturated soil is a three-phase material that is ubiquitous on the Earth's surface and exhibits complex behaviour, which becomes more complex in response to the Earth's changing climate and increasing engineering activities. This is because the former affects its moisture and temperature conditions significantly and the latter governs its stress state and suction condition. This book is designed to meet the increasing challenges of climate change and engineering activities by covering the mechanics and engineering of unsaturated soil in a logical manner. It comprises four major parts:

(1) water retention and flow characteristics;
(2) shear strength and stiffness at various temperatures;
(3) state-dependent elasto-plastic constitutive modelling; and
(4) field monitoring and engineering applications.

This second edition uniquely covers fundamental topics on unsaturated soil that are not covered in other similar books, including:

- the state-dependency of soil-water retention behaviour and water permeability functions, such as dependence on engineering activities;
- small strain stiffness considering the influence of wetting-drying cycles and recent suction history, such as that due to climate change;
- suction effects on dilatancy and peak shear strength;
- cyclic thermal effects on soil behaviour;
- state-dependent elastoplastic constitutive modelling of monotonic and cyclic behaviour; and
- engineering applications such as the South-to-North Water Transfer Project; an earthen landfill cover system devoid of geomembrane in the Xiaping landfill, Shenzhen; and a 15-m-deep multi-propped excavation in Tianjin, China.

Charles W.W. Ng is Chair Professor and the CLP Holdings Professor of Sustainability at The Hong Kong University of Science and Technology. He is also a Fellow of the Royal Academy of Engineering, an Editor-in-Chief of the Canadian Geotechnical Journal and the immediate past-President of the International Society for Soil Mechanics and Engineering (2017–2022).

Chao Zhou is Associate Professor at The Hong Kong Polytechnic University.

Junjun Ni is Professor at Southeast University in Nanjing, China.

Advanced Unsaturated Soil Mechanics

Theory and Applications

Second Edition

Charles W.W. Ng, Chao Zhou, and Junjun Ni

CRC Press is an imprint of the
Taylor & Francis Group, an **informa** business

Cover images:
1. Left-top: with courtesy from Geo-Experts
2. Left-middle: photo taken by the research team of the first author, Professor Charles Ng
3. Left-bottom: with courtesy from GDS Instruments
4. Right-top: taken by the first author, Professor Charles Ng
5. Right-middle: www.nxnews.net/zt/2018/liushi/dierqi/201806/t20180625_5981494.html
6. Right-bottom: with courtesy from Professor Yu Diao in Tianjin University, China

Second edition published 2025
by CRC Press
4 Park Square, Milton Park, Abingdon, Oxon, OX14 4RN

and by CRC Press
2385 NW Executive Center Drive, Suite 320, Boca Raton FL 33431

© 2025 Charles W.W. Ng, Chao Zhou, and Junjun Ni

First edition published by CRC Press 2007

CRC Press is an imprint of Informa UK Limited

The right of Charles W.W. Ng, Chao Zhou, and Junjun Ni to be identified as authors of this work has been asserted in accordance with sections 77 and 78 of the Copyright, Designs and Patents Act 1988.

All rights reserved. No part of this book may be reprinted or reproduced or utilised in any form or by any electronic, mechanical, or other means, now known or hereafter invented, including photocopying and recording, or in any information storage or retrieval system, without permission in writing from the publishers.

For permission to photocopy or use material electronically from this work, access www.copyright.com or contact the Copyright Clearance Center, Inc. (CCC), 222 Rosewood Drive, Danvers, MA 01923, 978-750-8400. For works that are not available on CCC please contact mpkbookspermissions@tandf.co.uk

Trademark notice: Product or corporate names may be trademarks or registered trademarks, and are used only for identification and explanation without intent to infringe.

British Library Cataloguing-in-Publication Data
A catalogue record for this book is available from the British Library

Library of Congress Cataloging-in-Publication Data
Names: Ng, C.W.W., author. | Zhou, Chao, 1987– author. | Ni, Junjun, author.
Title: Advanced unsaturated soil mechanics and engineering / Charles W.W. Ng, Chao Zhou and Junjun Ni.
Description: Second edition. | Abingdon, Oxon : CRC Press, 2025. |
Includes bibliographical references and index.
Identifiers: LCCN 2024007062 | ISBN 9781032298320 (hardback) | ISBN 9781032769585 (paperback) |
ISBN 9781003480587 (ebook)
Subjects: LCSH: Soil mechanics. | Swelling soils. | Zone of aeration. | Soil moisture.
Classification: LCC TA710 .N475 2025 | DDC 624.1/5136–dc23/eng/20240404
LC record available at https://lccn.loc.gov/2024007062

ISBN: 978-1-032-29832-0 (hbk)
ISBN: 978-1-032-76958-5 (pbk)
ISBN: 978-1-003-48058-7 (ebk)

DOI: 10.1201/9781003480587

Typeset in Sabon
by Newgen Publishing UK

Contents

Preface to the second edition	xiii
About the authors	xv
Acknowledgements	xvii
List of notations	xxv
List of nomenclature	xxxv

1 Introduction   1

Overview 1
    The vadose zone 2
    Climate changes and the vadose zone 3
    Some common unsaturated soils 5
    Definitions of suctions 6
    Stress state variables 7
    The stress state variable for saturated soil 8
    Possible sets of stress state variables for unsaturated soil 8
    Summary 13
Cavitation 13
    Overview 13
    Definitions 13
    Cavitation inception and nuclei 14
    Physics of pore water under tension in unsaturated soils 15
    Avoiding cavitation 16

2 Measurement and control of suction   19

Suction measuring devices 19
    Overview 19
    Psychrometer 20
    The filter paper method 22
    Porous block (e.g., gypsum block) 23
    Thermal conductivity sensor 24
    Suction plate and pressure plate 27
    Jet-fill tensiometer 30
    Mercury manometer 31
    Vacuum gauge 32

vi Contents

Pressure transducer 32
Small tip tensiometer 33
High suction tensiometer probe 33
Squeezing technique 34
Summary 35
Suction control and measurement methods 36
  Overview 36
  Axis-translation technique 36
  Osmotic technique 47
  Humidity control techniques 51
  Influence of suction equalisation 55
Comparisons of different suction control techniques 58
  Introduction 58
  Theoretical considerations of soil moisture transfer 59
  Water retention curves determined using different suction control
    techniques 61
  Summary 64

**3 Retention characteristics and permeability functions for water and
gas flows**   **65**
Introduction 65
Soil-water characteristic curve (SWCC) and water permeability function 68
  Introduction 68
  Descriptions of water content 69
  Terminology used to quantify a WRC 71
  Mathematical forms of SWCC or WRC 71
A new and simple model for SDSWCC or SDWRC 73
  Introduction 73
  Development of a new water retention model 75
  Summary 85
Laboratory measurements of drying and wetting stress-dependent permeability
  functions 86
  Theoretical consideration 86
  Design of the apparatus and instrumentation 88
  Experimental results 90
  Summary 94
A field study of SDSWCC and permeability function 95
  Introduction 95
  The test site and soil properties 95
  Instrumentation and testing procedures 96
  Observed field performance 98
  Summary 108
Gas breakthrough and emission through landfill final cover 108
  Introduction 108
  Theoretical considerations 109
  Test program 110
  Experimental apparatus and instrumentation 110

Contents vii

Testing soil and sample preparation 112
Test procedures 113
Experimental results 114
Summary 117
Numerical investigation of gas emission from three types of landfill covers
under dry weather conditions 118
Introduction 118
Working principle of the alternative three-layer capillary barrier cover 120
Numerical modelling 120
Computed results 125
Summary 129
Gas permeability of biochar-amended clay 130
Introduction 130
Materials and methods 130
Results and discussion 131
Summary 131
Influence of biopolymer on gas permeability 133
Introduction 133
Material and methods 134
Results and discussion 135
Summary 139
Effects of grass roots on gas permeability 140
Introduction 140
Materials and methods 140
Interpretation of test results 140
Summary 142
Effects of biofilm on gas permeability 143
Introduction 143
Materials and methods 143
Interpretation of test results 143
Summary 145

4 Shear strength and behaviour 146

Extended Mohr-Coulomb failure criterion 146
  Relationships between φ^b and χ 148
Comparisons of axis-translation and osmotic techniques for shear
  testing of unsaturated soils 150
Introduction 150
Testing equipment 151
Testing material and specimen preparation 153
Testing programme 155
Testing procedures 156
Test results from using the axis-translation technique 157
Test results from using the osmotic control technique 159
Comparison between test results from the axis-translation and osmotic
  control techniques 159
Summary 161

viii Contents

Effects of soil suction on the dilatancy of an unsaturated soil 162
 Introduction 162
 Soil type and test procedures 163
 Evolution of dilatancy during shear 166
 Stress-dilatancy relationship 168
 Maximum dilatancy 168
 Summary 170
Shear strength characteristics of compacted loess at high suctions 170
 Introduction 170
 Test apparatus and measuring devices 170
 Soil type and its water retention and compression behaviour 171
 Summary 183
Effects of microstructure on dilatancy 183
 Introduction 183
 Pore-size distributions of recompacted and natural loess specimens 185
 Significance of macrovoid ratio 188
 Summary 189
Comparisons of the shear strength of weathered lateritic, granitic and
 volcanic soils: compressibility and shear strength 190
 Introduction 190
 Soil types and grain size distribution 191
 Mineralogy, chemical and microstructure analysis 191
 Stress-strain relationship during shearing 196
 Summary 198

5 **Shear stiffness and damping ratio** 200
 Overview 200
 Anisotropic shear stiffness at very small strains 200
 Introduction 200
 Theoretical considerations 201
 Testing apparatus and measuring devices 204
 Soil types and test specimen preparation 206
 Test programme and procedures 207
 Interpretations of experimental results 209
 Summary 213
 Effects of current suction ratio and recent suction history on shear stiffness 215
 Introduction 215
 Theoretical considerations 216
 Experimental study 217
 Interpretations of test results 219
 Summary 227
 Effects of wetting-drying and stress ratio on anisotropic stiffness 230
 Introduction 230
 Interpretations of experimental results 231
 Summary 236
 Resilient modulus under cyclic shearing 237
 Introduction 237

Contents   ix

Theoretical considerations 237
Experimental study 238
Interpretations of experimental results 241
Verification of the proposed equation 246
Summary 248
Shear modulus and damping ratio at high suction of 40 MPa of specimens
  with different initial microstructures 249
Introduction 249
Interpretations of experimental results 250
Summary 261

6 **Thermal effects on soil behaviour and properties**   262

Cyclic shear behaviour of silt at various suctions and temperatures 262
Introduction 262
Test apparatus 262
Test program and procedures 263
Interpretation of experimental results 265
Summary 270
Effects of soil microstructure on the shear behaviour of loess at different
  suctions and temperatures 272
Introduction 272
Test apparatus and program 272
Test procedure 273
Interpretations of experimental results 274
Summary 286
Effects of temperature and suction on secant shear modulus 287
Introduction 287
Interpretations of experimental results 288
Summary 291
Thermal effects on yielding and wetting-induced collapse of recompacted
  and intact loess 291
Introduction 291
Test programme 292
Test apparatus 292
Test procedures 293
Interpretations of experimental results 295
Summary 307
Volume changes of an unsaturated clay during heating and cooling 308
Introduction 308
Test programme and test apparatus 308
Test procedures 308
Interpretation of experimental results 309
Summary 313

7 **Constitutive modelling of state-dependent behaviour of soils**   316

Elastoplastic modelling of state-dependent saturated and unsaturated soil
  behaviour 316

x Contents

Introduction 316
A unified and simple framework for simulating the state-dependent
    behaviour of soils 316
Discussion on the simple and unified framework 343
Summary 344
Simulating the cyclic behaviour of soils at different temperatures using
    bounding surface plasticity approach 345
Introduction 345
Mathematical formulations 345
Experimental program to verify the SOA model 360
Determination of soil parameters 361
Comparisons between measured and computed results 367
Summary 370

8 Field monitoring and engineering applications   376

The South-to-North Water Diversion Project, China 376
    The test site 377
    Soil profile and properties 379
    Field instrumentation programme 381
    Artificial rainfall simulations 385
    Observed field performance 386
    Summary 398
Field monitoring of an unsaturated saprolitic hillslope on Lantau Island
    in Hong Kong 401
    Introduction 401
    Description of the study area 401
    Features of active landslide mass 401
    Geological–hydrogeological model 403
    Soil properties 404
    Instrumentation programme and monitoring results 404
    Rainfall characteristics 406
    Response of PWP and VWC 406
    Piezometric-level variation 410
    Subsurface total horizontal stress 411
    Subsurface horizontal deformation characteristic 414
    Investigation of hillslope behaviour 417
    Infiltration characteristic and groundwater condition 417
    Deformation characteristics and failure mode 418
    Summary 419
Field performance of non-vegetated and vegetated three-layer landfill cover
    systems using construction waste without geomembrane 420
    Introduction 420
    Theoretical considerations of newly proposed landfill cover 422
    Fundamental principle of reducing water infiltration and percolation 423
    Long-term field test of the new vegetated cover system using recycled
        concretes without geomembrane 425
    Numerical analysis 433

Field monitoring results 435
Summary 450
Use of unsaturated small-strain soil stiffness in the design of wall deflection
  and ground movement adjacent to deep excavations 451
  Introduction 451
  The excavation site 452
  Theoretical basis 453
  Design analysis 455
  Comparison of analyses with and without considering suction effects
    on soil stiffness 457
  Summary 460

References	461
Author index	493
Subject index	499

Preface to the second edition

Since the publication of the first edition of *Advanced Unsaturated Soil Mechanics and Engineering* by CRC Press in 2007, the unsaturated soil research group at The Hong Kong University of Science and Technology has continued to investigate major scientific and engineering challenges associated with the coupling effects of suction state and stress state on unsaturated soil. These investigations have been carried out to meet the increasing demand for solutions to engineering problems under a changing climate. These challenges have included determining the effects of stress state on water retention characteristics and seepage behaviour, characterising the mechanism whereby suction paths influence shear stiffness at small strains under various stress states, and understanding the coupling effects of suction and stress states on shearing-induced contraction or dilation and strength. This second edition reports the result of these investigations and describes cyclic thermal effects on the behaviour of unsaturated soil and elastoplastic modelling of state-dependent behaviour under monotonic and cyclic loads. Furthermore, this new edition reports new engineering case histories, namely two successful applications of unsaturated soil mechanics. The first was the design and execution of all-weather earthen three-layer landfill cover systems without the use of geomembrane in Shenzhen, China, and the second was a 15-m-deep excavation in soils that had been unsaturated due to de-watering in Tianjin, China.

We are indebted to many postgraduate students and post-doctoral fellows, especially Mr Chang Guo, Mr Qi Zhang and Ms Qianyu Zhou, and Dr Jinquan Liu, Dr Pui San So and Dr Yuchen Wang, who have helped us to proofread the text of this book and to prepare the acknowledgements, figures, tables and references.

Charles W.W. Ng, Chao Zhou and Junjun Ni

About the authors

Charles W.W. Ng is a Chair Professor and the CLP Holdings Professor of Sustainability at The Hong Kong University of Science and Technology. He is also a Fellow of the Royal Academy of Engineering, an Editor-in-Chief of the Canadian Geotechnical Journal and the immediate past-President of the International Society for Soil Mechanics and Engineering (2017–2022). He has published over 400 SCI journal articles, many of which are related to unsaturated soil mechanics and engineering. He has received many international awards, such as the 2022 Vernes Medal from the International Consortium on Landslides, the 2022 Fredlund Award from the Canadian Geotechnical Society, the 2017 Telford Premium Prize from the Institution of Civil Engineers, and the 2016 R. M. Quigley Award from the Canadian Geotechnical Society. In addition, he has received awards in mainland China, namely the prestigious 2022 Ho Leung Ho Lee Foundation Prize for Scientific and Technological Progress, the 2020 National Natural Science Award and the 2015 Scientific Technological Advancement Award from the State Council, China.

Drs Chao Zhou and Junjun Ni are former PhD students of Professor Ng and have conducted extensive research on unsaturated soil mechanics and engineering during and after their PhD studies. Dr Zhou is currently an Associate Professor at The Hong Kong Polytechnic University, and Dr Ni is a Professor at Southeast University in Nanjing, China.

Acknowledgements

We are grateful to Mr Qi Zhang and Miss Qianyu Zhou (postgraduate research students), Dr Pui San So (research assistant professor) and Dr Yuchen Wang (postdoctoral fellow) at the Hong Kong University of Science and Technology for their significant efforts in proofreading this book to enhance its accuracy and quality. We also extend our thanks to Mr Chang Guo (postgraduate research student) and Dr Jinquan Liu (postdoctoral fellow) at The Hong Kong Poplytechnic University for their assistance in preparing the figures and tables.

We are also grateful for the following research grants, which financially supported some of the research work described in this book:

- Area of Excellence project (AoE/E-603/18) provided by the Research Grants Council of Hong Kong Special Administrative Region (HKSAR), China
- Research grant ECWW19EG01 awarded by the Environment and Conservation Fund of HKSAR, China
- Research grants U20A20320, 51778166 and 52022004 awarded by the National Natural Science Foundation of China
- Research grant 2012CB719805 from the National Basic Research Program (973 Program) administered by the Ministry of Science and Technology of China
- Research grant HKUST6/CRF/12R awarded by the Research Grants Council of HKSAR

Furthermore, we acknowledge permission from publishers to include verbatim extracts from the following published works. We declare that we do not intend to claim any credit from the authors of these extracts and any omitted acknowledgements are unintentional.

0.1 A.A. BALKEMA PUBLISHERS

- Fredlund, D. G. (1987). The stress state for expansive soils. *Proceedings of the 6th International Conference on Expansive Soils*, 1–9.
- Fredlund, D. G. (1996). The scope of unsaturated soil mechanics: An overview. *Proceedings of the 1st International Conference on Unsaturated Soils*, 140–156.
- Fredlund, D. G. (2002). Use of soil-water characteristic curve in the implementation of unsaturated soil mechanics. *Proceedings of the 3rd International Conference on Unsaturated Soils*, 57–80.

xviii Acknowledgements

- Fredlund, D. G., Fredlund, M. D. and Zakerzadeh, N. (2001). Predicting the permeability function for unsaturated soils. *Proceedings of International Symposium on Suction, Swelling, Permeability and Structured Clays*, 215–221.
- Fredlund, D. G., Rahardjo, H., Leong, E. C. and Ng, C. W. W. (2001). Suggestions and recommendations for the interpretation of soil-water characteristic curves. *Proceedings of the 14th Southeast Asian Geotechnical Conference*, 503–508.
- Marinho, F. A. M. and Chandler, R. J. (1995). Cavitation and the direct measurement of soil suction. *Proceedings of the 1st International Conference on Unsaturated Soils*, 623–630.
- Marinho, F. A. M. and de Sousa Pinto, C. (1997). Soil suction measurement using a tensiometer. *Symposium on Recent Developments in Soil and Pavement Mechanics*, 249–254.
- Ridley, A. M. and Wray, W. K. (1996). Suction measurement: A review of current theory and practices. *Proceedings of the 1st International Conference on Unsaturated Soils*, 1293–1322.
- Wheeler, S. J. and Karube, D. (1995). State-of-the-art report – constitutive modelling. *Proceedings of the 1st International Conference on Unsaturated Soils*, 1323–1356.

0.2 ACADEMIC PRESS, HARCOURT BRACE JOVANOVICH PUBLISHERS

- Hillel, D. (1982). *Introduction to Soil Physics*. Academic Press, London, UK.
- Hillel, D. (1998). *Introduction to Environmental Soil Physics*. Academic Press, San Diego, USA.

0.3 AMERICAN SOCIETY FOR CIVIL ENGINEERS

- Dong, Y., Lu, N. and McCartney, J. S. (2016). Unified model for small-strain shear modulus of variably saturated soil. *Journal of Geotechnical and Geoenvironmental Engineering*, 142(9), 04016039.
- Leong, E. C. and Rahardjo, H. (1997). Review of soil-water characteristic curve equations. *Journal of Geotechnical and Geoenvironmental Engineering*, 123(12), 1106–1117.
- Ng, C. W. W. and Chiu, A. C. F. (2001). Behavior of a loosely compacted unsaturated volcanic soil. *Journal of Geotechnical and Geoenvironmental Engineering*, 127(12), 1027–1036.
- Ng, C. W. W. and Chiu, A. C. F. (2003). Laboratory study of loose saturated and unsaturated decomposed granitic soil. *Journal of Geotechnical and Geoenvironmental Engineering*, 129(6), 550–559.
- Ng, C. W. W., Coo, J. L., Chen, Z. K. and Chen, R. (2016). Water infiltration into a new three-layer landfill cover system. *Journal of Environmental Engineering*, 142(5), 04016007.
- Ng, C. W. W. and Leung, A. K. (2012). Measurements of drying and wetting permeability functions using a new stress-controllable soil column. *Journal of Geotechnical and Geoenvironmental Engineering*, 138(1), 58–68.
- Ng, C. W. W. and Pang, Y. W. (2000). Influence of stress state on soil-water characteristics and slope stability. *Journal of Geotechnical and Geoenvironmental Engineering*, 126(2), 157–166.

Acknowledgements xix

- Sawangsuriya, A., Edil, T. B. and Bosscher, P. J. (2009). Modulus-suction-moisture relationship for compacted soils in postcompaction state. *Journal of Geotechnical and Geoenvironmental Engineering*, 135(10), 1390–1403.

0.4 AMERICAN SOCIETY FOR TESTING AND MATERIALS

- Likos, W. J. and Lu, N. (2003). Automated humidity system for measuring total suction characteristics of clay. *Geotechnical Testing Journal*, 26(2), 179–190.
- Ng, C. W. W., Lai, C. H. and Chiu, C. F. (2012). A modified triaxial apparatus for measuring the stress path-dependent water retention curve. *Geotechnical Testing Journal*, 35(3), 490–495.

0.5 CAMBRIDGE UNIVERSITY PRESS

- Fisher, R. A. (1926). On the capillary forces in an ideal soil; correction of formulae given by WB Haines. *The Journal of Agricultural Science*, 16(3), 492–505.

0.6 CHINESE SOCIETY OF CIVIL ENGINEERING

- Ng, C. W. W. and Chen, R. (2006). Advanced suction control techniques for testing unsaturated soils. *Yantu Gongcheng Xuebao/Chinese Journal of Geotechnical Engineering*, 28(2), 123–128.

0.7 EARTH PRODUCTS CHINA LIMITED

- www.epccn.com/pro_view-4874.html

0.8 ELSEVIER

- Cheng, Q., Zhou, C., Ng, C. W. W. and Tang, C. S. (2020). Effects of soil structure on thermal softening of yield stress. *Engineering Geology*, 269, 105544.
- Delage, P., Howat, M. D. and Cui, Y. J. (1998). The relationship between suction and swelling properties in a heavily compacted unsaturated clay. *Engineering Geology*, 50(1–2), 31–48.
- Ng, C. W. W., Akinniyi, D. B., Zhou, C. and Chiu, C. F. (2019). Comparisons of weathered lateritic, granitic and volcanic soils: Compressibility and shear strength. *Engineering Geology*, 249, 235–240.
- Ng, C. W. W., Chen, Z. K., Coo, J. L., Chen, R. and Zhou, C. (2015). Gas breakthrough and emission through unsaturated compacted clay in landfill final cover. *Waste Management*, 44, 155–163.
- Ng, C. W. W. and Shi, Q. (1998). A numerical investigation of the stability of unsaturated soil slopes subjected to transient seepage. *Computers and Geotechnics*, 22(1), 1–28.
- Ng, C. W. W., So, P. S., Lau, S. Y., Zhou, C., Coo, J. L. and Ni, J. J. (2020). Influence of biopolymer on gas permeability in compacted clay at different densities and water contents. *Engineering Geology*, 272, 105631.

- Ng, C. W. W., Zheng, G., Ni, J. and Zhou, C. (2020). Use of unsaturated small-strain soil stiffness to the design of wall deflection and ground movement adjacent to deep excavation. *Computers and Geotechnics*, 119, 103375.
- Ni, J. J. and Ng, C. W. W. (2019). Long-term effects of grass roots on gas permeability in unsaturated simulated landfill covers. *Science of the Total Environment*, 666, 680–684.
- Zhou, C. and Ng, C. W. W. (2014). A new and simple stress-dependent water retention model for unsaturated soil. *Computers and Geotechnics*, 62, 216–222.

0.9 JAPANESE GEOTECHNICAL SOCIETY

- Ng, C. W. W., Cui, Y., Chen, R. and Delage, P. (2007). The axis-translation and osmotic techniques in shear testing of unsaturated soils: A comparison. *Soils and Foundations*, 47(4), 675–684.

0.10 JOHN WILEY & SONS, INC.

- Brown, R. W. and Collins, J. M. (1980). A screen-caged thermocouple psychrometer and calibration chamber for measurements of plant and soil water potential. *Agronomy Journal*, 72(5), 851–854.
- Childs, E. (1969). *An Introduction to the Physical Basis of Soil Water Phenomena.* John Wiley, London and New York.
- Fredlund, D. G. and Rahardjo, H. (1993). *Soil Mechanics for Unsaturated Soils.* John Wiley & Sons, Inc., New Jersey, USA.
- Ng, C. W. W., Liu, J. and Chen, R. (2015). Numerical investigation on gas emission from three landfill soil covers under dry weather conditions. *Vadose Zone Journal*, 14(9), vzj2014.12.0180.
- Ng, C. W. W., Zhou, C. and Leung, A. K. (2015). Comparisons of different suction control techniques by water retention curves: Theoretical and experimental studies. *Vadose Zone Journal*, 14(9), vzj2015-01.
- Zhou, C., Ng, C. W. W. and Chen, R. (2015). A bounding surface plasticity model for unsaturated soil at small strains. *International Journal for Numerical and Analytical Methods in Geomechanics*, 39(11), 1141–1164.

0.11 LIPPINCOTT WILLIAMS AND WILKINS

- Zur, B. (1966). Osmotic control of the matric soil water potential: I. Soil-water system. *Soil Science*, 102, 394–398.

0.12 MCGRAW-HILL BOOK COMPANY

- Young, F. R. (1989). *Cavitation.* McGraw-Hill Book Company, London, UK.

0.13 MILL PRESS SCIENCE PUBLISHERS/IOS PRESS

- Lee, S. R., Kim, Y. K. and Lee, S. J. (2005). A method to estimate soil-water characteristic curve for weathered granite soil. *Proceedings of the 16th International Conference on Soil Mechanics and Geotechnical Engineering*, 543–546.

- Ng, C. W. W. and Zhou, R. (2005). Effects of soil suction on dilatancy of an unsaturated soil. *Proceedings of the 16th International Conference on Soil Mechanics and Geotechnical Engineering*, 559–562.

0.14 NATIONAL RESEARCH COUNCIL OF CANADA (CANADIAN SCIENCE PUBLISHING)

- Al-Mukhtar, M., Qi, Y., Alcover, J. F. and Bergaya, F. (1999). Oedometric and water-retention behavior of highly compacted unsaturated smectites. *Canadian Geotechnical Journal*, 36(4), 675–684.
- Chiu, C. F. and Ng, C. W. W. (2012). Coupled water retention and shrinkage properties of a compacted silt under isotropic and deviatoric stress paths. *Canadian Geotechnical Journal*, 49(8), 928–938.
- Delage, P. and Lefebvre, G. (1984). Study of the structure of a sensitive Champlain clay and of its evolution during consolidation. *Canadian Geotechnical Journal*, 21(1), 21–35.
- Feng, M. and Fredlund, D. G. (2003). Calibration of thermal conductivity sensors with consideration of hysteresis. *Canadian Geotechnical Journal*, 40(5), 1048–1055.
- Fredlund, D. G. (2000). The 1999 R.M. Hardy Lecture: The implementation of unsaturated soil mechanics into geotechnical engineering. *Canadian Geotechnical Journal*, 37(5), 963–986.
- Fredlund, D. G., Xing, A. and Huang, S. (1994). Predicting the permeability function for unsaturated soils using the soil-water characteristic curve. *Canadian Geotechnical Journal*, 31(4), 533–546.
- Gan, J. K. M., Fredlund, D. G. and Rahardjo, H. (1988). Determination of the shear strength parameters of an unsaturated soil using the direct shear test. *Canadian Geotechnical Journal*, 25(3), 500–510.
- Leung, A. K., Sun, H. W., Millis, S. W., Pappin, J. W., Ng, C. W. W. and Wong, H. N. (2011). Field monitoring of an unsaturated saprolitic hillslope. *Canadian Geotechnical Journal*, 48(3), 339–353.
- Ng, C. W. W., Baghbanrezvan, S., Sadeghi, H., Zhou, C. and Jafarzadeh, F. (2017). Effect of specimen preparation techniques on dynamic properties of unsaturated fine-grained soil at high suctions. *Canadian Geotechnical Journal*, 54(9), 1310–1319.
- Ng, C. W. W., Chen, R., Coo, J. L., Liu, J., Ni, J. J., Chen, Y. M., Zhan, L. T., Guo, H. W. and Lu, B. W. (2019). A novel vegetated three-layer landfill cover system using recycled construction wastes without geomembrane. *Canadian Geotechnical Journal*, 56(12), 1863–1875.
- Ng, C. W. W., Cheng, Q. and Zhou, C. (2018). Thermal effects on yielding and wetting-induced collapse of recompacted and intact loess. *Canadian Geotechnical Journal*, 55(8), 1095–1103.
- Ng, C. W. W., Leung, E. H. and Lau, C. K. (2004). Inherent anisotropic stiffness of weathered geomaterial and its influence on ground deformations around deep excavations. *Canadian Geotechnical Journal*, 41(1), 12–24.
- Ng, C. W. W., Mu, Q. Y. and Zhou, C. (2016). Effects of soil structure on the shear behaviour of an unsaturated loess at different suctions and temperatures. *Canadian Geotechnical Journal*, 54(2), 270–279.

- Ng, C. W. W. and Pang, Y. W. (2000). Experimental investigations of the soil-water characteristics of a volcanic soil. *Canadian Geotechnical Journal*, 37(6), 1252–1264.
- Ng, C. W. W., Sadeghi, H., Hossen, S. B., Chiu, C. F., Alonso, E. E. and Baghbanrezvan, S. (2016). Water retention and volumetric characteristics of intact and re-compacted loess. *Canadian Geotechnical Journal*, 53(8), 1258–1269
- Ng, C. W. W., Sadeghi, H. and Jafarzadeh, F. (2017). Compression and shear strength characteristics of compacted loess at high suctions. *Canadian Geotechnical Journal*, 54(5), 690–699.
- Ng, C. W. W., Sadeghi, H., Jafarzadeh, F., Sadeghi, M., Zhou, C., and Baghbanrezvan, S. (2020). Effect of microstructure on shear strength and dilatancy of unsaturated loess at high suctions. *Canadian Geotechnical Journal*, 57(2), 221–235.
- Ng, C. W. W. and Xu, J. (2012). Effects of current suction ratio and recent suction history on small-strain behaviour of an unsaturated soil. *Canadian Geotechnical Journal*, 49(2), 226–243.
- Ng, C. W. W., Xu, J. and Yung, S. Y. (2009). Effects of wetting-drying and stress ratio on anisotropic stiffness of an unsaturated soil at very small strains. *Canadian Geotechnical Journal*, 46(9), 1062–1076.
- Ng, C. W. W., Zhan, L. T. and Cui, Y. J. (2002). A new simple system for measuring volume changes in unsaturated soils. *Canadian Geotechnical Journal*, 39(3), 757–764.
- Ng, C. W. W., Zhou, C., Yuan, Q. and Xu, J. (2013). Resilient modulus of unsaturated subgrade soil: Experimental and theoretical investigations. *Canadian Geotechnical Journal*, 50(2), 223–232.

0.15 SPRINGER VERLAG

- Ng, C. W. W., Zhou, C. and Chiu, C. F. (2020). Constitutive modelling of state-dependent behaviour of unsaturated soils: An overview. *Acta Geotechnica*, 15, 2705–2725.
- Wong, J. T. F., Chen, Z., Ng, C. W. W. and Wong, M. H. (2016). Gas permeability of biochar-amended clay: Potential alternative landfill final cover material. *Environmental Science and Pollution Research*, 23(8), 7126–7131.

0.16 TAYLOR & FRANCIS BOOKS UK

- Ng, C. W. W. and Menzies, B. (2007). *Advanced Unsaturated Soil Mechanics and Engineering*. Taylor and Francis, London and New York.

0.17 THE HONG KONG UNIVERSITY OF SCIENCE AND TECHNOLOGY

- Chiu, C. F. (2001). *Behaviour of unsaturated loosely compacted weathered materials*. PhD Thesis, The Hong Kong University of Science and Technology, Hong Kong, China.
- Wang, B. (2000). *Stress effects on soil-water characteristics of unsaturated expansive soils*. MPhil Thesis, The Hong Kong University of Science and Technology, Hong Kong, China.

Acknowledgements   xxiii

- Zhan, L. (2003). *Field and laboratory study of an unsaturated expansive soil associated with rain-induced slope instability*. PhD Thesis, The Hong Kong University of Science and Technology, Hong Kong, China.

0.18 THOMAS TELFORD LTD (ICE PUBLISHING)

- Cui, Y. J. and Delage, P. (1996). Yielding and plastic behaviour of an unsaturated compacted silt. *Géotechnique*, 46(2), 291–311.
- Han, Z. and Vanapalli, S. K. (2016). Stiffness and shear strength of unsaturated soils in relation to soil-water characteristic curve. *Géotechnique*, 66(8), 627–647.
- Mair, R. J. (1993). Unwin Memorial Lecture 1992: Developments in geotechnical engineering research: Application to tunnels and deep excavations. *Proceedings of the Institution of Civil Engineers – Civil Engineering*, 97(1), 27–41.
- Ng, C. W. W., Cheng, Q., Zhou, C. and Alonso, E. E. (2016). Volume changes of an unsaturated clay during heating and cooling. *Géotechnique Letters*, 6(3), 192–198.
- Ng, C. W. W., Guo, H., Ni, J., Chen, R., Xue, Q., Zhang, Y., Feng, Y., Chen, Z., Feng, S. and Zhang, Q. (2022). Long-term field performance of non-vegetated and vegetated three-layer landfill cover systems using construction waste without geomembrane. *Géotechnique*, 74(2), 155–173.
- Ng, C. W. W., So, P. S., Coo, J. L., Zhou, C. and Lau, S. Y. (2019). Effects of biofilm on gas permeability of unsaturated sand. *Géotechnique*, 69(10), 917–923.
- Ng, C. W. W., Wong, H. N., Tse, Y. M., Pappin, J. W., Sun, H. W., Millis, S. W. and Leung, A. K. (2011). A field study of stress-dependent soil-water characteristic curves and permeability of a saprolitic slope in Hong Kong. *Géotechnique*, 61(6), 511–521.
- Ng, C. W. W. and Yung, S. Y. (2008). Determination of the anisotropic shear stiffness of an unsaturated decomposed soil. *Géotechnique*, 58(1), 23–35.
- Ng, C. W. W., Zhan, L. T., Bao, C. G., Fredlund, D. G. and Gong, B. W. (2003). Performance of an unsaturated expansive soil slope subjected to artificial rainfall infiltration. *Géotechnique*, 53(2), 143–157.
- Ng, C. W. W. and Zhou, C. (2014). Cyclic behaviour of an unsaturated silt at various suctions and temperatures. *Géotechnique*, 64(9), 709–720.
- Ridley, A. M. and Burland, J. B. (1995). A pore pressure probe for the in situ measurement of soil suction. *Proceedings of Conference on Advances in Site Investigation Practice*, 510–520.
- Wheeler, S. J., Sharma, R. S. and Buisson, M. S. R. (2003). Coupling of hydraulic hysteresis and stress-strain behaviour in unsaturated soils. *Géotechnique*, 53(1), 41–54.
- Wheeler, S. J. and Sivakumar, V. (1995). An elasto-plastic critical state framework for unsaturated soil. *Géotechnique*, 45(1), 35–53.
- Zhou, C. and Ng, C. W. W. (2016). Simulating the cyclic behaviour of unsaturated soil at various temperatures using a bounding surface model. *Géotechnique*, 66(4), 344–350.
- Zhou, C., Xu, J. and Ng, C. W. W. (2015). Effects of temperature and suction on secant shear modulus of unsaturated soil. *Géotechnique Letters*, 5(3), 123–128.

0.19 UNIVERSITY OF SASKATCHEWAN

- Shuai, F., Yazdani, J., Feng, M. and Fredlund, D. G. (1998). Supplemental report on the thermal conductivity matric suction sensor development (Year II). University of Saskatchewan, Saskatoon, Canada.

0.20 US PATENTS

- Ng, C. W. W. (2010). High suction double-cell extractor. US Patent No. US 7,793,552 B2; Granted on 14 September 2010.
- Ng, C. W. W. (2014). Humidity and osmotic suction-controlled box. US Patent No. US 8,800,353 B2; Granted on 12 August 2014.

0.21 ZHEJIANG UNIVERSITY PRESS

- Ng, C. W. W. and Chen, R. (2005). Keynote lecture: Advanced suction control techniques for testing unsaturated soils (in Chinese). *Proceddings of the 2nd National Conference on Unsaturated Soils*, 144–167.

Notations

β	a statistical factor of the same type of contact area in Jennings (1960)
β'	holding or bonding factor in Croney, Coleman and Black's (1958) effective stress equation
M_{b}	peak stress ratio attainable at the current state
M_{d}	dilation stress ratio
$(u_{\mathrm{a}} - u_{\mathrm{w}})_{\mathrm{f}}$	matric suction at failure
$(\gamma_{\mathrm{d}})_{\mathrm{max}}$	maximum dry unit weight
$(\rho_{\mathrm{d}})_{\mathrm{max}}$	maximum dry density
$(\sigma_1 - u_{\mathrm{a}})_{\mathrm{f}}$	net major stress at failure
$(\sigma_1 - \sigma_3)_{\mathrm{max}}$	maximum deviator stress
$(\sigma_3 - u_{\mathrm{a}})_{\mathrm{f}}$	net minor stress at failure
$(\sigma_{eqv})_{max}$	maximum equivalent effective stress
$(\sigma_{\mathrm{f}} - u_{\mathrm{a}})_{\mathrm{f}}$	net vertical normal stress at failure
$[C^*]$	compliance matrix
$[T_{\mathrm{a}}]$ and $[T_{\mathrm{b}}]$	matrices with values depending on the constitutive stress variables
$\{d\varepsilon\}$	strain increment
$\{d\sigma^*\}$	stress increment
$\{\varepsilon\}$	constitutive strain variables
$\{\sigma^*\}$	constitutive stress variables
A_{c}	attractive pore fluid stress due to chemistry in Lambe's (1960) effective stress equation
a	a parameter in Van Genuchten's (1980) equation
a_{tc}	thermal conductivity sensor reading at zero suction on the main hysteresis loop (Equation 2.1)
A	cross-sectional area
ae	empirical void ratio exponent (Equation 5.3)
a_{a}	fraction of the total area that is air-air contact in Lambe (1960) effective stress equation
a_{m}	mineral particle contact area in Lambe (1960) effective stress equation
a_{w}	water phase contact area in Lambe (1960) effective stress equation
b_{tc}	experimental parameters for thermal conductivity sensor (Equation 2.1)
b	soil parameter describing the effects of pore structure (Equation 3.21)
bs	empirical exponent reflecting the influence of matric suction on shear wave velocity (Equation 5.7)

xxvi Notations

c_{tc}	thermal conductivity sensor reading when the ceramic tip is in a dry condition (Equation 2.1)
c	total cohesion intercept (Equation 4.3)
c'	effective cohesion (Equation 4.1)
$C(\Psi)$	the correction function that causes the SWCC to pass through a suction of 1,000,000 kPa at zero water content
C_0 and C_s	soil parameters for describing shear modulus of unsaturated soils (Equation 7.68)
C_{ij}	material constant related to the pore and particle structure with a dimension of velocity related to the i-j plane
c_m	microstructural state variable at zero net mean stress (Equation 3.21)
d_{tc}	experimental parameters for thermal conductivity sensor (Equation 2.1)
d_{pc}	depth of penetration of the porous cap in Figure 2.10 (Equation 2.4)
d	dilatancy (Equation 4.9), equal to $d\varepsilon^p_v / d\varepsilon^p_s$
D_{10}	grain diameter at 10% passing
D_{30}	grain diameter at 30% passing
D_{50}	grain diameter at 50% passing
D_{60}	grain diameter at 60% passing
D_a	degree of aggregation
de	change in void ratio
d_m	dominant pore size of micropores
d_M	dominant pore size of macropores
d_{max}	maximum dilation angle
dp	change in mean net stress
dP^0_c	incremental yield stress at zero suction
dq	increment in deviatoric stress
D_q	dilatancy associated with plastic mechanism of shearing
$D_{q(c)}, D_{w(c)}$	dilatancy factors due to compression
ds	change in suction
D_s	dilatancy during drying/wetting
dS_r	increment in degree of saturation
$dS^p_{r(c)}$	plastic incremental degree of saturation due to compression
$dS^p_{r(h)}$	plastic incremental degree of saturation due to suction change
$dS^p_{r(s)}$	plastic incremental degree of saturation due to shearing
dS^e_r	elastic incremental degree of saturation
dt	time interval for measurements
$D_{v(h)}, D_{q(s)}$	dilatancy factors due to suction change
$D_{v(s)}, D_{w(s)}$	dilatancy factors due to shearing
$d\varepsilon^p_{q(c)}$	plastic shear strain due to compression
$d\varepsilon^p_{q(h)}$	plastic shear strain due to suction change
$d\varepsilon^p_{q(s)}$	plastic shear strain due to shearing
$d\varepsilon^p_s$	plastic shear strain
$d\varepsilon^p_v$	plastic volumetric strain
$d\varepsilon^p_{v(c)}$	plastic volumetric strain due to compression
$d\varepsilon^p_{v(h)}$	plastic volumetric strain due to suction change
$d\varepsilon^p_{v(s)}$	plastic volumetric strain due to shearing

$d\varepsilon_q$	deviatoric strain increment
$d\varepsilon_q^e$	elastic shear strain
$d\varepsilon_v$	volumetric strain increment
$d\varepsilon_v^e$	elastic volumetric strain
$d\Psi_w$	variation in Ψ_w
e	void ratio
E	Young's modulus
e_*	void ratio of the corresponding reconstituted soil at the same stress state
e_0	initial void ratio
e_{100}	void ratio at suction of 100 kPa
e^a	exponential void ratio function
e_{cr}	current void ratio and void ratio at the critical state
e_i	void ratio at given suction value during wetting process
e_m	microstructural void ratio
e_M	macrostructural void ratio
e_{MIP}	void ratio estimated from MIP
e_{nd}	non-detected pores
f	free energy per unit mass
$f(e)$	void ratio function relating the dependency of shear wave velocity to void ratio
f_0	free energy per unit mass at reference state
F_h	third bounding surface
F_n	normal force induced by meniscus water between two spherical particles
F_N	stabilizing interparticle normal force
F_s	second bounding surface
g	gravitational acceleration
G	shear modulus
$G(\eta)$	parameter related to the root distribution $\eta(z)$
G_0	stress-dependent small-strain soil stiffness (Equation 8.4)
$G_{0(hh)}$	elastic shear modulus of a wave propagating horizontally with a horizontal polarization
$G_{0(hv)}$	elastic shear modulus of a wave propagating horizontally with a vertical polarization
$G_{0(ij)}$	elastic shear modulus at a very small strain along the polarization plane i-j
G_0^{ref}	reference shear stiffness at given stress level
G_{max}	maximum shear modulus
G_s	specific gravity
G_{sec}	secant shear modulus
h_{ms}	height from the ground surface to the mercury reservoir bulb in Figure 2.10 (Equation 2.4)
h	total hydraulic head
h_m	a component of hydraulic head related to matric suction
h_s	a component of hydraulic head related to solute/osmotic suction
h_{wp} and h_{fc}	total hydraulic heads that correspond to the permanent wilting point and field capacity, respectively (Feddes et al., 2001)
I	unit matrix

xxviii Notations

I_{ij}	state-dependent variables for a given soil
$i\text{-}j$	principal stress directions (polarization plane)
$i_{zB,\text{tave}}$	hydraulic gradient at any depth zB for any average elapsed time
J_{pq}	moduli coupling shearing effects
J_{qp}	moduli coupling volumetric effects
k	water permeability
k_{in}	intrinsic permeability of the soil (Equation 3.27)
k_g	gas permeability (Equation 3.29)
K	bulk modulus
k_L	parameter related to the radiation interception by plant leaves (Equation 8.2)
K_0	at rest earth pressure
k_{u1}, k_{u2}, k_{u3}	regression coefficients in Equation 5.15
k_1, k_2, k_3	regression coefficients in Equation 5.16
k_4	regression coefficients in Equation 8.6
k_a	air coefficient of permeability
K_p	plastic modulus
k_{ra}	relative permeability of air
k_s	saturated permeability
K_w	bulk modulus for the water phase
k_x	water permeability in the x-direction
k_y	water permeability in the y-direction
$k_{zB,\text{tave}}$	unsaturated permeability at any depth zB for any average elapsed time
l	penultimate suction path
L	thickness of the soil sample
L_i	tip-to-tip distance
m	fitting parameter related to the results near the residual water content (Equation 3.9)
M	stress ratio at the critical state
M_0	fitting parameter (Equation 5.17)
m_1	soil parameter (Equation 3.12)
M^1_R	resilient modulus from the 1st cycle
m_2	soil parameter (Equation 3.12)
m_3	soil parameters (Equation 3.13)
m_3^I, m_3^D, m_T	soil parameters (Equation 7.61 and 7.62)
m_4	soil parameters (Equation 3.13)
m_5	soil parameter equal to the product of mm and b
m_a	constant at a particular time step in Equation 3.2
m_m	soil parameter describing the effects of pore structure on the parameter a (Equation 3.17)
M_m	the maximum value of stress ratio in the stress history
M^N_R	resilient modulus from the Nth cycle
M_R	resilient modulus
m_v	coefficient of volume change
m_w	slope of the water retention curve in Equation 3.4

n_1	soil parameter related to the slope of the soil-water characteristic curve at the inflection point (Equation 3.9)
n_2	a model parameter in Equation 3.28 typically equals to 3
n_s	empirically derived constant in Equation 5.1
N	number of cycles
n	porosity
$N(T)$	intercept of the NCL at a temperature T
N_a	normal component of intergranular force due to external stress
n_b *and* n_d	soil parameters (Equations 7.48 and 7.49)
n_{inf}	slope at the inflection point,(rate of desorption (drying) and absorption wetting)
N_0	intercept of the NCL at a reference temperature T_0
N_s	inter-granular force due to suction
p	mean net stress
p'	effective mean stress
p''	pore water deficiency in Aitchison (1961) effective stress equation
p^*	mean Bishop's stress (Equation 7.41)
p^*_k	mean effective stress proposed by Khalili and Khabbaz (1998)
p_0	pre-consolidation stress
P_1	atmospheric pressure (Equation 3.29)
P_2	applied gas pressure (Equation 3.29)
p_{atm}	atmospheric pressure
P_d	the area below the GSD measured using the dry-sieving method
P^0_c	initial yield stress at zero suction
p_r	reference pressure
p^t	mean total stress (Equation 7.41)
P_w	the area below the GSD measured using the wet sieving method
Q	applied boundary flux
q	deviatoric stress
Q_g	gas flow rate (Equation 3.29)
q/p	stress ratio
q_{cyc}	cyclic shear stress
R_{con}	universal gas constant
R_c	repulsive pore fluid stress due to chemistry in Lambe (1960) effective stress equation
r	the height of mercury in the tube in Figure 2.10 (Equation 2.4)
R_p	the radius of spherical particle (Equation 6.2)
R	similarity ratio
R^2	coefficient of determination
r_N	soil constant describing thermal effects on the location of NCL
R_p *and* R_s	parameters introduced to simulate the reduction of plastic modulus as the distance between state and bounding surfaces decreases.
r_Γ	soil parameter describing thermal effects on the location of CSL
s	matric suction
s'	half of the sum of the major and minor stress minus air pressure.

xxx Notations

s^*	modified suction
s^*	scaled suction (Gallipoli et al., 2015)
$s_{current}$	current soil suction
S_e	residual standard deviation
s_e	suction value marking the transition between saturated and unsaturated states (Equation 7.7)
$S_{g(h,z)}$	sink term the water volume transpired by grass from the entire root zone during a given time interval (Feddes et al., 1976)
s_{max}	maximum historical suction experienced by soil
S_r	degree of saturation
S_r^e	effective degree of saturation
S_r^m	microstructural degree of saturation
S_{rr}	residual degree of saturation
S_s	specific storage
$s_{s\alpha}$	soil suction obtained from the main drying/wetting curve at the current degree of saturation
$\overline{S_w}$	specific entropy of pore water (Equation 2.5)
S_y	sample standard deviation
T	temperature
t	elapsed time
T_a	the tangential component of inter-granular stress due to external stress
t_{ave}	average time $(t_1 + t_2)/2$
t_{ij}	measured travel time of a shear wave propagating in the polarization plane i-j
T_s	surface tension
u_a	pore air pressure
u_a-u_w	matric suction
u_v	the partial pressure of pore water vapour in equilibrium with soil water or vapour pressure
u_v/u_{v0}	relative humidity
u_{v1}	partial pressure of pore water vapour in equilibrium with a solution identical in composition to the water vapor
u_{v0}	saturation pressure of water vapour over a flat surface of pore water at the same temperature
u_w	pore-water pressure
$V(\Psi)$	sensor output voltage at suction ψ along the main drying or wetting curves
$V_d (\psi,\psi_1)$	output voltage at suction on the drying scanning curve that starts at a suction value of ψ_1
$V_{s(hh)}$	shear wave propagating horizontally with a horizontal polarization
$V_{s(hv)}$	shear wave propagating horizontally with a vertical polarization
$V_{s(ij)}$	shear velocity in the direction of wave propagation, i, and in the direction of particle motion, j.
$V_{s(vh)}$	shear wave propagating vertically with a horizontal polarization
V_t	total volume
V_w	output voltage at suction ψ along the main wetting curve
V_w	volume of water

Notations   xxxi

$V_w (\psi, \psi_1)$	output voltage at suction on the wetting scanning curve that starts at a suction value of ψ_1
$\overline{V}_w$	partial specific volume of water (Equation 2.5)
v_{w0}	specific volume of water or the inverse of the density of water
$v_{zB,tave}$	water flow rate at any depth zB at any average elapsed time
$v_{ze,tave}$	boundary water flow rate at any depth zB at any average elapsed time
w	gravimetric water content
w_{inf}	inflection point water content
w_L	liquid limit
w_{opt}	optimum water content
w_P	plastic limit
W^p	plastic work input
w_r	gravimetric water content at the start of the residual conditions
w_s	gravimetric water content at saturation
ω_v	molecular mass of water vapour
x	directional axis (horizontal direction)
y	directional axis (vertical direction)
z	depth
α_{tc}	an empirical fitting parameter that controls the degree of curvature of the scanning curves (Equations 2.2 and 2.3)
αe	material parameter for determining effective degree of saturation in Equation 4.8
α	curvature parameter (Equation 7.27)
$\alpha(h,z)$	transpiration reduction function which ranges from 0-1 (Feddes et al., 2001)
α_p	represents soil compressibility with respect to net mean stress
$\alpha_p(s)$	compressibility dependent on suction
α_s	compressibility of unsaturated soil upon a change in suction
α_{sk}	isotropic thermal coefficients for soil skeleton
β	parameter that controls the increasing rate of G_0 with increased suction (Mancuso et al., 2002)
$\gamma_{0.7}$	reference shear strain at which shear stiffness is 70% of G_0^{ref}.
γ_e	threshold shear strain corresponding to the limiting of elastic range (Equation 7.90)
γ_T	material parameter in Equation 6.9
γ_w	unit weight of water
Δe	change in void ratio
Δe	void ratio change
Δe_i	void ratio change at initial yield stress of a structured soil
δH	increments of horizontal displacement
ΔN	suction-dependent inter-particle normal force
δV	increments of vertical displacement
δx	incremental horizontal displacement
Δx	horizontal displacement
δy	incremental vertical displacement
$\delta y/\delta x$	dilatancy
ε	soil skeleton strain
ε_e	elastic threshold strain

xxxii Notations

ε^p_a	plastic axial strain
ε_q	deviatoric strain
ε_{qe}	elastic threshold strain
ε_{qref}	characteristic reference strain
ε^r_a	reversible axial strain
ε_s	shear strain
ε_v	volumetric strain
η	mean constant stress ratio
$\eta(z)$	root distribution with depth z
η/M	normalised stress ratio
η_r	stress ratio
θ	angle of rotation
Θ_d	dimensionless water content
Θ_n	normalised water content
θ_ω	volumetric water content
κ	elastic compressibility index
$\kappa(s)$	suction-dependent elastic compressibility index
κ_p	elastic stiffness parameter
κ_s	elastic stiffness parameter
κ_s	swelling index
λ	slope of the normal compression line/plastic compressibility indices
$\lambda(0)$	plastic compressibility index at zero suction
$\Lambda_{(c)},\ \Lambda_{(s)},\ \Lambda_{(h)},$	loading indices for compression, shearing and suction change (Equation 7.65)
$\lambda(s)$	suction-dependent compressibility index
λ_p	stiffness parameter with regard to stress at normally consolidated state
λ_s	stiffness parameter with regard to suction at normally consolidated state
μ	dynamic viscosity
μ_a	absolute (dynamic) viscosity
μm	micrometre
ν	Poisson's ratio
υ	specific volume
ξ	state parameter (Equation 7.46)
ξ_m	microstructural state variable
$\xi_m^{\,ref}$	microstructural state variable in a reference state
π	osmotic suction
π_{salt}	osmotic suction related to soil salts
π_{solute}	osmotic suction related to solute in solution
ρ	bulk density
ρ_a	density of air
ρ_d	dry density
ρ_{H20}	density of water
ρ_{HG}	density of mercury
ρ_w	density of water
σ	total stress or confining pressure
$\bar{\sigma}$	mineral interparticle stress in Lambe (1960) effective stress equation
σ^*	Bishop's stress

σ_a	axial stress
σ_{eqv}	equivalent effective stress
σ_i	principal stress in the direction of wave propagation, i.
σ_i	intergranular stress of the unsaturated packing
σ_i^b	inter-granular stress from bulk water
σ_{ij}	total stress (Equation 6.1)
σ_i^m	inter-granular stress from meniscus water
σ_j	principal stress in the direction of particle motion, j.
σ_r	radial stress
σ^s	suction stress
$\sigma\text{-}u_a$	net normal stress
$\sigma\text{-}u_w$	effective stress
σ_v	vertical stress
$\sigma_v - u_a$	net vertical stress
τ	shear stress
τ_f	shear stress at failure
τ_p	peak shear strength
φ'	friction angle
φ^b	angle with respect to changes in matric suction
φ_{cr}	friction angle at the critical state
φ_{cs}	effective critical friction angle (Equation 4.10)
φ_m	mobilised friction angle
χ	parameter to relate the degree of saturation to the shear strength of the soil in Bishop's (1959) effective stress equation
χ_m	effective stress parameter for matric suction ((Richards, 1966)
χ_s	effective stress parameter for solute suction (Richards, 1966)
ψ	total suction
ψ	angle of dilation (Equations 4.9 and 6.4)
ψ_a	air-entry value
ψ_{inf}	suction at the inflection point
ψ_{max}	maximum angle of dilation (Equation 4.11)
ψ_r	residual suction
ψ_v	energy state of pore vapour
ψ_w	energy state of pore water
ψ_{wev}	water-entry value
ω	slope of the CSL at temperature T
Γ	intercept of the CSL at temperature T
Γ_0	intercept of the CSL at temperature T_0
E^{ref}_{50}	triaxial loading Young's modulus when shear stress is 50% of shear strength
E^{ref}_{oed}	oedometric loading modulus
E^{ref}_{ur}	unloading-reloading Young's modulus
$\bar{K}$	suction modulus in Equation 5.12

Nomenclature

1D	One-dimensional
2D	Two-dimensional
AASHTO	American Association of State, Highway and Transportation Officials
AEV	Air-entry value
ASTM	American Society for Testing and Materials
atm	Atmospheres
ATT	Axis translation technique
BAC	Biochar amended clay
BBM	Barcelona basic model
BH	Borehole
BS	British Standards
BSI	British Standards Institution
CC	Coarsely crushed concrete
CCBE	Cover with capillary barrier effect
CD	Consolidated drained
CDG	Completely decomposed granite
CDT	Completely decomposed tuff
CDV	Completely decomposed volcanic
CEDD	Civil Engineering and Development Department
CFI	Carbon Farming Initiative
CL	Low plastic clay
CP	Casagrande-type piezometers
CRC	Coarse recycled concrete
CSL	Critical state line
CSR	Current suction ratio
CU	Consolidated undrained with pore pressure measurements
CVR	Cumulative void ratio
CW	Constant water content
DAVI	Diffused air volume indicator
DG	Decomposed granite
DMT	Dilatometer tests
DOC	Degree of compaction
DPT	Differential pressure transducer

DPT N	Number of blows per 100 mm penetration in dynamic probe tests (Figure 3.18)
DTI	Digital transducer interface
DV	Decomposed volcanic
DVLO	Derjaguin, Landau, Verwey and Overbeek
e.m.f	Electromotive force
EC	Electrical conductivity
ENPC	Ecole Nationale des Ponts et Chaussees
EPA	Environmental Protection Agency
EPC	Vibrating wire earth pressure cell
EPS	Exopolysaccharides
ET	Evapotranspiration
FOS	Factor of safety
GDS	Global Digital Systems Ltd
GEO	Geotechnical Engineering Office
GSD	Grain size distribution
GWC	Gravimetric water content
HDT	Highly decomposed tuff
HET	Hall-effect transducers
HKO	Hong Kong Observatory
HKSAR	Hong Kong Special Administrative Region
HKUST	The Hong Kong University of Science and Technology
HP	Hewlett Packard
IC	Integrated circuit
IEEE	Institute of Electrical and Electronics Engineers
IP	Instantaneous profile
IPI	In-place inclinometer
IPM	Instantaneous profile method
JFT	Jet fill tensiometer
JGS	Japanese Geotechnical Society
LAI	Leaf area index
LAT	Lateritic soil
LC	Loading collapse
LL	Liquid limit
LSL	Limit state line
LVDT	Linear variable differential transducer
LY	Loading yield
MCCM	Modified Cam Clay model
MDD	Maximum dry density
MDT	Moderately decomposed tuff
MIP	Mercury intrusion porosimetry
mPD	Meters above the Principal Datum
MSW	Municipal solid waste
MWCO	Molecular weight cut off
NA	Not applicable

NCL	Normal compression line
NG	Non-grassed landfill
OCR	Over-consolidation ratio
OD	Ordinance datum
OMC	Optimum moisture content
OMT	Osmotic technique
PC	Personal computer
PEG	Polyethylene glycol
PET	Potential evapotranspiration
PI	Plasticity index
PL	Plastic limit
PSD	Pore-size distribution/pore-size density
PVC	Polyvinyl chloride
PWP	Pore-water pressure
RC	Relative compaction
REV	Representative element volume
RG	Rain gauge
RH	Relative humidity
S.D	Standard deviation
SC	Stress-controlled soil column
SD	Suction decrease
SDSWCC	Stress-dependent soil water characteristic curve
SDSWRC	Stress-dependent soil water retention curve
SDT	Slightly decomposed tuff
SDWRC	Stress-dependent water retention curve
SEM	Scanning electron microscopy
SI	Suction increase
SNWDP	South-to-North Water Diversion Project
SOA	State-of-the-art
SP	Standpipe
SPT	Standard penetration tests
SWCC	Soil water characteristic curve
TC	Thermal conductivity- heat dissipation matric water potential sensors
TD	Temperature decreases
TDR	Time-domain reflectometry
TT	Trial trench
UC	Unconfined compression
UH	Unified hardening
URL	Unloading-reloading line
USCS	Unified Soil Classification System
USEPA	The United States Environmental Protection Agency
UU	Unconsolidated undrained
VET	Vapour equilibrium technique
VWC	Volumetric water content
WIP	Wish-in-place

WRC	Water retention curve
XRD	X-ray diffraction
XRF	X-ray fluorescence
YRSRI	Yangtze River Scientific Research Institute
ZY	Zao-yang

Chapter 1

Introduction

OVERVIEW

Soil above the groundwater table is generally unsaturated in nature. Saturated and dry soils are only two special cases of unsaturated soil when the degree of saturation is equal to 100% and 0%, respectively. In many textbooks and classes, soil mechanics and geotechnical engineering have often been taught with an implicit assumption that soil is either dry (0% saturation) or saturated (100% saturation) for simplicity. Soil behaviour can then be taught using Terzaghi's effective stress principle (Terzaghi, 1936). However, soil is neither saturated nor dry in many engineering problems, such as the instability of onshore slopes and settlement of embankments. Relatively little research has been conducted on unsaturated soils and only a handful of textbooks have been published on the subject. Clearly, there is an urgent need to improve the understanding of the behaviour and mechanics of unsaturated soil.

The general field of classical soil mechanics is often subdivided for convenience into the portion dealing with saturated soils and the portion dealing with unsaturated soils. Although this artificial division between saturated and unsaturated soils can be shown to be unnecessary, it may still be helpful to make use of the knowledge gained from saturated soils as a reference and then to extend it to the broader unsaturated world, as shown in Figure 1.1, which provides a visual aid for the generalised world of soil mechanics (Fredlund, 1996). For simplicity, this world of soil mechanics is divided by the water table. Below the water table, soil behaviour is governed by effective stress $(\sigma - u_w)$, whereas the unsaturated soil above the water table is governed by at least two stress variables (e.g., the net normal stress $(\sigma - u_a)$ and the matric suction $(u_a - u_w)$) (Fredlund and Morgenstern, 1977; Alonso et al., 1987; Houlsby, 1997; Gens, 2010; Ng et al., 2020a). Focusing on the soil above the water table, it may be useful to categorise the soil according to its degree of saturation, as shown in Figure 1.2. Instead of being a two-phase material (i.e., solid and water) when a soil is saturated, an unsaturated soil is recognised by Fredlund and Morgenstern (1977) as having four phases (i.e., solid, water, air and an air-water interface called the contractile skin (Padday, 1969; Fredlund and Rahardjo, 1993; Ng and Menzies, 2007)). It is obvious that the behaviour of an unsaturated soil is more complex than that of a saturated soil. It is the intention of this book to simplify the complexities of an unsaturated soil to a digestible level mainly for postgraduate students and practising engineers.

DOI: 10.1201/9781003480587-1

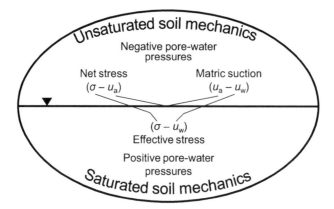

Figure 1.1 A visualisation aid for the generalised world of soil mechanics.

Source: Fredlund, 1996.

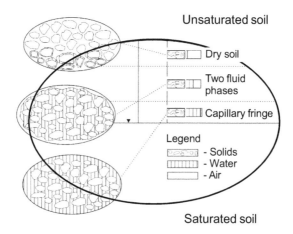

Figure 1.2 Categorisation of soil above the water table based on the variation in degree of saturation.

Source: Fredlund, 1996.

The vadose zone

According to the representation in Figure 1.1, the hypothetical geotechnical world is divided by a horizontal line representing the groundwater table. Below the water table, the pore-water pressures are positive, and the soil is, in general, saturated. Above the water table, the pore-water pressures will, in general, be negative with respect to the atmospheric pressure (i.e., gauge pressure). The entire soil zone above the water table is called the vadose zone (see Figure 1.3). Immediately above the water table, there is a zone called

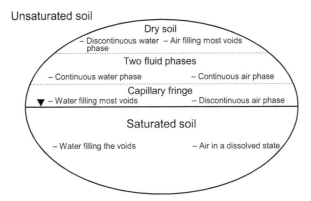

Figure 1.3 A visualisation of saturated/unsaturated soil mechanics based on the nature of the fluid phases.
Source: Fredlund, 1996.

the capillary fringe where the degree of saturation approaches 100%. This zone may range from less than one metre to approximately 10 metres in thickness, depending upon the soil type (Fredlund, 1996) and climate condition. Within this capillary fringe, the water phase can be assumed to be continuous, while the air phase is generally discontinuous. Above this capillary fringe, a two-phase zone (i.e., two fluid phases) can be identified in which both the water and air phases may be idealised as continuous. Within this zone, the degree of saturation may vary from about 20% to 90%, depending on the soil type and state. Terzaghi's effective stress principle cannot be directly applied. Bishop (1959), therefore, proposed a "χ" parameter to relate the degree of saturation to shear strength of soil. The pros and cons of using this "χ" parameter have been debated among leading researchers ever since. Above this two-phase zone, the soil becomes dryer, and the water phase will be discontinuous while the air phase remains continuous (Fredlund and Rahardjo, 1993; Ng and Menzies, 2007). For simplicity, Terzaghi's effective stress principle may be applied with the assumption that pore-water pressure is zero for calculation or interpretation of shear strength.

Climate changes and the vadose zone

As expected, the location of the groundwater table is strongly influenced by climatic conditions in a region. If the region is arid or semi-arid, the groundwater table slowly lowers with time (i.e., may be on a geological time scale). If the climate is temperate or humid, the groundwater table may remain quite close to the surface of the ground. It is the difference between the downward flux (i.e., precipitation) and the upward flux (i.e., evaporation or evapotranspiration) which determines the location of the groundwater table (refer to Figure 1.4).

Regardless of the degree of saturation of the soil, the pore-water pressure profile will come to equilibrium in a hydrostatic condition when there is zero net flux from the ground

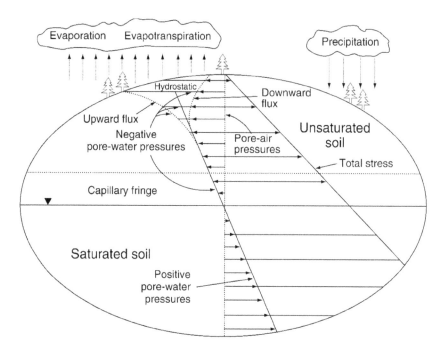

Figure 1.4 A visualisation of soil mechanics showing the role of the surface flux boundary condition.
Source: Fredlund, 1996.

surface. If moisture is extracted from the ground surface (e.g., by evaporation), the pore-water pressure profile will move to the left. If moisture enters the ground surface (e.g., by infiltration), the pore-water profile will move to the right.

A net upward flux produces a gradual drying, cracking, and desiccation of the soil mass, whereas a net downward flux eventually saturates the soil mass. The depth of the water table is influenced, amongst other things, by the net surface flux. A hydrostatic line relative to the groundwater table represents the equilibrium condition when there is no net flux at the ground surface. During dry periods, the pore-water pressures become less than those represented by the hydrostatic line, and greater during wet periods.

Grass, trees, and other plants growing on the ground surface dry the soil by applying a tension to the pore water through evapotranspiration (Dorsey, 1940; Ng et al., 2019a). Most plants can apply 1–2 MPa (10–20 atm) of tension to the pore water before reaching their wilting point (Taylor and Ashcroft, 1972). The tension applied to the pore water acts in all directions and can readily exceed the lateral confining pressure in the soil. When this happens, desiccation cracking commences. Evapotranspiration also results in the consolidation and desaturation of the soil mass.

Over the years, a soil deposit is subjected to varying environmental conditions. These produce changes in the pore-water pressure distribution, which in turn results in shrinkage and swelling of the soil deposit. The pore-water pressure distribution with depth as shown in Figure 1.4 can take on a wide variety of shapes as a result of environmental changes (Fredlund, 1996).

There are many complexities associated with the vadose zone because of its fissured and fractured nature. Conventionally, the tendency in geotechnical engineering has been to avoid or grossly simplify the analysis of this zone, if possible. However, in many cases it is an understanding of this zone which holds the key to the performance of an engineered structure. Historically, classic seepage problems involved a saturated soil where the boundary conditions are either a designated head or zero flux. However, the real world for the engineer often involves the ground surface, where there may be a positive or negative flux. Geo-environmental problems have done much to force the engineer to consider saturated/unsaturated transient seepage analyses with flux boundary conditions. The improved computing capability available to the engineer has helped to accommodate these changes (Fredlund, 1996).

Most man-made structures are located on the surface of the earth and as such will have an environmental flux boundary condition. This is the case for a highway where the soil of the embankment and subgrade have an initial set of conditions or stress states. These conditions will change with time primarily because of environmental and climatic (or surface moisture flux) changes. The foundations for light structures are likewise generally placed well above the groundwater table where the pore-water pressures are negative. In fact, most light-engineered structures of the world are placed within the vadose zone which is affected strongly by the climate.

One of the characteristics of the upper portion of the vadose zone is its ability to release water vapour slowly to the atmosphere at a rate depending on the permeability of the intact portions of soil. At the same time, downward flow of water can occur through fissures with a gradient of unity. There appears to be no hindrance to the inflow of water until the soil swells and the mass becomes intact, or until the fissures and cracks fill with water (Fredlund, 1996).

Some common unsaturated soils

A large portion of the world's population is found in the arid regions of the world where the groundwater table is deep because the annual evaporation from the surface in these regions exceeds the annual precipitation. Of the many types of unsaturated soils, some are notorious and problematic for engineers. Some examples are listed as follows:

- Medium to highly plastic clays containing a substantial amount of expansive minerals such as Montmorillonite which swell when subject to an increased moisture content, producing the category of materials known as swelling soils. The shrinkage of the soils may be equally severe. Expansive plastic clays are commonly found in Colorado, Texas and Wyoming in the USA (Chen, 1988), in Hubei, Guangxi and Shandong in China (Shi et al., 2002; Ng et al., 2003), in Alberta and Saskatchewan in Canada (Chen, 1988; Fredlund and Rahardjo, 1993), in Madrid in Spain and in Gezara, Blue Nile and Kasalla in Sudan (Chen, 1988). A summary of some problematic unsaturated soils is given in Table 1.1.
- Loess soils often undergo collapse when wetted, and possibly also in a loading environment. They are commonly found in Missouri, Nebraska and Wisconsin in the USA (Dudley, 1970; Handy, 1995), in Gansu, Ningxia and Shaanxi in China (Liu, 1988), and in Kent, Sussex, Hampshire in the United Kingdom (Jefferson et al., 2001).

6 Advanced Unsaturated Soil Mechanics

Table 1.1 Distribution of some expansive soils and loess soils worldwide

Expansive soils	Loess soils
China: Hubei, Guangxi, Shandong	China: Gansu, Ningxa, Shaanxi
United States: Colorado, Texas, Wyoming	United Kingdom: Kent, Sussex, Hampshire
Canada: Alberta, Saskatchewan, Manitoba	United States: Missouri, Nebraska, Wisconsin
Spain: Andalucia, Madrid	Romania: Galati, Constanta, Tulcea
Sudan: Gezara, Blue Nile, Kasalla	

Source: Ng and Menzies, 2007.

- Residual and saprolitic soils located above the groundwater table, particularly on many hillsides in Brazil, Portugal and in the Far East, such as in Hong Kong in China and Malaysia.

Apart from natural and geological processes, human activities such as excavation, remoulding, and recompacting may also lead to the desaturation of saturated soils and hence the formation of unsaturated soils. These natural and man-made materials are difficult to understand and analyse, particularly where volume changes are concerned, within the framework of classical saturated soil mechanics.

The drier climatic regions have become increasingly aware of the uniqueness of their regional soil mechanics problems. In recent years there has also been a shift in emphasis in the developed regions from the behaviour of engineered structures to the impact of developments on the natural world. This shift in emphasis has resulted in a greater need to deal with the vadose zone. There has been an ongoing desire to expand the science of soil mechanics to embrace the behaviour of unsaturated soils. An all-encompassing saturated and unsaturated soil mechanics has emerged in a number of countries worldwide (Fredlund, 1996).

Definitions of suctions

Soil suction is commonly referred to as the free energy state of soil water (Edlefsen and Anderson, 1943), which can be measured in terms of its partial vapour pressure. From a thermodynamic standpoint, total suction can be described quantitatively by Kelvin's equation (Sposito, 1981) as follows:

$$\psi = -\frac{R_{con} T}{v_{w0} \omega_{v}} \ln\left(\frac{u_v}{u_{v0}}\right) \tag{1.1}$$

where ψ is total suction (kPa); R_{con} is the universal gas constant [J/(mol·K)]; T is the absolute temperature (K); v_{w0} is the specific volume of water or the inverse of the density of water (m³/kg); ω_v is the molecular mass of water vapour (g/mol); u_v is the partial pressure of pore-water vapour (kPa); u_{v0} is the saturation pressure of water vapour over a flat surface of pure water at the same temperature (kPa). The term (u_v/u_{v0}) is called relative humidity, RH (%).

Equation (1.1) can be simply re-written as follows:

$$\psi = -\frac{RT}{v_{w0}w_v}\left[\ln\left(\frac{\bar{u}_v}{\bar{u}_{v1}}\right) + \ln\left(\frac{\bar{u}_{v1}}{\bar{u}_{v0}}\right)\right]$$

(1.2)

It can be seen that the first term $\left(\dfrac{\bar{u}_v}{\bar{u}_{v1}}\right)$ in the above equation represents the ratio of the partial pressure of the water vapour in equilibrium with the soil water, relative to the partial pressure of the water vapour in equilibrium with a solution identical in composition with the soil water. Hence, this is the matric or capillary component of total free energy (Fredlund and Rahardjo, 1993).

The second term in Equation (1.2), $\left(\dfrac{\bar{u}_{v1}}{\bar{u}_{v0}}\right)$, denotes the partial pressure of the water vapour in equilibrium with a solution identical in composition with the soil water, relative to the partial pressure of the water vapour in equilibrium with free pure water. Hence, this is the osmotic (or solute) component of total free energy.

Thus, the total suction has two components: matric suction $(u_a - u_w)$ and osmotic suction (π), i.e.,

$$\psi = \left(u_a - u_w\right) + \pi,$$

(1.3)

where u_a and u_w are the pore air and water pressures, respectively.

A change in total suction is generally caused by a change of relative humidity in the soil in geotechnical engineering. Relative humidity can be reduced by the presence of a curved water surface produced by capillary phenomena, i.e., contractile skin (Fredlund and Rahardjo, 1993). The radius of curvature of the curved water surface is inversely proportional to the difference between air pressure (u_a) and water pressure (u_w) across the surface, which is called matric suction.

Osmotic suction is a function of the amount of dissolved salts in the pore fluid and is expressed in terms of pressure. Alternatively, a reduction in the relative humidity in a pore, caused by the presence of dissolved salts in pore water, is referred to as the osmotic suction.

Stress state variables

What are stress state variables? According to the International Dictionary of Physics and Electronics (Michels, 1961), state variables are defined as follows:

A limited set of dynamical variables of the system, such as pressure, temperature, volume, etc., which are sufficient to describe or specify the state of the system completely for the considerations at hand.

8 Advanced Unsaturated Soil Mechanics

Fung (1965) describes the state of a system as the "information required for a complete characterisation of the system for the purpose at hand". Typical state variables for an elastic body are those that describe the strain field, the stress field, and its geometry. The state variables must be independent of the physical properties of the material involved.

The stress state variable for saturated soil

An understanding of the meaning of effective stress proves valuable when considering the stress state description for unsaturated soil (Fredlund, 1987). Terzaghi's 1936 statement regarding effective stress defined the stress state variables necessary to describe the behaviour of saturated soils. The statement is shown as follows:

> The stress in any point of a section through a mass of soil can be computed from the total principal stresses, σ_1, σ_2, σ_3, which act at this point. If the voids of the soil are filled with water under a stress, u, the total principal stresses consist of two parts. One part, u, acts in the water and in the solid in every direction with equal intensity ... The balance $\sigma_1' = \sigma_1 - u$, $\sigma_2' = \sigma_2 - u$, $\sigma_3' = \sigma_3 - u$, represents an excess over the neutral stress, u, and it has its seat exclusively in the soil phase of the soil. All the measurable effects of a change in stress, such as compression, distortion and a change in shearing resistance are exclusively due to changes in the effective stress σ_1', σ_2' and σ_3'.

Possible sets of stress state variables for unsaturated soil

Numerous attempts have been made to extend stress concepts useful for saturated soil into the unsaturated soil range. The proposed equations have taken the form of a single-valued effective stress equation. Table 1.2 contains a summary of the more common forms which have been proposed. All the equations in the table propose a single-valued equation. Soil properties can also be identified in all equations. It could be argued that these equations are constitutive relations and as such "fall short" of meeting the conditions for a state variable. The difficulties are primarily conceptual in nature, and their adoption gives rise to a deviation from classical continuum mechanics. Practical difficulties have also been encountered in the use of these effective stress equations.

Morgenstern (1979) amply summarised the difficulties associated with the use of the first equation listed in Table 1.2, by saying that the Bishop's effective stress equation

> proved to have little impact or practice. The parameter χ when determined for volume change behaviour was found to differ when determined for shear strength. While originally thought to be a function of the degree of saturation and hence bounded by 0 and 1, experiments were conducted in which χ was found to go beyond these bounds. As a result, the fundamental logic of seeking a unique expression for the effective stress independent of degree of saturation has been called into question (Fredlund, 1973). The effective stress is a stress variable and hence related to equilibrium considerations alone.

Table 1.2 Proposed effective stress equations for unsaturated soils

Equation	Symbol	Reference
$\sigma' = \sigma - u_a + \chi(u_a - u_w)$	χ = parameter related to degree of saturation u_a = the pressure in gas and vapour phase u_w = the pressure of water	Bishop (1959)
$\sigma' = \sigma - \beta' u_w$	β' = holding or bonding factor which is a measure of the number of bonds under tension effective in contributing to soil strength	Croney et al. (1958)
$\sigma = \bar{\sigma} a_m + u_a a_a + u_w a_w + R_c - A_c$	a_a = fraction of total area that is air-air contact $\bar{\sigma}$ = mineral inter-particle stress a_m = mineral particle contact area a_w = water phase contact area R_c = repulsive pore fluid stress due to chemistry A_c = attractive pore fluid stress due to chemistry	Lambe (1960)
$\sigma' = \sigma + \psi p''$	ψ = parameter with values ranging from zero to one p'' = pore-water pressure deficiency	Aitchison (1961)
$\sigma' = \sigma + \beta p''$	β = statistical factor of same type as contact area; should be measured experimentally in each case	Jennings (1960)
$\sigma' = \sigma - u_a + \chi_m(h_m + u_a) + \chi_s(h_s + u_a)$	χ_m = effective stress parameter for matric suction h_m = matric suction h_s = solute suction χ_s = effective stress parameter for solute suction	Richards (1966)

Source: Fredlund, 1987.

And that it

contains a parameter, χ, that bears on constitutive behaviour. This parameter is found by assuming that the behaviour of a soil can be expressed uniquely in terms of a single effective stress variable and by matching unsaturated behaviour with saturated behaviour in order to calculate χ. Normally, we link equilibrium considerations to deformations through constitutive behaviour and do not introduce constitutive behaviour directly into the stress variable.

Of course, Morgenstern's comments above would also apply to the other single-valued equations listed in Table 1.2.

Many subsequent authors have found that while it is relatively easy to relate the shear strength of unsaturated soil to a single stress parameter involving σ, u_a and u_w, the volumetric behaviour is not controlled by the same stress parameter or by any other single stress variable. In particular, it has proved impossible to represent the complex pattern of wetting-induced swelling and collapse in terms of a single effective stress (Wheeler and Karube, 1996), as shown in Figure 1.5.

It seems unlikely that it will ever be possible to devise a satisfactory definition of a single effective stress for unsaturated soil. The reason for this is that suction within the pore water and external stress applied to the boundary of a soil element act in qualitatively different ways on the soil skeleton, as noted by Jennings and Burland (1962), and hence these two stresses cannot be combined into a single effective stress parameter. This

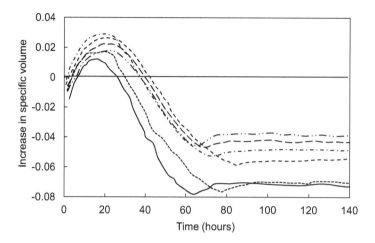

Figure 1.5 Volume change observed during wetting tests on compacted kaolin.
Source: Wheeler and Sivakumar, 1995.

qualitative difference in the mode of action of suction and external stress is not taken into account in the method of mixtures, which has been used by some authors in attempts to derive a single effective stress expression (Wheeler and Kurabe, 1995).

As an example of the different modes of action of suction and external stress, consider the idealised case shown in Figure 1.6 of an unsaturated soil made up of spherical soil particles, with the pore air at atmospheric pressure and the pore water at negative pressure within menisci at the particle contacts (Wheeler and Karube, 1996). An external total stress, σ, applied to the boundary of a soil element containing many particles will produce both normal and tangential forces at particle contacts, even if the external stress state is isotropic. Of course, the effectiveness of σ will be influenced by the presence of bulk water inside soil pores. If the external stress is increased sufficiently, the tangential forces at particle contacts will cause inter-particle slippage and plastic strains (this is why soils, unlike most metals, undergo plastic volumetric strains if loaded beyond a preconsolidation pressure). By contrast, the capillary effect caused by meniscus water arising from suction within the menisci only increases the normal forces at particle contacts.

For any stress variable(s) to capture the essential features of unsaturated soil behaviour, two forms of liquid water and two different influences of suction on the mechanical behaviour (refer to Figure 1.7) must be fully recognised:

- Suction modifies both the normal and tangential skeleton stresses of an unsaturated soil by changing the average bulk pore fluid pressure within its pores;
- Suction provides an additional normal bonding force (stabilising effect) at the particle contacts, attributed to capillary phenomena occurring in the water menisci or contractile skin.

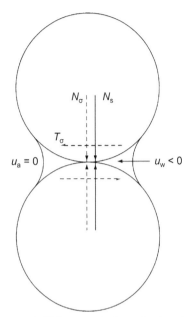

N_σ = normal component of inter-granular force due to external force
T_σ = tangential component of inter-granular force due to external force
N_s = inter-granular force due to suction

Figure 1.6 Influence of external stress and suction on inter-particle forces.

Source: Wheeler and Kurabe, 1995.

It is important to realise that for these two different mechanisms, two independent stress variables are therefore required. It is known that the effects of suction are influenced by the degree of saturation of the soil. The relative area over which the water and air pressures act depends directly on the degree of saturation (the percentage of pore voids occupied by water), but the same parameter also affects the number and intensity of capillary-induced inter-particle forces (Gallipoli et al., 2003).

According to Fredlund (1987), criticism of a single-valued effective stress equation for unsaturated soils can be summarised as follows: Coleman (1962) suggested the use of "reduced" stress variables $(\sigma_1 - u_a)$, $(\sigma_3 - u_a)$ and $(u_a - u_w)$ to represent the axial, confining and pore-water pressure, respectively, in triaxial tests. Constitutive relations for volume change were then formulated using the "reduced" stress variables. Bishop and Blight (1963) re-evaluated the use of the single-valued effective stress equation and stated that a change in matric suction did not always result in the same change in effective stress. It was also suggested that the laboratory data of volume change could be plotted in terms of the independent stress variables, $(\sigma_1 - u_a)$ and $(u_a - u_w)$. This appears to have initiated a transition towards using the stress variables independently. This approach was further reinforced by Blight (1965) and Burland (1964, 1965).

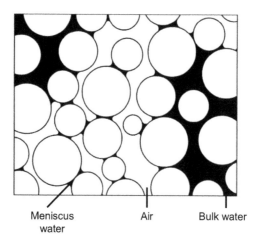

Figure 1.7 Two forms of liquid water in an unsaturated soil.
Source: **Wheeler and Karube, 1995; Wheeler et al., 2003.**

In 1977, Fredlund and Morgenstern suggested the use of any two of three possible stress variables, $(\sigma - u_a)$, $(\sigma - u_w)$ and $(u_a - u_w)$, to describe the mechanical behaviour of unsaturated soils. The possible combinations are:

(1) $(\sigma - u_a)$ and $(u_a - u_w)$,
(2) $(\sigma - u_w)$ and $(u_a - u_w)$,
(3) $(\sigma - u_a)$ and $(\sigma - u_w)$.

The most common choice is to use net stress $(\sigma - u_a)$ and matric suction $(u_a - u_w)$ as the two independent stress state variables. Houlsby (1997) reported perhaps the most convincing theoretical derivations and justifications of the need of two stress state variables to describe the behaviour of unsaturated soils. The theoretical derivations and justifications are based on the consideration that the rate of work input to the soil is equal to the sum of the products of the stresses with their corresponding strain rates. It should be noted that if the finite compressibilities of the soil grains and pore fluid are to be included, more than two stress state variables may be needed to model the behaviour of unsaturated soils fully.

It is worth considering the merits of selecting $(\sigma - u_a)$ and $(u_a - u_w)$ as stress state variables, rather than $(\sigma - u_w)$ and $(u_a - u_w)$. The former combination has the advantage that the pore air pressure u_a is zero in many practical situations, so that net stress and matric suction simplify to total stress and negative pore-water pressure, respectively. In addition, the pore-water pressure, which is commonly negative, is often very difficult to measure. This leads to uncertainty in the value of only one stress state variable if $(\sigma - u_a)$ and $(u_a - u_w)$ are selected, but uncertainty in the values of both stress state variables if $(\sigma - u_w)$ and $(u_a - u_w)$ are chosen. However, a counterargument in favour of the combination of $(\sigma - u_w)$ and $(u_a - u_w)$ is that this choice leads to a slightly easier transition to fully saturated conditions (although it does not solve all the problems associated with this transition (Wheeler and

Kurabe, 1995). Thus, $(\sigma - u_a)$ and $(u_a - u_w)$, are chosen to be the most satisfactory combination from the standpoint of practical analysis (Fredlund, 1987).

Several new combinations of two independent stress state variables have been proposed recently, as modifications to $(\sigma - u_a)$ and $(u_a - u_w)$. They are mainly in an attempt to account directly for the influence of the degree of saturation on soil behaviour. This is vital because of hydraulic hysteresis in the soil-water characteristic curve (SWCC) or the water retention curve (WRC) during drying and wetting processes. The merits and shortcomings of these new stress state variables are introduced and discussed by Wheeler et al. (2003) and Gallipoli et al. (2003).

Summary

To summarise, a suitable set of independent stress state variables is one which produces no distortion or volume change in an element when the individual components of the stress state variables are modified but the stress state variables themselves are kept constant. Thus, the stress state variable for each phase should produce an equilibrium in that phase when a stress point in space is considered. In spite of the many ingenious attempts to find a general expression for effective stresses, it is time to conclude that describing the full range of behaviour of unsaturated soils requires the simultaneous use of at least two independent stress state variables. However, in order to avoid too much complexity for engineering applications, the use of the two simple independent stress state variables, $(\sigma - u_a)$ and $(u_a - u_w)$, may hold more promise for geotechnical engineers. There are convincing experimental verifications and theoretical derivations (Houlsby, 1997) in support of these stress state variables in general. Formulations for some engineering analyses using these stress state variables have been published (Chiu and Ng, 2003).

CAVITATION

Overview

Young (1989) points out that the pore water in soils can sustain very high negative pressures (suctions), and it is possible to estimate this pressure using many different indirect methods (Marinho and Chandler, 1995). However, attempts to measure suction greater than one atmosphere directly are often unsuccessful until recently, because of cavitation in the measuring system. It is recognised that water has a high tensile strength, and this characteristic conflicts with the phenomenon of cavitation observed in measuring systems. The lack of inter-communication between the various sciences involved in measuring negative gauge pressure in liquids has delayed the development of direct suction measurement system for soils and understanding of the behaviour of unsaturated soils, which generally have negative gauge pore-water pressures.

Definitions

Young (1989) reports that cavitation is the formation and activity of bubbles (or cavities) in a liquid. Here the word "formation" refers, in a general sense, both to the creation of a new cavity or to the expansion of a pre-existing one to a size where macroscopic effects can be observed. These bubbles may be suspended in the liquid or may be trapped in tiny cracks either at the liquid's boundary surface or in solid particles suspended in the liquid.

The expansion of the minute bubbles may be affected by reducing the ambient pressure by static or dynamic means. The bubbles then become large enough to be visible to the unaided eye. The bubbles may contain gas or vapour or a mixture of both gas and vapour. If the bubbles contain gas, then the expansion may be by diffusion of dissolved gases from the liquid into the bubble, or by pressure reduction, or by temperature rise. If, however, the bubbles contain mainly vapour, reducing the ambient pressure sufficiently at essentially constant temperature causes an "explosive" vaporisation into the cavities which is the phenomenon that is called *cavitation*, while raising the temperature sufficiently causes the main vapour bubbles to grow continuously producing the effect known as *boiling*. This means that "explosive" vaporisation or boiling does not occur until a threshold is reached.

There are thus four ways of inducing bubble growth (Young, 1989):

- For a gas-filled bubble, by pressure reduction or an increase in temperature. This is called *gaseous cavitation*.
- For a vapour-filled bubble, by pressure reduction. This is called *vaporous cavitation*.
- For a gas-filled bubble, by diffusion. This is called *degassing* as gas comes out of the liquid.
- For a vapour-filled bubble, by sufficient temperature rise. This is called *boiling*.

The situation is complicated because the bubble usually contains a mixture of gas and vapour.

Cavitation inception and nuclei

Pearsall (1972) observed that in theory, a liquid will vaporise when the pressure is reduced to its vapour pressure. In practice, the pressure at which cavitation starts is greatly dependent on the liquid's physical state. If the liquid contains much dissolved air, then as the pressure is reduced the air comes out of the solution and forms cavities within which the pressure is greater than the vapour pressure of the liquid. Even if there are no visible air bubbles, the presence of sub-microscopic gas bubbles may provide nuclei which cause cavitation at pressures above the vapour pressure. Each cavitation bubble grows from a nucleus to a finite size and collapses again, the entire cycle taking place within perhaps a few milliseconds. Bubbles may follow each other so rapidly that they appear to the eye to form a single continuous bubble.

In the absence of nuclei, the liquid may withstand negative pressures or tensions without undergoing cavitation. In theory, a liquid should be able to withstand tensions equivalent to thousands of atmospheres. It is estimated that water, for example, will withstand a tension ranging from 500 to 10,000 atmospheres (Harvey et al., 1947). In practice, even when water has been subjected to rigorous filtration and pre-pressurised to several hundred atmospheres, it has ruptured at tensions of 300 atmospheres. According to Plesset (1969), water without nuclei will theoretically withstand tensions of 15,000 atmospheres, but the probability of this happening is low unless the bubbles are of molecular dimensions. However, when solid non-wetted nuclei of size 10^{-8} cm are present, it is likely that water will rupture at tensions of the order of tens of atmospheres. It follows from this that it should be possible to raise water under vacuum to a height greater than that corresponding to atmospheric pressure. This has been achieved in practice by pre-pressurising the liquid (Hayward, 1970).

Physics of pore water under tension in unsaturated soils

Marinho and Chandler (1995) observed that measurement of matric suction, using a tensiometer, requires the water inside the measuring system to be at the same value of tensile stress as the soil pore water, and this may induce cavitation within the system. In soil suction measurement, cavitation is the result of pressure reduction and has the effect that, after cavitation has occurred, the pressure measured will be approximately equal to the pressure of gas in the system. This probably gives rise to the belief that it is impossible to measure a suction greater than 1 atmosphere.

The conditions under which a liquid with a flat liquid-gas interface (free water) will vaporise are given by the vapour pressure-temperature curve for that particular liquid. However, if the liquid-gas interface is not flat, as in the case of a vapour cavity (bubbles) in a liquid, it is possible to reach a state normally associated with the vapour phase without the vapour phase developing (Apfel, 1970).

A phase diagram for a simple substance, as shown in Figure 1.8 indicates the regions where under a given combination of pressure and temperature, the substance is in the solid, liquid or vapour phase. The substance at point A would be a vapour. However, under certain circumstances it is possible, starting from A_1, to reduce the pressure and reach A, at a pressure lower than the vapour pressure, and for the substance to remain in liquid. In the same way, it is possible to go from A_2 to A while remaining liquid. When the phase boundary is crossed without the substance changing phase, the phase boundary is said to have been transgressed. If transgression has occurred, the system is said to be in a

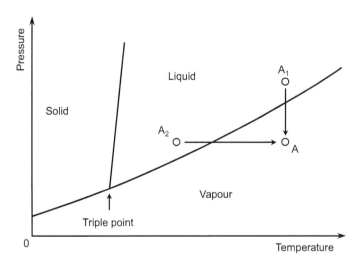

Figure 1.8 Phase diagram for a simple substance.

Source: After Marinho and Chandler, 1995.

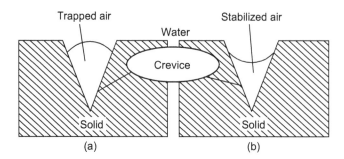

Figure 1.9 Cavitation nuclei.

Source: After Marinho and de Sousa Pinto, 1997.

metastable state (Apfel, 1970). Water under tension is in a metastable condition and this metastability can be destroyed if nucleation occurs. Nucleation is the formation of vapour cavities within the liquid itself or at their boundaries (Trevena, 1987).

Cavitation in a metastable liquid can result from two types of nucleation: nucleation in the pure liquid, and nucleation caused by impurities in the liquid. Impurities can be other pure substances, solid impurities, or even radiation acting on the liquid. Nucleation within a pure liquid is called "homogeneous" nucleation and nucleation due to impurities is called "heterogeneous" nucleation. Heterogeneous nucleation is more common, and it is responsible for most cavitation which occurs in suction measurement systems.

Avoiding cavitation

Marinho and de Sousa Pinto (1997) observed that although water can sustain tension, attempts to measure soil suction greater than 1 atm, using tensiometers, have failed. The reason is associated with the entrapment of air inside micro crevices in the system. There are many theories that attempt to explain the entrapment of air between liquid and a solid container. Harvey et al. (1944) presented the most accepted model to justify the presence of air and how the air nuclei can be stabilised.

Figure 1.9 presents a schematic representation of air trapped during the usual saturation procedure of a tensiometer. The air trapped is called cavitation nuclei. In order to dissolve this air, high positive water pressure should be applied. However, due to particular aspects of the geometry of the crevice, the air may not dissolve. In this case, the application of positive pressure may "stabilise" the cavitation nuclei (i.e., trapped air), as shown in Figure 1.9b.

The stabilisation of the cavitation nuclei increases the level of the suction that can be applied. It is not clear how the process occurs, and thus it cannot be controlled precisely (Marinho and de Sousa Pinto, 1997).

Usually, the pressure required to stabilise the cavitation nuclei is higher than 5 MPa (Ridley and Burland, 1993). Marinho and de Sousa Pinto (1997) suggested a chemical procedure to be used with the usual technique for stabilising the cavitation nuclei inside a suction probe (see Figure 1.10).

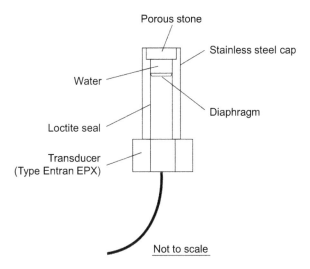

Figure 1.10 Schematic representation of the suction probe.
Source: After Marinho and de Sousa Pinto, 1997.

They reported that the chemical procedure used reduced the level of positive pressure to be applied. The system used by Marinho and de Sousa Pinto (1997) was pressurised to a maximum pressure of 3.5 MPa by a hand pump as shown in Figure 1.11. The pressure was maintained for 24 hours. The system was then cycled 10 times from 3.5 MPa to zero pressure. After that, it could sustain suction up to 650 kPa. The porous stone used has a nominal air-entry value of 500 kPa. Marinho and de Sousa Pinto (1997) believed that chemical action can help to eliminate cavitation nuclei.

Therefore, to reduce the likelihood of trapping permanent undissolved air in the system, the following measures were recommended by Marinho and Chandler (1995):

- The use of de-aired water is important to avoid air saturation (boiling the water is an appropriate method).
- The water and all surfaces within the measurement system must be extremely pure and clean (Henderson and Speedy, 1980).
- The surfaces in contact with the water must be as smooth as possible to avoid or reduce the number and size of crevices. The smaller the surface area, the easier it is to avoid cavitation.
- The system should be evacuated by vacuum application in order to remove the maximum amount of air entrapped in the crevices, though it is unlikely that all the air will be removed (Jones et al., 1981).
- The system should be cycled from positive to zero (or negative) pressure. This may help to dissolve persistent bubbles (Chapman et al., 1975; Richards and Trevena, 1976).
- Pre-pressurisation of the system to high pressure is required in order to dissolve all the free air (Harvey et al., 1944).

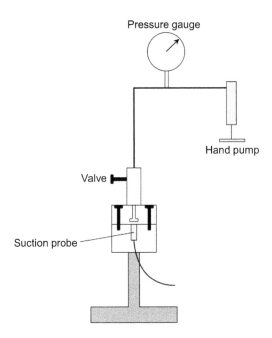

Figure 1.11 System for saturating the tensiometer using high positive pressure.
Source: After Marinho and de Sousa Pinto, 1997.

It is important to point out that cavitation can be delayed if the nuclei are stabilised against tension. The procedures above probably only stabilise the nuclei, which should allow the measurement of a certain level of suction before cavitation occurs. It should be noted that cavitation will not occur if no cavitation nuclei are present. Examples of the success and failure of measuring soil suction higher than one atmosphere using tensiometers are given and discussed in detail by Take et al. (2003) and Zhou et al. (2006).

SOURCES

Overview (Fredlund, 1996)
Cavitation (Young, 1989; Marinho and Chandler, 1995; Marinho and de
    Sousa Pinto, 1997)

Chapter 2

Measurement and control of suction

SUCTION MEASURING DEVICES (FREDLUND AND RAHARDJO, 1993; RIDLEY AND WRAY, 1996; FENG AND FREDLUND, 2003)

Overview

It is generally recognised that unsaturated soil consists of solids, voids and pore fluids, typically water and air. If the voids are filled up with water, the soil is called "saturated", which has a 100% degree of saturation. On the other hand, if the voids are full of air, then the soil is called dry soil (i.e., 0% degree of saturation). When the voids are partly filled with water and partly filled with air, the soil is unsaturated. The interface between water and air in voids is called "contractile skin", which behaves like an elastic membrane due to surface tension (Fredlund and Rahardjo, 1993).

There are two distinct levels at which suction can be measured. The first involves the measurement of the energy required to move a water molecule within the soil matrix – this is termed the matric suction, i.e., $(u_a - u_w)$. The second is a measure of the energy required to remove a soil-water molecule from the soil matrix into the vapour phase – it is termed the *total suction* as defined by Equation (1.1)).

In a salt-free granular material, the total suction and the matric suction are equal (refer to Equation (1.3)). If the pore water contains ions (as is generally the case for clay soils), the vapour pressure (u_v) is lowered and the energy required to remove a water molecule from the water phase of the soil (i.e., the total suction) increases. The additional suction caused by the dissolved salts is called *osmotic suction*. Ideally, if the vapour pressure of a sample of soil water (extracted from the soil) is measured, the stress implied by Equation (1.2) will be equal to the difference between the total suction and the matric suction of the soil. However, the composition of the soil water must not change during the extraction process. At present, the only satisfactory method of estimating the osmotic suction is by subtracting the matric suction from the total suction.

As suggested by Ridley and Wray (1996), in the absence of a truly semipermeable membrane, the presence of dissolved salts within the pore water should not result in any appreciable movement of the pore water. Therefore, dissolved salts do not change the energy required to move a water molecule (while maintaining its state) within the soil matrix and the matric suction is independent of the osmotic suction.

In deciding whether an instrument is measuring total or matric suction, it is necessary to determine whether the instrument is making direct contact with the pore water in the soil. If no contact is established between the measuring instrument and the soil, dissolved

DOI: 10.1201/9781003480587-2

20 Advanced Unsaturated Soil Mechanics

salts cannot move from the pore fluid to the measuring instrument. When the instrument being used to measure suction makes contact with the pore water, the dissolved salts are free to move between the measuring instrument and the soil. Therefore, unless it is certain that the concentration of dissolved salts is everywhere the same, their effect on the suction measured in this way is unquantifiable. However, the effect is generally believed to be negligible. If this is so, the matric suction will be measured in this way. Moreover, at high values of suction, the meniscus may recede completely into the soil and an instrument that is making contact with the soil will not necessarily be making close contact with the soil water. Therefore, contact between soil and instrument does not guarantee that matric suction will be measured (Ridley and Wray, 1996).

Generally, devices to measure suction fall into two categories depending on whether the measurement is direct or indirect. A direct measurement is one that measures the relevant quantity under scrutiny (e.g., the pore-water energy). With indirect measurements, another parameter (e.g., relative humidity, resistivity, conductivity or moisture content) is measured and related to the suction through calibration against known values of suction. These devices will now be introduced.

Psychrometer

A psychrometer is an instrument that measures humidity and can therefore be used to measure the total suction. In its simplest form, it consists of a thermometer that has a wet bulb whose temperature is lowered below ambient temperature by evaporation into the adjacent air. When evaporation ceases and an equilibrium with the ambient vapour pressure is reached, the measured temperature is compared with the temperature of a dry bulb thermometer placed in the same environment. The difference between the temperatures of the dry bulb and the wet bulb is related to the relative humidity and can be calibrated using salt solutions of known concentrations.

Electrical devices now exist to measure the heat (which is a function of the electrical current) generated in an electrical junction when water evaporates from or condenses onto the junction. They may be subdivided into two groups:

- thermistor/transistor psychrometers (see Figure 2.1) and
- thermocouple psychrometers (see Figure 2.2).

Thermistor/transistor psychrometers

A thermistor is a temperature-sensitive resistor. In the thermistor psychrometer, two identical (matched) thermistors are employed. A drop of water is placed on the first (the wet bulb) and the other is left dry (the dry bulb). If both thermistors are then exposed to an enclosed environment, evaporation from or condensation onto the wet thermistor will result in an electromotive force (e.m.f.) being generated in the thermistor. The voltage can be related to the relative humidity of the environment. Richards (1965) gave a detailed description of the theory, construction and calibration of thermistors psychrometer.

Measurement and control of suction 21

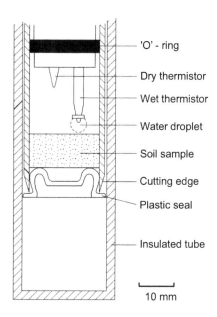

Figure 2.1 A thermistor psychrometer.
Source: After Ridley and Wray, 1996.

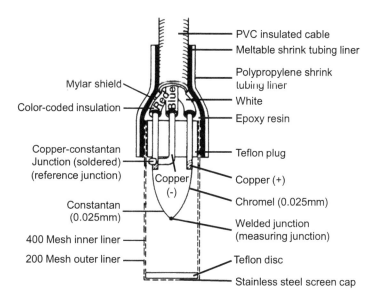

Figure 2.2 Screen-caged single-junction Peltier thermocouple psychrometer.
Source: From Brown and Collins, 1980; after Fredlund and Rahardjo, 1993.

22   Advanced Unsaturated Soil Mechanics

Thermocouple psychrometer

The development of the thermocouple psychrometer is attributed to Spanner (1951). Working in the field of plant physiology, Spanner (1951) recognised the potential of using the Peltier and Seebeck effects to measure the suction pressure in plant leaves.

In 1834, Peltier discovered that when passing an electrical current across the junction between two different metal wires, there is a change of temperature, the sign of which is dependent on the direction of flow of the current. If such a junction is placed in an atmosphere of humid air and a current is passed in the direction which causes the junction to cool sufficiently to cause water to condense on it, then it essentially becomes a "wet bulb" thermometer. When the circuit is broken, the water will evaporate and the change in temperature of the wet junction will generate an e.m.f. (the Seebeck effect) between it and a reference junction which is at the ambient temperature (i.e., a "dry bulb"). This e.m.f. can be measured with a sensitive galvanometer. The characteristic output curve for this type of psychrometer has a plateau voltage at the "wet bulb" temperature corresponding to the vapour pressure of the ambient air. Unfortunately, the evaporation continues, and the plateau only lasts for a period determined by the relative humidity (and therefore by the suction) being measured. At suctions less than 500 kPa, the plateau is difficult to detect. Details of the operating principles and procedures for this type of psychrometer are given by Fredlund and Rahardjo (1993).

Thermocouple psychrometers are seriously affected by temperature fluctuations (Rawlins and Dalton, 1967; Fredlund and Rahardjo, 1993). When the measuring junction is cooled, heat is generated in the reference junction. It is important for the correct measurement of relative humidity that the reference junction is held at the ambient temperature. Temperature gradients between the cooling junction and the reference junction at the start of the test can result in a large error in the measured humidity. Measuring the initial potential difference between the two junctions will give an indication of any temperature gradient. It is recommended that a zero offset exceeding 1 µV represents an excessive temperature gradient (Ridley and Wray, 1996). In order to obtain satisfactory performance from a psychrometer, a constant temperature water bath which can be regulated to +/- 0.001°C may be required to measure total suctions as low as 100 kPa (Fredlund and Rahardjo, 1993). The useful range over which the total suction can be measured lies between about 100 kPa and 8000 kPa.

The efficiency of the thermocouple psychrometer is also temperature dependent and the calibration needs to be adjusted relative to the temperature at which the probe was calibrated. The effect of the temperature on the calibration of thermocouple psychrometers was investigated by Brown and Bartos (1982). They found that the calibration was more sensitive at higher temperatures.

The filter paper method

According to Fredlund and Rahardjo (1993), the filter paper method for measuring soil suction was developed in the soil science discipline and has since been used primarily in soil science and agronomy (Gardner, 1937; Fawcett and Collis-George, 1967; McQueen and Miller, 1968; Al Khafaf and Hanks, 1974). Attempts have also been made to use the filter paper method in geotechnical engineering (Ching and Fredlund, 1984; Gallen,

1985; McKeen, 1985; Chandler and Gutierrez, 1986; Crilly et al., 1991; Gourley and Schreiner, 1995).

McQueen and Miller (1968) proposed to use the filter paper method to measure either the total or the matric suction of a soil. The filter paper is used as a sensor. The filter paper method is classified as an "indirect method" for measuring soil suction, based on the assumption that a filter paper will come into equilibrium (with respect to moisture flow) with a soil having a specific suction. Equilibrium can be reached by either liquid or vapour moisture exchange between the soil and the filter paper. When a dry filter paper is placed in direct contact with a soil specimen, it is assumed that water flows from the soil to the paper until equilibrium is achieved. When a dry filter paper is suspended above a soil specimen (i.e., no direct contact with the soil), a vapour flow of water will occur from the soil to the filter paper until equilibrium is achieved. Having established equilibrium conditions, the water content of the filter paper is measured. Once the water content is known, soil suction can be obtained from a calibration curve. This is a standard method of measuring soil suction (ASTM, 2016).

It should be emphasised that the filter paper technique is highly user-dependent, and great care must be taken when measuring the water content of the filter paper (Fredlund and Rahardjo, 1993). The balance to determine the water content of the filter paper must be able to weigh to the nearest 0.0001 g. Each dry filter paper has a mass of about 0.52 g, and at a water content of 30%, the mass of water in the filter paper is about 0.16 g.

Porous block (e.g., gypsum block)

The electrical resistance of an absorbent material changes with its moisture content (Ridley and Wray, 1996). This is the principle of the porous block method of measuring soil matric suction. The block (which is placed inside the soil sample) consists of two concentric electrodes buried inside a porous material (Figure 2.3). If the suction in the soil is higher than the suction in the block, the latter will lose moisture to the soil until the soil and the block have the same matric suction. Alternatively, if the block is initially dry, it will absorb moisture from the soil until equilibrium is reached. The electrical resistance of the block can then be related to the suction. The calibration is performed by burying each block in a soil sample that is subsequently subjected to known suctions in a pressure plate apparatus.

Gypsum (in the form of plaster of Paris) is found to be the best medium for the measurement of electrical resistance. It takes the shortest time to saturate, which is the quickest to respond when placed in the ground and has the most stable electrical properties. However, it has the distinct disadvantage of softening when saturated (Ridley and Wray, 1996).

The gypsum block is a low-cost device that is easy to handle and install. Measurements are simple and correction factors, for the effects of temperature and salt content, can be applied where necessary. However, the blocks are susceptible to hysteresis and their response to a change in suction can be slow (2–3 weeks) which precludes using them as accurate indicators of absolute suction values in a rapidly changing moisture environment. Therefore, it is recommended that only a small portion of the calibration curve should be used (i.e., the instruments should not be used in situations with large fluctuations in soil suction).

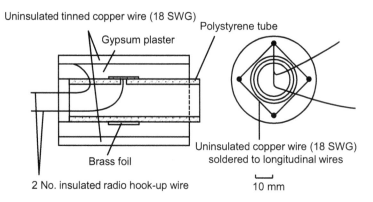

Figure 2.3 A typical gypsum block.
Source: After Ridley and Wray, 1996.

The ideal conditions for the use of gypsum blocks, are in a soil that is relatively non-saline, having a suction in the range 50 kPa to 3000 kPa, and that is subjected to small and unidirectional moisture changes. Conversely, the most adverse conditions for the use of the blocks are in relatively saline soils subjected to rapid changes in water content. These rather narrow constraints have led to a reduced use of porous blocks for measuring soil suction.

Thermal conductivity sensor (Feng and Fredlund, 2003)

A thermal conductivity soil suction sensor consists of a cylindrical porous tip containing a miniature heater and a temperature-sensing element (Phene et al., 1971). Figure 2.4 shows the structure of a thermal conductivity sensor developed at the University of Saskatchewan (Shuai et al., 1998). The porous tip is a specially designed and manufactured ceramic with a pore-size distribution appropriate for the range of soil suctions to be measured. The heater at the centre of the ceramic tip converts electrical energy to thermal energy. A portion of the thermal energy will be dissipated within the ceramic tip. The undissipated thermal energy results in the temperature rising at the centre of the ceramic tip. The temperature sensor (i.e., the integrated circuit IC in Figure 2.4) measures the temperature rise with respect to time in terms of output voltage. Since water has a much higher thermal conductivity than air, the rate of dissipation of the thermal energy within the ceramic tip increases with the water content of the ceramic. Higher water contents result in a lower rise in temperature at the centre of the ceramic and, consequently, a lower output voltage from the temperature sensor. Since the suction in the sensor is equal to the suction in the surrounding soil, the voltage output of the temperature sensor (i.e., the output of the thermal conductivity suction sensor) can be calibrated against matric suction in the surrounding soil.

Calibration is the first step in using the thermal conductivity sensor for field measurements of soil suction. The calibration should reproduce the actual field conditions as closely as possible. The sensors are conventionally calibrated following a drying process. The

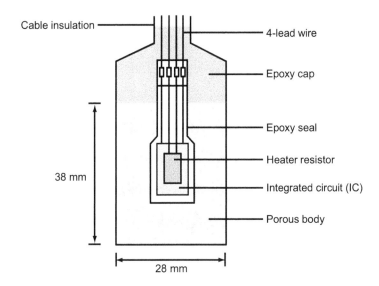

Figure 2.4 A cross-sectional diagram of the newly developed thermal conductivity sensor.

Source: From Shuai et al., 1998; after Feng and Fredlund, 2003.

calibration curve represents the relationship between the sensor output and the applied matric suction for the specific drying process. It is recognised, however, that the water content (and consequently the output voltage) versus matric suction relationships for a porous material exhibit hysteresis between the wetting and drying processes (Fredlund and Xing, 1994).

Several hypothetical hysteresis models have been proposed in the literature (Feng, 1999). An examination of some of the hysteresis models using measured hysteresis data from the ceramic sensors shows that the models either require too much calibration data or cannot provide a reasonable prediction of hysteresis (Feng, 1999).

The experimental results show that the hysteresis curves are consistent from one suction sensor to another. If a model for prediction can be developed that fits the measured calibration curves of the sensors with known hysteresis characteristics, the model can be used to predict the hysteresis curves of other sensors of the same type. The following equation is proposed to fit the main drying and wetting curves:

$$V(\psi) = \frac{a_{tc} b_{tc} + c_{tc} \psi^{d_{tc}}}{b_{tc} + \psi^{d_{tc}}} \qquad (2.1)$$

where a_{tc} is the sensor reading at zero suction on the main hysteresis loop; c_{tc} is the sensor reading when the ceramic tip is dry. Parameters a_{tc} and c_{tc} are easy to measure and remain unchanged for the main wetting and drying curves, respectively. With one branch of the main hysteresis loop measured, only two parameters, b_{tc} and d_{tc}, remain unknown for the other branch. If two points on the unknown branch are measured, this branch can be estimated using Equation (2.1).

26 Advanced Unsaturated Soil Mechanics

With one branch of the main hysteresis loop measured and the other branch estimated, the following equations are used to fit the scanning curves:

$$V_d\left(\psi, \psi_1\right) = V_d + \left(\frac{\psi_1}{\psi}\right)^{\alpha_{tc}}\left(V_w - V_d\right)$$ (2.2)

$$V_w\left(\psi, \psi_1\right) = V_w + \left(\frac{\psi_1}{\psi}\right)\left(V_w - V_d\right)$$ (2.3)

where $V_d\left(\psi, \psi_1\right)$ is the output voltage at suction ψ on the drying scanning curve starting at a suction value ψ_1; ψ_1 is the soil suction at which the scanning curve starts; V_w and V_d are the output voltages at suction ψ on the main wetting and drying curves, respectively; and α_{tc} is an empirical parameter that controls the degree of curvature of the scanning curves and is the only unknown parameter in Equations (2.2) and (2.3).

The measured hysteresis curves for the six sensors were fitted using Equations (2.1), (2.2) and (2.3). Similar prediction results were obtained for the six sensors. The predicted curves for the main hysteresis loop and primary scanning curves for sensors 1, 2, and 3 are shown in Figures 2.5, 2.6 and 2.7, respectively.

A best-fit value of α_{tc} equal to 1.8 was used for both the primary drying scanning curves and the primary wetting scanning curves of all six sensors. As shown in Figures 2.5 to 2.7, the predicted curves are close to the measured curves. The errors between the predicted and measured values are less than 5%. The α_{tc} value of 1.8 appears to be reasonable to predict the primary scanning curves of the newly developed suction sensors. It should be noted that the value of α_{tc} could be different for sensors other than the newly developed suction sensor. It is necessary to investigate the hysteresis properties using typical sensors to estimate the α_{tc} value when calibrating other types of thermal conductivity sensors.

Feng and Fredlund (2003) recommend the following procedure to calibrate the thermal conductivity suction sensors:

- saturate the ceramic sensor tip by submerging it in water for 2 days or more;
- place the sensor ceramic tip in the pressure plate cell and apply a suction of 50–100 kPa;
- when equilibrium has been reached, reduce the applied suction to zero;
- after equilibrium at zero suction, increase the applied suction in increments following the conventional calibration procedure to measure the main drying curve; and
- rewet the ceramic sensor to obtain two points on the main wetting curve.

The main wetting curve is estimated using Equation (2.1). The primary scanning curves are estimated using Equations (2.2) and (2.3), assuming a α_{tc} value of 1.8.

A similar procedure can be used to calibrate other types of thermal conductivity suction sensors. However, a study of the hysteretic properties of the ceramic sensor output voltage versus matric suction should be carried out to establish the value of the α_{tc} parameter for Equations (2.2) and (2.3).

Measurement and control of suction

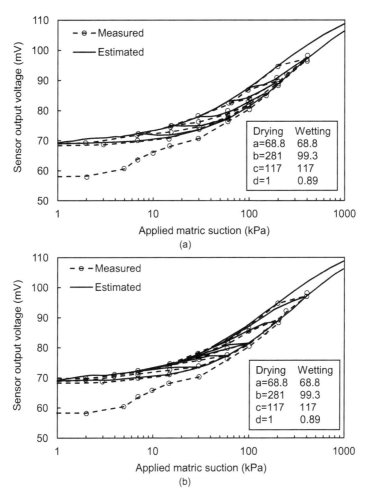

Figure 2.5 Measured and predicted primary scanning curves for sensor 1: (a) primary drying scanning curves; (b) primary wetting scanning curves.

Source: After Feng and Fredlund, 2003.

Suction plate and pressure plate (Ridley and Wray, 1996)

In its simplest form, the suction plate consists of a saturated flat porous ceramic filter disc that separates a soil specimen from a water reservoir and a mercury manometer. Soil which, by virtue of its suction, is deficient in water will absorb water from the porous disc, causing a drop in the water pressure in the reservoir which is measured using the manometer. When the pore-water pressure in the soil and the tension in the reservoir water are in equilibrium, there will be no further flow of water and the soil suction can be interpreted from the manometer reading. As a result of the water exchange, the measured suction will probably be less than the actual suction in the soil prior to the test.

28 Advanced Unsaturated Soil Mechanics

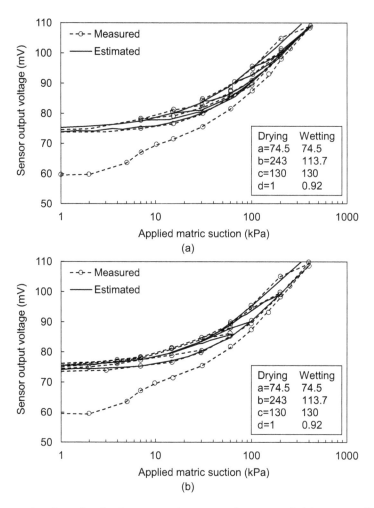

Figure 2.6 Measured and predicted primary scanning curves for sensor 2: (a) primary drying scanning curves; (b) primary wetting scanning curves.

Source: After Feng and Fredlund, 2003.

The main limitation of the suction plate is its range of usefulness; it fails to prevent the formation of air in the reservoir at suctions greater than one atmosphere (100 kPa). When air is present in the reservoir, the tension measured by the manometer will remain less than or equal to one atmosphere and any further moisture exchange will only result in a change in the volume of air present. Schofield (1935) recognised this limitation and proposed extending the useful range by enclosing the specimen in a chamber and applying air pressure inside the chamber.

The pressure plate apparatus (Figure 2.8) consists of a base plate with a porous ceramic filter set into it. The air-entry pressure of the ceramic should be higher than the maximum suction

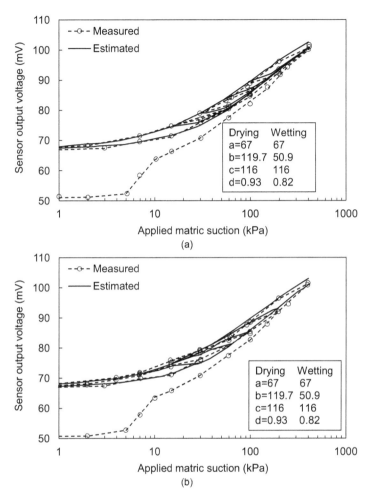

Figure 2.7 Measured and predicted primary scanning curves for sensor 3: (a) primary drying scanning curves; (b) primary wetting scanning curves.

Source: After Feng and Fredlund, 2003.

to be measured (usually 500 kPa but 1500 kPa ceramic filters are available). Beneath the filter is a water reservoir connected to a pressure measurement device and a drainage system. An airtight chamber into which compressed air can be piped is placed on the base plate.

When a soil specimen is placed on the ceramic filter and the air pressure in the chamber is increased, the water pressure in the soil specimen rises by an amount roughly equal to the difference between the air pressure in the chamber and atmospheric pressure. This technique is called axis translation (Hilf, 1956). Therefore, if the difference between the air pressure in the chamber and the original atmospheric pressure is greater than the suction in the soil, the final water pressure in the soil specimen will be positive and can be measured

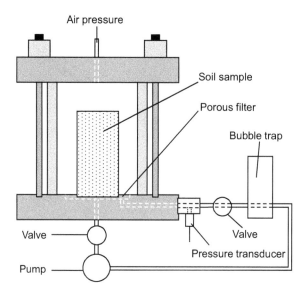

Figure 2.8 A typical pressure plate apparatus.

Source: After Ridley and Wray, 1996.

using standard pore pressure measuring equipment. Table 2.1 summarises and compares the characteristics of various measuring devices.

Jet-fill tensiometer

A conventional tensiometer measures the absolute negative pore-water pressure in a similar manner to the suction plate. It is principally used in the field but some of its applications have been found in the laboratory (Tadepalli and Fredlund, 1991; Chiu et al., 1998). It works by allowing water to be extracted from a reservoir in the tensiometer, through a porous ceramic filter and into the soil, until the stress holding the water in the tensiometer is equal to the stress holding the water in the soil (i.e., the soil suction). At equilibrium, no further water flow will occur between the soil and the tensiometer. The suction will then manifest itself in the reservoir as a tensile stress in the water and can be measured using any stress measuring instrument. Stannard (1992) has reviewed the theory, construction and usage of common tensiometers.

All the commercially available tensiometer devices designed for use in the field consist of a porous ceramic cup, a fluid reservoir and a pressure measuring device. Figure 2.9 shows a typical jet fill-type tensiometer. The jet fill type is an improved variant of the regular tensiometer. A water reservoir is provided at the top of the tensiometer tube to remove air bubbles. The jet fill mechanism is like the action of a vacuum pump. The accumulated air bubbles are removed by pressing the button at the top to activate the jet fill action. The jet fill action causes water to be injected from the water reservoir into the tube of

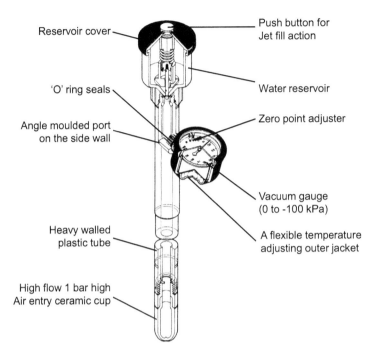

Figure 2.9 Jet-fill tensiometer (courtesy Soilmoisture Equipment Corp.; after Fredlund and Rahardjo, 1993).

the tensiometer, and air bubbles move upward to the reservoir. Three types of measuring instruments are commonly used with a tensiometer (see Figure 2.10):

- a mercury manometer,
- a vacuum gauge, and
- an electronic pressure transducer.

Mercury manometer

The manometer type uses a water-filled tube to connect the porous cup to a mercury reservoir (Figure 2.10a). The suction at the porous cup causes the mercury to rise in the tube above the level of the free surface. Using the free mercury surface as a datum, the suction is related to the rise of the mercury and the depth of the porous cup below the free surface. Referring to Figure 2.10a, this suction may be expressed as:

$$\text{suction} = \left(\rho_{Hg} - \rho_{H_2O}\right)r - \rho_{H_2O}\left(h_{ms} + d_{pc}\right) \tag{2.4}$$

where ρ_{Hg} and ρ_{H_2O} are the density of mercury and water, respectively; d_{pc} and h_{ms} are the vertical distance from the ground surface to the porous cup and free surface of mercury, respectively; r is the capillary rise of mercury.

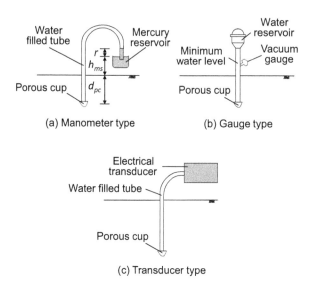

Figure 2.10 Typical tensiometer arrangements.

Source: After Ridley and Wray, 1996.

Vacuum gauge

A vacuum gauge is the device most commonly used in commercially available tensiometers. In the vacuum gauge tensiometer (Figure 2.10b), the system is sealed, and measurements are made on a standard vacuum gauge. The porous cup is connected to the vacuum gauge by a rigid pipe that extends above the level of the gauge to help in refilling the system when air penetrates the reservoir. The advantage of this approach is that provided water continuity is maintained between the vacuum gauge and the porous cup, leading to the direct reading of soil suction in the porous cup by the gauge. However, the range of suction over which the tensiometer can be used is limited when long lengths of tube are required (e.g., the correction required is 10 kPa per metre between the tip and the gauge). The vacuum gauge will normally have an adjustable zero that is used to compensate for the difference in height between the gauge and the porous cup. The adjustment is made with the cup immersed in water to its mid-height. Without adjustment, the gauge can be used to measure small positive pressures (should they occur) but this practice will also reduce the range over which suction can be recorded.

Pressure transducer

Tensiometers employing electronic pressure transducers (Figure 2.10c) are similar to the vacuum gauge type except that pressure measurements are converted to a voltage measurable by a digital voltmeter. This means that data can be logged automatically by a computer or chart recorder. The sensitivity of the transducer can make it susceptible to the thermal expansion/contraction of the water in the system and this can result in errors when continuous monitoring is necessary. The sensor used in electronic transducers is normally a semiconducting resistor located inside or on the surface of a diaphragm. The resistance of these devices is (a) strain dependent; (b) a nonlinear function of pressure and (c) highly sensitive to changes in temperature. Connecting the circuit to a Wheatstone bridge arrangement smooths the nonlinearity and reduces the temperature sensitivity. However, the zero and occasionally the sensitivity may change with time (a condition known as drift) and therefore frequent re-calibration may be required. Using an electronic transducer means that the sensor can be located as close to the porous cup as is practical, negating the need for a depth correction.

To measure the suction accurately and quickly, it is necessary to de-air the tensiometer. The porous filter should be de-aired by immersing it in water under a vacuum before assembling the equipment. Once the tube is full of water, a handheld vacuum pump can be used to remove any large air bubbles which become trapped during the filling process. If the tube is carefully filled with water that has been previously de-aired, the operation is made considerably easier. If the sensor is buried (as in the case of electronic sensors), a purging system will be required to flush out any trapped air in the reservoir.

The response time of tensiometers is affected by (a) the sensitivity of the measuring gauge; (b) the volume of water in the system; (c) the permeability of the porous filter and (d) the amount of undissolved air in the system.

In the tensiometers manufactured by *Soil Moisture Equipment Inc. USA*, the gauge is either a standard Bourdon type or an electronic transducer. The volume of water varies depending on the tube diameter and length. The porous ceramic filter has an air-entry value of 100 kPa. Response times in these tensiometers are about a few minutes but they are limited by the air-entry value of the porous cup and cavitation to measuring suctions less than 100 kPa.

Small tip tensiometer

A small tip tensiometer with a flexible coaxial tubing is shown in Figure 2.11. The tensiometer is prepared for installation using a procedure similar to the description for the regular tensiometer tube. A vacuum pump can be used initially to remove air bubbles from the top of the tensiometer tube. Subsequent removal of air bubbles can be performed by flushing through the coaxial tube.

High suction tensiometer probe

Ridley and Burland (1993) introduced a new high suction tensiometer (see Figure 2.12) to measure suctions greater than 100 kPa directly in the laboratory. By pre-pressurising the water in the reservoir to inhibit the release of air within the tensiometer and taking advantage of the high tensile strength of water (Harvey et al., 1944; Marinho and Chandler, 1995; Guan and Fredlund, 1997), matric suction of up to 1800 kPa can be measured. Subsequently, Ridley and Burland (1995) also extended the use of this type of high-suction tensiometer probe in the field, together with a technique for excavating a borehole with a flat, horizontal bottom in compacted London clay. Subsequently, Take and Bolton (2003), Jacobsz (2018) and Ng et al. (2023) successfully developed miniature tensiometer suction transducers for use in centrifuge model tests.

34 Advanced Unsaturated Soil Mechanics

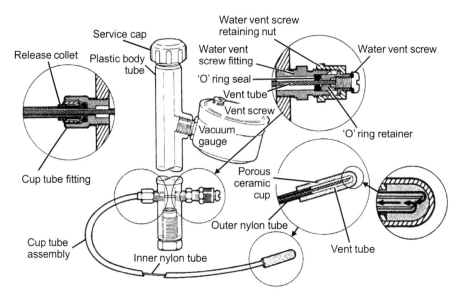

Figure 2.11 Small tip tensiometer with flexible coaxial tubing (courtesy Soilmoisture Equipment Corp.; after Fredlund and Rahardjo, 1993).

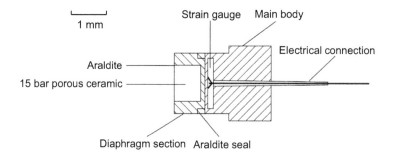

Figure 2.12 The Imperial College tensiometer.

Source: After Ridley and Burland, 1995; Ridley and Wray, 1996.

Squeezing technique

The osmotic suction of a soil can be estimated indirectly by measuring the electrical conductivity of the pore water from the soil (Fredlund and Rahardjo, 1993). Pure water has a lower electrical conductivity by comparison with pore water which contains dissolved salts. The electrical conductivity of the pore water can be measured and used to indicate the total concentration of dissolved salts, which is related to the osmotic suction of the soil.

Pore water can be extracted from the soil using a pore fluid squeezer which consists of a heavy-walled cylinder and piston squeezer. The electrical resistivity (or electrical conductivity) of the pore water is then measured. A calibration curve can be used to relate

Measurement and control of suction   35

the electrical conductivity to the osmotic pressure of the soil. The results of squeezing technique measurements appear to be affected by the magnitude of the extraction pressure applied. Krahn and Fredlund (1972) used an extraction pressure of 34.5 MPa in osmotic suction measurements on glacial till and Regina clay.

Summary

Ridley and Wray (1996) point out that before choosing a particular instrument, it is important for the engineers to decide whether to measure the total suction, the matric suction or both. The direct measurement of osmotic suction can be misleading, and it is probably sufficient to derive a value from the difference between the measured total and matric suctions. An appropriate device can then be chosen based on its range of measurement, simplicity (primarily for installation) and expense. Table 2.1 summarises the principal methods of suction measurement in terms of the type of suction they measure, the range over which they are useable and the response time. In general, those instruments that make contact with the soil usually measure matric suction, while those instruments which do not make contact usually measure total suction.

The measurement of total suction requires special attention to the calibration procedure and equipment design. The psychrometer and the out-of-contact filter paper method are the only techniques available for measuring total suction, but both present difficulties if they are used in the low suction range.

It is in the interests of the engineer to measure soil suction directly. However, there are some difficulties associated with doing this at atmospheric pressure coupled with uncertainties regarding the effect of raising the ambient pressure. They have resulted in the adoption of many indirect methods (e.g., filter paper, psychrometer and thermal conductivity sensor, particularly for *in situ* measurements). However, the indirect methods are not always accurate, and a large number of tests are often necessary to produce a reasonable degree of confidence in the estimate of suction. In addition, indirect techniques such as the filter paper method and psychrometer rely on another indirect method (i.e., salt solutions) for calibration.

Table 2.1 Methods for measuring soil suction

Device	Measurement mode	Range: kPa	Approximate equilibrium time
Thermocouple psychrometer	Total	100 to 7500	Minutes
Thermistor/Transistor psychrometer	Total	100 to 71000	Minutes
Filter paper (in contact)	Matric	30 to 30000	7 days
Filter paper (no contact)	Total	400 to 30000	7–14 days
Porous block	Matric	30 to 3000	Weeks
Thermal conductivity probe	Matric	0 to 300	Weeks
Pressure plate	Matric	0 to 1500	Hours
Standard tensiometer	Matric	0 to 100	Minutes
Tensiometer suction probe	Matric	0 to 1800	Minutes

Source: After Ridley and Wray, 1996.

Tensiometers are the only devices that can truly measure the matric suction directly, but theoretically, the maximum suction which standard tensiometers can measure is 100 kPa. Recent advances in direct measurement of soil suction have extended the range of such measurements to about 1800 kPa in the laboratory environment.

Soil suction is quantitatively defined as the difference between soil pore air pressure and soil pore-water pressure. Although the pressure plate apparatus is regarded as a direct measurement of matric suction because it measures the absolute difference between ambient air pressure and soil pore-water pressure, it does not measure negative water pressures directly. The axis-translation technique is the most common method of measuring and/or controlling matric suction in the laboratory. In the field, however, it is impractical to introduce a stable raised air pressure down a borehole.

SUCTION CONTROL AND MEASUREMENT METHODS (NG AND CHEN, 2005, 2006; NG ET AL., 2007)

Overview

In laboratory studies, an important issue is the control or measurement of suction in an unsaturated soil specimen. Generally, total suction can be controlled by using the humidity control technique (Esteban and Saez, 1988). Matric suction can be controlled using the axis-translation technique (Hilf, 1956) and the osmotic technique (Zur, 1966). Osmotic suction can be controlled using different solutions as pore fluids or changing the solute concentrations of the pore fluid in the soil. In most geotechnical engineering applications, the chemistry of pore fluids in the soil is not changed and soil-water content varies over a range in which the concentrations of pore fluids do not change significantly, so osmotic suction appears to be insensitive to change in soil-water content.

In this section, the working principles, development and applications of the three suction control techniques used in the laboratory are introduced and reviewed. Experimental data using these techniques are compared and discussed. No matter which technique is used, suction equalisation is a vital stage in testing unsaturated soil. To illustrate the influence of suction equalisation on subsequent shearing behaviour, two direct shear tests were performed on a compacted expansive soil applying different durations of suction equalisation under the same applied vertical stress and matric suction.

Axis-translation technique

Working principle

Matric suction is considered to be an important variable in defining the state of stress in an unsaturated soil. Therefore, it is necessary to control or measure matric suction in laboratory studies on unsaturated soils. However, difficulties associated with the measurement and control of negative pore-water pressure present an important practical limitation. Water is normally assumed to have little tensile strength and may therefore start to cavitate when the gauge pressure approaches –1 atmosphere. With suitable conditioning (see Chapter 1), water can withstand tensions of the order of 40–300 atmospheres (Temperley and Chambers, 1946; Young, 1989). As cavitation occurs, the water phase

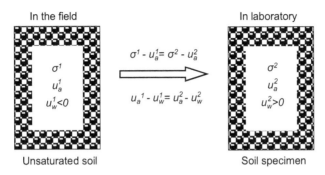

Figure 2.13 Schematic diagram illustrating the axis-translation principle.

Source: Ng and Chen, 2005.

becomes discontinuous, making the measurements unreliable or impossible. Because it is necessary to control the matric suction over a range far greater than 1 atmosphere for many soil types and their applications, alternatives to the measurement or control of negative water pressure are desirable.

Hilf (1956) introduced the axis-translation technique of raising the pore air pressure u_a thereby raising the pore-water pressure above zero, preventing cavitation in the water drainage system. Total stress σ is increased with air pressure by the same amount so the net stress $(\sigma - u_a)$ remains unchanged. As shown in Figure 2.13, the stresses on an unsaturated soil in the field are total stress σ^1, pore air pressure u_a^1 (generally equal to atmospheric pressure) and pore-water pressure u_w^1 (generally negative gauge pressure). When applying the axis-translation technique, total stress is increased from σ^1 to σ^2, pore air pressure is increased from u_a^1 to u_a^2 and pore-water pressure is increased from u_w^1 to u_w^2 (generally positive gauge pressure). The net stress $(\sigma - u_a)$ and matric suction $(u_a - u_w)$ remain unchanged. This process is referred to as "axis translation". Based on the axis translation principle, the matric suction variable $(u_a - u_w)$ can be controlled over a range far greater than the cavitation limit for water under negative pressure. Axis translation is accomplished by separating the air and water phases of the soil through porous material with a high air-entry value. When saturated, these materials allow the passage of water, but prevent the free flow of air when the applied matric suction does not exceed the air-entry value of the porous material. Its air-entry value can be as high as 1500 kPa for sintered ceramics or 15 MPa for special cellulose membranes (Zur, 1966).

Applications

Measurements of SWCC (or WRC) and SDSWCC (or SDWRC)

The axis-translation technique has been applied successfully by numerous researchers to study the soil-water characteristic or water retention ability of unsaturated soils (Fredlund and Rahardjo, 1993; Ng and Pang, 2000a; Ng and Pang, 2000b), as well as the volume change and shear strength properties of unsaturated soils (Fredlund and Rahardjo, 1993; Gan et al., 1988; Ng and Chiu, 2001; Ng and Chiu, 2003; Ng and Zhou, 2005; Ng and Menzies, 2007; Ng et al., 2016a, 2020b).

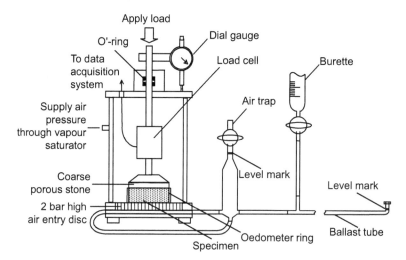

Figure 2.14 A new total stress controllable one-dimensional volumetric pressure plate extractor at HKUST.
Source: After Ng and Pang, 2000b; Ng and Chen, 2005.

The soil-water characteristic curve (SWCC) or water retention curve (WRC) is the relationship between suction and water content or degree of saturation for an unsaturated soil. It is generally accepted that the behaviour of unsaturated soil is governed by at least two stress state variables, i.e., matric suction and net stress (Fredlund and Morgenstern, 1977; Gens et al., 2006; Ng and Menzies, 2007). Therefore, it is necessary to consider the influence of not only the matric suction but also the net stress on the SWCC. However, the SWCC of a soil is conventionally measured by means of a pressure plate extractor with no external stress applied and the volume change of the soil specimen is assumed to be zero. To investigate the influence of net stress as well as matric suction on SWCC, Ng and Pang (2000b) developed a total stress controllable one-dimensional volumetric pressure plate extractor based on the axis translation principle at the Hong Kong University of Science and Technology (HKUST) (see Figure 2.14 and Plate 2.1a). Subsequently, the design of this SDSWCC apparatus was commercialised by Geo-Experts of Earth Products China Limited (refer to Plate 2.1b).

This apparatus can be applied to measure the SWCCs at various vertical stresses under the K_0 condition. An oedometer ring equipped with a high air-entry ceramic plate at its base is located within an airtight chamber. Vertical stress is applied to a soil specimen through a loading frame inside the oedometer ring. To eliminate the error arising from side friction of the loading piston, a load cell is attached near the end of the piston inside the airtight chamber to determine the actual vertical load applied to the soil specimen. Because the radial deformation is zero for the K_0 condition, the total volume change of the specimen is measured from the vertical displacement of the soil specimen using a dial gauge. Using this apparatus, the state-dependent soil-water characteristics curve (SDSWCC) or stress-dependent water retention curve (SDWRC) can be measured and

Measurement and control of suction 39

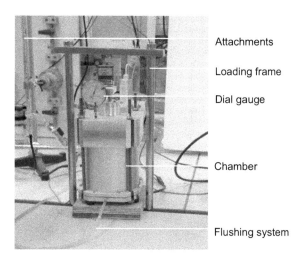

Plate 2.1a A total stress controllable one-dimensional volumetric pressure plate extractor at HKUST.
Source: After Ng and Pang, 2000b; Ng and Chen, 2005.

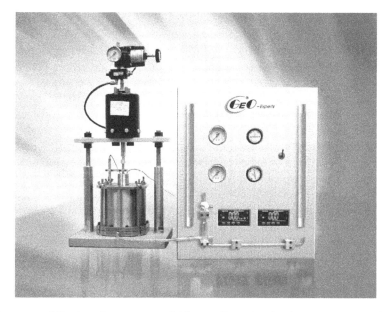

Plate 2.1b A commercialised total stress controllable one-dimensional volumetric pressure plate extractor by Geo-Experts (with the courtesy of Earth Products China Limited).

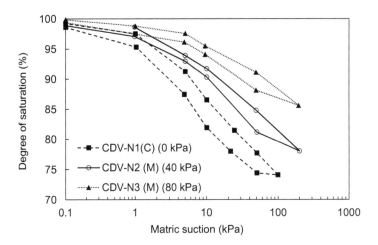

Figure 2.15 Influence of stress state on soil-water characteristics of natural CDV specimens.
Source: After Ng and Pang, 2000a; Ng and Chen, 2005.

it is no longer necessary to assume zero volume change. As with the conventional volumetric pressure plate extractor, the pore air pressure u_a is controlled through a coarse porous stone together with a coarse geotextile located at the top of the specimen. Pore-water pressure u_w is controlled at atmospheric pressure through the high air-entry ceramic plate mounted on the base of the specimen. In addition, some attachments are used to study the hysteresis of the SWCCs associated with drying and wetting of the soil. These are a vapour saturator, air trap, ballast tube, and burette. The vapour saturator is used to saturate the air flowing into the airtight chamber to prevent the soil from drying by evaporation. The air trap is attached to collect air that may diffuse through the high air-entry disc. The ballast tube serves as a horizontal storage for water flowing in or out of the soil specimen. The burette is used to store or supply water and to measure the change of water volume in the soil specimen. A commercial version of the apparatus is shown in Plate 2.1b.

Figure 2.15 shows the influence of stress state on the soil-water characteristics of natural completely decomposed volcanic (CDV) specimens (Ng and Pang, 2000a). The size of the hysteresis loops does not seem to be governed by the applied stress level. The specimens subjected to higher applied stresses possess a slightly higher air-entry value and lower rates of desorption and adsorption as a result of smaller pore sizes. These experimental results demonstrate that the stress state has a substantial influence on the water retention behaviour of unsaturated soils. Ng and Pang (2000b) adopted the measured wetting SDSWCCs to perform numerical analyses on slope stability. They found that during a prolonged low-intensity rainfall, the FOS (factor of safety) predicted using the SDSWCC is substantially lower than that predicted using the conventional drying SWCC. Under highly intense but brief rainfall, however, the two predicted FOS values are close. Therefore, when studying the soil-water retention behaviour of unsaturated soils and their applications, the effects of two independent stress state variables, i.e., net stress and suction, should be considered simultaneously.

Stress path testing in the triaxial apparatus

The triaxial test and the direct shear test (using shear box) are two commonly used shear strength tests. Plate 2.2 illustrates a triaxial system using the axis-translation technique to test unsaturated soils at HKUST (Zhan, 2003).

The triaxial cell is a Bishop and Wesley stress path cell for testing up to 100 mm diameter specimens. In terms of the four GDS pressure controllers, two are automatic pneumatic controllers controlling cell and pore air pressure, and the other two are digital hydraulic pressure/volume controllers controlling back pressure and axial stress. With these four GDS controllers, cell pressure, axial stress, pore air pressure and pore-water pressure can be controlled independently. The six transducers consist of an internal load cell (measuring axial force), a linear variable difference transformer (LVDT) (measuring axial displacement), a differential pressure transducer (DPT) (a component of the total volume change measuring system, see Plate 2.3) and three pressure transducers (monitoring cell pressure, pore-water pressure and pore air pressure).

Figure 2.16 shows the setup of the total volume measuring system (Ng et al., 2002). The basic principle of the double-cell total volume measuring system is that the overall volume change in a specimen is measured by recording the differential pressures between the water in an open-ended, bottle-shaped inner cell and the water in a reference tube using a high-accuracy DPT. The inner cell is sealed onto the pedestal of the outer cell in the triaxial apparatus. The high-accuracy DPT is connected to the inner cell and to a reference tube. The aim is to record changes in differential pressure between them. During a

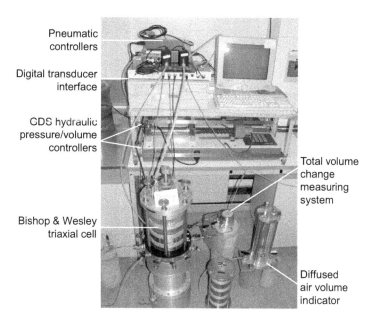

Plate 2.2 A computer-controlled triaxial system for unsaturated soils based on the axis-translation principle at HKUST.

Source: After Ng and Chen, 2005, 2006.

42 Advanced Unsaturated Soil Mechanics

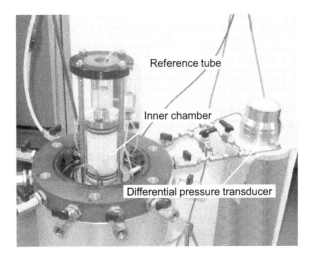

Plate 2.3 A new double-cell total volume change measuring system for unsaturated soils at HKUST.

Source: After Ng et al., 2002.

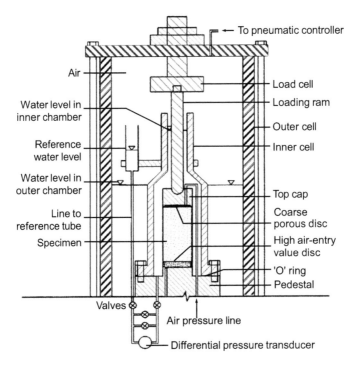

Figure 2.16 Schematic diagram of a new double-cell total volume change measuring system for unsaturated soils at HKUST.

Source: After Ng et al., 2002.

test, the water pressure in the inner cell changes upon a volume change in the specimen, while the water pressure in the reference tube remains constant. Detailed calibrations have been carried out to account for apparent volume changes caused by changes in cell pressure, fluctuations in the ambient temperature, creep in the inner cell wall and relative movement between the loading ram and the inner cell (Ng et al., 2002). The estimated accuracy of the volume change measuring system is of the order of 31.4 mm^3 if the system is properly calibrated. For a test specimen with 100 mm in diameter and 200 mm in height, this corresponds to a volumetric strain of 0.002%. In this system, apparent volume changes caused by changes in cell pressure and fluctuations due to variations of ambient temperature and creep are all smaller than other existing double-cell volume change measuring systems. More detailed comparisons are made by Ng and Chiu (2003). This double-cell total measuring device has been adopted and recommended by an internationally well-known laboratory equipment company – GDS as its Method B (HKUST inner cell) and Method C (double-walled cell) for measuring the volume change of an unsaturated soil specimen. Details are given on their website: www.gdsinstrume nts.com/gds-products/unsaturated-triaxial-testing-of-soil. Moreover, this device is also recommended by Earth Products on their website: www.epccn.com/pro_view.php?id= 3367&language=Ch.

Using the axis translation principle, Ng and Chiu (2001) conducted triaxial stress path tests on a loosely compacted unsaturated completely decomposed volcanic (CDV). Figure 2.17 shows the results of field stress path (wetting) tests at constant deviator stress (i.e., field stress path tests which simulate rainfall infiltration in a slope element). As shown in Figure 2.17a, when the suction falls below its initial value (i.e., 150 kPa), there is a threshold suction above which only small axial strain is mobilised. As the suction drops below this threshold value, the rate of increase in axial strain increases towards the end of the test. The threshold suction increases with applied net mean stress. Figure 2.17b shows the variation of volumetric strain with suction. Like the variation of axial strain with suction, there also exists a threshold suction above which only a small volumetric strain is mobilised. As the suction drops below this threshold suction, contraction is observed for the two specimens (u_a^1 and u_a^2) compressed to net mean stress less than 100 kPa. By contrast, the other two specimens (u_a^3 and u_a^4) show dilation.

Shearing test in the direct shear box

Compared to the triaxial test, the direct shear test is simpler to perform and requires shorter test durations because of the shorter drainage paths. Figure 2.18 shows a direct shear apparatus for unsaturated soils based on the axis-translation principle located in HKUST. It was modified from a conventional direct shear box for testing saturated soils (Gan et al., 1988; Zhan, 2003). The cylindrical pressure chamber was made of stainless steel and designed for pressures up to 1000 kPa. Three holes were drilled to provide the necessary housing for a vertical loading ram and two horizontal pistons (refer to Plate 2.4). Teflon ring seals were installed to ensure an airtight seal around the loading ram and pistons. A high air-entry value ceramic disk was installed in the lower portion of the shear box. The water chamber beneath the ceramic disk was designed to serve as a water compartment as well as a channel for flushing air, as in the pedestal of the tri-axial apparatus for testing unsaturated soils. The desired matric suction is applied to a

44 Advanced Unsaturated Soil Mechanics

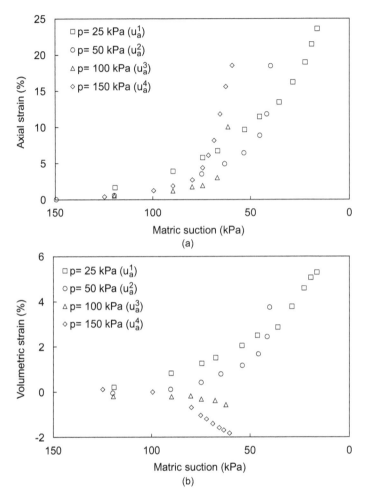

Figure 2.17 Stress path tests on unsaturated CDV triaxial test specimens showing the relationships between (a) axial strain and suction; (b) volumetric strain and suction.

Source: After Ng and Chiu, 2001.

soil specimen by maintaining a constant air pressure in the air pressure chamber and a constant water pressure in the water chamber below the ceramic disk. The pore air pressure and pore-water pressure in the soil are then allowed to reach equilibrium with these applied pressures. Matric suction in the soil is equal to the difference between the applied air and water pressures. The modified direct shear apparatus has five measuring devices. There are two LVDTs for horizontal and vertical displacements, a load cell for shear force, a pressure transducer for water pressure, and a water volume indicator for water volume changes in the soil specimen. All these transducers are connected to a data logger for data acquisition.

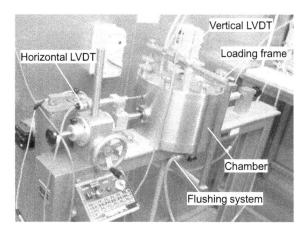

Plate 2.4 A direct shear apparatus for unsaturated soil test specimens.
Source: Ng and Chen, 2005, 2006.

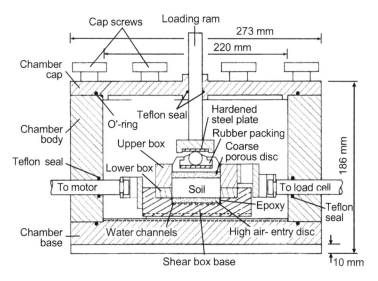

Figure 2.18 Schematic diagram of a direct shear apparatus for unsaturated soil test specimens (after Zhan, 2003). Interpreting the results uses the axis-translation principle.
Source: Ng and Chen, 2005, 2006.

Figure 2.19 shows direct shear results for a completely decomposed granite (CDG) test specimen (Ng and Zhou, 2005). In this figure, the stress ratio is defined as the ratio of shear stress (t) to net vertical stress ($\sigma_v - u_a$). Dilatancy is defined as the ratio of incremental vertical displacement (δy) to incremental horizontal displacement (δx). A negative value (or negative dilatancy) means dilative behaviour. As shown in Figure 2.19a, at zero

46 Advanced Unsaturated Soil Mechanics

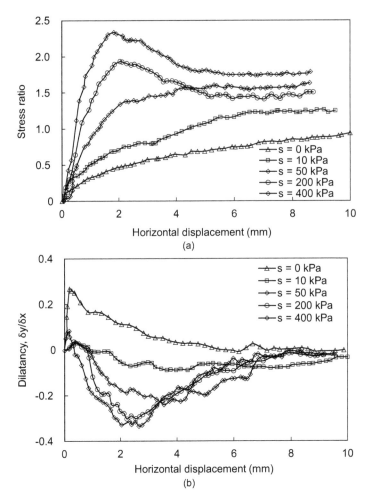

Figure 2.19 Evolution of (a) stress ratio and (b) dilatancy of CDG subjected to direct shear under different controlled suctions.

Source: After Ng and Zhou, 2005.

suction and suctions of 10 kPa and 50 kPa, the stress ratio-displacement curve displays strain hardening behaviour. As suction increases, strain softening behaviour is observed at suctions of 200 kPa and 400 kPa. Generally, measured peak and ultimate stress ratios increase with suction, except for the ultimate stress ratio measured at a suction of 200 kPa. Figure 2.19b shows the effects of suction on dilatancy of the CDG in the direct shear tests. Under saturated conditions, the soil specimen shows contractive behaviour (i.e., positive dilatancy). On the other hand, under unsaturated conditions, all soil specimens exhibit contractive behaviour initially but then dilative behaviour as the horizontal displacement continues to increase. The measured maximum negative dilatancy is increased by an increase in suction. The measured peak stress ratio in each test does not correspond with the maximum negative dilatancy.

Advantages and limitations

In the axis-translation technique, both pore-water pressure and pore air pressure are controlled and measured independently. This enables control of the variation in suction. When a feedback system is used, the suction can be controlled automatically. The majority of experimental results for unsaturated soils have been obtained by applying the axis-translation technique because it is easy to measure and control the suction.

One limitation of the axis-translation technique relates to the maximum value of suction that can be applied. It is limited by the maximum cell pressure (for safety reasons) and the air-entry value of the porous material. Hence, this technique is generally used to control suctions on the order of several hundred kPa.

Another disadvantage of the axis-translation technique is that by elevating the pore-water pressure from a negative to a positive value, the possibility of cavitation is prevented not only within the measuring system but also within soil pores. This implies that any influence of cavitation on the pore water by altering the desaturation mechanism of a soil *in situ* is not accounted for in laboratory tests using the axis-translation technique (Dineen and Burland, 1995). It is then fundamental to understand whether experimental results obtained using the axis-translation technique can be extrapolated to unsaturated soils under atmospheric conditions in the field. Some researchers (Zur, 1966; Williams and Shaykewich, 1969; Ng and Chen, 2005; Ng et al., 2007) have compared experimental data using the axis-translation technique and the osmotic technique where the pore air pressure is the atmospheric pressure. They found that there were some differences in the experimental results using the two techniques. These comparisons will be discussed later in this chapter.

Osmotic technique

Working principle

Due to the limitations of the axis-translation technique, an alternative method, the osmotic technique, has been used when testing unsaturated soils to control matric suction. Delage et al. (1998) reported that this technique was developed initially by biologists (Lagerwerff et al., 1961) and then adopted by soil scientists (Williams and Shaykewich, 1969) and geotechnical researchers (Kassif and Ben Shalom, 1971; Komornik et al., 1980; Cui and Delage, 1996; Delage et al., 1998; Ng and Chen, 2005).

Figure 2.20 illustrates the principle of the osmotic technique for controlling suction in a soil specimen. The specimen and an osmotic solution are separated by a semipermeable membrane. The membrane is permeable to water and ions in the soil but impermeable to large solute molecules and soil particles (Zur, 1966). Therefore, at equilibrium, the component of osmotic suction related to soil salts is the same on both sides of the membrane (i.e., $\pi_{salt}^1 = \pi_{salt}^2$) and the component of osmotic suction related to the solute is zero in the soil. Then the difference in osmotic suction on both sides of the membrane is equal to the

48 Advanced Unsaturated Soil Mechanics

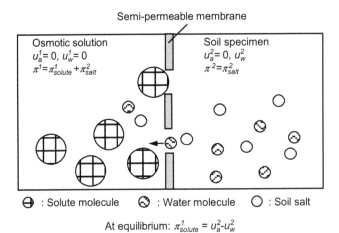

Figure 2.20 Schematic diagram illustrating the osmotic technique.
Source: Ng and Chen, 2005.

component of osmotic suction related to the solute in the solution (i.e., π_{solute}^1). The pore air pressure u_a in the soil is at atmospheric pressure, as in natural field conditions. Zur (1966) presented the principle of the osmotic technique in a review of the energy analysis of soil water. When water exchange through the membrane is in equilibrium, the energy potential in soil water is equal to that in solution water, i.e., total suction in the soil is equal to that in the solution. Therefore, the difference in osmotic suction is equal to the difference in matric suction on both sides of membrane. Since the matric suction in the solution is zero, the matric suction $(u_a^2 - u_w^2)$ in the soil is equal to the difference in osmotic suction on both sides of the membrane, i.e., equal to the component of osmotic suction related to the solute π_{solute}^1. Because of its safety and simplicity, polyethylene glycol (PEG) is the solute most commonly used in biological, agricultural and geotechnical testing. The concentration of PEG in the PEG solution determines the osmotic suction (or osmotic pressure). The maximum value of osmotic pressure in PEG solution was reported to be over 10 MPa (Delage et al., 1998).

Applications

In geotechnical testing, the osmotic technique has been successfully adapted to an oedometer (Dineen and Burland, 1995; Kassif and Ben Shalom, 1971), a hollow cylinder triaxial apparatus (Komornik et al., 1980) and a modified triaxial apparatus (Cui and Delage, 1996; Ng and Chen, 2005).

Collaborative research between HKUST and the Ecole Nationale des Ponts et Chaussees (ENPC) investigated the reliability of the osmotic technique and compared the axis-translation and osmotic techniques (Ng and Chen, 2005). Figure 2.21a illustrates a modified triaxial apparatus using the osmotic technique at ENPC. A soil specimen is put in contact with semipermeable membranes on both the top and bottom surfaces. Concentric

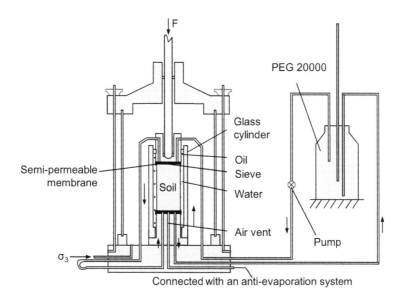

Figure 2.21 Schematic diagram of the ENPC system using the osmotic technique.
Source: After Cui and Delage, 1996.

grooves are machined in the base pedestal and top cap to circulate PEG solution. A thin mesh sieve is placed over the grooves and covered by the membranes, which are glued with epoxy resin. The top cap is connected to the base of the cell by two flexible tubes. A closed circuit comprising the serial connection of the base of the cell, the top cap, a reservoir and a pump contains the PEG solution. The reservoir is large enough to ensure a fairly constant concentration in spite of water exchanges through the membranes between the specimen and the solution. The reservoir is closed with a rubber cap, pierced by three glass tubes. Two of these tubes allow the solution to flow into and out of the triaxial cell; the third is connected to a graduated capillary tube to monitor water exchanges. Stabilisation of the level of the solution in the tube indicates that an equilibrium suction has been reached in the whole specimen. An air vent is machined on the base of the cell in order to ensure a constant pore air pressure equal to the atmospheric pressure within the specimen. At HKUST, the triaxial apparatus is similar to that at ENPC, except that a double-cell volume change measuring system (see Figure 2.16) and an electronic balance for measuring water volume change is used. Plate 2.5 shows a photograph of the triaxial apparatus at HKUST.

Figure 2.22 shows the stress point failure envelopes resulting from triaxial shear using the osmotic technique at HKUST and ENPC on an expansive soil (Ng et al., 2007). In the figure, $s = (\sigma_1 + \sigma_3)/2 - u_a$ and $s = (\sigma_1 - \sigma_3)/2$. Under the same net confining pressure and suction, tests OH4 and OE3 exhibit similar shear strengths. Comparison between tests OH5 and OE6 shows the same conclusion. The consistency of the results from HKUST and ENPC confirms the reliability of the osmotic technique. If the tests at HKUST and ENPC are regarded as one test series, four stress point failure envelopes at different suctions can

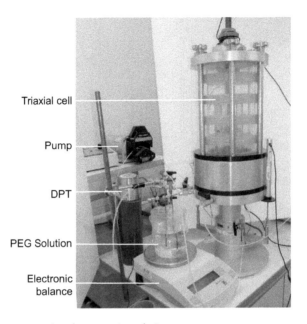

Plate 2.5 Triaxial test systems using the osmotic technique.

Source: Ng and Chen, 2005, 2006.

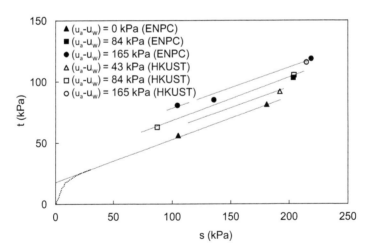

Figure 2.22 Triaxial test results using the osmotic technique at ENPC and HKUST.

Source: After Ng et al., 2007.

be presented. Parallel stress envelopes suggest a constant internal frication angle ϕ' within a suction range from 0 kPa to 165 kPa. The increasing intercepts on the t axis indicate that apparent cohesion increases with suction. There also appears to be a small intercept of 19.4 kPa on the t axis when the failure envelope at zero suction is extended linearly. This apparent intercept is probably caused by experimental error or a curved failure envelope within the low stress range. The comparison of test results between the axis-translation and osmotic techniques will be discussed later.

Advantages and limitations

Compared with the axis-translation technique, the main advantage of the osmotic technique is that the pore-water pressure within the soil is maintained at a negative value. A second advantage is that high values of suction can be applied without using very high cell pressure. The maximum applied suction in the osmotic technique is determined by the maximum osmotic pressure of PEG solution, which has been reported to be above 10 MPa (Delage et al., 1998). However, an evaluation of the performance of three different semipermeable membranes by Tarantino and Mongiovi (2000) indicated that all membranes experience a chemical breakdown when the osmotic pressure of the PEG solution exceeds a threshold value, which was found to depend on the type of membrane. Beyond this value, solute molecules were no longer retained by the semipermeable membrane and passed into the soil specimen, resulting in a reduced concentration gradient and a decay of soil suction. Accordingly, the maximum applied suction in the osmotic technique is further limited by the performance of semipermeable membranes.

A further limitation is that the osmotic technique in its current form cannot be used to control suction continuously because in existing technology, suction changes are applied in steps by manually changing the PEG solutions with ones of different concentrations.

Moreover, the osmotic technique requires calibration of the osmotic pressure against concentration of PEG solution. Calibration of PEG solutions can be carried out with a psychrometer (Williams and Shaykewich, 1969), an osmotic tensiometer (Peck and Rabbidge, 1969), a high suction probe (Dineen and Burland, 1995) or an osmotic pressure cell (Slatter et al., 2000; Ng and Chen, 2005). All these methods involve a semipermeable membrane except the psychrometer method, which requires strict temperature control. It has been found that the relationship between osmotic pressure and the concentration of PEG solution is affected significantly by the calibration method (Dineen and Burland, 1995; Slatter et al., 2000; Ng and Chen, 2005). Dineen and Burland (1995) suggested a need for a means to measure negative pore-water pressure in soil specimens directly when using the osmotic technique. Therefore, the crucial issues in the osmotic technique are not only the calibration of osmotic pressure against PEG concentration but also direct measurement of the applied suction in the soil.

Humidity control techniques

Working principle

The thermodynamic definition in Equation (1.1) shows that total suction can be imposed on an unsaturated soil specimen by controlling the relative humidity of the atmosphere

52 Advanced Unsaturated Soil Mechanics

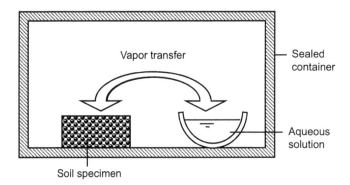

Figure 2.23 Schematic diagram illustrating the principle of humidity control with solutions.
Source: Ng and Chen, 2005.

surrounding the soil. Humidity can be controlled using aqueous solutions (Esteban and Saez, 1988; Delage et al., 1998; Oteo Mazo et al., 1995; Al-Mukhtar et al., 1999; Ng et al., 2016a, 2020b) or by mixing vapour-saturated gas with dry gas via a feedback system (Kunhel and van der Gaast, 1993).

Delage et al. (1998) reported that humidity control using solutions was initially developed by soil scientists and was firstly applied to geotechnical testing by Esteban and Saez (1988). This technique was generally used to study the combined effects of high suction and stress on the mechanical properties of various expansive soils (Lloret et al., 2003; Oteo Mazo et al., 1995; Al-Mukhtar et al., 1999).

Figure 2.23 shows a schematic diagram illustrating the working principle of humidity control by solutions. A soil specimen is placed in a closed thermodynamic environment containing an aqueous solution of a given chemical compound. Depending on the physico-chemical properties of the compound, a given relative humidity is forced on the sealed environment. Water exchanges occur by vapour transfer between the solution and the specimen, and the given suction is applied to the specimen when vapour equilibrium is achieved. The solution can be the same substance at various concentrations (i.e., unsaturated solutions, sulphuric acid or sodium chloride for instance – Oteo Mazo et al., 1995), or various saturated saline solutions (Delage et al., 1998). Saturated saline solutions have the practical advantage over unsaturated solutions. The saturated saline solutions are able to liberate or adsorb relatively large quantities of water without the equilibrium relative humidity being significantly affected.

Kunhel and van der Gaast (1993) reported a system for X-ray Diffraction (XRD) analyses under controlled humidity by mixing vapour-saturated gas with dry gas. Figure 2.24 shows a feedback system to control humidity in a soil specimen. Humidity is controlled by proportionate mixing of vapour-saturated nitrogen gas and desiccated nitrogen gas in a closed environmental chamber. The vapour-saturated and desiccated gas streams are combined in a three-neck flask where the resulting gas stream has a relative humidity that is a direct function of the flow ratio of wet to dry gas. The humid gas stream is routed into an acrylic environmental chamber containing a soil sample. An effluent gas vent on the

Measurement and control of suction 53

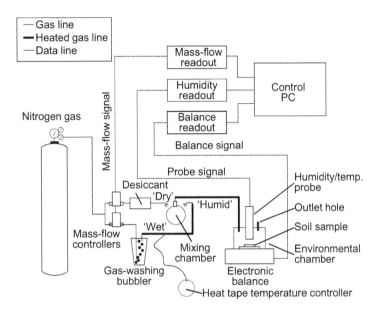

Figure 2.24 General layout diagram for an automatic humidity control system.
Source: After Likos and Lu, 2003.

top cap of the chamber allows the inflowing humid gas to escape after flowing around the soil. Relative humidity and temperature in the chamber are continuously monitored by a polymer capacitance probe. Signals from the probe form a feedback loop with a control computer for automated regulation of the ratio of wet to dry gas, enabling control of the relative humidity (RH). The variation of humidity variation is controlled to approximately 0.6% RH by this system.

Applications

The humidity control technique with solutions is extremely time-consuming because vapour exchanges are quite slow due to the very low kinetics of vapour transfer. Oedometers are often adopted to achieve a small drainage path of soil specimens when applying this technique. Figure 2.25 shows oedometer results for a highly compacted smectite (Na-laponite) under constant relative humidity (Al-Mukhtar et al., 1999). The maximum applied total suction was 298 MPa and the maximum applied vertical stress was 10 MPa. The combined high suction and stress on volume change properties of the expansive soil is illustrated in the figure. The results show that

- void ratio increases as suction decreases (RH increases);
- increasing axial stress reduces the difference in the void ratio of the tested samples;
- the slope of compression curve decreases as suction increases, indicating that suction stiffens the soil.

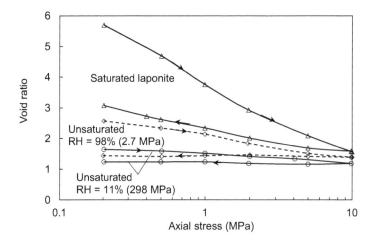

Figure 2.25 Oedometer results for Na-laponite tested under controlled humidity by using solutions.
Source: After Al-Mukhtar et al., 1999.

Likos and Lu (2003) used the system described by Al-Mukhtar et al. (1999) to determine total suction characteristic curves over a high suction range for four types of clay, ranging from highly expansive smectite to nonexpansive kaolinite. Figure 2.26 shows the measured total suction characteristic curves for the expansive smectite and nonexpansive kaolinite. Both drying and wetting paths can be performed using the testing system. The values of total applied suction encompass a large range (7 MPa to 700 MPa). The expansive smectite contains more water than the kaolinite under the same applied total suction, exhibiting a larger capacity for water adsorption. The expansive smectite shows larger rate of change in water content due to suction and larger hysteresis than the kaolinite.

Advantages and limitations

Compared with the osmotic technique, humidity control techniques can maintain the pore-water pressure within a soil specimen at a negative value when applying high suction values without using a very high cell pressure. When using solutions to control humidity, very high values of suction (up to 1000 MPa) can be applied. However, Delage et al. (1998) pointed out that uncertainties in this technique limit applications to less than 10 MPa suction. When using a feedback system to control humidity, the suction control range is determined by the measurement range and accuracy of the humidity probe. For the results given in Figure 2.26, the suction control range was from 7 MPa to 700 MPa (Likos and Lu, 2003).

When using solutions to control humidity, only fixed values of suction can be applied and cannot be used to vary suction continuously. If the humidity is controlled by mixing vapour-saturated gas with dry gas via a feedback system, suction can be controlled automatically. The accuracy of the method depends on both the accuracy of the humidity probe and the resolution of the automated control loop regulating the relative humidity.

When humidity is controlled using solutions, the temperature must be strictly controlled during testing as the activity of the solutions is very sensitive to thermal fluctuations. The

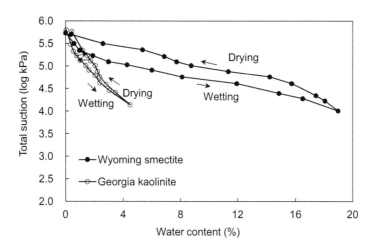

Figure 2.26 Measured total suction characteristic curves for Wyoming smectite and Georgia kaolinite under controlled humidity by using a feedback system.

Source: After Likos and Lu, 2003.

control of suction by this technique is much slower than in the techniques involving liquid transfer (i.e., axis-translation and osmotic techniques) because of the very low kinetics of vapour transfer. When humidity is controlled by proportioning vapour-saturated gas with dry gas, test durations can be reduced because of active gas circulation. Likos and Lu (2003) reported that the equilibrium water content of several clays was reached within 12 hours when changes in humidity in the soils were controlled by the system described in Figure 2.24. The short suction equalisation duration (within 12 hours) is very questionable, since it has been reported that testing durations of months were required when humidity was controlled using solutions (Oteo Mazo et al., 1995).

Characteristics of the three suction control techniques (i.e., axis-translation, osmotic and humidity control) are summarised in Table 2.2.

Influence of suction equalisation

No matter which method is used to control suction, a vital issue in testing unsaturated soils is suction equalisation. To investigate the influence of suction equalisation on subsequent shearing behaviour, two direct shear tests were performed on an expansive soil subjected to different durations of suction equalisation.

The expansive soil was taken from Zaoyang, Hubei province of China. It has 3% sand, 58% silt and 39% clay. Therefore, it is classified as a silty clay (BSI, 1990). The liquid limit and plasticity index of the soil are 50.5% and 31%, respectively. Soil specimens for direct shear tests were compacted statically to a dry density of 1.56 g/cm^3 with an initial degree of saturation of 69%. The initial matric suction of the specimens was about 540 kPa, measured by a high suction probe (Zhan, 2003).

The direct shear apparatus using the axis-translation principle as shown in Figure 2.18 and Plate 2.4 was used to perform direct shear tests. Zero suction was applied to two

56 Advanced Unsaturated Soil Mechanics

Table 2.2 Characteristics of three suction control techniques

	Axis-translation	Osmotic	Humidity control	
			Using solution	Using feedback system
Controlled suction	Matric	Matric	Total	Total
Automatisation	Automatic	Manual	Manual	Automatic
Suction range	Zero to hundreds of kPa	Zero to 10 MPa (maximum is limited by performance of semipermeable membrane)	10 MPa to 1000 MPa	Determined by measuring range and accuracy of humidity probe
Similarity with field condition	Positive pore-water pressure	Negative pore-water pressure (similar to field condition)	Negative pore-water pressure (similar to field condition)	Negative pore-water pressure (similar to field condition)
Verification	Valid (continuous air phase) Controversial (occluded air phase)	NA	NA	NA
Requirement	Continuous air and water phases	Calibration of PEG solution or direct measurement of negative pore-water pressure	Strict temperature control; time-consuming	Accurate humidity probe and feedback loop

Source: Ng and Chen, 2005, 2006.

specimens and their suction equalisation durations were 2 hours for Test A and 1 day for Test B. Then the two specimens were sheared at a rate of 0.0019 mm/min under the same applied suction (i.e., 0 kPa) and vertical stress (i.e., 50 kPa).

Figure 2.27 shows the results of direct shear tests with different durations of suction equalisation. In Figure 2.27a, the measured relationships between shear stress and horizontal displacement are reported. The curves of the two tests show distinct behaviours. The curve of Test A exhibits strain softening behaviour. The curve of Test B shows strain hardening behaviour. The influence of duration of suction equalisation on the shear stress-displacement behaviour is similar to that of suction as shown in Figure 2.19a. Each curve reaches a steady value at a displacement in excess of 2.5 mm. The steady value decreases with increasing duration of suction equalisation. The difference in the steady values of shear stress is about 12%. The difference in the maximum values of shear stress is about 18%. These differences may be caused by nonequilibrium of suction at the end of the suction equalisation period and will be discussed later.

Figure 2.27b shows the relationships between vertical displacement (positive value denotes soil contraction) and horizontal displacement. The curves of the two tests show opposite behaviours in terms of volume change. When the suction equalisation duration increases from 2 hours to 1 day, the volume change behaviour changes from dilative to contractive. Again, the influence of duration of suction equalisation on volume change is similar to that of suction as shown in Figure 2.19b. The volume change of the two tests

Measurement and control of suction 57

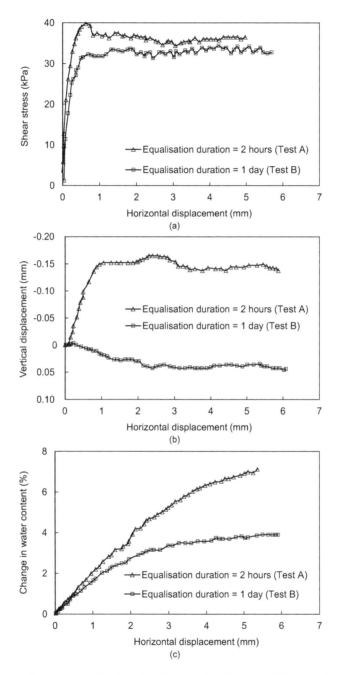

Figure 2.27 Influence of suction equalisation duration on shearing tests: (a) stress-displacement curves; (b) vertical displacement versus horizontal displacement; (c) water content change versus displacement. Test results from the HKUST direct shear apparatus using the axis-translation principle.

Source: Ng and Chen, 2005, 2006.

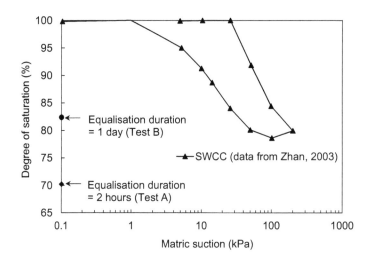

Figure 2.28 Degree of saturation before shearing in direct shear tests.

Source: Zhan, 2003; Ng and Chen, 2005, 2006.

tends to a steady volume at a horizontal displacement in excess of 2.5 mm. Figure 2.27c shows the change in gravimetric water content with horizontal displacement. Both tests showed a continuous adsorption of water during shearing. The magnitude of water absorption decreases as the applied duration of suction equalisation increases.

Figure 2.28 shows the degree of saturation of Tests A and B just before shearing. Wetting and drying SWCCs for this expansive soil are also shown in this figure. Since the value of matric suction after specimen preparation was relatively large (i.e., 540 kPa), the specimens in Tests A and B were subjected to a wetting path at an applied suction of 0 kPa. The degree of saturation of the two specimens increases with the applied duration of suction equalisation and is less than the value obtained from the wetting SWCC at the suction of 0 kPa. The degree of saturation is considered to indicate actual suction in the specimens. The actual suction decreases with the applied duration of suction equalisation, however, and is larger than the applied suction (i.e., 0 kPa) for the same degree of saturation. This is similar to the influence of suction as shown in Figure 2.19. The higher value of actual suction in Test A resulted in larger strength, more dilative behaviour and stronger water adsorption during the shearing stage as compared with Test B. These tests illustrate that it is vital to reach equalisation of suction before shearing.

COMPARISONS OF DIFFERENT SUCTION CONTROL TECHNIQUES (NG ET AL., 2007; NG ET AL., 2015A)

Introduction

Various techniques to control suction have been developed including the axis-translation technique (Hilf, 1956) and the osmotic technique (Zur, 1966) to control matric suction, and the humidity control technique (Tessier, 1984) to control total suction (Agus and

Schanz, 2005; Ng et al., 2007; Tarantino et al., 2011; Laloui et al., 2013; Ng et al., 2016a, 2020b). The axis-translation technique is the most widely used to determine unsaturated soil properties, but it has been criticised for several reasons. First, this technique artificially elevates u_a and hence it is not representative of field conditions where u_a is generally at atmospheric pressure (Dineen and Burland, 1995; Delage et al., 2008). The elevation of u_a increases the pore-water pressure (u_w) and eliminates the possibility of cavitation in the pore water (Or and Tuller, 2002; Baker and Frydman, 2009). Second, the axis-translation technique may not be valid in the nearly saturated state because applying u_a will compress occluded air bubbles (Bocking and Fredlund, 1980). Third, some researchers have postulated that matric suction is attributed not only to capillary but also to adsorptive forces (Philip, 1977; Tuller et al., 1999; Baker and Frydman, 2009). The influence of the air pressurisation process on u_w is not fully understood when water is held by adsorption (Baker and Frydman, 2009). Thus, it is important to compare different suction control techniques.

Theoretical considerations of soil moisture transfer

Thermodynamic theory was introduced by Edlefsen and Anderson (1943) to study moisture transfer in unsaturated soil. In their study, the energy state of soil water is represented by free energy per unit mass (f) [SI unit: J/kg]. f controls water movement and phase transformation in unsaturated soil. Water flows from a region of higher f to a region of lower f, and transforms from a phase with a higher f to a phase with a lower f. Free energy is equivalent to other widely used terminologies, for example, potential (ψ) [SI unit: J/kg] (Sposito, 1981). ψ is defined as $f - f_0$, where f_0 is free energy per unit mass at a reference state. For simplicity, potential is used to describe the energy state of soil water in this study.

Figure 2.29a shows an idealised configuration of a representative unsaturated soil subjected to a confining pressure (σ). Water in unsaturated soil can be present as liquid and gaseous phases, the latter being contained in pore air. The liquid and gaseous phases of soil water are called pore-water and pore vapour. Their potentials are denoted by ψ_w and ψ_v, respectively. To interpret suction effects on the mechanical behaviour of unsaturated

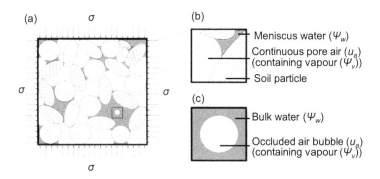

Figure 2.29 Idealised configuration of an unsaturated soil: (a) Representative elementary volume of a soil-water air system; (b) Air water interface with continuous air phase; (c) Air water interface with discrete air bubble.

60 Advanced Unsaturated Soil Mechanics

soil, pore-water in this configuration was classified into meniscus water (see Figure 2.29b) and bulk water (see Figure 2.29c) (Wheeler et al., 2003). The energy state (ψ_w) of these two types of water should be identical at thermodynamic equilibrium, otherwise there would be water movement within the soil specimen. This implies that it is unnecessary to differentiate bulk water and meniscus water from an energy state point of view.

Figures 2.29b and 2.29c show the air water interfaces in unsaturated soil when pore air is continuous and discrete, respectively. At thermodynamic equilibrium, there should not be any net mass transfer of soil water across the air water interface between liquid and gaseous phases and ψ_v is equal to ψ_w. The potential of soil water can thus be controlled through either ψ_w (axis-translation technique and osmotic technique or ψ_v (humidity control technique). These two methods are equivalent as long as thermodynamic equilibrium is achieved.

As shown in Figure 2.29, the energy states of pore vapour and pore water (both bulk water and meniscus water) should be the same in the thermodynamic state. Equation (1.1) suggests that ψ_v is uniquely related to relative humidity under isothermal conditions. This is the principle of the humidity control technique, which imposes soil suction by controlling the relative humidity of pore air in unsaturated soil.

For the energy state of pore water, the variation in ψ_w can be presented in the following incremental form (Sposito, 1981):

$$d\psi_w = -\overline{S_w}dT + \overline{V_w}d\sigma + \left(\frac{1}{\rho_w} - \overline{V_w}\right)du_a + \left(\frac{\partial\psi_w}{\partial w}\right)_{T,\sigma,u_a} dw \tag{2.5}$$

where $\overline{S_w}$ is the specific entropy of pore water; σ is total mean stress; u_a is pore air pressure; and $\overline{V_w}$ is the partial specific volume of pore water, defined as the ratio of incremental soil volume to incremental water mass in the soil specimen. For a nondeformable soil specimen, $\overline{V_w}$ is equal to 0. For a deformable soil specimen, $\overline{V_w}$ is equal to $1/\rho_w$ in the saturated state and decreases as the degree of saturation decreases.

Based on the thermodynamic theory, Houlsby (1997) showed that a complete description of unsaturated soil behaviour requires at least two stress state variables. To enable rigorous yet simple interpretation of experimental data, this study adopts the most commonly used two independent stress state variables (Jennings and Burland, 1962), namely net stress ($\sigma - u_a$) and matric suction ($u_a - u_w$). Equation (2.5) can thus be rearranged as follows:

$$d\psi_w = \frac{1}{\rho_w}du_a + \overline{V_w}d(\sigma - u_a) + \left(\frac{\partial\psi_w}{\partial w}\right)_{T,\sigma-u_a,u_a} dw - \overline{S_w}dT \tag{2.6}$$

The first term on the right-hand side of Equation (2.6) suggests that ψ_w increases linearly with u_a. This term is equivalent to the gas potential defined by Aitchison (1965). The second term represents the effects of net stress ($\sigma - u_a$) on ψ_w and hence WRC of unsaturated soil. It should be pointed out some recent studies have shown that WRC is not only dependent upon matric suction ($u_a - u_w$) but it is also governed by net stress, which is one of the stress state variables governing unsaturated soil behaviour (Coleman,

1962). The water retention ability of unsaturated soil generally increases with an increase in net stress (Ng and Pang, 2000b; Ng and Menzies, 2007; Ng and Leung, 2012a; Zhou and Ng, 2014). The third term describes the change in ψ_w with water content and is governed by the water retention ability of soil specimen. It is important to note that this term is expressed using $\sigma - u_a$ instead of σ to align with the framework of two stress state variables. The last term represents thermal effects on the energy state of soil water. Equation (2.6) can be rewritten to describe the incremental water content of unsaturated soil:

$$dw = \frac{1}{\left(\dfrac{\partial \psi_w}{\partial w}\right)_{T,\sigma-u_a,u_a}} \left[d\psi_w - \frac{1}{\rho_w}du_a + \overline{V}_w d\left(\sigma - u_a\right) + \overline{S}_w dT \right] \tag{2.7}$$

At constant net stress and temperature, Equation (2.7) can be simplified as follows:

$$dw = \frac{1}{\left(\dfrac{\partial \psi_w}{\partial w}\right)_{T,\sigma-u_a,u_a}} \left[d\psi_w - \frac{1}{\rho_w}du_a \right] \tag{2.8}$$

The numerator on the right-hand side of this equation is a function of ψ_w and u_a, where the denominator $\left(\partial \psi_w / \partial w\right)_{T,\sigma-u_a,u_a}$ depends on the relationship between ψ_w and w. Equation (2.8) is derived to describe soil moisture exchange in soil pores. It is applied later to analyse the WRCs obtained using different suction control techniques.

Water retention curves determined using different suction control techniques

Figures 2.30 and 2.31 compare the WRCs of several soils obtained using the axis-translation technique (ATT) and the osmotic technique (OMT). Some researchers have argued that the artificial elevation of u_a in the axis-translation technique may alter the water retention capacity of unsaturated soil (Or and Tuller, 2002; Baker and Frydman, 2009). However, it is clear in the two figures that the WRCs obtained using the axis-translation technique and the osmotic technique are remarkably consistent. The maximum percentage difference in equilibrium w at any given suction is less than 5%. This implies that the artificial increase of u_a in the axis-translation technique does not affect the equilibrium w noticeably at a given suction. Hence, the term $\left(\partial \psi_w / \partial w\right)_{T,\sigma-u_a,u_a}$ in Equation (2.8) can be simplified to $\left(\partial \psi_w / \partial w\right)_{T,\sigma-u_a}$. Equation (2.8) can be reduced further to the following:

$$dw = \frac{1}{\left(\partial \psi_w / \partial w\right)_{T,\sigma-u_a}} \left[d\psi_w - \frac{1}{\rho_w}du_a \right] \tag{2.9}$$

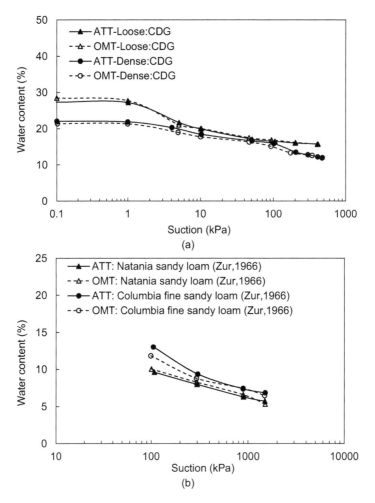

Figure 2.30 Comparisons of drying WRCs determined using the ATT and the OMT of coarse-grained soils: (a) new data from this study; (b) data from Zur (1966).

This equation suggests that the final w at thermodynamic equilibrium must be the same regardless of the suction control technique used. The observed difference at higher w is probably caused by experimental errors. As suggested by Tarantino et al. (2011), in the nearly saturated state, occluded air bubbles may be present in the unsaturated soil. These air bubbles would be compressed when u_a is artificially increased in the axis-translation technique, resulting in an overestimate of w during transient states (Marinho et al., 2008). When thermodynamic equilibrium is reached, however, the equilibrium w should not be affected significantly by the artificial elevation of u_a. To minimise experimental errors, the test duration should be long enough for the soil specimen to achieve thermodynamic equilibrium (Gee et al., 2002).

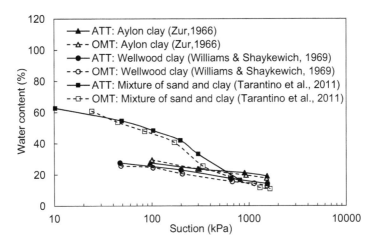

Figure 2.31 Comparisons of drying WRCs obtained using the ATT and the OMT of fine-grained soils.

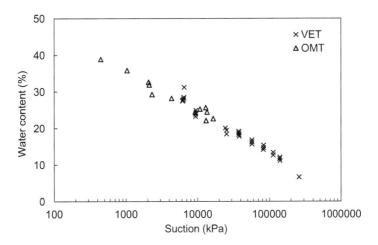

Figure 2.32 Comparisons of drying WRCs of Foca7 clay determined using the OMT and the VET.

Source: Data from Delage et al., 1998.

Figure 2.32 compares the WRCs of one clay measured using the osmotic technique (OMT) and the humidity control technique (VET, Delage et al., 1998; Tarantino et al., 2011). The WRCs measured using the two techniques show smooth continuity and consistency. At a given suction, the equilibrium w obtained using the osmotic technique is almost identical to that obtained using the humidity control technique. The difference in measured water content is not more than 5%. Moisture exchange between the soil specimen and the solution generally occurs by liquid water transfer in the osmotic technique and by vapour transfer in the humidity control technique. Experimental evidence suggests

64 Advanced Unsaturated Soil Mechanics

that these two techniques produce remarkably consistent WRCs for unsaturated soil at zero osmotic suction. This is because the energy states of pore water (liquid phase) and pore vapour (gaseous phase) should be identical in the thermodynamic equilibrium, as illustrated in Figure 2.29. The suction of unsaturated soil can be controlled through either the liquid phase or the gaseous phase. The change in w is independent of the form of moisture exchange (see Equation (2.9)) as long as thermodynamic equilibrium is achieved.

Summary

Water retention curves and shear strengths obtained using the axis-translation technique, the osmotic technique and the humidity control technique have been compared and analysed over a wide range of suctions. Considering possible experimental errors such as suction calibration in the osmotic technique, analysis of the results reveals that the experimental results obtained from the three techniques are generally consistent. The consistency between the experimental results obtained by different suction-controlled techniques can be explained theoretically from the derived thermodynamic equations. This is because the final soil-water content in the thermodynamic equilibrium state must be the same regardless of the suction control technique used, even though the techniques involve different moisture exchange processes.

SOURCES

Theory of soil suction (Fredlund and Rahardjo, 1993)
Suction devices (Fredlund and Rahardjo, 1993; Ridley and Wray, 1996; Feng and Fredlund, 2003)
Suction control methods (Ng and Chen, 2005, 2006)
Comparisons of different suction control techniques: theoretical and experimental studies (Ng et al., 2007; Ng et al., 2015a)

Chapter 3

Retention characteristics and permeability functions for water and gas flows

INTRODUCTION (NG AND SHI, 1998)

Water flow through unsaturated soils is governed by the same physical law – Darcy's law – as fluid flow through saturated soils. The major difference between water flow in saturated and unsaturated soils is that the coefficient of permeability (or hydraulic conductivity), which is conventionally assumed to be a constant in saturated soils, is a function of degree of saturation or matric suction in unsaturated soils. The pore-water pressure generally has a negative gauge value in the unsaturated region, whereas the pore-water pressure is positive in the saturated zone. Despite the differences, the formulation of the partial differential flow equation is similar in the two cases.

The governing differential equation (Lam et al., 1987) for water flow through a two-dimensional unsaturated soil element is as follows:

$$\frac{\partial}{\partial x}\left(k_x\frac{\partial h}{\partial x}\right)+\frac{\partial}{\partial y}\left(k_y\frac{\partial h}{\partial y}\right)+Q=\left(\frac{\partial\theta_w}{\partial t}\right) \tag{3.1}$$

where h is the total hydraulic head; k_x and k_y are the water permeability in the x-direction and y-direction, respectively; Q is the applied boundary flux; θ_w is the volumetric water content. The equation illustrates that the sum of the rates of change of flows in the x-direction and y-direction plus an externally applied flux is equal to the rate of change of the volumetric water content with respect to time.

The amount of water stored within the soil depends on water retention characteristics of the soil and hence a water retention curve (see Figure 3.1). The slope of the curve represents the retention characteristics of a soil, i.e., it represents the rate of water taken or released by the soil as a result of a change in the pore-water pressure. At 100% saturation, the volumetric water content is equivalent to the soil porosity.

Different from saturated soil, the hydraulic conductivity or water permeability is highly dependent on the water content of an unsaturated soil. This is because of the heterogeneous volume distribution of the pore-water (i.e., water content or matric suction) within the soil mass. Figure 3.2 illustrates the desaturation of a soil by the withdrawal of the air-water interface at different stages of matric suction or degree of saturation (i.e., stages 1–5) (from Childs, 1969; after Fredlund and Rahardjo, 1993). Water flows along a web of interconnected but continuous conduits, and as the water content increases,

DOI: 10.1201/9781003480587-3

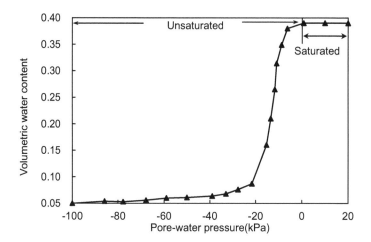

Figure 3.1 A typical water retention curve.

the size and number of conduits increase, thereby enhancing the capability to conduct water through the soil, as shown in Figure 3.3. The higher the water content, the shorter the water flow path and hence the larger the water permeability. On the contrary, at low water content, water has to flow through a torturous flow path leading to low water permeability of soil.

If a unique relationship between water content and pore-water pressure can be assured (i.e., no hysteresis of a WRC), the water permeability is thus also a function of pore-water pressure as shown in Figure 3.4.

For an isotropic unsaturated soil, the constitutive equation for the water phase (Lam et al., 1987) is

$$\partial \theta_w = m_a \partial(\sigma - u_a) + m_w \partial(u_a - u_w) \qquad (3.2)$$

where m_a and m_w can be regarded as constants for a particular time step during a water transient process. Under saturated conditions, m_a is equivalent to the conventional coefficient of volume change, m_v. For transient seepage analysis, it can be assumed that the total stress in the soil mass and the pore-air pressure remain constant. This means that $(\sigma - u_a)$ does not have any effects on the change in volumetric water content. In addition, no hysteresis is assumed between drying and wetting paths in the water retention curve (as shown in Figure 3.1). Then, a change in volumetric water content can be related to a change in pore-water pressure by

$$\partial \theta_w = m_w \partial u_w \qquad (3.3)$$

Substituting Equation (3.3) into (3.1) leads to the following governing differential equation for water flow in unsaturated soils:

Retention characteristics and permeability functions 67

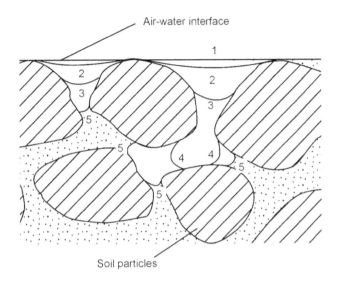

Figure 3.2 Development of an unsaturated soil by the withdrawal of the air-water interface at different stages of matric suction or degree of saturation (i.e., stages 1–5).

Source: From Childs, 1969; after Fredlund and Rahardjo, 1993.

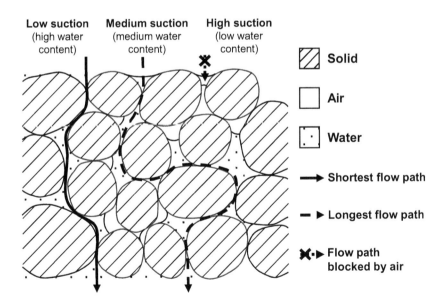

Figure 3.3 Relationships between water flow paths and distributions of water content (suction).

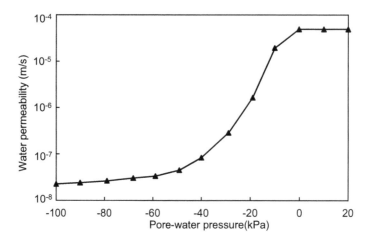

Figure 3.4 A typical nonlinear water permeability function.

$$\frac{\partial}{\partial x}\left(k_x \frac{\partial h}{\partial x}\right) + \frac{\partial}{\partial y}\left(k_y \frac{\partial h}{\partial y}\right) + Q = m_w \left(\frac{\partial u_w}{\partial t}\right) \quad (3.4)$$

where m_w is equal to the slope of the water retention curve (see Figure 3.1), which can readily be determined by experiments (Fredlund and Rahardjo, 1993). If the soil permeability is homogenous and isotropic, i.e., $k_x = k_y = k$ and $Q = 0$, Equation (3.4) can be reduced to the following equation:

$$\nabla^2 h = \left(\frac{\gamma_w m_w}{k}\right)\left(\frac{\partial h}{\partial t}\right) \quad (3.5)$$

From Equation (3.5), it can be deduced that $S_s = \gamma_w m_w$. For water flow in saturated soils, $u_a = u_w$, then Equation (3.2) becomes:

$$\partial \theta_w = m_a \partial (\sigma - u_w) \quad (3.6)$$

where m_a is equivalent to the coefficient of volumetric change, m_v, which is common to saturated soil mechanics.

SOIL-WATER CHARACTERISTIC CURVE (SWCC) AND WATER PERMEABILITY FUNCTION (FREDLUND ET AL., 2001A, 2002)

Introduction

What is a soil-water characteristic curve (SWCC) or a water retention curve (WRC)? It is the relationship between water content and suction of a soil. Hydraulically and physically, it means how much equilibrium water a soil can accept at a given original suction. Although SWCCs are taking on an increasingly important role in the application of unsaturated soil mechanics to geotechnical and geoenvironmental engineering, there is a lack of consistency in the terminology used to describe the relationship between the amount of water in a soil and soil suction (Fredlund et al., 2001b). The lack of a consistent terminology is primarily the result of numerous disciplines being involved with measuring and using the data. The desire, however, is to encourage greater consistency in the terminology of the geotechnical and geoenvironmental engineering disciplines. Some common terms used when referring to the relationship between the amount of water in the soil and soil suction, are as follows (Fredlund et al., 2001b):

- soil-water characteristic curve (SWCC)
- water retention curve (WRC)
- suction-water content relationship
- moisture retention curve
- soil moisture retention curve and
- numerous other terms.

Conventionally, a SWCC is determined from a drying path without considering the effects of volume changes and stress state. To address this problem, Ng and Pang (2000a, 2000b) developed a novel stress-controllable volumetric pressure plate. Plate 3.1 shows a commercial product manufactured by Geo-Expert in recent years. Ng and Pang (2000a, 2000b) have demonstrated experimentally and numerically the relevance and importance of considering these two effects and introduced the concept of stress-dependent soil-water characteristic curve (SDSWCC) and stress-dependent water retention curve (SDWRC). In this book, SWCC and WRC are reserved for a drying curve without including the effects of volume changes and stress state as usual, unless stated otherwise. On the other hand, the terms "SDSWCC" and "SDWRC" are used exclusively for a stress-dependent soil-water characteristic curve which incorporates the effects of two stress variables (i.e., suction and net mean stress) and volume changes. Details of the measurement of an SDSWCC are described in Chapter 2.

Descriptions of water content

Numerous observations have been used as a measure of the amount of water in the soil. The three most basic measures are:

- volumetric water content, θ_w,
- gravimetric water content, w, and
- degree of saturation, S_r.

All three of the above variables convey similar information to the engineer provided the structure of the soil is essentially incompressible. Clayey soils may undergo volume change as a result of a soil suction increase. In this case, it is the degree of saturation variable, S_r, that indicates the air-entry value of the soil.

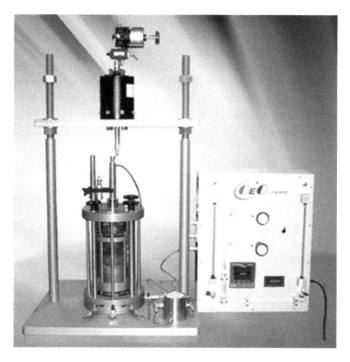

Plate 3.1 A photograph showing a triaxial stress-controllable volumetric pressure plate extractor equipped with an inner cell for measuring volume change of a specimen.

Source: Ng, 2010.

Volumetric water content is most commonly used in agriculture-related disciplines when plotting soil-water characteristic curve data. Volumetric water content is defined as instantaneous water content in the sense that the volume of water in the soil is referenced to the present total volume. However, it is generally not the instantaneous volumetric water content that is being plotted since the volume of water is usually referred back to the original volume of soil, and volume changes are not considered generally. Consequently, both the gravimetric water content and the volumetric water content representations portray similar information.

It is recommended that gravimetric water content be used when plotting the WRC when there are no continuous volume measurements. At the same time, it is recognised that volumetric water content appears in the formulation of transient flow processes when using a referential element. Care must be exercised when using volumetric water content to ensure that continuous volume measurements are made, and the correct reference volume is used in both the mathematical formulations and the laboratory measurements. If so, θ_w, or, S_r, can be used when plotting the WRC. If the volume of water in the soil is always referenced back to the original total volume, then it is better that this variable simply be referred to as V_w/V_t, where V_w is the present volume of water and V_t is the initial total volume of the soil specimen.

If the WRC data is to be used to estimate unsaturated soil properties, it is beneficial to normalise or dimensionalise the water content scale. The dimensionless water content, Θ_d, can be defined as follows:

$$\Theta_d = \frac{w}{w_s} \tag{3.7}$$

where w = any gravimetric water content and w_s = gravimetric water content at saturation.

The total range of water contents from the saturated state under zero confining pressure to zero water content, are reduced to a scale of 1.0 to 0.0, respectively. The normalised water content, Θ_n, can be defined as follows:

$$\Theta_n = \frac{w - w_r}{w_z - w_r} \tag{3.8}$$

where w_r = gravimetric water content at the start of residual conditions.

Terminology used to quantify a WRC

Variables that should be defined from the WRC are, if possible:

- saturated water content, w_s
- air-entry value, ψ_a
- inflection point suction, ψ_{inf}
- inflection point water content, w_{inf}
- slope at the inflection point, n_{inf} (rate of desorption (drying) and absorption (wetting)),
- residual suction, ψ_r,
- residual water content, w_r, and
- water-entry value, ψ_{wev}

Figure 3.5 illustrates an ideal WRC along with the definition of key variables to describe its drying and wetting components. Straight lines on a semi-log plot are necessary for the constructions. Along the drying path, the first line is horizontal through saturation water content. The second line passes tangent to the steepest portion of the WRC. The third line passes through 1000000 kPa and extends back through data points at high suctions (provided the data are available). Similarly, water-entry value and adsorption rate along the wetting branch are defined and illustrated in the figure.

Mathematical forms of SWCC or WRC

Fredlund (2000) reviews several mathematical equations that have been proposed to describe the SWCC. The equation of Gardner (1958) was originally proposed to define the unsaturated coefficient of permeability function, and its application to the SWCC is inferred. The mathematical equations proposed by Burdine (1953) and Mualem (1976) are two-parameter equations that become special cases of the more general three-parameter

72 Advanced Unsaturated Soil Mechanics

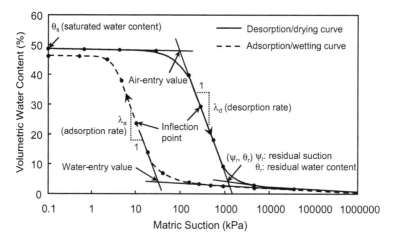

Figure 3.5 A typical water retention curve.

equation proposed by van Genuchten (1980). These equations are asymptotic to horizontal lines at low soil suction and a suction beyond residual conditions. Therefore, these equations are not forced through zero water content at 1000000 kPa of suction. A correction factor, C_r, has been applied to the equation proposed by Fredlund and Xing (1994). The correction factor forces the SWCC function through a water content of zero at a suction of 1000000 kPa. A complete summary of proposed equations can be found in Sillers (1997). All the proposed equations provide a reasonable fit to SWCC data in the low- and intermediate-suction ranges (Leong and Rahardjo, 1997). In all cases, the parameter a bears a relationship to the air-entry value of the soil and usually refers to the inflection point along the curve. The parameter n_1 corresponds to the slope of the straight-line portion of the main desorption (i.e., drying) or adsorption portion of the SWCC.

The Fredlund and Xing (1994) function applies over the entire range of soil suctions from 0 to 1000000 kPa. The relationship is essentially empirical and, as with earlier models, assumes that soil consists of a set of interconnected pores that are randomly distributed. Subsequent discussions of the SWCC are restricted to the Fredlund and Xing's equation.

The Fredlund and Xing (1994) equation, written in terms of gravimetric water content, w, is as follows:

$$w = C(\psi)\frac{w_z}{\left[\ln\left(2.718+\left(\frac{\psi}{\psi_{\text{inf}}}\right)^{n_1}\right)\right]^m} \quad (3.9)$$

where w_s is the saturated gravimetric water content;

> ψ_{inf} is a suction value corresponding to the inflection point on the curve and is somewhat greater than the air-entry value;

n_1 is a soil parameter related to the slope of the soil-water characteristic curve at the inflection point;

ψ is the soil suction (i.e., matric suction at low suctions and total suction at high suctions);

m is a fitting parameter related to the results near the residual water content; and

$C(\psi)$ is the correction function that causes the SWCC to pass through a suction of 1,000,000 kPa at zero water content.

The correction function $C(\psi)$ is defined as

$$C(\psi) = \left[1 - \frac{\ln\left(1 + \dfrac{\psi}{\psi_r}\right)}{\ln\left(1 + \dfrac{1000000}{\psi_r}\right)} \right] \tag{3.10}$$

where ψ_r is the suction value corresponding to the residual water content, w_r. Residual suction can be estimated as 1500 kPa for most soils, unless the actual value is known.

Equation (3.10) can be written in a dimensionless form by dividing both sides of the equation by the saturated gravimetric water content (i.e., $\Theta_d = w/w_s$, where Θ_d is the dimensionless water content):

$$\Theta_d = C(\psi) \frac{1}{\left[\ln\left(e + \left(\dfrac{\psi}{a}\right)^{n_1}\right)\right]^m} \tag{3.11}$$

The above equation can be used to provide a good fit of the desorption or adsorption branches of the water retention curve over the entire range of suctions. The fitting parameters (i.e., a, n_1, and m values) can be determined using a nonlinear regression procedure such as the one proposed by Fredlund and Xing (1994).

A NEW AND SIMPLE MODEL FOR SDSWCC OR SDWRC (ZHOU AND NG, 2014)

Introduction

As illustrated above, WRC is an important hydraulic parameter for seepage analysis in unsaturated soils. It has been found to be dependent on many factors such as net stress (Lloret and Alonso, 1985; Vanapalli et al., 1999; Ng et al., 2000). Two idealised configurations of unsaturated granular soils with the same average void ratio are shown in Figure 3.6. Although the particles can be arranged in many configurations that result in the same average void ratio, the pore structures of the two configurations, including pore-size distribution, pore shape and pore orientations, are almost always different. Thus they should have different water retention abilities, as the experimental results shown in Figure 3.7 (Ng and Pang, 2000b). In this figure, the measured WRCs of a recompacted (CDV-R2(M)) specimen and a natural (CDV-N2(M)) specimen are compared. The specimen CDV-R2(M) was recompacted to the same average density at the same initial moisture contents as the natural specimen. The hysteresis loop in the recompacted soil is considerably larger than that in the natural specimen. The soil recompacted at wet of optimum (i.e., CDV-R2(M)) is generally believed to be more homogenous, whereas the natural soil specimen (CDV-N2(M)) has relatively non-uniform pore-size distributions due to various geological processes such as leaching in the field. CDV-N2(M) has a slightly lower air-entry value and a higher rate of desorption than CDV-R2(M)) for suctions up to 50 kPa. The rates of desorption of the two soil specimens appear to be the same for high suctions. On the other hand, the rates of adsorption for the two specimens are considerably different. The rate of wetting curve obtained from the natural soil specimen is substantially higher than that of the recompacted specimen.

Ng and Pang (2000b) found that WRC of unsaturated soil alters when stress changes (see Figure 2.15). This is because applying stress affects not only soil density (or void ratio) but also pore structure, as revealed in scanning electronic microscopy conducted by Delage and Lefebvre (1984). Through 1D consolidation tests on saturated clay, Delage and Lefebvre (1984) found that platy clay particles and hence soil pores show preferential orientations during the loading process. Such a rearrangement of the pore structure obviously cannot be completely captured by the soil density (or void ratio) alone. In addition, Ng and Xu (2012) carried out a series of suction-controlled triaxial compression tests on unsaturated compacted silt and measured shear modulus reduction curves. They found that the elastic shear range is very small (less than 0.003%). Beyond this very small elastic range, continuous loading induced the rearrangement of soil particles and pore structures (i.e., plastic behaviour). This suggests that even a small increase in net stress induces plastic

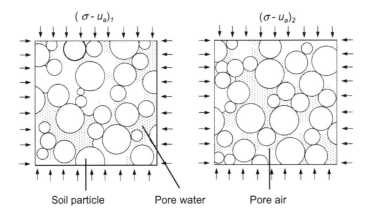

Figure 3.6 Idealised configurations of two representative unsaturated soil specimens having the same void ratio but different pore structures.

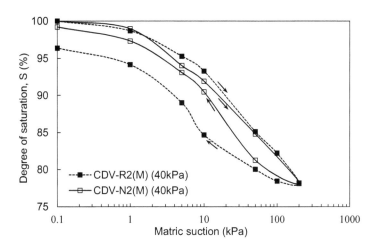

Figure 3.7 Influences of soil structure on WRC.
Source: Ng and Pang, 2000a.

deformation and alters the pore structure and hence affects the water retention behaviour. Similarly, Khosravi and McCartney (2012) observed plastic changes in shear modulus during drying under constant net stress. For simplicity, this may be explained by Bishop's effective stress. As the stress increases during drying, it induces plastic deformation and a change in pore-size distribution. Bishop's effective stress is defined as $\sigma + S_r (u_a - u_w)$, where σ, S_r, u_a and u_w are net stress, degree of saturation, pore-air pressure and pore-water pressure. The Bishop's effective stress becomes Terzaghi's effective stress when the soil specimen is saturated.

Development of a new water retention model

Van Genuchten (1980) proposed a simplified equation to describe the relationship between matric suction and volumetric water content. Assuming that the degree of saturation S_r tends to unity at zero suction, the van Genuchten equation can be expressed as follows:

$$S_r = \left[1 + \left(\frac{s}{a}\right)^{m_2}\right]^{-m_1} \tag{3.12}$$

where a, m_1 and m_2 are soil parameters. Of these parameters, a is closely related to the air-entry value of unsaturated soil. It is widely recognised that the air-entry value of unsaturated soil increases with decreasing e. To obtain the relationship between a and e, Gallipoli et al. (2003a) proposed the following semi-empirical equation:

$$a = m_3 e^{-m_4} \tag{3.13}$$

76 Advanced Unsaturated Soil Mechanics

where m_3 and m_4 are soil parameters. S_r can be obtained by combining Equations (3.12) and (3.13):

$$S_r = \left[1 + \left(\frac{se^{m_4}}{m_3} \right)^{m_2} \right]^{-m_1}$$

(3.14)

From Equation (3.14), it can be seen that S_r is a function of s and e. However, previous experimental results revealed that the soil-water content (w) is almost independent of e at high suction. Based on this evidence, Gallipoli (2012) showed that the product $m_1 m_2 m_4$ should equal 1.

Although Equation (3.14) can describe the density effects on the water retention behaviour, it fails to consider the stress effects on the pore structure (including the pore-size distribution, pore shape and pore orientation). To account for changes in the pore structure induced by the application of net stress, results from microstructural analysis such as Mercury Intrusion Porosimeter (MIP) and Environmental Scanning Electron Microscope (ESEM), which measure the variation in pore-size distribution (PSD), may be used. Delage and Lefebvre (1984) investigated the pore structures of saturated natural clay during consolidation. Figure 3.8a shows the PSD curves they measured for soil consolidated at different effective mean stresses (p'). As shown by the figure, soil pores may be classified as macropores (or inter-aggregate pores) having a radius greater than 0.125 µm and micropores (or intra-aggregate pores) having a radius less than 0.125 µm. During consolidation, the macropores became significantly compressed, while the micropores remained almost unchanged. Considering the obvious differences between micropores and macropores, it is reasonable to conclude that the average e is the sum of the microstructural void ratio e_m (i.e., the ratio of micropore volume to solid volume) and the macrostructural void ratio e_M (i.e., the ratio of macropore volume to solid volume) (Romero et al., 2011). To describe the pore structure, Alonso et al. (2013) introduced a simple microstructural state variable ξ_m:

$$\xi_m = \frac{e_m}{e}$$

(3.15)

Following this definition, it can be shown that

$$\frac{1}{\xi_m} = 1 + \frac{e_M}{e_m}$$

(3.16)

It can be seen from Equations (3.15) and (3.16) that ξ_m, which ranges from 0 to 1, provides information about the pore-size distribution of soil specimens. As shown in Figure 3.8, e_m of the clay tested by Delage and Lefebvre (1984) is almost independent of consolidation pressure $(e_m = 0.8)$, while e decreases from 2.4 to 1.1 as the clay specimen is compressed from 20 to 1500 kPa. Therefore ξ_m increases from 0.36 to 0.78 during the compression process. Such an increase of ξ_m suggests a significant increase in the presence of micropores

Retention characteristics and permeability functions 77

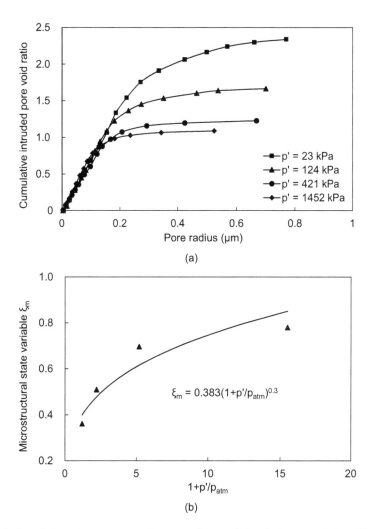

Figure 3.8 Evolution of the microstructure of saturated natural clay during consolation: (a) pore-size distribution at different stress levels; (b) influence of effective mean stress on the microstructural state variable.

Source: Data from Delage and Lefebvre, 1984.

in the soil specimen. Soils with a greater presence of micropores exhibit a larger air-entry value (AEV) and a higher water retention ability (Romero et al., 2011). To consider the effects of pore structure, Equation (3.13) may be modified by including ξ_m:

$$a = m_3 e^{-m_4} \left(\frac{\xi_m}{\xi_m^{ref}} \right)^{m_m} \tag{3.17}$$

78 Advanced Unsaturated Soil Mechanics

where m_m is a soil parameter describing the effects of pore structure on parameter a; ξ_m^{ref} is a microstructural state variable in a reference state. In this study, ξ_m^{ref} is taken to be ξ_m at zero net mean stress. Substituting Equation (3.17) into Equation (3.12) yields

$$S_r = \left[1 + \left(s\frac{e^{m_4}}{m_3}\left(\frac{\xi_m}{\xi_m^{ref}}\right)^{-m_m}\right)^{m_2}\right]^{-m_1} \tag{3.18}$$

It can be seen from this equation that the degree of saturation is dependent not only on s but also on e and ξ_m/ξ_m^{ref}. Both e and ξ_m/ξ_m^{ref} are related to the stress history. To describe the stress effects on the average void ratio and hence the water retention behaviour, one of the simplest formulations used to compute the volume change is:

$$de = -\alpha_p \frac{dp}{p_r + p} - \alpha_s \frac{ds}{p_{atm} + s} \tag{3.19}$$

where p is net mean stress; p_r is a reference pressure, taken as 1 kPa; p_{atm} is atmospheric pressure; α_p represents soil compressibility with respect to the net mean stress; and α_s represents soil compressibility with respect to suction. Equation (3.19) is widely included in the elastoplastic modelling of unsaturated soil (for example, Alonso et al., 1990). The two terms on the right-hand side of this equation describe the soil deformation induced by changes in mean net stress and suction, respectively. The values of α_p and α_s are expected to depend on the overconsolidation ratio. For normally consolidated soil, α_p and α_s should correspond to the stiffness parameters λ_p and λ_s used to calculate elastoplastic soil deformation. For overconsolidated soils, α_p and α_s are equal to the elastic stiffness parameters κ_p and κ_s because the soil response is essentially elastic. By integrating Equation (3.19), the void ratio under any given net stress and suction conditions can be obtained. If a soil specimen is compressed in the saturated state and then dried at constant net stress, the void ratio may be expressed as

$$e = e_0 - \alpha_p \ln\left(1 + \frac{p}{p_r}\right) - \alpha_s \ln\left(1 + \frac{s}{p_{atm}}\right) \tag{3.20}$$

where e_0 is the initial void ratio before changes in net stress and suction.

To explore the relationship between ξ_m and mean stress, the data published by Delage and Lefebvre (1984) are reinterpreted and ξ_m at each consolidation pressure is calculated as shown in Figure 3.8b. The relationship between ξ_m and p' can be approximated by a power function:

$$\xi_m = c_m\left(1 + \frac{p'}{p_{atm}}\right)^b \tag{3.21}$$

where c_m is a microstructural state variable at zero net mean stress; and b is a soil parameter describing the stress effects on the pore structure. The least squares method was used to fit Equation (3.21) to the measured relationship between ξ_m and p'. The fitted values of c_m

Retention characteristics and permeability functions 79

and b are 0.38 and 0.3, respectively. The coefficient of determination (R^2) is found to be 0.9. The equation fits the data very well when the effective stress is less than 500 kPa, but it does not capture the nonlinearity equally well when the stress is greater than 500 kPa. The maximum difference between measured and predicted ξ_m is about 15%. For simplicity, this equation is assumed to be equally applicable to unsaturated soil. Using two independent stress-state variable (i.e., net stress and matric suction) approach (Coleman, 1962), effective mean stress p' may be replaced by net mean stress p. The semi-empirical Equation (3.21) may be used to estimate the ratio ξ_m/ξ_m^{ref} in Equation (3.18):

$$\frac{\xi_m}{\xi_m^{ref}} = \left(1 + \frac{p'}{p_{atm}}\right)^b \tag{3.22}$$

Substituting Equations (3.20) and (3.22) into (3.18) leads to

$$S_r = \left[1 + \left(\frac{\left(e_0 - \alpha_p \ln\left(1 + p/p_r\right) - \alpha_s \ln\left(1 + s/p_{atm}\right)\right)^{m_4}}{m_3 \left(1 + p/p_{atm}\right)^{m_5}} s\right)^{m_2}\right]^{-m_1} \tag{3.23}$$

where m_5 is a new soil parameter equal to the product of m_m and b. Equation (3.23) takes into account not only the effects of stress on the void ratio but also its effects on the pore structure. The stress effects on void ratio and pore structure are taken into account through the semi-empirical equations (i.e., Equations (3.18) and (3.21)). It should be pointed out that the approach introduced in this study gives the flexibility to define and use other semi-empirical equations to incorporate the effects of stress. For example, it is possible that the pore structure may be related to other variables such as the memory variable of a mechanical model, instead of stress. Compared to Equation (3.14), this equation reflects the stress effects on the water retention behaviour more fully. Its usefulness will be evaluated using published experimental data in the next section.

Note that Equation (3.23) requires seven soil parameters: α_p, α_s, m_1, m_2, m_3, m_4 and m_5. Given that the product m_1, m_2, m_4 should be equal to 1, there are six independent parameters. Two of these parameters (α_p and α_s) can be calibrated using the void ratios obtained during isotropic compression and drying processes. The other four parameters (m_1, m_2, m_3 and m_5) can be calibrated using measured WRCs. Values of m_1, m_2 and m_3 can be obtained by three-dimensional least-square fitting in S_r:s:e space. The new parameter m_5 can be calibrated from two WRCs at different levels of net stress. Table 3.1 summarises all the parameters derived and used in the comparisons. It should be noted that pore structure is simplified as a microstructural state variable ξ_m in this section. Equation (3.23) provides a simple approach to consider stress effects on WRC. Its validity is evaluated using experimental results in the next section. Soil deformation was monitored in water retention tests. The measured average void ratio at each suction and stress state was used in the Gallipoli et al. (2003a) model to calculate the corresponding degree of saturation. For the new model, only the initial void ratio before the application of net stress and drying is required. The initial void ratios of different soils are summarised in Table 3.1.

80 Advanced Unsaturated Soil Mechanics

Table 3.1 Summary of the input parameters in Equations (3.14) and (3.23) for three soils

Reference	Soil	Void ratio	α_p	α_s	m_1	m_2	m_3 : kPa	m_4	m_5
Ng and Pang (2000a)	Silt of high plasticity (MH)	0.78	0.012	0.04	0.26	0.7	13.3	5.52	0.5
Lee et al. (2005)	Silty sand (SM)	0.56	0.018	0.04	0.04	7.6	0.3	3.3	0.5
Ng et al. (2012)	Silt (ML)	0.80	0.02	0.005	0.99	1.88	179.6	0.54	0.5

Predictability of the new model

Water retention behaviour under K_0 condition

To explore the capability of the new water retention model and existing ones, Equations (3.14) and (3.23) were fitted to the WRCs of completely decomposed volcanic (CDV) soil. CDV soil is classified as silt of high plasticity (MH) (ASTM, 2006a). Its drying WRCs have been measured at different levels of initial density and vertical net stress (Ng and Pang, 2000a). Soil deformation was monitored in each stress and suction-controlled water retention test. The measured void ratios at various stresses and suctions were used to calibrate the parameters of Equation (3.20), and α_p was found to be 0.012 and α_s was found to be 0.04. These values were used in the new water retention model (i.e., Equation (3.23)).

Figure 3.9 shows that the water retention capacity increases with increasing density. Based on the measured data, the soil parameters in Equations (3.14) and (3.23) were calibrated as follows: $m_1 = 0.26, m_2 = 0.7, m_3 = 13.3$ kPa, $m_4 = 5.52$. It should be noted that the other parameter (m_5) in Equation (3.23) is not needed to calculate WRCs at zero stress. Using these parameters, WRCs for different initial dry densities were calculated using these two equations and are shown in Figures 3.9a and 3.9b, respectively. As expected, the WRCs calculated using these two equations are very similar since both incorporate the void ratio in the formulations. Moreover, the calculated WRCs match the measured ones well. This implies that inclusion of the void ratio in the water retention model can capture the density effects effectively. This set of calibrated parameters (i.e., $\alpha_p = 0.012, \alpha_s = 0.04, m_1 = 0.26, m_2 = 0.7, m_3 = 13.3$ kPa, $m_4 = 5.52$) was then used to calculate the WRC of CDV soil under different stress conditions.

Figure 3.10a compares the measured and calculated (Equation (3.14)) WRCs of CDV soil under 1 D stress. Before the drying process, three specimens with the same average initial void ratio (0.78) were compressed to net vertical stress levels of 0, 40 and 80 kPa. From the measured WRCs, it can be seen that the water retention ability is strongly dependent on net stress. The AEV of unsaturated soil increases with increasing net stress. Above the AEV, the equilibrium degree of saturation at a given suction is greater at a higher net stress than at a lower net stress. The influence of net stress on the water retention behaviour may be explained qualitatively using density effects on WRCs. As net stress increases, the void ratio and hence the water retention ability increases. However, it is interesting to observe from the figure that the influence of net stress on WRCs is underestimated quantitatively by Equation (3.14). This may be because Equation (3.14) considers the effects of stress on the void ratio, but ignores its effects on the pore structure.

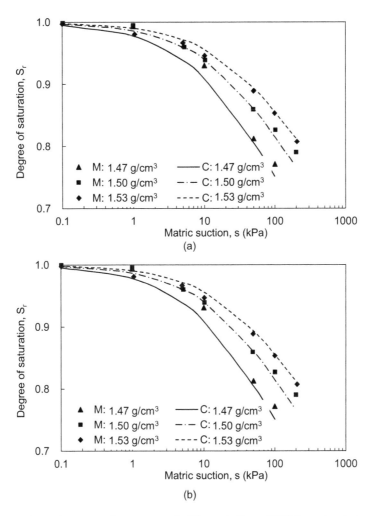

Figure 3.9 Comparisons between the measured (Ng and Pang, 2000a) and calculated WRCs of CDV soil (silty clay) at different dry densities under zero stress: (a) Equation (3.14) and (b) Equation (3.23).

Equation (3.23) was also applied to fit the stress-dependent WRCs of CDV soil. The six calibrated parameters (i.e., $\alpha_p = 0.012$, $\alpha_s = 0.04$, $m_1 = 0.26$, $m_2 = 0.7$, $m_3 = 13.3$ kPa, $m_4 = 5.52$) were kept constant, while the new parameter m_5 was determined to be 0.5 using the least-square method. Using this set of parameters, WRCs at different stress levels were calculated. Figure 3.10b compares the measured and calculated WRCs. The measured WRC is reasonably consistent with the calculated WRC at each level of net stress. A comparison between Figures 3.10a and 3.9b reveals that including the net mean stress in the water retention model does indeed improve the modelling of stress effects on WRCs.

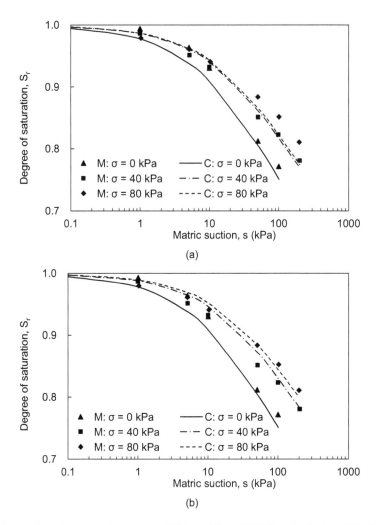

Figure 3.10 Comparisons between the measured (Ng and Pang, 2000a) and calculated WRCs of CDV soil (silty clay) under 1D stress: (a) Equation (3.14) and (b) Equation (3.23).

The newly introduced parameter m_s in Equation (3.23) is equal to 0.5 for CDT. It is assumed that this value is equally applicable to other soils. In the remaining two sections, this assumption is evaluated using the measured stress-dependent WRCs of different soils.

Water retention behaviour under an isotropic state of stress

Lee et al. (2005a) measured the WRCs of weathered granite, which is classified as silty sand (SM), at four levels of isotropic net stress (0, 100, 200 and 300 kPa). Four specimens were compacted to a relative compaction of 90% (with an initial void ratio of 0.56) at a gravimetric water content of 10%. It should be noted that even for weathered granite, a bimodal pore-size distribution forms during soil compaction (Li and Zhang, 2009). In the water retention test, each specimen was subjected to predefined levels of isotropic stress and then to drying. Based on the measured void ratios at various suctions and temperatures, α_p and α_s in Equation (3.20) were found to be 0.018 and 0.04, respectively.

Figure 3.11 shows the measured WRCs of the silty sand. As expected, the water retention capacity increases with increasing net stress. Based on the measurement, soil parameters in Equations (3.14) and (3.23) were calibrated using the least-square method as follows: $m_1 = 0.04$, $m_2 = 7.6$, $m_3 = 0.3$ kPa, $m_4 = 3.3$. As discussed above, the other parameter m_5 in Equation (3.23) was assumed to be 0.5.

The WRCs calculated using Equation (3.14) are also shown in Figure 3.11 for comparison. It can be seen that the difference in water retention ability induced by stress is only partially captured. This is consistent with observations from Figure 3.10 that the stress effects on WRCs cannot be simplified as density effects alone. On the other hand, Figure 3.11 shows a comparison between the measured WRCs and those calculated from Equation (3.23). It can be seen that Equation (3.23) predicts the WRC reasonably well at each stress level.

Water retention behaviour at different stress ratios

Ng et al. (2012) investigated the influence of net stress on the water retention behaviour of CDT, which is classified as silt (ML). They prepared all soil specimens at the optimum water content for a void ratio of 0.8. Before drying, soil specimens were compressed to different levels of net mean stress (0, 40 and 80 kPa) at a constant stress ratio. For each level of net mean stress, they considered two different stress ratios q/p (0 and 1.2). During compression and subsequent drying, void ratios were measured under various stress and suction conditions. Based on experimental results, parameters α_p and α_s in Equation (3.20) were found to be 0.02 and 0.005, respectively.

The WRCs measured at different levels of isotropic stress are shown in Figure 3.12a. The measured water retention ability increases with increasing stress. Based on these measurements, the parameters in Equation (3.14) were determined as follows: $m_1 = 0.99$, $m_2 = 1.88$, $m_3 = 179.6$ kPa, $m_4 = 0.54$. Thus the soil parameters in Equation (3.22) were $\alpha_p = 0.02$, $\alpha_s - 0.005$, $m_1 = 0.99$, $m_2 = 1.88$, $m_3 = 179.6$ kPa, $m_4 = 0.54$ and $m_5 = 0.5$. These parameters were used to calculate the WRCs under both isotropic and anisotropic stress states. The WRCs under isotropic stress calculated using Equation (3.14) are also included in Figure 3.12a for comparison. It can be seen that the calculated WRCs for different stress levels are almost identical, probably because the change in e is quite small during the application of net stress. Figure 3.12b shows a comparison between the measured WRC and the one calculated using Equation (3.23). The influence of net stress on WRC is captured well.

Figures 3.13a and 3.13b show a comparison between the measured and calculated WRCs under an anisotropic state of stress ($q/p = 1.2$), where Equations (3.14) and (3.23) were used in the calculations. It can be seen that Equation (3.23) provides a much better prediction of stress-dependent WRCs than Equation (3.14). Figures 3.13a and 3.13b reveal that at different stress ratios, the newly proposed model gives predictions that are very similar to the measured values. This suggests that although the stress ratio may affect the pore structure, its influence is not as important as that of net mean stress.

The comparisons shown in Figures 3.9 through 3.13 demonstrate that including net mean stress in the water retention model definitely improves the modelling of stress-dependent

84 Advanced Unsaturated Soil Mechanics

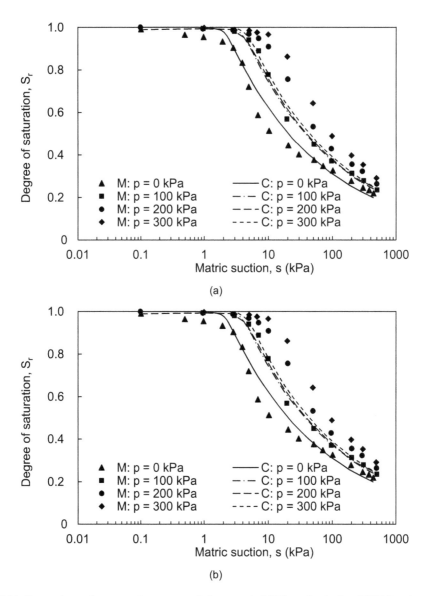

Figure 3.11 Comparisons between the measured (Lee et al., 2005) and calculated WRCs of weathered granite (silty sand) under an isotropic state of stress: (a) Equation (3.14) and (b) Equation (3.23).

WRCs. This improvement is mainly because the new model incorporates the effects of stress on pore structure through a microstructural state variable ξ_m. Further, the stresses used in the water retention tests of this study are all less than 500 kPa. Within this stress range, the model provides a good fit to the measured ξ_m (see Figure 3.8). In addition, the new parameter (m_s) in the model is not sensitive to soil parameters and is approximately equal to 0.5.

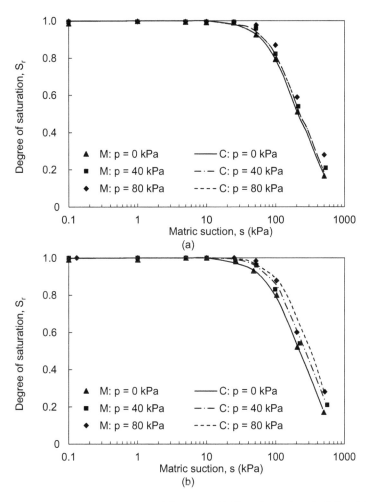

Figure 3.12 Comparisons between the measured (Ng et al., 2012) and calculated WRCs of CDT (silty clay) under an isotropic state of stress: (a) Equation (3.14) and (b) Equation (3.23).

Summary

Net stress affects not only the void ratio but also the pore structure including the pore-size distribution, pore shape and pore orientation. Thus it is not sufficient to include the void ratio alone in a water retention model that captures the effects of stress on the water retention of unsaturated soil. A new model was developed in this study to predict the stress effects on the WRC by incorporating an additional parameter specifically to describe the effects of stress on the pore structure. This new model can provide predictions of the water retention behaviour that are an improvement over several existing water retention models over a wide range of stress conditions, including isotropic and anisotropic stress conditions, as comparison with experimental data has convincingly demonstrated.

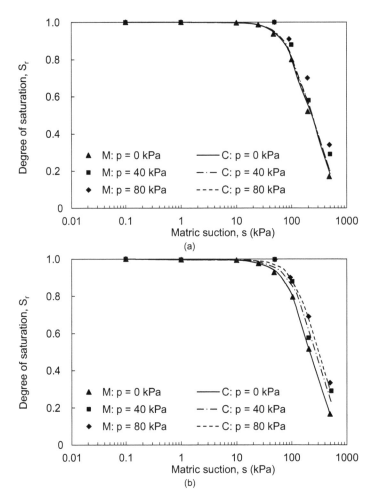

Figure 3.13 Comparisons between the measured (Ng et al., 2012) and calculated WRCs of CDT (silty clay) under an anisotropic state of stress ($q/p = 1.2$): (a) Equation (3.14) and (b) Equation (3.23).

LABORATORY MEASUREMENTS OF DRYING AND WETTING STRESS-DEPENDENT PERMEABILITY FUNCTIONS (NG AND LEUNG, 2012B)

Theoretical consideration

In this study, the Instantaneous Profile Method (IPM) is adopted to determine the $k(\psi)$ of an unsaturated soil directly at different vertical net normal stresses. This is a transient-state method, where a cylindrical soil specimen is subjected to a continuous flow of water from one end of the specimen to the other. During the flow process, both water content and pore-water pressure (PWP) distributions along the specimen are measured instantaneously to obtain the water flow rate and the hydraulic gradient. Assuming

Darcy's law is applicable to water flow in unsaturated soils (Buckingham, 1907), the unsaturated water permeability at any location and time can be determined by dividing the measured water flow rate by the hydraulic gradient. The calculation procedures are briefly summarised below.

Figure 3.14 shows two arbitrary profiles of volumetric water content (VWC, θ_w) and hydraulic head at elapsed time $t = t_1$ and t_2 along a one-dimensional soil column. The values of VWC and PWP can be measured using different types of instruments at z_A (Row A), z_B (Row B), z_C (Row C) and z_D (Row D). The measured VWC profiles can be extrapolated to the surface and the bottom of the soil column to determine the water flow rate. According to the one-dimensional (1 D) continuity equation, the water flow rate, $v_{zB,tave}$, at any depth z (z_B in this example) for any average elapsed time $t_{ave} = (t_1 + t_2)/2$ can be determined by estimating the change of total water volume between the depth under consideration, i.e., z_B, and one end of the soil column, z_E:

$$v_{zB,tave} = \frac{d}{dt}\int_{zE}^{zB} \theta_w(z,t)\cdot dz = \frac{\Delta V}{t_2 - t_1} + v_{ze,tave} \tag{3.24}$$

where $\theta_w(z, t)$ is the VWC profile as a function of depth z at specific time t; ΔV is the shaded area between the $\theta_w(z, t)$ at $t = t_1$ and t_2 as shown in Figure 3.14; dt is the time interval the measurements, i.e., $t_2 - t_1$; and $v_{ze,tave}$ is the boundary water flow rate. In this new apparatus, the $v_{ze,tave}$ can be measured or controlled during a test. The location of z_e and the corresponding $v_{ze,tave}$ are discussed and reported in the test procedures later.

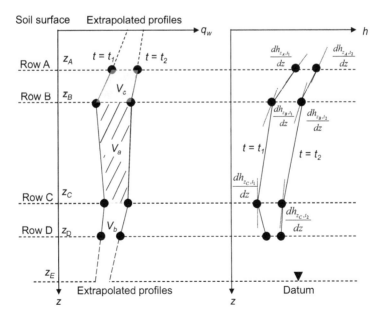

Figure 3.14 Arbitrary profiles of volumetric water content, θ_w, and hydraulic head, h, at elapsed time, $t = t_1$ and t_2, along a one-dimensional soil column.

88   Advanced Unsaturated Soil Mechanics

On the other hand, the hydraulic head gradient, $i_{zB,tave}$, at any depth z_B for any average elapsed time t_{ave} can be obtained by estimating the slope of the hydraulic head profile, which is the summation of the measured PWP head profile and the gravitational head profile. It can be expressed mathematically as follows:

$$i_{zB,tave} = \frac{1}{2}\left(\frac{dh_{zB,t1}}{dz} + \frac{dh_{zB,t2}}{dz} \right)$$

$$= \frac{1}{4}\left[\left(\frac{h_{zA,t1} - h_{zB,t1}}{z_A - z_B} + \frac{h_{zB,t1} - h_{zC,t1}}{z_B - z_C} \right) + \left(\frac{h_{zA,t2} - h_{zB,t2}}{z_A - z_B} + \frac{h_{zB,t2} - h_{zC,t2}}{z_B - z_C} \right) \right] \quad (3.25)$$

where $h_{zi,tj}$ is the hydraulic head at depth z_i (i = A, B, C and D) for elapsed time t_j (j = 1, 2).

Hence, the unsaturated permeability, $k_{zB,tave}$, at any depth z_B for any average elapsed time t_{ave} can be calculated by dividing the water flow rate by the corresponding hydraulic head gradient:

$$k_{zB,tave} = \frac{v_{zB,tave}}{i_{zB,tave}} \quad (3.26)$$

Design of the apparatus and instrumentation

Figure 3.15 shows a schematic diagram of the newly developed stress-controllable soil column (SC). It consists of a 1 m high acrylic tube, a constant head water supply system and a loading system. The inner and outer diameters of the tube are 150 mm and 160 mm, respectively, i.e., 5 mm wall thickness. The aspect ratio, i.e., height-to-diameter ratio, of this SC is designed to be greater than 5. This is to ensure 1 D flow conditions when determining water permeability using the IPM. It has been reported by many researchers that the aspect ratio of most soil column apparatuses reported in the literature should usually be greater than 5 when studying 1 D infiltration problems (Richards and Weeks, 1953; Watson, 1966; Wendroth et al., 1993; Choo and Yanful, 2000; Yang et al., 2004a; Chapuis et al., 2006; Li et al., 2009).

The newly developed apparatus is designed to control and measure the top and the bottom boundary flow conditions. At the upper boundary, the apparatus is designed in such a way that vertical stress and boundary flux can be applied independently. A 150 mm-diameter, 20 mm-thick, circular, perforated, stainless-steel plate is placed on top of the compacted soil column. As the vertical stress is applied to the centre of the plate through a loading ram, water can infiltrate or evaporate through the twenty-four perforated holes, which are each 10 mm in diameter (see Figure 3.15). Upon ponding, constant head infiltration can be achieved by using a constant head water supply system. The principle of this system is similar to that of Mariotte's bottle (McCarthy, 1934). A copper tube, 2 mm in diameter, is sealed and inserted into the water storage tank. The bottom part of the copper tube is submerged in water while the top is open to the atmosphere. The elevation of the bottom part of the tube is carefully adjusted to the same level as that of the ponding head. Since the connection between the copper tube and the tank is tightly sealed, the total head at point A has to be identical to that at points B and C. Any lowering of the ponding head

Retention characteristics and permeability functions 89

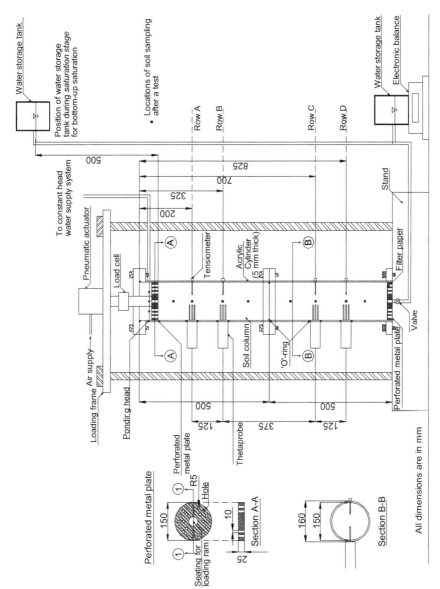

Figure 3.15 Schematic diagram of the stress-controllable soil column.

90 Advanced Unsaturated Soil Mechanics

(i.e., decrease of total head at point B) will cause water to flow from the storage tank to the pond until the total head at A and B is equalised. Therefore, the infiltration rate can be measured by continuously recording the change of water weight in the storage tank. Since the openings of the system are all sealed by o-rings, loss of water by evaporation from the storage tank is prevented. As shown in Figure 3.15, another 150 mm diameter, 20 mm thick, circular, perforated, stainless-steel plate is placed at the bottom of the cylinder. The perforated plate can be filled completely with water to form a water compartment allowing for uniform drainage. To control the bottom drainage condition, a valve is installed. Under drained conditions, the volume of outflow water at any time can be measured by continuously recording the weight of the water storage tank.

Regarding the loading system, a vertical load (or net stress) can be applied using a pneumatic actuator. The applied load is transmitted to the centre of the stainless-steel plate (and hence the soil column) through a loading ram. The magnitude of the applied load is recorded by a load cell. To estimate the lateral deformation of the 1 m high acrylic cylinder caused by an applied vertical load, cavity expansion theory, which is an elastic solution, is used. Under a maximum vertical load of 100 kPa, the estimated radial strain of the 5 mm-thick acrylic cylinder is less than 0.03%. This value is 40% less than that stated in the Japanese Geotechnical Society (JGS) standard (JGS-0525), i.e., 0.05%, where the K_0 loading condition is considered to be satisfied when using triaxial apparatus (JGS, 1999). In other words, the K_0 loading condition can be assumed when using this apparatus and any change of VWC and PWP due to the lateral deformation of the 1 m high acrylic cylinder may be neglected. Plate 3.2 shows a commercial product manufactured by Geo-Expert in recent years.

Experimental results

Permeability-matric suction relationship, $k(\psi)$

As described previously, the unsaturated permeability at the evaporation and ponding stage of each soil column can be determined using the IPM (refer to Equations (3.24) to (3.26)). For clarity, the measured permeability function, which relates the permeability to its corresponding matric suction, $k(\psi)$, at average vertical net normal stresses of 4, 39 and 78 kPa is shown in Figures 3.16a, 3.16b and 3.16c, respectively. The $k(\psi)$ measured at the four rows in each test are denoted using four different symbols separately. The solid and open symbols refer to drying and wetting $k(\psi)$, respectively. Because of the small variation of σ_v' along each test, it is not surprising that the overall trends of both drying and wetting $k(\psi)$ at any average vertical net normal stress are consistent between the four rows (see Figures 3.16a, 3.16b and 3.16c). At a given suction, the maximum variation in the measured permeability is less than half an order of magnitude. The influences of interface friction and self-weight do not appear to affect the investigation of stress dependency of $k(\psi)$ significantly. The saturated permeability, k_s, at each vertical net normal stress was independently measured in a triaxial apparatus. As the effective stress increases from 0 to 80 kPa, the void ratio of the specimen decreases from 0.69 to 0.62 and the measured k_s thus decreases by up to one and half orders of magnitude.

At any applied average vertical net normal stress, the measured drying permeability decreases log-linearly as matric suction increases. For an example, the reduction in drying

Retention characteristics and permeability functions 91

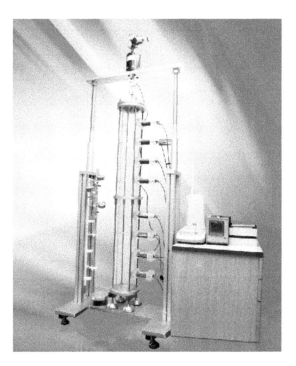

Plate 3.2 One-dimensional stress-controllable soil column (with courtesy from Geo-Expert).

permeability at constant zero vertical net normal stress is up to two orders of magnitude (see Figure 3.16a) as matric suction increases from 0 to 80 kPa. An increase in matric suction may induce more air bubbles within each soil column. These induced air bubbles block some hydraulic paths and thus increase the contortion of the water flow paths and reduce the permeability. At a matric suction of 6 kPa, the maximum reduction in drying permeability is about one order of magnitude (from 2×10^{-8} to 9×10^{-10} m/s) when the average vertical net normal stress increases from 4 to 78 kPa (see Figures 3.16a, 3.16b and 3.16c). Moreover, at a given change of matric suction (from 4 to 80 kPa), the decreasing rate of drying permeability reduces as the average vertical net normal stress increases, particularly when the average vertical net normal stress increases from 39 to 78 kPa. A summary of the decreasing rates estimated in this study is given in Table 3.2. As compared to other experimental data, the decreasing rate of $k(\psi)$ in silty clay at zero net stress as measured by Meerdink et al. (1996) and Li et al. (2009) is estimated to be about $1.84(\log_{10} \text{m/s}) \cdot (\log_{10} \text{kPa})^{-1}$ and $2.17(\log_{10} \text{m/s}) \cdot (\log_{10} \text{kPa})^{-1}$, respectively. The observed stress dependency of the drying $k(\psi)$ is probably attributable to the substantial increase in the dry density of the soil column when an average vertical net normal stress is applied. This supposition is also supported by the microstructure analysis of a Bloom clay carried out by Romero et al. (1999).

At a given average vertical net normal stress, the magnitude of the wetting permeability is always lower than that of the drying permeability at any matric suction considered in the three tests (see Figures 3.16a, 3.16b and 3.16c). The observed consistent lower

92 Advanced Unsaturated Soil Mechanics

Table 3.2 Summary of the estimated decreasing and increasing rates of each $k(\psi)$ and $k(\theta_w)$

	$k(\psi)$		$k(\theta_w)$	
	Decreasing rate	Increasing rate	Decreasing rate	Increasing rate
Test ID	$(\log m/s)(\log kPa)^{-1}$		$(\log m/s)$	
SC0	1.392	0.982	8.730	15.544
SC40	1.116	0.869	14.507	27.881
SC80	0.628	0.734	14.818	26.138

Note:
Each decreasing/increasing rate is determined by averaging the constant portion of a drying/wetting curve.

wetting permeability at all stress levels is probably because the VWCs along wetting paths are less than those along the drying path at any given suction, possibly leading to fewer hydraulic flow paths and thus lowering the wetting permeability. As summarised in Table 3.2, as the average vertical net normal stress increases from 4 to 78 kPa, the increasing rate of wetting permeability reduces relatively slowly by comparison with the reduction in the decreasing rate of drying permeability. Remarkable hysteresis loops are identified between drying and the wetting $k(\psi)$ at each average vertical net normal stress. The measured hysteretic behaviour is probably because of the difference in the receding and advancing contact angles of the soil-water interface (Hillel et al., 1972) during a drying and wetting cycle in the soil column. However, the size of the hysteresis loop of $k(\psi)$ decreases with an increase in average vertical net normal stress. This may be caused by the smaller difference between the receding and advancing contact angles of the soil-water interface in denser soil when an average vertical net normal stress is applied (Ng and Pang, 2000a).

Permeability-volumetric water content relationship, $k(\theta_w)$

The permeability function can also be expressed in another form, which relates the measured permeability to its corresponding VWC, $k(\theta_w)$. Figure 3.17 shows the measured $k(\theta_w)$ at average vertical net normal stresses of 0, 40 and 80 kPa. The solid and open symbols refer respectively to a drying and a wetting $k(\theta_w)$. The k_s at each vertical net normal stress is also shown for reference. At a given average vertical net normal stress, the drying permeability decreases log-linearly as VWC decreases upon evaporation. At a given change of VWC, the decreasing rate of the drying permeability almost doubles when the average vertical net normal stress increases from 0 to 40 kPa (refer to Table 3.2). However, a smaller increase in the decreasing rate of the drying permeability is recorded when the net stress increases from 40 to 80 kPa.

As shown in Figure 3.17, the wetting permeability increases with an increase in VWC but at a higher rate than that of the drying $k(\theta_w)$ at a given average vertical net normal stress. At an average vertical net normal stress of 4 kPa, the maximum difference between the drying and the wetting permeability is about a half order of magnitude at a VWC of 21%. Although there is some hysteresis in the measured $k(\theta_w)$, the effects of a/the drying-wetting

Retention characteristics and permeability functions 93

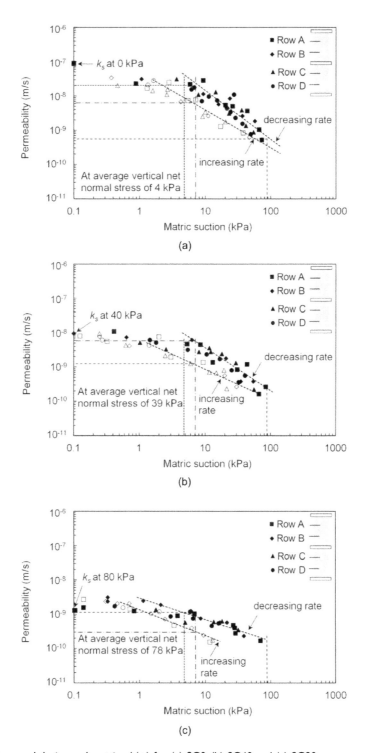

Figure 3.16 Measured drying and wetting $k(\psi)$ for (a) SC0; (b) SC40 and (c) SC80.

94 Advanced Unsaturated Soil Mechanics

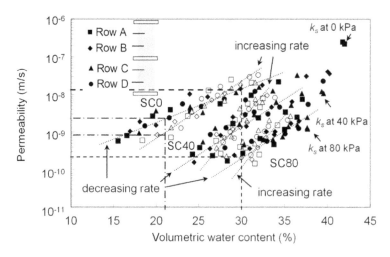

Figure 3.17 Permeability function, $k(\theta_w)$, at average vertical net normal stress of 0 kPa (SC0), 40 kPa (SC40) and 80 kPa (SC80).

cycle on $k(\theta_w)$ appear to be less significant when compared to those observed from $k(\psi)$ (see Figures 3.16a, 3.16b and 3.16c).

On the other hand, it can be seen from Figure 3.17 that when the average vertical net normal stress increases from 4 to 78 kPa, the maximum reduction in permeability (either drying or wetting) is up to two orders of magnitude at a VWC of 30%. This implies that when a soil column is subjected to different average vertical net normal stresses, any changes in its pore-size distribution can cause different distributions of water and hydraulic flow paths even at the same VWC.

Summary

A new 1 m high stress-controllable soil column (SC) apparatus was developed in this study and was found to provide a useful direct measure of a stress-dependent permeability function (i.e., $k(\psi)$ and $k(\theta_w)$) of an unsaturated soil. Based on measurements of matric suctions and volumetric water contents (VWC) in this SC, $k(\psi)$ can be determined using the Instantaneous Profile Method. In this new apparatus, the effects of the drying-wetting cycle and the two stress-state variables (i.e., matric suction and net normal stress) on $k(\psi)$ (or $k(\theta_w)$) can also be investigated. The $k(\psi)$ of a compacted silty clay (tuff) was measured and some conclusions are drawn as follows:

- The measured permeability decreases log-linearly by up to two orders of magnitude when the matric suction increases from the AEV to 80 kPa, depending upon the stress level.
- A noticeable hysteresis loop is observed between the measured drying and wetting $k(\psi)$. At a given matric suction, the measured drying permeability is generally larger than the wetting one by about a half order of magnitude.

- When the average vertical net normal stress increases from 4 to 78 kPa, the maximum decrease in permeability is up to two orders of magnitude at a given matric suction and the size of the hysteresis loop decreases.
- The observed hysteresis between the drying and wetting curves appears to not be very significant at any vertical net normal stress.
- As the average vertical net normal stress increases from 4 to 78 kPa, the measured permeability decreases by two orders of magnitude at a given VWC whereas the rate of change of permeability almost doubles for a given change of VWC.

A FIELD STUDY OF SDSWCC AND PERMEABILITY FUNCTION (NG ET AL., 2011)

Introduction

This section presents measured results obtained from the field monitoring of variations in volumetric water content, θ_w, with matric suction ($u_a - u_w$, where u_a and u_w are pore-air and pore-water pressure respectively) in response to wetting-drying cycles on a hillside in Tung Chung, Hong Kong. The field monitoring project was undertaken as part of a pilot natural hillside assessment and field instrumentation project commissioned by the Geotechnical Engineering Office (GEO), Civil Engineering and Development Department (CEDD), the Government of the Hong Kong SAR. VWC and PWP were monitored using time-domain reflectometry (TDR) moisture probes and jet-fill tensiometers respectively, throughout a 48-day monitoring period. The SDSWCCs obtained from the field and laboratory measurements and the SWCC predicted from PSD using Fredlund et al. (1997) approach were compared. The $k(\psi)$ values of the soil profile were estimated using the instantaneous profile (IP) method. The field-determined $k(\psi)$ values were then compared with predictions given by Fredlund et al. (1994) using deduced SWCCs obtained from PSDs. The comparisons are reported and discussed.

The test site and soil properties

The test site is located on a sloping hillside above the North Lantau Expressway on Lantau Island, Hong Kong, as shown in Plate 3.3. According to the GEO (1994), the site is underlain by undivided rhyolite lava and tuff of the Lantau Formation. A ground investigation was conducted in the vicinity of the test plot comprising trial trench 1 (TT1), trial trench 2 (TT2) and trial pit 1 (TP1). As shown in Figure 3.18, the ground consists of about 1 m of loose colluvial deposit accumulated through the action of gravity. The colluvium is thus anticipated to have large inter-pores, while its thickness may vary from place to place. A layer of about 2 m of completely decomposed tuff (CDT, clayey silt) is then encountered, and some relict joints with silty clay infill were identified. The completely decomposed tuff at this site is a saprolitic soil, which is commonly found in Hong Kong. The decomposed tuff overlies moderately and slightly decomposed tuff at further depths. It was found that the water content of the ground decreases with depth but the dry density increases with depth, as expected in a typical unsaturated weathered-rock soil profile. Borehole records reveal that the groundwater table is located about 3 m below the ground surface. Block samples of colluvium and CDT were taken in the vicinity of the test plot at 0.9 m and

2.9 m, respectively. Figure 3.19 shows the particle size distributions from the samples. Using the Unified Soil Classification System (USCS), both the colluvium and the CDT found at this site are classified as inorganic silty clay of low to medium plasticity (CL). Selected index parameters of the soils are summarised in Table 3.3.

Prior to the installation of instruments for the IP test, a flat test plot 3.5 m × 3.5 m was formed by cutting into the ground at the test location. To achieve the one-dimensional downward flow assumption, a trench 3 m deep and 1.2 m wide was excavated on the uphill side of the test plot to install 0.06 mm-thick polythene sheeting, which prevented lateral groundwater flows (see Figure 3.20). After installing the polythene sheeting, the trench was backfilled. Subsequently, a circular steel test ring, 3 m in diameter, was installed at the ground surface and embedded 100 mm into the flattened plot to retain water for the IP test.

Instrumentation and testing procedures

Figure 3.20 shows the instrumentation scheme at the site. Ten tensiometers (JFT_1 to JFT_10) were installed at depths of 0.36 m, 0.77 m, 0.95 m, 1.17 m, 1.54 m, 1.85 m, 2.13 m, 2.43 m, 2.6 m and 2.99 m to monitor the PWP, while, because of budget constraints, only four TDRs (TDR_1 to TDR_4) could be installed at depths of 0.84 m, 1.85 m, 2.50 m and 3.59 m to measure θ_w in the ground. The consequent need to interpolate the measurements of θ_w is not ideal but is inevitable. The measuring range of the tensiometers was from 0 to 90 kPa. The sampling interval for each tensiometer and TDR was 5 min.

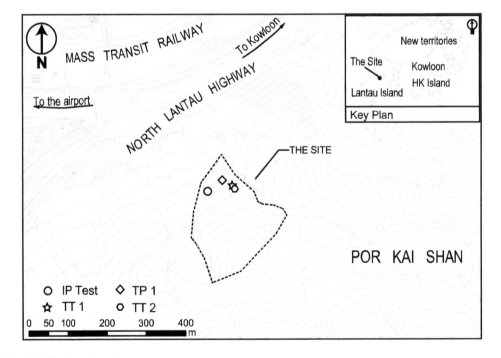

Plate 3.3 Site location plan.

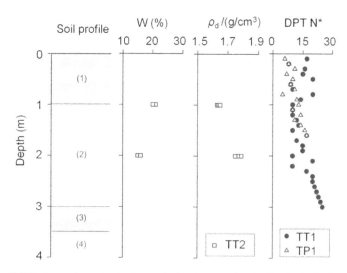

*DPT=dynamic probe tests carried out according to Geotechnical control office (1987); N=number of blows per 100mm penetration. Similar to the standard BS EN ISO 22476-2:2005 DPL, the hammer mass is 10 kg, except that the height of fall is modified to be 300 mm instead of 100 mm by the Geotechnical control office (1987).
(1) Firm, light brown, slightly sandy silt with occasional angular to sub-angular fine to coarse gravel and sub-angular cobbles of moderately decomposed tuff (colluvium).
(2) Very weak (Grade V), light grey, dappled light brown, completely decomposed coarse ash crystal tuff (clayey silt with occasional angular to sub-angular fine gravel. Joints are closely spaced, rough planar, extremely narrow and dipping at 10°, 20°, 30° and 40° to the horizontal.
(3) Moderately strong (Grade III), grey, dappled brown, moderately decomposed fine ash crystal tuff.
(4) Strong to very strong ((Grade II), light grey, spotted dark grey and white, slightly decomposed tuff.

Figure 3.18 Soil profiles and properties revealed by ground investigation at test plot (from TT1, TT2 and TP1).

To investigate the influence of wetting-drying cycles on *in situ* water permeability and SDSWCCs, the test programme consisted of two wetting-drying cycles, which were divided into four stages. Details of the test programme are summarised in Table 3.4. This field experiment was intentionally conducted after the wet season. According to rainfall data obtained from Hong Kong Observatory, negligible daily rainfall was recorded during the entire testing and monitoring period. Daily rainfall of 2.7 mm, 2.4 mm and 2.1 mm was recorded on 30 and 31 October and 1 November, respectively. For each wetting period, water about 0.1 m deep was applied to the ground surface inside the ring, and the water level was checked and refilled to the same level every 12 h. For each drying period, the test plot was allowed to dry under natural evaporation.

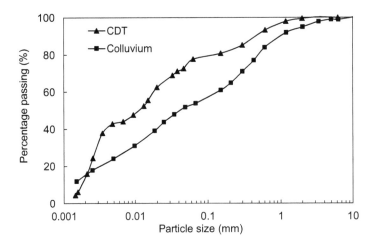

Figure 3.19 Particle size distribution of colluvium and CDT.

Observed field performance

Distributions of pore-water pressure and volumetric water content with depth

Figure 3.21 shows the PWP profiles measured on several key days. The hydrostatic line is also included for ease of reference, assuming that the initial groundwater table is located at about 3 m depth. The PWPs at all depths, except on day 28, are greater than the hydrostatic line, indicating a net down-flux of water and the overly wet nature of the plot. Upon first wetting, the PWP profile remains almost unchanged (day 0 to day 4) as a result of the wet ground caused by the preceding rainfall in September. This field observation is consistent with numerical simulations of the effects of previous rainfall on subsequent changes of PWP distribution (Ng and Shi, 1998). When the water source is removed and drying starts (day 4 to day 28), the PWPs generally decrease at all depths, because of continuous loss of water content, and the PWP profile moves towards the hydrostatic line. At the end of the drying period, the PWP at the colluvial stratum (i.e., 0.36 m below the ground surface) reached its minimum value of −42 kPa. By contrast, the measured PWPs remain greater than the hydrostatic pressure at depths between 0.77 m and 2.13 m, whereas they are hydrostatic at deeper levels (i.e., 2.43 m, 2.60 m and 2.99 m). The fairly constant suction of about 5 kPa at depths between 0.77 m and 2.13 m may be attributed to the comparable magnitudes of the water flow rate and the unsaturated permeability (i.e., unit gradient downward seepage) in the CDT stratum.

Figure 3.22 shows the θ_w profiles measured on several key days. The intersection between the θ_w profiles on days 0 and 4 indicates that θ_w increased because of the downwards flux in the ground above 2.50 m, while the water drained downwards below this level because of gravity. The initial high θ_w values measured near the ground surface were caused by the antecedent rainfall prior to the IP test. As expected, the entire θ_w profile generally increases

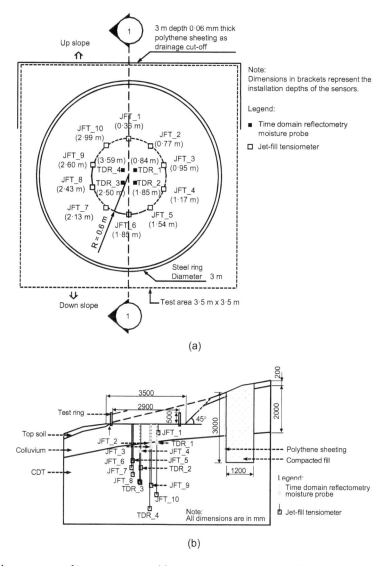

Figure 3.20 Arrangement of instrumentation: (a) instrumentation plan and (b) section I–I along slope.

during the wetting period from day 0 to day 4, and from day 28 to day 32. By contrast, the θ_w profiles typically decreased during each drying cycle from day 4 to day 28, and from day 32 to day 48. Between depths of 0.85 m to 1.85 m, the reduction in θ_w caused by drying was much more significant than that at greater depths. This suggests that the depth of the influence of evaporation (or drying) was shallow (less than 2 m), which is consistent with the measurements from the tensiometers.

Table 3.3 Index properties of the colluvium and CDT

Property	Colluvium	CDT
Specific gravity	2·73	2.68
Bulk density: kg/m³	1806	1877
Maximum dry density: kg/m	1580	1777
Optimum moisture content: %	15.2	17.2
Moisture content: %	20.1	17.3
Gravel content (≥2 mm): %	5.0	0.0
Sand content (2 mm–63 μm): %	40.0	25.0
Silt content (63 μm–2 μm): %	40.0	60.0
Clay content (≤ 2 μm): %	15.0	15.0
Liquid limit: %	32	34
Plastic limit: %	17	20
Plasticity index: %	15	14
Calculated dry density: kg/m³	1504	1600
Calculated porosity	0.45	0.40

Table 3.4 Test schedule

Stage	Wetting/ drying	Date	Duration: days
1 (day 0–4)	Wetting	2–6 Oct 2007	4
2 (day 4–28)	Drying	6–30 Oct 2007	24
3 (day 28–32)	Wetting	30 Oct–3 Nov 2007	4
4 (day 32–48)	Drying	3–19 Nov 2007	16

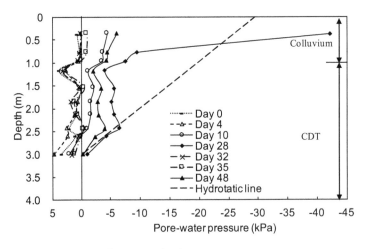

Figure 3.21 Pore-water pressure profile against depth.

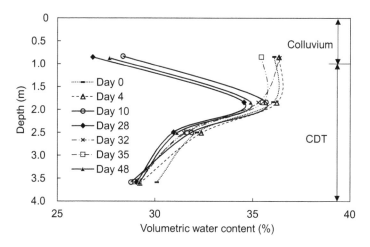

Figure 3.22 Volumetric water content profile against depth.

In situ SDSWCCs

In field conditions, the air phase is generally assumed to be continuous and to remain at atmospheric pressure. Thus the magnitude of negative PWP is equivalent to matric suction. Figures 3.23a to 3.23j show the relationships between measured θ_w and measured matric suction (i.e., in situ SDSWCCs) at the 10 locations of the tensiometers. Table 3.5 summarises the air-entry values, desorption rates, adsorption rates and hysteretic loop sizes of the hysteretic SDSWCCs at 0.36 m, 0.77 m, 0.95 m, 1.17 m and 1.54 m. The results seem to suggest that the measured SDSWCCs are depth-dependent and that the hysteretic loop size decreases with depth.

Within the colluvial strata (0.36–0.95 m), the soil appears to desaturate when the matric suction is greater than 1 kPa (i.e., air-entry value,1 kPa), as shown in Figures 3.23a–3.23c. At a depth of 0.36 m (see Figure 3.23a), the reduction in θ_w is negligible when the matric suction increases beyond 5.6 kPa in the first drying curve. It should be noted that the influence of the inevitable in situ stratification, discontinuities (relict joints) and heterogeneities may substantially overwhelm the stress effect, which is already low within 1 m of the ground surface.

Noticeable hydraulic hysteretic loops during wetting-drying cycles are seen in Figures 3.23a to 3.23f. These may be caused by the difference in contact angle during a wetting-drying cycle. The contact angle and the radius of curvature are reported to be greater for an advancing meniscus than for a receding one (Hillel, 1998). As shown in Figure 3.23, the size of the hysteretic loop decreases with depth within the colluvium layer. Figures 3.23d–3.23j show the measured SDSWCCs of CDT at depths of 1.17 m, 1.54 m, 1.85 m, 2.13 m, 2.43 m, 2.60 m and 2.99 m, respectively. Below 1.85 m depth the measured volumetric water content shows only a small reduction with an increase in matric suction up to a few kPa. Considering the fact that the ground has been subjected to countless wetting and drying cycles in the past, it is not surprising to find that the characteristics of in situ SDSWCCs obtained from the first wetting-drying cycle are comparable to those from

102 Advanced Unsaturated Soil Mechanics

Table 3.5 Summary of air-entry values, desorption rates, adsorption rates and hysteretic loop sizes of hysteretic SDSWCCs at 0.36 m, 0.77 m, 0.95 m, 1.17 m and 1.54 m

Depth: m	AEV: kPa		Desorption rate: (log kPa)$^{-1}$		Adsorption rate: (log kPa)$^{-1}$		Hysteretic loop size: kPa	
	1st cycle	2nd cycle	1st cycle	2nd cycle	1st cycle	2nd cycle	1st cycle	2nd cycle
0.36	1.4	1.0	−0.23	−0.24	−0.12	−0.08	0.20	0.12
0.77	1.1	0.8	−0.13	−0.17	−0.08	−0.06	0.14	0.09
0.95	0.9	0.9	−0.28	−0.34	−0.06	−0.03	0.05	0.05
1.17	0.5	NA*	−0.04	NA*	NA**	NA*	NA*	0.02
1.54	0.8	1.0	−0.06	−0.07	−0.02	−0.03	0.01	0.01

Notes:
*   Not available because air-entry value not exceeded.
** Not available because first wetting curve unavailable.

the second cycle (Ng and Pang, 2000a) for the colluvium strata. Below 2.13 m, the change in θ_w with matric suction and the hysteresis phenomenon of soil-water characteristics are essentially negligible. Apart from the possible influence of ink-bottle effects, this could be because the soil was subjected to low matric suction, which was possibly within the air-entry value and thereby resulted in little change in θ_w with suction during the test. It should be noted that the degree of weathering varied within the CDT layer, and moderately to slightly decomposed rock was present below about 3 m deep.

Comparisons between field and laboratory SDSWCCs

SDSWCCs of the natural colluvium and CDT specimens, which were obtained from the block samples taken at the site, were measured using the modified one-dimensional stress-controllable volumetric pressure plate extractor (Ng and Pang, 2000a, 2000b). The tests were conducted under net normal stresses of 0 and 40 kPa for colluvium and CDT, respectively. Figure 3.24a compares the laboratory- and field-measured SDSWCCs obtained from the first wetting-drying cycle. Field measurements at 0.36 m and 2.13 m correspond approximately to net vertical stresses of about 7 kPa and 41 kPa, respectively. For the colluvium, it can be seen that the corresponding laboratory values and field measurements are generally consistent, in terms of the air-entry value, desorption rate, adsorption rate and the shape of the hysteretic loops respectively. In the first wetting-drying cycle, the laboratory-measured SDSWCCs appear to capture the overall field situation satisfactorily. By contrast, the field-measured SDSWCC of CDT appears to have a higher air-entry value, negligible hydraulic hysteretic loop and smaller saturated θ_w than those of laboratory-measured ones. As discussed previously, these observations may probably be attributed to the smaller and more uniform pore-size of the *in situ* soils, resulting from different degrees of weathering at different depths. The flat shape of the *in situ* SDSWCCs indicates that the CDT is within the boundary effect zone (Fredlund et al., 2001b), where the soil remains saturated at a given suction.

To investigate the effects of the wetting-drying cycles, laboratory- and field-measured SDSWCCs from the second wetting-drying cycles are compared in Figure 3.24b. Relatively

Retention characteristics and permeability functions 103

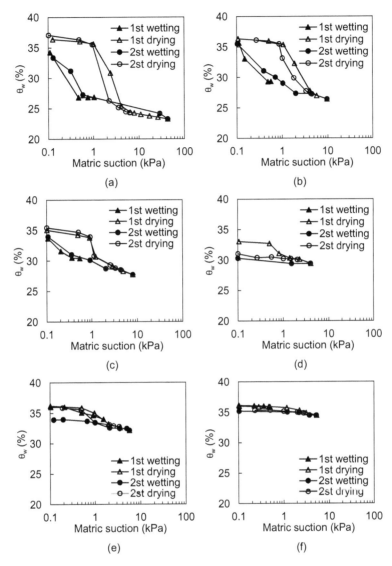

Figure 3.23 In situ SDSWCCs at: (a) 0.36 m; (b) 0.77 m; (c) 0.95 m; (d) 1.17 m; (e) 1.54 m; (f) 1.85 m; (g) 2.13 m; (h) 2.43 m; (i) 2.60 m and (j) 2.99 m.

speaking, the correspondence between the laboratory and field measurements of colluvium obtained in the second cycle is not as good as that measured in the first cycle, especially at low suctions. We attribute this mainly to the smaller hysteretic loop obtained for the laboratory specimen in the second cycle than that obtained in the first cycle. Given the consistent field measurements of SDSWCCs between the first and second cycles obtained at depths of 0.36 m and 2.13 m (see Figures 3.23a and 3.23g), this seems to suggest that there might have been a change in the microstructure or PSD of the laboratory soil specimen after the first wetting-drying cycle, leading to different measurements being obtained during the second cycle. It is reported that the size of the hysteretic loop can be reduced

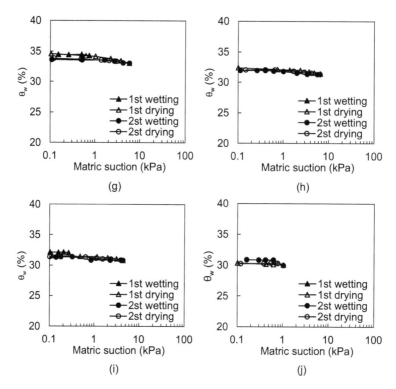

Figure 3.23 (Continued)

significantly in a recompacted CDT between the first and the second wetting-drying cycles (Ng and Pang, 2000a), owing to the collapse of the soil structure during the first wetting cycle. Similarly, it may be not unreasonable to expect a collapse in the soil structure in natural soil specimens if the soil has been disturbed during sampling. Using the Young-Laplace equation and assuming that T_s is equal to 72 mN/m at 25°C, the corresponding calculated pore radius is 1.4 mm and 0.072 mm for matric suctions of 0.1 kPa and 2 kPa, respectively. It is postulated that these relatively large pores at low suctions might have closed (or collapsed) after the first wetting-drying cycle, leading to a different SDSWCC in the second cycle.

Comparisons between field-measured SDSWCCs and predictions from PSDs

Since it may be time-consuming and resource-intensive to measure SWCC and $k(\psi)$ either in the laboratory or in the field, some researchers (Gupta and Larson, 1979; Fredlund et al., 1996, 1997) have suggested using PSDs to predict SWCCs. Figure 3.25 shows a comparison of field-measured SDSWCCs at 0.84 m and 1.85 m depth with empirical predictions based on the corresponding PSDs of colluvium and CDT using the predictive

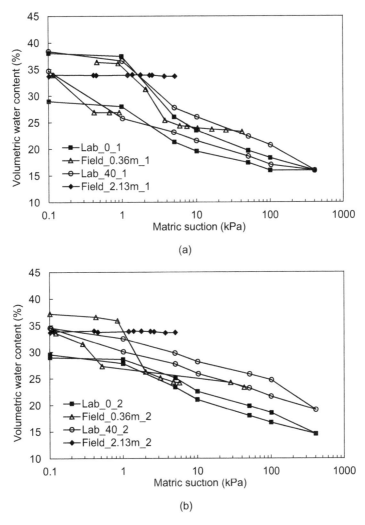

Figure 3.24 Comparisons of SDSWCCs from *in situ* and laboratory measurements of colluvium and CDT in Tung Chung from: (a) first wetting-drying cycle and (b) second wetting-drying cycle.

method proposed by Fredlund et al. (1997). Since the predictive method is based on calibrating drying SWCCs with PSDs, only one drying curve can be obtained from each soil type. As shown in the figure, the predicted drying curves for the colluvium and CDT are found to be in satisfactory agreement with the drying curves obtained from the field in terms of the air-entry values and desorption rates. However, it is obvious that this method may not be appropriate when predicting wetting SWCCs, and thus cannot account for soil-water hysteresis.

106 Advanced Unsaturated Soil Mechanics

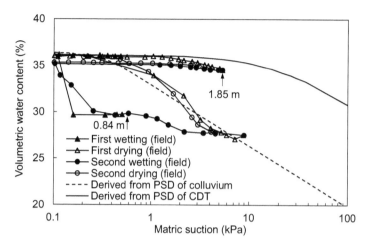

Figure 3.25 Soil-water characteristic curves at depths of 0.84 m and 1.85 m obtained from field monitoring and predicted from particle size distributions using Fredlund et al. (1997).

Comparisons between field-measured permeability functions and predictions from PSDs

In the literature, different studies have suggested various ways to predict $k(\psi)$ from SWCCs. These methods usually require information about the SWCC and the saturated soil permeability. Here, the drying SWCC for colluvium and CDT is first predicted from the corresponding PSDs (see Figure 3.19) using the method proposed by Fredlund et al. (1997). GCO (1982) reported that the range of saturated permeability in colluvium and CDT within 5 m of the ground surface is 2×10^{-6} to 9×10^{-4} m/s and 3×10^{-6} to 9×10^{-6} m/s, respectively. These reported values were obtained from constant head tests in boreholes and double-ring infiltration tests in the mid-level area of Hong Kong Island. Using the method proposed by Fredlund et al. (1994), the $k(\psi)$ of both colluvium and CDT can now be predicted. Figure 3.26 compares the predicted $k(\psi)$ with field measurements at depths of 0.84 m and 1.85 m for the colluvium and CDT, respectively. Although the measured data are fairly scattered, the water permeability along the two wetting paths is generally higher than that of the drying paths. For a given matric suction, however, θ_w is higher along the drying path than along the wetting path (see Figure 3.26), and thus a higher permeability may be expected along the drying path. This unusual $k(\psi)$ might be the result of *in situ* features such as cracks, fissures and relict joints in the soil, or might be due to enhanced water connectivity in the soil structure as a result of structural collapse in the soil upon wetting.

The measured $k(\psi)$ along the second wetting path is higher than that along the first wetting path by nearly one order of magnitude at a matric suction of about 0.5 kPa. This may be due to errors introduced by the limited resolution of a tensiometer at low matric suction. At a given suction, the range of the measured permeability appears to be significant (two orders of magnitude) during the two wetting-drying cycles This may be attributed to the heterogeneity of colluvium resulting from the presence of cracks, fissures

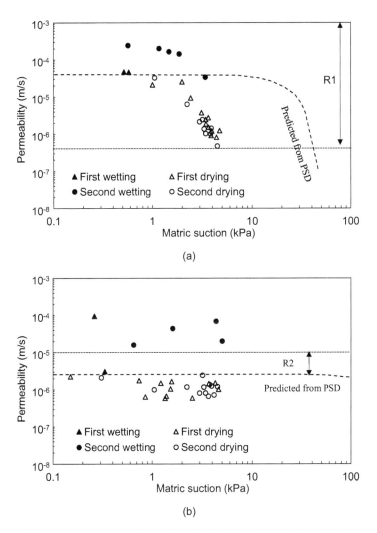

Figure 3.26 Permeability-matric suction relationships obtained from field and predicted from particle size distribution using Fredlund et al. (1994): (a) colluvium and (b) CDT.

and rootlets in the field (e.g., Basile et al., 2006). Wendroth et al. (1993) reported that the measured permeability of a silty loam decreased from 3×10^{-7} to 6×10^{-8} m/s over 1 to 10 kPa matric suction, even under laboratory conditions.

In the CDT (see Figure 3.26b) the measured data are quite scattered, probably as a result of the wide spacing of the installed instruments. Nevertheless, the measured water permeability along the wetting paths is higher than that along the drying paths at a given suction. For the range of suctions measured, the permeability varies from 3×10^{-6} to 1×10^{-4} m/s along the wetting paths and from 4×10^{-7} to 3×10^{-6} m/s along the drying paths. As with those measured in the colluvium, measurements from the first and second drying paths are generally consistent. At a given suction, the measured $k(\psi)$ in colluvium is generally higher than that in CDT along all the wetting-drying paths, indicating the less permeable nature of CDT, which might have led to the formation of a perched water table at the colluvium/CDT interface (see Figure 3.21).

Generally speaking, the predictions of $k(\psi)$ using PSDs are overestimated in the drying cycles and underestimated in the wetting cycles, for both the colluvium and CDT. It is obvious that the predictions are neither qualitatively nor quantitatively in agreement with the measured results, except that the predicted $k(\psi)$ of CDT is quantitatively in agreement with the measured drying permeability, within one order of magnitude. Predicting a $k(\psi)$ directly from a PSD of the two soils seems to be unsatisfactory, at least over the range of matric suction studied.

Summary

Above 1.85 m, noticeable hydraulic hysteretic loops were observed during the wetting-drying cycles. The size of the hysteretic loop in the second wetting-drying cycle was slightly less than that in the first cycle. However, below 1.85 m, the change in θ_w with matric suction and the hysteresis phenomenon of soil-water characteristics are essentially negligible, and no significant hysteretic loop was obtained at any depth during the wetting-drying cycles. These differences may be attributed to the smaller range of matric suctions, the stiffer nature and larger dry density (or smaller porosity) of CDT, and an increase in the net normal stress with depth.

The laboratory-measured SDSWCCs under a net normal stress of 0 kPa (colluvium) seem to be able to capture the overall field characteristics (i.e., air-entry values, desorption rates, adsorption rates and shapes of the hysteretic) satisfactorily for the first wetting-drying cycle but not for the second, especially in the low suction zone. This may be caused by the presence of preferential flow paths arising from the presence of cracks, fissures and relict joints in the field that cannot be captured in small specimens tested in the laboratory. For the CDT, there is also clear inconsistency between the laboratory and field measurements of SDSWCCs.

It has been demonstrated that an unsaturated $k(\psi)$ is dependent on the current matric suction, θ_w and suction history (wetting-drying cycles). The values of $k(\psi)$ for wetting paths are generally higher than those for drying paths, in both the colluvium and the CDT. Along each wetting/drying path, the $k(\psi)$ of colluvium is generally above that of CDT. A comparison of the predictions of $k(\psi)$ using PSDs shows that the predicted $k(\psi)$ values in both colluvium and CDT along the drying paths are overestimated, but they are underestimated along the wetting paths. Although the predicted SWCCs along the drying paths are fairly consistent with corresponding laboratory measurements, there is a clear qualitative and quantitative inconsistency between predictions of $k(\psi)$ from PSD and field measurements.

GAS BREAKTHROUGH AND EMISSION THROUGH LANDFILL FINAL COVER (NG ET AL., 2015B)

Introduction

The main functions of a final landfill cover, which is unsaturated in nature, are to reduce rainfall infiltration and gas emission from underlying municipal waste (Ng et al., 2015b,

2015c; Ng et al., 2016b). The biodegradation (i.e., anaerobic activities) of municipal waste produces a large amount of landfill gas, namely CO_2 and CH_4. The landfill gas may discharge into the atmosphere and hence endanger the lives of nearby residents. Although gas collection systems have been used widely, landfill gases cannot entirely be collected (McBean et al., 1995; EPA, 1999). To minimise the emission of landfill gas to the atmosphere, a common approach is to construct an engineered cover over landfills to isolate municipal waste from the environment. Compacted clay has been used widely for landfill final cover systems because of its relatively low saturated water permeability (Albright et al., 2004; Benson et al., 2007; Barnswell and Dwyer, 2012). Although saturated permeability is a key parameter for assessing the feasibility of using compacted clays as hydraulic barriers in landfill final covers, it is not sufficient to evaluate the performance of compacted clay barriers as gas barriers in terms of gas breakthrough pressure and emission rates under different weather conditions. This is because landfill covers are unsaturated in service (Weeks and Wilson, 2005; Zhan et al., 2014a). Infiltrating rainfall water can increase the degree of saturation of soil but is unlikely to render it fully saturated. This implies that it would be non-conservative to consider using a saturated compacted clay to impede gas flow since unsaturated soils typically favour gas flows to a greater extent than saturated soils.

Two series of gas breakthrough tests over a pressure range of 0 to 50 kPa and two series of gas emission rate tests over a pressure range of 0 to 20 kPa were carried out on unsaturated compacted kaolin clay using a one-dimensional soil column. Three degrees of clay saturation (40%, 60% and 80%) and two thicknesses of clay (0.4 m and 0.6 m) were considered. The results improve our understanding of gas breakthrough and emission in unsaturated compacted clay and provide useful insights into the optimum design of landfill cover systems.

Theoretical considerations

According to Gallé (2000), the gas breakthrough pressure of a soil cover can be determined from the relationship between landfill gas pressure and gas emission rate. Figure 3.27 shows a conceptual diagram illustrating the relationship between gas emission rate and gas pressure for unsaturated soils. It can be seen that when gas pressure is low (i.e., below point A shown in the figure), the gas emission rate is almost negligible. This may be because, for unsaturated clay, the soil pores occupied by gas are probably not interconnected to form a pathway for gas flow. Only gas diffusion takes place in this pressure range (Graham et al., 2002). When the gas pressure increases over a threshold value (point A), the gas emission rate increases dramatically. The sudden increase in gas emission rate is mainly because isolated pores may be forced into communication, resulting in a much higher gas flow. Point A is referred to as the breakthrough point and the corresponding pressure as the gas breakthrough pressure.

It should be noted that the gas breakthrough pressure of unsaturated soil is not the same as the well-known air-entry value of soil. The entry of gas into the soil pores does not necessarily induce a dramatic increase of gas flow rate.

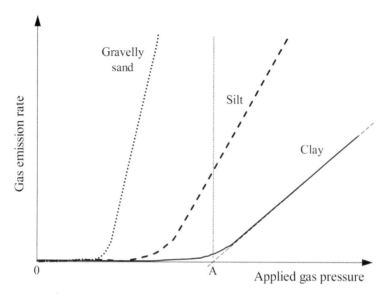

Figure 3.27 Schematic diagram showing the relationships between gas emission rate and gas pressure of silt, gravelly sand and clay.

Test program

Two types of tests (gas breakthrough and gas emission rate test) were carried out using a 1 D soil column. The first type was conducted to determine the gas breakthrough pressure of unsaturated clay at the ultimate limit state over a high gas pressure ranging from 0 to 50 kPa. In each test, the gas pressure was increased in a stepwise manner at constant time intervals until gas breakthrough occurred. For this test, two series (BSr and BTh) were carried out to determine the gas breakthrough pressures taking into consideration three degrees of saturation (S_r = 60%, 80% and 100%) and two thicknesses of clay (H = 0.4 and 0.6 m). Another type of test was carried out to investigate the gas emission rate at the serviceability limit state with a low gas pressure ranging from 0 to 20 kPa. In each test, a constant gas pressure was applied at specific time intervals until the gas flow rate reached a steady state. Details of the test program are given in Table 3.6.

Experimental apparatus and instrumentation

A 1-D soil column testing system comprising a transparent acrylic cylinder and a loading system was used. A schematic diagram and photograph of this system are shown in Figure 3.28 and Plate 3.4, respectively. To investigate gas flow in unsaturated compacted soil, the upper and lower boundaries of the soil sample are designed to be controllable. At the base of the soil column, inflow gas was controlled using a high-precision gas pressure regulator. The pressure of the inflow gas was double-checked using an air-pressure transducer with an accuracy of 0.2 kPa. Further, to avoid the drying effect of gas on the soil sample, the gas was passed through a gas saturator before entering the soil sample.

Table 3.6 Test program for investigating gas breakthrough and emission

Series	Degree of saturation at compaction: %	Thickness, H: m	Gas Pressure, Pg: kPa
BSr	60, 80, 100	0.4	0–50
BTh	80	0.6	0–50
CSr	40, 60, 80	0.4	1, 5, 10, 20
CTh	40, 60, 80	0.6	1, 5, 10, 20

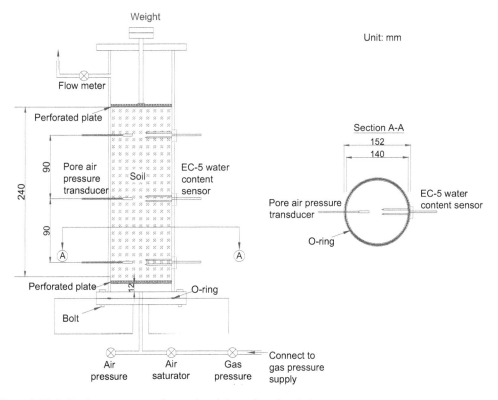

Figure 3.28 Soil column apparatus for gas breakthrough and emission rate tests.

A gravel layer with a height of 20 mm was used to ensure uniformity of gas pressure at the bottom boundary. On the other hand, the top of the soil column was open to the atmosphere, with the rate of gas outflow monitored by a flow meter. In addition, a constant load corresponding to a pressure of 20 kPa was applied at the top of the soil sample to simulate the weight of the soil layers overlying the compacted clay layer in landfill final covers.

The soil column testing system was equipped with various instruments, including pore-air pressure transducers, EC-5 water content sensors and an air flow meter. Pore-air pressure transducers were installed at 75, 225 and 375 mm above the bottom of the soil column to monitor pore-air pressure distribution in the unsaturated compacted soil sample. The measurement range was 0 to 100 kPa with an accuracy of ±0.2%. EC-5 water

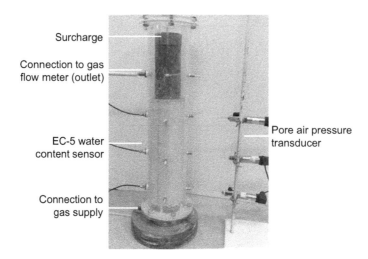

Plate 3.4 Soil column apparatus for gas breakthrough and emission rate tests.

content sensors were used to measure the volumetric water content of the soil and located at the same elevation as the pore-air pressure transducers. Further, depending on the gas emission rate in each test, two types of air flow meter with different ranges (0 to 100 ml/min and 0 to 30 L/min) and accuracies (0.5 and 5 ml/min) were used. When high gas emission is expected, the device with a wider measuring range was used. Otherwise, the one with a narrower measuring range but higher accuracy was used.

Testing soil and sample preparation

The test soil is kaolin clay. Some basic properties (including Atterberg limits, specific gravity, compaction curve) of the soil were determined following ASTM standard test methods (ASTM D 4318-10, D 2487-11, D 698-12). The saturated permeability was measured using a flexible wall permeameter according to the ASTM standard (ASTM 5084-10). The test results are summarised in Table 3.7. According to the findings of Benson et al. (1999), the geotechnical properties of the selected clay barrier material fall within the typical range of fine-grained soils used as landfill barriers.

Figure 3.29 shows the WRC of the compacted clay during the process of drying from the fully saturated state. It was measured using a modified pressure plate apparatus (Ng and Pang, 2000b). From the relationship between volumetric water content and suction, the air-entry value of the compacted clay is estimated to be about 70 kPa. Further, the measured WRC was fitted using van Genuchten's equation (van Genuchten, 1980). The fitted values of all parameters are summarised in Table 3.8.

Two methods were used to prepare soil samples, depending on their degree of saturation. For tests at degrees of saturation of 40%, 60% and 80%, each sample (140 mm in diameter and 400/600 mm in height), each of 20 sub-layers, was dynamically compacted at the predefined water content. Based on the design guideline EPA (1999) for the compaction requirements of fine-grained soil in landfill final cover, a degree of compaction of

Table 3.7 Basic properties of the kaolin clay

Property	Value
Unified soil classification system	CH
Specific gravity	2.52
Liquid limit: %	59
Plastic limit: %	32
Plasticity index: %	27
Maximum dry density: kg/m^3	1264
Optimum moisture content: %	36.18
Saturated water coefficient of permeability: m/s	5.2 × 10^{-9}
Water content at saturation: %	42.06

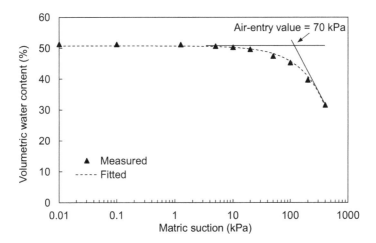

Figure 3.29 Measured and fitted WRC.

90% was chosen and used. To achieve this degree of compaction, 15 hammer drops were required for each sub-layer. It should be noted that a thin layer of vacuum grease was placed on the inner surface of the soil column to minimise preferential flow paths through the gap between the soil and the sidewall of the column. For tests at a degree of saturation of 100%, the soil samples were first prepared at the optimum water content which is equivalent to a degree of saturation of 80%, following the methods described above. Then, each sample was saturated from the bottom up to achieve a degree of saturation of about 100%.

Test procedures

For gas breakthrough tests (series BSr and BTh), after soil preparation, a constant gas pressure of 5 kPa was applied to the base of the soil sample for 3 days. If no gas breakthrough occurred, the gas pressure was increased from 5 to 10 kPa and then maintained

114 Advanced Unsaturated Soil Mechanics

Table 3.8 Result of soil-water characteristic curve and van Genuchten (1980) fitting parameters of kaolin clay

Parameter	Symbol	Value
Saturated volumetric water content: %	θ_s	51.22
Residual volumetric water content: %	θ_r	9
van Genuchten fitting parameter	a	120 kPa
	m	0.33
	n	1.49

at a constant value for another 3 days. Subsequently, gas pressures of 15, 20, 30, 40 and 50 kPa were applied to the soil sample until gas breakthrough occurred.

In the gas emission rate tests (series CSr and CTh), constant gas pressures were applied to the bottom of the soil after soil preparation until a stable gas flow was achieved. When the step at a gas pressure of 1 kPa was complete, the gas pressure was then increased to 5 kPa and then kept constant until a stable gas flow was achieved. 10 and 20 kPa gas pressures were applied following a similar procedure.

During the tests, pore-air pressure distributions were measured using pore-air pressure transducers along the depth of the soil column. The gas emission rate was monitored at the top of the soil sample using a flow meter. Further, all tests were conducted in a temperature-controlled room (20 ± 1°C) where the density and viscosity of gas can be considered to be constant.

Experimental results

Gas breakthrough pressure at different degrees of saturation

Figure 3.30 shows the relationship between applied gas pressure and gas emission rate in clay at three different degrees of saturation (S_w = 60%, 80% and 100%). The figure clearly reveals two types of soil response. At a degree of saturation of 60% and 80%, the gas emission rate is almost zero at lower gas pressures. When the gas pressure reaches a threshold value, the gas emission rate suddenly increases. According to the discussion in the section on "theoretical considerations", the threshold value is defined as the gas breakthrough pressure. From the figure it is estimated that the gas breakthrough pressure is 22 and 38 kPa at degrees of saturation of 60% and 80%, respectively. On the other hand, for the test at 100% saturation, the gas emission rate is almost zero over the whole range of gas pressure (0–50 kPa). This suggests that the gas breakthrough pressure is higher than 50 kPa under this testing condition. Comparisons between the three curves illustrate that the gas breakthrough pressure increases with an increasing degree of saturation. The influence of the degree of saturation is probably because more pores are occupied by soil-water in the compacted sample at higher degrees of saturation. Higher pressure is therefore required to create an interconnected air-filled channel for gas flow.

Based on field measurements, McBean et al. (1995) reported that landfill gas pressure in the field is generally less than 10 kPa. Under typical landfill gas pressures, a compacted clay layer with a thickness of 0.4 m or above can prevent gas breakthrough when the clay's degree of saturation is greater than 60%, as illustrated in Figure 3.30. This finding

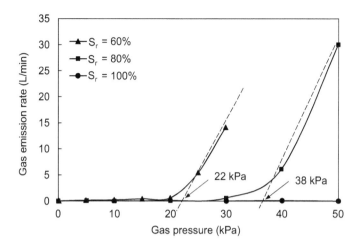

Figure 3.30 Relationship between applied gas pressure and gas emission rate at three degrees of saturation (H = 0.4 m).

suggests that in humid regions where the degree of soil saturation is generally higher than 60% (Weeks and Wilson, 2005; Zhan et al., 2014a), a 0.4 m-thick layer of clay is sufficient to satisfy the design requirements for the ultimate limit state.

Influence of clay thickness on gas breakthrough pressure

Figure 3.31 shows the results of gas breakthrough tests at two thicknesses of clay (H = 0.4 and 0.6 m). A gas breakthrough pressure of about 38 kPa was estimated for 0.4 m-thick clay. When the clay thickness increases to 0.6 m, the gas emission rate still remains almost zero even for gas pressures up to 50 kPa. This suggests that the gas breakthrough pressure can be higher than 50 kPa for a 0.6 m-thick clay layer. The observed increase of gas breakthrough pressure with increasing clay thickness is mainly because as the clay thickness increases, the potential to form interconnected air-filled flow paths across the whole sample depth decreases. Similar observations have been reported by Hildenbrand et al. (2002). Based on tests over a limited range of sample thickness (less than 0.03 m), they also found that gas breakthrough pressure increases as soil thickness increases for the same gas pressure. These results suggest that the capacity of compacted clay to prevent gas breakthrough can be improved significantly by increasing the thickness of clay. According to the results in Figures 3.30 and 3.31, in arid regions where the soil's degree of saturation may drop below 60%, a 0.4 m-thick clay layer may fail to prevent gas breakthrough. One possible solution to this problem is to increase the thickness of the clay layer in the landfill cover.

Gas emission rates at different degrees of saturation

Figure 3.32 shows the relationship between gas emission rate and gas pressure at three degrees of saturation (40%, 60% and 80%). For 40% and 60% degrees of saturation, the

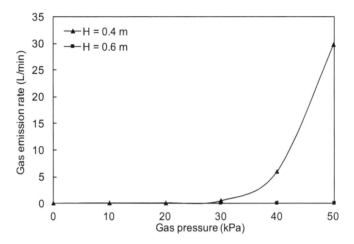

Figure 3.31 Thickness effect on gas breakthrough pressure (S_w = 80%).

gas emission rate increases almost linearly with increasing gas pressure. For a degree of saturation of 80%, the gas emission rate is almost zero even for gas pressures up to 20 kPa. This is consistent with the result of gas breakthrough tests. As illustrated in Figure 3.30, the gas breakthrough pressure at a degree of saturation of 80% is up to 38 kPa and the gas emission rate is almost negligible at gas pressures less than 20 kPa. Comparisons between the three curves (corresponding to degrees of saturation of 40%, 60% and 80%) clearly reveal that at the same gas pressure, the gas emission rate is significantly lower at higher degrees of saturation.

In Figure 3.32, the methane emission limit proposed by the Australian government guideline (CFI, 2013) is included for comparison. According to this guideline, the methane emission rate in a landfill should be less than 63 ml/m²/min. To meet the gas emission rate limit set by the Australian guideline, a clay layer with a thickness of 0.4 m at degrees of saturation of 60% and 80% is sufficient over the range of typical landfill gas pressures (less than 10 kPa) observed in the field (McBean et al., 1995). However, at a degree of saturation of 40%, a 0.4 m-thick clay layer is only sufficient if the gas pressure is less than 5 kPa. A possible solution to this problem is to increase the thickness of the clay layer. The gas emission rate would decrease with an increase in clay thickness (Ng et al., 2015c). More discussions of effects of clay thickness are given later. It should be noted that although this guideline is proposed for methane emission, it should be applicable to some other gases (such as N_2). This is because the physical properties, including viscosity, of methane and nitrogen are similar.

Influence of clay thickness on gas emission rate

Figure 3.33 shows the influence of clay thickness on the gas emission rate at a degree of saturation of 60%. The maximum gas emission rate allowed by the Australian guideline

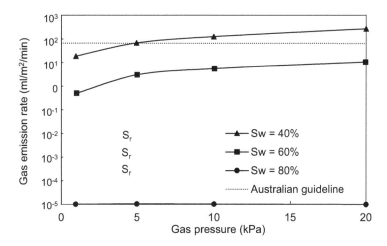

Figure 3.32 Measured gas emission rates at different degrees of saturation (H = 0.4 m).

(CFI, 2013) is also included for comparison. As expected, the gas emission rate increases with an increase in gas pressure for both clay thicknesses. Moreover, over the full range of gas pressures (0 to 20 kPa), the gas emission rate decreases significantly with increasing clay thickness, which suggests that the gas emission rate is inversely proportional to the thickness of the soil sample.

The observations from Figures 3.32 and 3.33 illustrate that the gas emission rate through compacted unsaturated clay is strongly affected by the thickness and degree of saturation of the clay. To meet the requirement of the gas emission rate limit set by the Australian guideline, landfill designers should require that the clay layer is at the bottom of the landfill cover to eliminate moisture loss. Moreover, a thicker clay layer should be adopted in drier areas in order to minimise the gas emission rate.

Summary

The gas breakthrough pressure of unsaturated compacted clay increases as the degree of saturation and thickness of the clay increase. Under a gas pressure of 10 kPa (the upper bound of typical landfill gas pressure), a 0.4 m or thicker clay layer can prevent gas breakthrough at a degree of saturation of about 60% (in humid regions). A thicker clay layer is required if the clay's degree of saturation is below 60%.

Gas emission rate tests show that a clay layer with a thickness of 0.4 m at a degree of saturation of 60% or above (i.e., in relatively humid regions) is sufficient over the range of typical landfill gas pressures (less than 10 kPa) to meet the requirement of the gas emission rate limit set by the Australian guideline. When the gas pressure is below 5 kPa, the gas emission rate through a 0.4 m-thick layer of clay at a degree of saturation of 40% (i.e., in semi-arid regions) can still satisfy the above requirement.

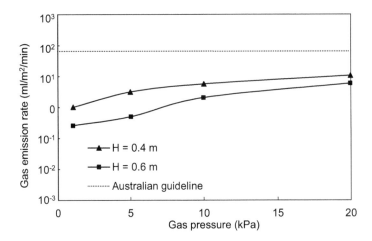

Figure 3.33 Measured gas emission rates at different clay thicknesses (S_w = 60%).

NUMERICAL INVESTIGATION OF GAS EMISSION FROM THREE TYPES OF LANDFILL COVERS UNDER DRY WEATHER CONDITIONS (NG ET AL., 2015B, C; NG ET AL., 2016B)

Introduction

Single compacted clay layer covers (Albright et al., 2006) and covers with capillary barrier effect (CCBE) (Ross, 1990; Aubertin et al., 2009), have been used widely as final covers for MSW landfills. Figure 3.34a shows the configuration of a single clay layer cover. We expect that a low coefficient of air permeability at a high degree of saturation in clay will minimise gas emission, particularly in saturated conditions. However, the coefficient of air permeability for compacted clay increases significantly as its water content decreases (Moon et al., 2008) and hence the single clay layer cover may not act as a satisfactory gas barrier in extremely dry weather conditions. On the other hand, a CCBE, as shown in Figure 3.34b, basically consists of a layer of fine-grained soil overlying a layer of coarse-grained soil. Bussière et al. (2003) pointed out that high saturation at the interface in a CCBE will result in a low coefficient of air permeability and could prevent oxygen intrusion. Because of the potential for malfunction of these two types of covers when minimising gas emission, particularly under extremely dry conditions, it is necessary to add an underlying geomembrane to minimise gas emissions. However, a geomembrane is generally thin and is susceptible to construction damage and puncture (Wilson-Fahmy et al., 1996; Koerner et al., 2010). Moreover, the shear resistance at the interface of geomembrane and overlying soil is usually lower than that of soil-soil interfaces. There have been notorious cases of landfill instability and failure caused by the weak interface at the geomembrane (Stark et al., 2012).

In order to increase the effective minimisation of gas emission and hence eliminate the use of a geomembrane, a three-layer capillary barrier cover as an alternative soil final landfill cover is proposed, as shown in Figure 3.34c. This cover consists of a fine-grained

Retention characteristics and permeability functions

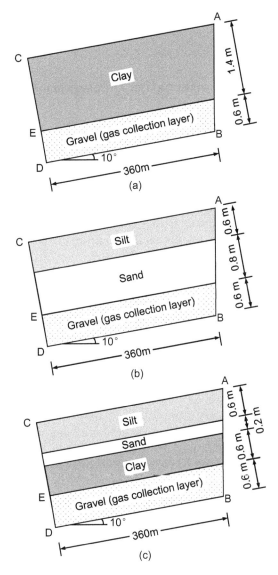

Figure 3.34 Configuration of (a) single clay layer cover; (b) CCBE and (c) three-layer capillary barrier cover (not to scale).

soil layer overlying a coarse-grained layer, which in turn overlies a fine-grained soil such as clay. Its working principle is described below.

The feasibility and effectiveness of a three-layer capillary barrier cover regarding minimisation of gas emissions is explored by carrying out a series of numerical simulations. For comparison, a single clay layer cover and a CCBE without geomembrane are also simulated. Since the critical case for gas emission is dry weather, the longest duration of

120 Advanced Unsaturated Soil Mechanics

evaporation (maximum continuous no-rainfall days) in a humid climate, e.g., Hong Kong, is considered in numerical simulations. The numerical results for the three types of soil covers are compared to explore and assess their gas emission performance and to investigate the underlying mechanism.

Working principle of the alternative three-layer capillary barrier cover

In essence, the three-layer cover consists of a two-layer CCBE and an underlying clay layer. The main purpose of the upper two-layer CCBE is to protect the underlying clay layer from desaturation by virtue of its capillary effects. During evaporation, the suction of the cover increases, which results in a significant decrease in the coefficient of water permeability for coarse-grained soil. Hence, this coarse-grained soil minimises the upward flow of water (i.e., evaporation) in clay and the coefficient of air permeability in the clay layer barely changes. It indicates that gas emissions from waste can then be minimised by a clay layer at high saturation in the three-layer capillary barrier cover.

Numerical modelling

Numerical parametric scheme

As stated earlier, dry weather conditions are the worst case for a landfill cover with respect to minimising gas emission. Table 3.9 summarises the continuous no-rainfall days over the past 10 years (2004–2014) for Hong Kong, which has a humid climate. The data were collected from the daily total rainfall data from the Hong Kong International Airport Station and published online by the Hong Kong Observatory (2015). The maximum continuous days without rainfall vary from 10 to 34 days. Therefore, the maximum continuous duration without rainfall (i.e., 34 days) in Hong Kong was selected as the extreme dry condition for numerical simulations. During these 34 days, the average potential evaporation rate of 2.91 mm/d in Hong Kong was applied to the cover surface. This value is derived from the annual potential evapotranspiration between 1981 and 2010 provided by Hong Kong Observatory (2012), i.e., 1062.4 mm/year.

Finite element software and mesh

As pointed out by Zeng et al. (2011), air flow is strongly coupled to moisture transport in soil near the land surface. In other words, coupled water and air flow should be considered when assessing the performance of soil covers with regard to gas emission. Computer software capable of taking coupled water and air flow into account was used to carry out the finite element analyses. The software is called CODE_BRIGHT (COupled DEformation BRIne, Gas and Heat Transport), developed by the Technical University of Catalonia (UPC) (Olivella et al., 1994).

Figure 3.35 shows the finite element mesh used to simulate the proposed three-layer capillary barrier cover. The mesh describes a 0.6 m-thick silt layer overlying a 0.2 m-thick sand layer and a 0.6 m-thick clay layer. The use of 0.6 m of silt was considered to be the minimum thickness required to allow a large storage capacity under humid

Table 3.9 Maximum continuous no-rainfall days in Hong Kong during 2004–2014

Year	Start	End	Maximum continuous no-rainfall days
2004–2005	Nov. 25, 2004	Dec. 22, 2005	28
2005–2006	Nov. 29, 2005	Dec. 27, 2005	29
2006–2007	Jan. 24, 2007	Feb. 7, 2007	15
2007–2008	Nov. 10, 2007	Dec. 13, 2007	34
2008–2009	Jan. 5, 2009	Jan. 26, 2009	22
2009–2010	Oct. 22, 2009	Nov. 7, 2009	17
2010–2011	Nov. 17, 2010	Dec. 10, 2010	24
2011–2012	Dec. 8, 2011	Dec. 17, 2011	10
2012–2013	Dec. 30, 2012	Jan. 22, 2013	24
2013–2014	Dec. 18, 2013	Jan. 6, 2014	20

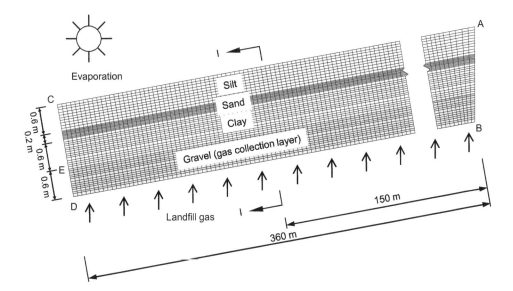

Figure 3.35 Finite element mesh used for three-layer capillary barrier cover.

climate conditions (Koerner and Daniel, 1997). 0.2 m was chosen for the sand layer acting as the coarse-grained soil layer in a typical CCBE, according to Morris and Stormont (1999). Based on some preliminary one-dimensional column tests (Ng et al., 2015b), 0.6 m thickness of compacted clay is required to minimise gas emission. In order to consider the influence of the gas collection layer underlying the cover system, a 0.6 m-thick gravel layer was also simulated beneath the three-layer capillary barrier cover.

The USEPA (2004) guideline was followed to select the slope angle. It suggests that the slope angle of a typical cover should lie between 3° (i.e., 20H:1 V) and 18° (i.e., 3H:1 V). Therefore, a 10° (i.e., 6H:1 V) slope was chosen here. In order to simulate a sufficiently large landfill, 360 m length was used.

122 Advanced Unsaturated Soil Mechanics

A single clay layer cover and a CCBE were also simulated, as shown in Figure 3.35. For ease of comparison, the total thickness of these two covers was set to be the same as that of the three-layer cover system (i.e., 1.4 m-thick cover and a 0.6 m-thick gravel layer as a gas collection layer). For CCBE, the thickness of fine-grained soil was identical to that of the three-layer cover system, to give the same water storage capability (Khire et al., 2000). Thus, the CCBE consists of a 0.6 m-thick silt layer and a 0.8 m-thick sand layer. It should be noted that a 0.6 m-thick gas collection layer is also simulated beneath the single clay layer cover and the CCBE.

Material properties

The measured WRCs of silt, sand, clay and gravel obtained from published experimental data were used for numerical simulation of transient seepage (see Figure 3.36a). It can be estimated from the figure that the air-entry values of silt, sand and clay are approximate 10, 1 and 100 kPa, respectively (Table 3.10). Based on these WRCs, the water permeability functions of the four soils are estimated using Mualem's (1976) model and are shown in Figure 3.36b.

The air coefficient of permeability, k_a, takes the form:

$$k_a = \frac{\rho_a g}{\mu_a} K_{in} k_{ra}$$

(3.27)

where k_{ra} is the relative permeability of air; ρ_a is the density of air; g is the gravitational acceleration; μ_a is the absolute (dynamic) viscosity of air and K_{in} is the intrinsic permeability of the soil. Based on the saturated coefficient of water permeability, the values of intrinsic permeability of sand, silt, clay and gravel are 4.91×10^{-12}, 1.03×10^{-10}, 1.03×10^{-16} and 1.03×10^{-8} m^2, respectively. A power relationship between relative air permeability and degree of saturation (Olivella and Alonso, 2008) was used as follows:

$$k_{ra} = \left(1 - S_r\right)^{n_2}$$

(3.28)

where S_r is the degree of saturation and n_2 is a model parameter (typically equal to 3). The air coefficients of permeability for the four soils were then calculated, as shown in Figure 3.36c.

Numerical simulation procedure

All three types of soil covers are subjected to the same modelling procedures. Before any transient analysis, the initial pore-water pressure and pore-air pressure at each node of the finite element grid were obtained from a steady-state seepage analysis. In this steady-state simulation, the three types of covers were subjected to the same boundary conditions.

Retention characteristics and permeability functions 123

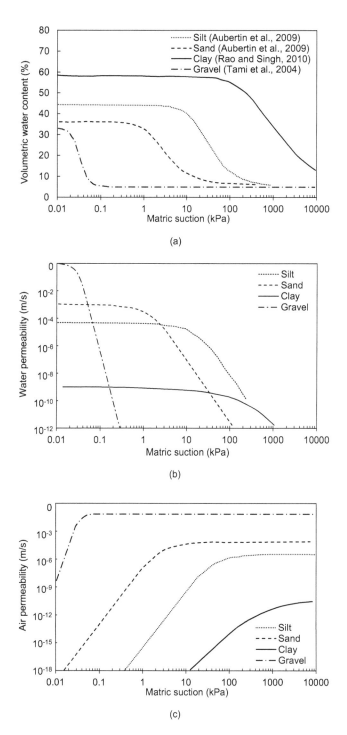

Figure 3.36 (a) Water retention curves; (b) water permeability functions and (c) air permeability functions for silt, sand, clay and gravel.

124 Advanced Unsaturated Soil Mechanics

Table 3.10 Main hydraulic properties of different soils

Soil	Air-entry value: kPa	Water-entry value: kPa	Saturated θ_w: %	Residual θ_w: %	Intrinsic permeability: m^2	n_2(Eq. (3.28))
Silt	10	200	44	5	4.91×10^{-12}	3
Sand	1	10	36	6	1.03×10^{-10}	3
Clay	100	>10000	57	3	1.03×10^{-16}	3
Gravel	0.02	0.06	33	5	1.03×10^{-8}	3

Note:

θ_w: volumetric water content.

(1) As shown in Figure 3.35, the right boundary AB is taken as the plane of symmetry and assumed to be impermeable. This implies that both water and air flux through AB are zero.

(2) According to the Hong Kong Observatory (2012), the average rainfall is 6.56 mm/d. On the upper exposed surface (AC), an average net infiltration rate (precipitation minus potential evaporation, i.e., 6.56 mm/d – 2.91 mm/d) of 3.65 mm/d is applied at the boundary AC. For the air phase, the pore-air pressure is set to 0 kPa at the boundary AC since it is in direct contact with the atmosphere. Therefore, the boundary condition at AC (surface boundary) is specified to have a constant water flux and a constant air pressure.

(3) In engineering practice, the cover is connected to an exit drain which allows percolated water to be removed (Koerner and Daniel, 1997). The exit drain is typically a perforated drainage pipe. In order to simulate the exit drain, a seepage surface is imposed at boundary CE. This suggests that water accumulates forming a drip face to seep out on boundary CE until its pore-water pressure increases to atmospheric pressure (i.e., 0 kPa). Regarding air flow, zero air flux is imposed at CE. For DE, both water and air fluxes are assumed to be zero.

(4) At the bottom boundary BD, air pressure is specified to be the landfill gas pressure. Walter (2003) and Gebert and Gröngröft (2006) reported that the maximum measured pressure of landfill gas under a cover system is about 0.5 kPa, whereas Wei (2007) found that the maximum value could be up to 10 kPa. To simulate the worst case, 10 kPa of landfill gas pressure was assumed in this study. Chen et al. (2013) pointed out that the degree of saturation of municipal solid waste may be as high as 70% in humid regions (e.g., southern China). Based on the water retention curve (WRC) of municipal solid waste (Ng et al., 2015c), a corresponding suction of about 20 kPa may be deduced for waste at 70% degree of saturation. Thus, a constant pore-water pressure of –10 kPa is specified at boundary BD to simulate the influence of the underlying wet waste. In other words, constant water and air pressures are both specified for the boundary conditions at BD (bottom boundary).

After obtaining the initial steady-state solutions for the three types of covers, an average potential evaporation rate of 2.91 mm/d was imposed on the boundary AC for 34 days to simulate extremely dry conditions (evaporation) in Hong Kong. Other boundary conditions are kept the same as those of the initial steady-state seepage analysis.

Computed results

The computed results from Section I-I shown in Figure 3.35 are chosen for detailed interpretation and comparison. This section is located 150 m upstream of the cover (i.e., boundary AB) or almost half way along the 360 m long mesh.

Variations of pore-water pressure

Figure 3.37a shows the computed pore-water pressure distribution along Section I-I for the initial state of the three types of soil cover. As previously stated, the initial distribution of pore-water pressure for each cover was obtained from steady-state flow analysis with an average net infiltration rate of 3.65 mm/d at the surface. For the single clay layer cover, the initial pore-water pressure decreases with depth. The initial pore-water pressure profile moves right from the hydrostatic line because of rainfall-induced infiltration. Since a constant suction is applied at the bottom boundary, the lowest pore-water pressure is almost −10 kPa at the bottom. For the CCBE, the initial pore-water pressure increases along a hydrostatic line in the silt layer and remains almost constant (i.e., −10 kPa) in the sand layer. For the three-layer capillary barrier cover, the initial pore-water pressure increases along a hydrostatic line in the silt and sand layers but decreases in the clay layer. The maximum pore-water pressure occurring at the interface between clay and sand layers is −3.2 kPa. The initial hydraulic gradient is almost 0 above this interface but becomes positive below the interface. Because of the high water storage capacity of the clay layer at the bottom, the three-layer cover is relatively wet under the initial conditions and thus the initial pore-water pressure at the surface of three-layer capillary barrier cover is higher than that of CCBE by 5.4 kPa.

On the 34th day, the pore-water pressure at the cover surface decreases to −1600 kPa in the single clay layer cover, −60.2 kPa in the CCBE, and −45.3 kPa in the three-layer capillary barrier cover. In the single clay layer cover, the suction is greater than the air-entry value from top to bottom. However, the suction of the clay layer in the three-layer cover is less than its air-entry value. This indicates that only the clay layer in the three-layer cover can be used as a gas barrier. This is because of the dry sand layer in the three-layer cover. In this sand layer, the pore-water pressure is less than −30 kPa, at which pressure the coefficient of water permeability of sand is below that of the underlying clay (see Figure 3.36b). This small coefficient of water permeability in the sand layer results in a hydraulic barrier between the silt and clay layers. Therefore, the decrease in pore-water pressure induced by water loss in the three-layer cover is much smaller than that in the cover with a single clay layer.

Variations of pore-air pressure

Figure 3.38 shows the computed air pressures in the three types of cover. As previously stated, the air pressure is set to 0 kPa at the cover surface and 10 kPa at the bottom.

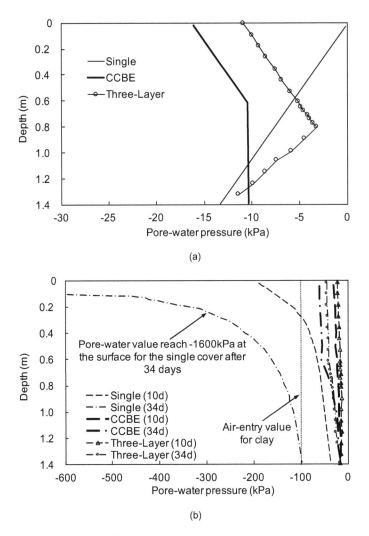

Figure 3.37 Pore-water pressure profiles in three types of covers: (a) at the initial state and (b) during drying.

For the cover with a single layer of clay, the initial pore-air pressure increases with depth. The initial air pressure decreases significantly in a zone near the surface of the cover, indicating that the air flow is minimised mainly by this zone with an initial rainfall of 3.65 mm/d. The depth of the gas-blocking zone is about 0.2 m. For CCBE, the initial air pressure increases with depth in the silt layer but maintains 10 kPa in the sand layer. The constant air pressure indicates that the sand layer cannot minimise landfill gas emission with an initial rainfall. The air pressure mainly dissipates within the silt layer of the CCBE, indicating that gas emission from the CCBE subjected to an initial rainfall is governed by the silt layer. For the three-layer capillary barrier cover,

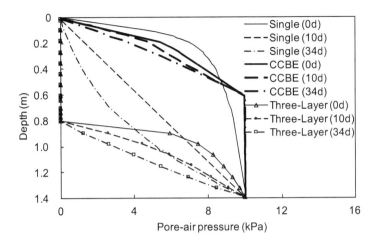

Figure 3.38 Pore-air pressure profiles in three types of covers.

the initial air pressure is zero in the silt and sand layers, while it increases with depth in the clay layer. This implies that the clay layer dominates the air flow through the three-layer cover.

After 34 days of drying, the pore-air pressures in the single layer of clay cover decrease, particularly near the surface. The higher the location, the closer to zero the pore-air pressure becomes. This is because of lower pore-water pressure (see Figure 3.37b) and thus a larger coefficient of air permeability (see Figure 3.36c) near the surface. On the 10th and 34th day, the air pressure gradients are similar and both increase from top to bottom. The maximum air pressure gradient lies at a depth between 1.0 m and 1.4 m, which indicates that gas emission is mainly blocked by the part near the bottom. For CCBE, by comparison with the initial condition, the pore-air pressure in the silt layer decreases but remains constant (i.e., 10 kPa) in the sand layer. The largest air pressure gradient occurs in the silt layer, indicating that the silt layer dominates gas emission through the CCBE. For the three-layer capillary barrier cover, the pore-air pressure in the clay layer decreases with time whereas those in the silt and sand layers remain constant (i.e., 0 kPa). On the 34th day, the pore-air pressure of the clay layer increases linearly with depth. This indicates that only the clay layer in the three-layer capillary barrier cover contributes to minimising gas emission.

Variations of gas emission rate

Figure 3.39 shows the development of the gas emission rate in the three types of cover during continuous drying. The gas emission rate is calculated from the total amount of gas emission over the area of the cover surface (assuming that the width of the cover is 1 m), as in the calculation of average water percolation by Morris and Stormont (1999). According to CFI (2013), the mass of landfill gas in a unit area and time (unit: $g/m^2/h$) is used for the gas emission rate.

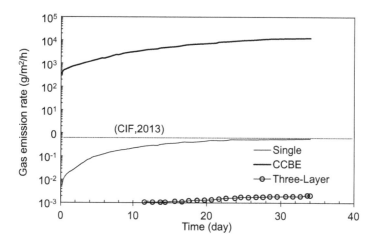

Figure 3.39 Development of gas emission rate during drying.

For a cover with a single layer of clay, the gas emission rate gradually increases with time caused mainly by the decrease in volumetric water content (Ng et al., 2015c) and hence the increase in the coefficient of air permeability in the clay layer. After 34 days of drying, the gas emission rate reaches an almost steady value of about 0.7 g/m²/h. As previously analysed, it is because the suction is greater than the air-entry value of the clay layer (see Figures 3.37 and 3.38). This gas emission rate exceeded the value allowed by CFI (2013) on the 22nd day.

For the CCBE, the gas emission rate increases to about 300 g/m²/h only after about 0.5 h of drying. Such a rapid increase in gas emission rate is because the suction of silt (see Figures 3.37 and 3.38) is greater than its air-entry value (see Figure 3.36a) and thus the coefficient of air permeability becomes large (see Figure 3.36). Subsequently, the gas emission rate increases with time and its value reaches 10000 g/m²/h at the end of continuous drying. On the 34th day, the gas emission rate is even larger than that of the single clay layer cover by about 4 orders of magnitude. This indicates that CCBE cannot serve as a gas barrier when subjected to drying.

For the three-layer capillary barrier cover, the gas emission rate is less than 0.001 g/m²/h for the first 12 days of drying. Subsequently, the gas emission rate increases slowly with time mainly because of the small increase in suction (see Figure 3.38) and hence in coefficient of air permeability for the clay layer. After 34 days of drying, the gas emission rate almost reaches a steady value of about 0.002 g/m²/h. As previously analysed, it is because the suction of the clay layer protected by the dry sand layer (see Figures 3.37 and 3.38) is smaller than its air-entry value. The relatively dry sand layer acts as a barrier, minimising water loss from the clay layer because of its low water permeability. In this way, the clay layer can maintain its high volumetric water content and low coefficient of air permeability. The gas emission rate is below the value allowed by CFI (2013), indicating that the three-layer capillary barrier cover can minimise gas emission effectively.

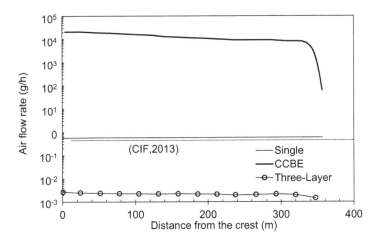

Figure 3.40 Distribution of gas emission rate along surface of three types of covers after 34 days of drying.

Distribution of gas emission rate

Figure 3.40 shows the distribution of the gas emission rate along the cover surface at the end of 34 days of continuous drying. For the single clay layer cover, the gas emission rate barely changes along the cover surface, mainly the distribution of suction and volumetric water content is uniform in the single clay layer cover. For the other two types of covers, the gas emission rate decreases with distance from the upstream boundary (i.e., AB). The gas emission rate reaches a maximum at the upstream boundary AB of the cover and then decreases gradually along the cover surface, finally dropping significantly at the downstream boundary CD. Such a highly uneven distribution of gas emission rate is mainly caused by lateral diversion of the water in the silt layer for the CCBE and three-layer capillary barrier covers. Relatively high water content appears near the downstream boundary CD of the cover and thus the gas emission rate is significantly reduced. For CCBE, because of the high coefficient of air permeability of the silt layer, even at the downstream boundary CD of the cover, the gas emission rate (100 g/m^2/h) is still greater than the criterion set by CFI (2013). For the three-layer capillary barrier cover, because of the lower coefficient of air permeability for the clay layer, the gas emission rate remains below the allowable value along almost all the cover surface. It indicates that the three-layer capillary barrier cover performs well when subjected to 34-day continuous drying.

Summary

Because of evaporation, the suction in the single clay layer cover increases until it exceeds its air-entry value, hence leading to an increase in the coefficient of air permeability. The gas emission rate exceeds the allowable gas emission set by CFI (2013).

For CCBE, the evaporation leads to water loss in the silt layer resulting in an increase in suction. Since the suction in the silt exceeds the air-entry value, it cannot serve as a gas barrier. After 34 days of drying, the gas emission rate of CCBE is even higher than that of the single clay layer cover by about 4 orders of magnitude.

130 Advanced Unsaturated Soil Mechanics

For the alternative three-layer capillary barrier cover, the relatively dry sand layer acts as a barrier, minimising water loss from the bottom clay layer because of its low water permeability. In this way, the clay layer can maintain its high volumetric water content and low coefficient of gas permeability even after 34 days of drying. The gas emission rate of the alternative three-layer cover is about 0.002 $g/m^2/h$, which is below the allowable gas emission set by CFI (2013).

Of the three types of soil cover, it is evident that the alternative three-layer capillary barrier cover has performed best in terms of minimising gas emission under continuous dry conditions.

GAS PERMEABILITY OF BIOCHAR-AMENDED CLAY (WONG ET AL., 2016)

Introduction

Biochar is a carbon-rich substance produced by heating biomass, such as leaves and wood chips, at high temperature in oxygen-deficient conditions (Lehmann and Joseph, 2009). It has been used in soil amendment to improve soil fertility and as a pollutant sorbent for soil and water remediation (Lehmann and Joseph, 2009; Ahmad et al., 2014; Mohan et al., 2014). The objective of the present study is to investigate the possibility of using compacted biochar-amended clay as an alternative soil layer in landfill final covers by measuring the gas permeability. Peanut shell biochar and kaolin clay were used to produce biochar-amended clay. The gas permeability of the biochar-amended clay was measured using a flexible wall gas permeameter and compared with pure kaolin clay.

Materials and methods

Preparation of biochar

Peanut shell biochar was produced by a slow pyrolysis process. The biomass was heated at 5–10°C/min and kept at about 500°C for 30–40 min. In general, 30% of the biomass was converted into biochar. Nitrogen gas was pumped into the chamber to maintain an inert environment during the cooling period. A water cooling system was installed to shorten the cooling period. The biochar produced was sieved through a 5-mm sieve. In the present study, the biochar was additionally sieved through a 425-µm sieve to obtain a more homogenised biochar-amended clay sample and prevent any possible preferential flow during the gas permeability tests that may interfere with the results.

Measurements of gas permeability

The gas permeability of the soil samples was measured using a flexible wall air permeameter. Biochar was air dried and sieved through a 425-µm sieve and mixed with clay at three different percentages (5, 10, 15%, w/w) to produce biochar-amended clay (BAC), which was then mixed with water to give an optimum 35% gravimetric water content. After homogenising the water content for 2 days, the BAC was compacted with different degrees of compaction (80, 85, and 90%) into a soil column. The compacted columns were covered with an elastic membrane (flexible wall) and placed in the gas permeameter. The confining pressure of the permeameter was then set to 20 kPa and air flows with varying base pressures were applied. The air outflow rate was recorded by a gas flow meter after the gas flow became stable, and the gas permeability was calculated according to the generalised Darcy's law for porous media (Bouazza and Vangpaisal, 2003):

$$k_g = \frac{2Q_g \mu L P_1}{A(P_2^2 - P_1^2)} \qquad (3.29)$$

where k_g is the gas permeability (m^2), Q_g is the gas outflow rate (m^3/s), μ is the gas dynamic viscosity (Pa·s), L is the thickness of the soil sample (m), P_1 is the atmospheric pressure (Pa), P_2 is the applied gas pressure (Pa), and A is the cross-section area of the sample (m^2).

Results and discussion

The physiochemical soil properties related to gas permeability of kaolin clay, peanut shell biochar, and 425-µm sieved biochar that are summarised in Table 3.11. The pH of the clay is acidic (5.20 ± 0.03) while biochar is alkaline (9.98 ± 0.01). The sieved biochar is less alkaline (9.71 ± 0.04, $p < 0.05$) when compared with the un-sieved one. This indicates that sieving alters the physiochemical properties of biochar. Similar results were shown in organic matter content, electrical conductivity, and ash content. Sieving biochar significantly increased ($p < 0.05$) its organic matter from 30.15 ± 3.07 to 51.91 ± 0.09, while significantly decreased ($p < 0.05$) its EC from 6.41 ± 0.36 to 1.80 ± 0.09 and ash content from 69.85 ± 3.07 to 48.09 ± 0.09. Biochar used in this study (425-µm sieved) had larger particle sizes than clay, which implies that clay would fill the pores between the biochar particles in the biochar-amended soil. The results of the tests show that the gas permeability decreases with increasing biochar content, especially at relatively high degrees of compaction (Figure 3.41). The gas permeability increases with the degree of compaction for a particular biochar content. As the biochar content increases, the rate of reduction of gas permeability increases. For instance, the gas permeability of 15% BAC decreased by two orders of magnitude (3×10^{-11} to 2×10^{-13} m^2) when the degree of compaction increased from 80 to 90%. However, at low biochar content (i.e., 5%), the gas permeability of BAC was almost constant for all degrees of compaction. This implies that gas permeability is mainly governed by inter-pores of clay aggregates at low biochar content.

Summary

The effects of the biochar content on the BAC gas permeability vary with degree of compaction (DOC). At high DOC (i.e., 90%), the gas permeability of the BAC decreases with increasing biochar content because of the combined effect of clay aggregation and biochar inhibition in the gas flow. However, at low DOC (i.e., 80%), the effects of biochar incorporation on gas permeability are negligible because it no longer acts as a filling material to retard gas flow (Sun et al., 2013). Results from the present study did not show similar trends to those studies using biochar in agricultural soils. The result revealed that the gas permeability of BAC was lowered by increasing the biochar content and degree of

Table 3.11 Physiochemical properties of kaolin clay and biochar

Material	Moisture content :%	pH	Organic matter :%	EC: ds m^{-1}	Ash content :%	Particle size distribution: %		
						> 0.02 mm	0.002– 0.02 mm	< 0.002 mm
Kaolin clay	0.87 ± 0.06a	5.20 ± 0.03a	31.26 ± 1.18a	6.49 ± 0.02a	68.74 ± 1.18a	0	0	100
Peanut shell biochar (5-mm sieved)	4.92 ± 0.34b	9.98 ± 0.01b	30.15 ± 3.07a	6.41 ± 0.36a	69.85 ± 3.07a	64	28	8
Peanut shell biochar (425-μm sieved)	4.73 ± 0.19b	9.71 ± 0.04c	51.91 ± 0.09b	1.80 ± 0.09b	48.09 ± 0.09b	0	72	28

Note:
Values are presented in mean ± SD (n = 3). Values followed by the same letter on the same column indicate no significant difference (at the 0.05 probability level) according to the Duncan's multiple range test.

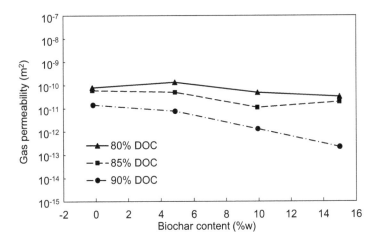

Figure 3.41 Gas permeability of the biochar-amended clay (BAC) with different biochar content and degrees of compaction (DOC) at 35% gravimetric water content.

compaction. This suggests that apart from being used as a soil amendment in agriculture, biochar-amended clay is a potential final cover material for landfill to control gas emission.

INFLUENCE OF BIOPOLYMER ON GAS PERMEABILITY (NG ET AL., 2020C)

Introduction

One way to condition soil is to incorporate it with biopolymers, which are long chains of natural polymers with high molecular mass. Biopolymers are widely present in nature, serving as the matrix for living bacterial cells (Flemming and Wingender, 2010). They can be produced by non-food parts of plants (Khatami and O'Kelly, 2013) and through bacterial fermentation (Giavasis et al., 2000). By contrast with chemical conditioners like cement and polyacrylamide, which cause the emission of carbon dioxide and are toxic to the environment, biopolymers are regarded as sustainable and environmentally friendly (Chen et al., 2015; Chang et al., 2016a). Because numerous functional groups such as hydroxyl groups are present, biopolymers can absorb water through hydrogen bonds to form viscous hydrated colloids. Biopolymers are commonly found in food, pharmaceutical and cosmetic products, and potentially in waste water treatment (More et al., 2014). In geotechnical engineering, investigation of soil with these sustainable materials has focused on controlling soil or dust erosion (Orts et al., 2000; Chen et al., 2015), enhancing mechanical properties (Karimi, 1998; Khatami and O'Kelly, 2013; Chang et al., 2016b) as well as hydrological properties which are a prerequisite in the performance of soil barriers.

In this section, different biopolymers were mixed with soil to measure their gas permeability at different water contents using a flexible wall permeameter. Also, the soil was prepared with various degrees of compaction to investigate biopolymer effects on soil density.

134 Advanced Unsaturated Soil Mechanics

Table 3.12 Test program for investigating gas permeability of biopolymer-amended soil

Soil type	Biopolymer type	Compaction water content	Degree of compaction and void ratio e_0
Compacted clay	None	25% 30% 35%	75% (e_0 = 1.66)
Compacted clay with gellan	Gellan gum	25% 30% 35%	85% (e_0 = 1.35)
Compacted clay with xanthan	Xanthan gum	25% 30% 35%	95% (e_0 = 1.10)

Note:
Three replicates were considered for the condition at degree of compaction of 95%.

Material and methods

Test program

Three series of tests were carried out to examine the effects of compaction water content, soil density and biopolymer type on the gas permeability of biopolymer-amended clay. Each series is defined according to the type of biopolymer added to the clay. These three test series, namely "Compacted clay", "With gellan" and "With xanthan", represent clay without biopolymers, clay with gellan gum and clay with xanthan gum respectively. First, investigating the soil at different compaction water contents evaluates changes in soil structure and clogging effects caused by hydration of the biopolymers with different water amounts. Second, different soil densities were studied in terms of degree of compaction (DOC). This determines the performance of biopolymers in soil having different pore volumes. A lower DOC may be sufficient when compacting an earthen barrier. Finally, different types of biopolymer were used to investigate whether their different properties significantly affect their performance and thus the gas permeability of unsaturated clay. Details of the test program are summarised in Table 3.12.

Soil and biopolymers

Kaolin clay in powder form and two types of biopolymer (xanthan gum and gellan gum) were used. Xanthan gum can be produced from waste (Yoo and Harcum, 1999) at a relatively low cost (Chen et al., 2015). Its effects in reducing the saturated water permeability of soil may lower the cost of improving soil barriers. Gellan gum has good thermal stability and is stable over a wide range of pH from 2 to 10 (Moslemy et al., 2003) so its performance is expected to be unaffected in soil contaminated with pollutants. These two biopolymers are both anionic polysaccharides with molecular weights up to an order of 1000 kDa (Hoefler, 2004). Since previous studies showed that an increasing concentration of biopolymers finally reaches a maximum reduction in the saturated water permeability (Bouazza et al., 2009; Chang et al., 2016b), a 3% concentration (w/w) of biopolymer was adopted to give optimal effects in this study. The basic properties of the clay with and

Table 3.13 Physical properties of kaolin clay with and without biopolymers

Property		Value	
Condition	Clay	With gellan	With xanthan
Specific gravity, G_s	2.52	-	-
Liquid limit, LL	80	70	69
Plastic limit, PL	35	40	36
Plasticity index, PI	45	37	33
Maximum dry density, ρ_d: kg/m³	1264	1287	1259
Optimum moisture content: %	36.2	34.5	34.3

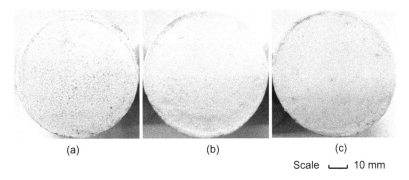

(a)   (b)   (c)

Scale ⊢—⊣ 10 mm

Plate 3.5 Photographs showing the surface of (a) the compacted clay; (b) clay with gellan gum and (c) xanthan gum.

without 3% of biopolymers were measured. They are summarised in Table 3.13. Standard test methods, BS 1377-2 (BSI, 1990a) and BS 1377-4 (BSI, 1990b), were followed to determine the Atterberg limits and compaction curve, respectively. Plate 3.5 shows the compacted clay, clay with gellan and clay with xanthan respectively at DOC of 95% after soil compaction.

Results and discussion

Reducing the gas permeability at different degrees of compaction (DOC) using biopolymers

Figures 3.42a and 3.42b show the gas permeability of the clay with and without biopolymers at different degrees of compaction for 25% and 35% compaction water content, respectively. As the variation in the soil gas permeability with DOC at a compaction water content of 30% was similar to those at 25% and 35%, the data are not reported to avoid repetition. Note that the logarithmic scale is adopted for all figures concerning the gas permeability of soils. By comparison with the compacted clay without biopolymers, the gas permeability of clay with either type of added biopolymer was always lower at any degree of compaction. For a given compaction water content, an increasing degree

136 Advanced Unsaturated Soil Mechanics

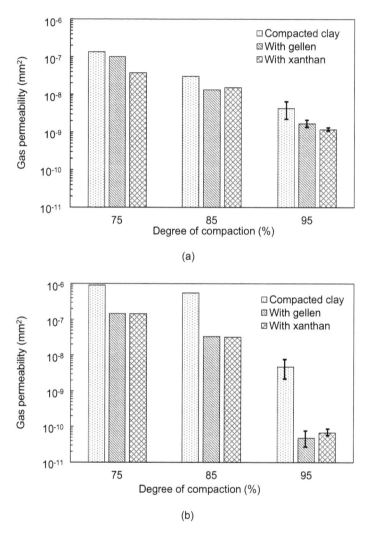

Figure 3.42 Gas permeability of the soil at various degrees of compaction with compaction water content (a) 25% and (b) 35%.

Note: At 95% of DOC, the value is average ± standard error, n = 3.

of compaction further increased the difference between the gas permeability of the clay with and without the biopolymer. At 25% water content, the gas permeability of the clay with the biopolymers was at least 25% and 60% lower than that of the compacted clay at 75% and 95% DOC, respectively. For the same compaction water content (i.e., 25%), the gas permeability of the clay with xanthan gum at a DOC of 75% was also significantly lower than that with gellan gum. The difference reached nearly 0.7 orders of magnitude. At 35% compaction water content, adding biopolymers to soil compacted at DOCs of 75% and 95% reduced the gas permeability by about 1 and 2 orders of magnitude, respectively. Biopolymers reduce the gas permeability mainly through their hydrated viscous biopolymer chains, which in turn affect the soil microstructure including the pore-size distribution and structure of the clay particles. A detailed discussion and explanation of the mechanisms of biopolymer effects on the performance of biopolymer at different compaction water contents are provided in Ng et al. (2020c). An increase in DOC indicates that soil particles are packed more closely in a unit volume. The soil particles cohere with each other to form bigger aggregates, thereby reducing the volume of soil pores (Gupta et al., 1989). In general, an increase in DOC decreases the gas permeability of soil. Also, an increasing DOC allows the biopolymers to be packed closely with the soil. The soil-biopolymer interaction affects the soil structure more significantly and clogs the pores. As a result, the gas permeability of the clay with biopolymers is lower than that of compacted clay at any degree of compaction. The addition of the biopolymer to the clay further reduces the gas permeability at higher DOCs.

Effects of compaction water content on the gas permeability with and without biopolymers

Figure 3.43 shows the variation in gas permeability of the compacted clay without biopolymers at different compaction water contents. Different lines represent clay compacted at different DOCs. At 75% DOC, the gas permeability of the compacted clay increases with increasing compaction water content. This trend is consistent with the gas permeability of the clay at 85% DOC at different compaction water contents. Until the clay was compacted to 95% DOC, the gas permeability was nearly constant with an increasing compaction water content. Once the clay was compacted at a higher degree of compaction, the increasing trend in gas permeability caused by higher compaction water contents lessened. Normally, an increasing water content in soil should lower the gas

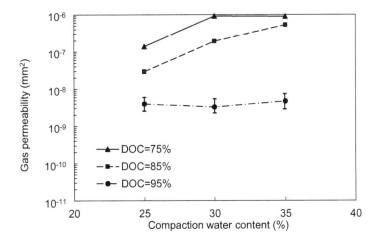

Figure 3.43 Variations of gas permeability of compacted clay with different compaction water contents and at various degrees of compaction.

Note: At 95% of DOC, the value is average ± standard error, $n = 3$.

138 Advanced Unsaturated Soil Mechanics

Table 3.14 Conditions of soil specimens at various DOC and water contents

		Water content at compaction		
Degree of saturation after compaction		*25%*	*30%*	*35%*
Degree of compaction	75%	38.0	45.6	53.2
	85%	46.8	56.2	65.6
	95%	57.3	68.8	80.3

permeability of the soil since the pores in the soil through which air flows are reduced and occupied by water (Fredlund et al., 2012). The increasing trends found in this study are caused by the formation of different soil fabrics, especially in fine-grained soils. The pore-size distribution at the same void ratio is different, potentially causing increasing gas permeability with an increase in the water content. When the compaction water content is below a certain value, pores filled with water do not affect the gas permeability significantly. As gas flows through well-connected macro pores, the gas permeability of the soil is nearly constant (Wickramarachchi et al., 2011; Zhan et al., 2014a). Model predictions by Wickramarachchi et al. (2011) show that the gas permeability of silty sand ultimately becomes almost constant with a decreasing soil-water content. In this study, the trend of increasing gas permeability with compaction water content has already been observed even at low DOC (i.e., 75%), including at the lowest degree of saturation (i.e., 38% shown in Table 99 3.14). This means that an increase in the number of water-filled pores at the higher water content affect the gas permeability. But it is outweighed by the effect of larger soil pores. On the other hand, Moon et al. (2008) prepared soil specimens with different compaction water contents on the dry side of the optimum using a modified Proctor method. They found that the gas permeability of the compacted soil first decreased and then increased with increasing water content even though the void ratio was lower. As their focus was to investigate the gas permeability of clayey soil prepared by different compaction methods at various water contents, they provided no interpretation of this finding.

Figure 3.44 shows the gas permeability of clay with biopolymers at different compaction water contents. Each line represents soil compacted at a certain DOC. At any compaction water content, the gas permeability of the clay with the biopolymers was always lower than that of the compacted clay without the biopolymers. As with the compacted clay shown in Figure 3.44, an increasing gas permeability with an increasing water content was also observed in the clay with either biopolymer for 75% or 85% DOC. Until the DOC was increased to 95%, the gas permeability of the clay with biopolymers decreased with increasing compaction water content. This in the opposite direction to the change in gas permeability with water contents of 75% and 85% DOC. Comparing the clay with and without biopolymers, adding biopolymers suppressed the increasing trend at both 75% and 85% DOC and finally reversed the trend at 95% DOC. Further, regardless of the degree of compaction, the reduction in permeability by the biopolymers at high compaction water content was always greater than that at low compaction water contents. For all test conditions except at 75% DOC and at 25% compaction water content, the maximum difference between the gas permeability of clay amended with xanthan gum and gellan gum was about 50%. This implies that the mechanisms by which these two

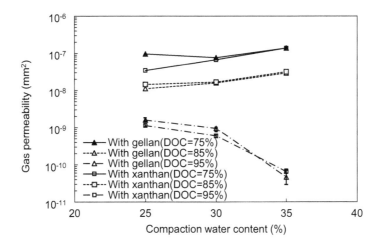

Figure 3.44 Variations of gas permeability of clay with the biopolymers with different compaction water contents and at different degrees of compaction.

Note: At 95% of DOC, the value is average ± standard error, $n = 3$.

hydrated biopolymers affect the gas permeability of the clay were similar. Although the present study did not consider different degrees of saturation or matric suctions in a given specimen, reducing the gas permeability by biopolymers may still be valid. For a specimen with a given soil structure, the gas permeability of the soil normally decreases with increasing saturation (Bouazza and Vangpaisal, 2003; Fredlund et al., 2012; Zhan et al., 2014a). The kaolin clay adopted in the current study is neither expansive nor collapsible so its volume change should not significantly affect the gas permeability upon wetting and drying. Therefore, compared with clay compacted under the same conditions, the presence of biopolymers in the clay studied is expected to reduce the gas permeability at different degrees of saturation.

Summary

When the biopolymers were introduced, the gas permeability of clay with biopolymers was always lower than that of compacted clay, regardless of the soil density and the compaction water content. The soil density was studied in terms of degree of compaction. Generally, there was no significant difference between the two biopolymers in reducing gas permeability. The mechanisms by which both gellan and xanthan gum reduce the gas permeability are based on reducing the number of larger soil pores as well as pore clogging. Since the hydrated biopolymer chains are highly viscous, they can stick the soil particles together and thus reduce the formation of larger soil pores. Also, after absorbing the water, the biopolymer can hydrate to block the soil pores.

Considering the effect of the compaction water content, the gas permeability of the clay compacted at DOC of 75% and 85% increased with increasing compaction water content. Then, it became nearly constant when the compaction water content reached 95% DOC.

140 Advanced Unsaturated Soil Mechanics

Such an increasing trend can be attributed to the formation of larger pores in the soil at a higher compaction water content. These larger pores led to heterogeneity of the soil and preferentially provided more flow paths for gas. Therefore, these larger pores dominated the effects of reducing pores filled with air at higher compaction water content.

As with the compacted clay, an increasing gas permeability with an increase in the water content was also found for clay with biopolymers at 75% and 85% DOC, except that at 95% DOC. However, the presence of the biopolymers suppressed the increasing trend and finally reversed it at 95%. Further, at a given DOC, the reduction in permeability resulting from the biopolymers was always greater at a higher compaction water content. An increasing compaction water content provides more water and enhances the hydration of the biopolymer chains. As a result, more soil pores are clogged at the higher water content.

When the performance of the biopolymers at different soil densities was considered, the reduction in permeability caused by the biopolymers was enhanced with increasing DOC. The increase in DOC indicates that soil particles and biopolymers are closely packed in a unit volume. At a higher DOC, the soil-biopolymer interaction affects the soil structure more significantly and blocks the soil pores. Therefore, an increase in DOC further reduces the gas permeability when using biopolymers. Overall, the use of biopolymers, which can be produced by non-food parts of plants and bacterial fermentation, can reduce substantially the gas permeability of the clay in unsaturated soil barriers such as landfill covers to minimise odour emission.

EFFECTS OF GRASS ROOTS ON GAS PERMEABILITY (NI AND NG, 2019)

Introduction

This section aims to investigate the evolution of gas permeability in unsaturated vegetated soil, which can form a final landfill cover. A common grass species (*Cynodon dactylon*) was selected for investigation. There are six replications for the grass case.

Materials and methods

There were 6 columns with grass-covered soil (see Figure 3.45 and Plate 3.6) and one with bare soil. An array of miniature-tip tensiometers (2100 F, Soil Moisture Equipment Cooperation) were installed at depths of 50, 130, 210 and 290 mm in each column to measure the negative pore-water pressure or matric suction up to 90 kPa. At depths of 130 and 210 mm, two gas pressure transducers (Chen, 2016) were installed to measure the gas pressure during testing.

Interpretation of test results

Figure 3.46 shows the relationship between gas permeability (k_g) and soil suction for both bare and grass-covered conditions after 6, 15 and 24 months, respectively. This relationship is also called the gas permeability function. Test data in each scenario was fitted

Retention characteristics and permeability functions 141

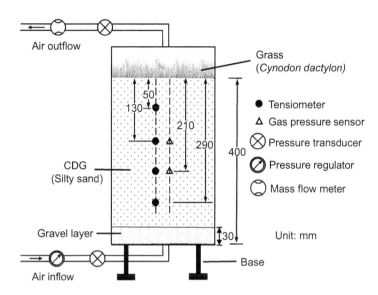

Figure 3.45 Overview of the columns vegetated with grass.

Plate 3.6 Setup for the gas permeability test.

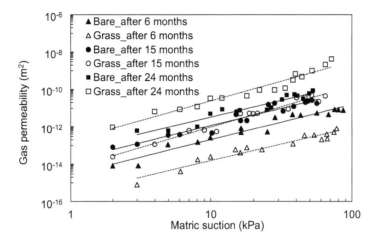

Figure 3.46 Comparison of gas permeability function in bare and grass-covered soils.

using least squares. It was found that for the suction range considered (2-86 kPa), there were linear relationships between k_g and suction for all the scenarios. k_g increased with increasing suction. Also, the slopes of fitted lines were almost the same. After 6 months of growth, k_g in grass-covered soil was about one order of magnitude less than that in bare soil. Similar observations were also captured in compacted silty clay with a mixture of *Ilex purpurea* Hassk, ryegrass and bluegrass (Zhan et al., 2016).

After continuous growth of grass, the relationship between k_g and suction for both bare and vegetated soils shifted upwards (Figure 3.46). That means that k_g increased with time at any specific suction after 6 months. However, the increase in the vegetated case was much greater than that in bare soil by up to two orders of magnitude. On one hand, the higher k_g in vegetated soils can increase the available CH_4 and O_2 in the soil and hence enhance methane oxidation in landfill cover. On the other hand, it may also increase gas emissions into the atmosphere. In order to balance ecological restoration using vegetation with the practical function of reducing landfill gas emission, materials with low permeability (such as membranes, compacted clay) should be included at the bottom of conventional landfill covers.

Summary

With increasing suction, k_g increased linearly on a semi-log scale for both bare and grass-covered soil. At an early stage (i.e., 6 months), k_g in grass-covered soil was lower than that in bare soil by 85%. This is because of the blockage of flow channels by plant roots. After that, the k_g of bare and vegetated soils increased, but the increase in grass-covered soil after 24 months was two orders of magnitude greater than that of bare soil over a wide suction range (2-86 kPa). Decayed roots associated with plant aging may cause preferential channels and hence increase gas migration through the soil.

EFFECTS OF BIOFILM ON GAS PERMEABILITY (NG ET AL., 2019B)

Introduction

Bacterial activity in soil may significantly affect soil properties (Rowe, 2005; DeJong et al., 2013; He et al., 2013; Dashko and Shidlovskaya, 2016). This can provide new opportunities for sustainable soil improvement such as using biofilm, which is estimated to form in 99% of bacterial populations (Dalton and March, 1998). Biofilms are aggregates of bacterial cells encased in a self-produced matrix of exopolysaccharides (EPS) (Flemming and Wingender, 2010). Typically, EPS comprises of polysaccharides, proteins, nucleic acids, humic acid and water to form hydrated slime, which facilitates bacterial attachment to material surfaces. For soil properties, EPS may cause bio-clogging in soil pores and reduce soil permeability. Reducing gas and water permeability is important for the performance of earthen structures such as cover systems for landfill and mining waste (Bouazza and Rahman, 2007; Ng and Coo, 2015). In this chapter, the effects of biofilm on the gas permeability of soil at different degrees of saturation were quantified using a flexible wall permeameter.

Materials and methods

Toyoura sand was adopted and its D10, D50 and D90 values are 0.12, 0.17 and 0.25, respectively. The void ratio is 0.635. The aerobic rhizobacterium *Bacillus subtilis* (NCIB 3610) is a non-pathogenic model bacterium typically used for biofilm studies (Vlamakis et al., 2013). It was chosen rather than other pathogenic model bacteria such as *P. aeruginosa* and *E. coli* (O'Toole et al., 2000) because it is commonly found to act as a biofertiliser in soil, promoting plant growth (Morikawa, 2006; Vlamakis et al., 2013). A bacterial solution with 0.7 optical density was used, corresponding to approximately 7×10^9 cells per ml (Hsueh et al., 2015). The nutrient was prepared by dissolving Lennox broth, 30 gL^{-1} glycerol and 6.3 gL^{-1} $MnCl_2$ in Milli-Q water to optimise the biofilm produced by *Bacillus subtilis* (Morikawa et al., 2006; Shemesh and Chai, 2013).

Interpretation of test results

Figure 3.47 shows the gas permeability for each condition at different degrees of saturation. The variation in experimental results is probably caused by the heterogeneity of biofilm formation (Flemming and Wingender, 2010) and nutrient precipitation. At a given degree of saturation, the gas permeability of "Combined" and "Nutrient" were consistently lower than that of "Water". When the soil was relatively dry (around 5% degree of saturation), the gas permeability of "Combined" and "Nutrient" were 22% and 14% lower than that of "Water", respectively. The lower permeability of "Nutrient" is attributed to nutrient precipitation as verified by Scanning Electron Microscope (SEM) analysis. When the soil specimens were dried, decreasing water content less nutrient, such as organic matter, dissolved in the soil. The nutrient precipitated and could block the pores. As indicated in the figure, the difference in gas permeability between "Nutrient" and "Water" increased with decreasing saturation. For "Combined", in addition to the precipitation of nutrient,

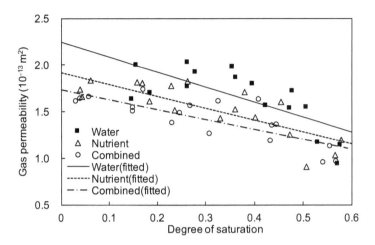

Figure 3.47 Variation of gas permeability of each condition at different degrees of saturation.

the further reduction of gas permeability was attributed to biofilm. When nutrient is present, bacterial cells secret attaches to surfaces and connect other cells (Lemon et al., 2008). The bacteria then become immobile and consume the nutrient to produce EPS with high molecular mass (e.g., 1000 kDa) to surround themselves. Normally, EPS is highly viscous (Flemming and Wingender, 2010) and more viscous than hydrated biopolymer gels such as guar gum (Isobe et al., 1992). The high viscosity of EPS with immobile cells ensures that biofilms attach firmly to the particle surface and cause bio-clogging. Consequently, the gas permeability of "Combined" was the lowest after the biofilm formed in soil. Combined with other beneficial effects such as decomposing harmful contaminants (Morikawa, 2006) and mitigating soil drying (Roberson and Firestone, 1992), biofilm can potentially be applied as soil barriers to reduce harmful gas emission.

The difference in gas permeability between "Combined" and "Water" increases as the degree of saturation decreased. This implies that dehydrated biofilm still maintained the function of reducing the permeability of unsaturated soil during drying. However, the difference in gas permeability between "Combined" and "Nutrient" was almost constant with degree of saturation. This suggests that biofilm formation was inhibited. Since the incubation chamber provided sufficient oxygen through air-ventilation, the slow rate of biofilm formation could mainly be attributed the limited amount of nutrient provided and the decrease in soil-water content. The consumption of nutrient by bacteria always involves oxidation and reduction reactions. Organic carbon in the nutrient is oxidised to provide energy for microbial activities. As suggested by Proto et al. (2016), the organic carbon can be estimated simply as 50% of the organic components in the nutrient by weight. The organic carbon content provided was only around 0.5% of the total soil dry mass. We note that some bacteria can use other substances such as carbon dioxide to oxidise organic carbon in an anaerobic environment (Thauer, 1998). But, the production of cell biomass is less than that produced by aerobic respiration (Lengeler et al., 1999), which may suggest that biofilm formation would be lower in an anaerobic environment. Typical examples of oxidation and reduction with regards to nutrient can be found in Nealson and Saffarini (1994) and Vepraskas and Craft (2016). More importantly, organisms also require water to transport chemicals and regulate the environment for metabolism (Mitchell and Santamarina, 2005).

Summary

The gas permeability of "Combined" and "Nutrient" were consistently lower than "Water" for all degrees of saturation. At around 5% degree of saturation, "Combined" and "Nutrient" were 22% and 14% lower than "Water", respectively. Precipitation of nutrient causes a reduction in permeability in "Nutrient" while both nutrient precipitation and biofilm formation in "Combined" reduce the permeability. Interactions between immobile bacterial cells and viscous EPS allows biofilm to adsorb firmly on particle surface to cause bio-clogging. The difference in gas permeability between "Combined" and "Water" increased with a decreasing degree of saturation, which implies that permeability reduction by biofilm is maintained during soil drying.

SOURCES

A numerical investigation of the stability of unsaturated soil slopes subjected to transient seepage (Ng and Shi, 1998)

Soil- water characteristic curves and water permeability functions (Fredlund et al., 2001a; Fredlund, 2002)

A new and simple stress- dependent water retention model for unsaturated soil (Zhou and Ng, 2014)

Laboratory measurements of drying and wetting permeability functions using a stress- controllable soil column (Ng and Leung, 2012b)

A field study of stress- dependent soil- water characteristic curves and permeability of a saprolitic slope in Hong Kong (Ng et al., 2011)

Gas breakthrough and emission through unsaturated compacted clay in landfill final cover (Ng et al., 2015b)

Numerical investigation on gas emission from three landfill soil covers under dry weather conditions (Ng et al., 2015b, c; Ng et al., 2016b)

Gas permeability of biochar- amended clay: potential alternative landfill final cover material (Wong et al., 2016)

Influence of biopolymer on gas permeability in compacted clay at different densities and water contents (Ng et al., 2020c)

Effects of grass roots on gas permeability in unsaturated simulated landfill covers (Ni and Ng, 2019)

Effects of biofilm on gas permeability of unsaturated sand (Ng et al., 2019b)

Chapter 4

Shear strength and behaviour

EXTENDED MOHR–COULOMB FAILURE CRITERION (FREDLUND AND RAHARDJO, 1993)

To investigate the stability and safety of many geotechnical structures such as bearing capacity of foundation, lateral earth pressure acting on retaining wall, man-made and natural slopes, it is necessary to determine the shear strength of a soil. The two shear strength tests commonly performed are the triaxial test (see Plates 4.1 and 4.2) and the direct shear test (refer to Plate 4.3) in the laboratory. For determining the shear strength of an unsaturated soil for engineering designs, the extended Mohr-Coulomb failure criterion may be used and formulated in terms of two independent stress state variables for simplicity (Fredlund et al., 1978; Fredlund and Rahardjo, 1993). Any two of the three possible stress state variables, including $(\sigma - u_a)$, $(\sigma - u_w)$ and $(u_a - u_w)$, can be used for the shear strength equation. The stress state variables, $(\sigma - u_a)$ and $(u_a - u_w)$, have been shown to be the most convenient combination in practice. Using these stress variables, the shear strength equation is written as follows:

$$\tau_f = c' + \left(\sigma - u_a\right)_f \tan \varphi' + \left(u_a - u_w\right)_f \tan \varphi^b \tag{4.1}$$

where

τ_f = shear stress at failure
c' = intercept of the "extended " Mohr-Coulomb failure envelope on the shear stress axis where the net normal stress and the matric suction at failure are equal to zero; it is also referred to as "effective cohesion"
$\left(\sigma - u_a\right)_f$ = net normal stress at failure
u_a = pore-air pressure
u_w = pore-water pressure
σ = normal total stress
ϕ' = angle of internal friction associated with the net normal stress state variable, $\left(\sigma - u_a\right)_f$
$\left(u_a - u_w\right)_f$ = matric suction at failure
φ^b = angle indicating the rate of increase in shear strength relative to the matric suction, $\left(u_a - u_w\right)_f$.

146   DOI: 10.1201/9781003480587-4

The shear strength equation for an unsaturated soil exhibits a smooth transition to the shear strength equation for a saturated soil. As the soil approaches saturation, the pore-water pressure, u_w, approaches the pore-air pressure, u_a, and the matric suction, $(u_a - u_w)$, goes to zero. The matric suction component vanishes, and Equation (4.1) reverts to the equation for a saturated soil:

$$\tau_f = c' + (\sigma - u_w)_f \tan \phi' \tag{4.2}$$

The failure envelope for a saturated soil is obtained by plotting a series of Mohr circles corresponding to failure conditions on a two-dimensional plot. The line tangent to the Mohr circles is called the failure envelope, as described by Equation (4.2). In the case of an unsaturated soil, the Mohr circles corresponding to failure conditions can be plotted three-dimensionally, as illustrated in Figure 4.1.

The three-dimensional plot has the shear stress, τ, as the ordinate and the two stress state variables, $(\sigma - u_a)$ and $(u_a - u_w)$, as abscissae. The frontal plane represents a saturated soil where the matric suction is zero. On the frontal plane, the $(\sigma - u_a)$ axis reverts to the $(\sigma - u_w)$ axis since the pore-air pressure becomes equal to the pore-water pressure at saturation. Thus it can be seen that a saturated soil is just a special case of an unsaturated soil.

The Mohr circles for an unsaturated soil are plotted with respect to the net normal stress axis, $(\sigma - u_a)$, in the same manner as the Mohr circles are plotted for saturated soils with respect to the effective stress axis, $(\sigma - u_w)$. However, the location of the Mohr circle plot in the third dimension is a function of the matric suction (Figure 4.1). The surface tangent to the Mohr circles at failure is referred to as the extended Mohr-Coulomb failure envelope for unsaturated soils. The extended Mohr-Coulomb failure envelope defines the shear strength of an unsaturated soil. The intersection line between the extended Mohr-Coulomb failure envelope and the frontal plane is the failure envelope for the saturated condition.

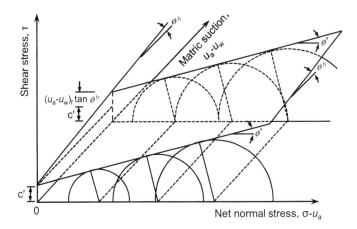

Figure 4.1 Extended Mohr–Coulomb failure envelope for unsaturated soils.

Source: After Fredlund and Rahardjo, 1993.

148 Advanced Unsaturated Soil Mechanics

The inclination of the theoretical failure plane is defined by joining the tangent point on the Mohr circle to the pole point. The tangent point on the Mohr circle at failure represents the stress state on the failure plane at failure. Clearly, the extended Mohr-Coulomb failure envelope is idealised as a planar surface. In reality, the failure envelope is somewhat curved (Gan and Fredlund, 1996).

As shown in the figure, a planar failure envelope intersects the shear stress axis, giving a cohesion intercept, c'. The envelope has slope angles of φ' and φ^b with respect to the $(\sigma - u_a)$ and $(u_a - u_w)$ axes, respectively. Both angles are assumed to be constant. The cohesion intercept, c', and the slope angles, φ' and φ^b, are the strength parameters used to relate the shear strength to the stress state variables. The shear strength parameters represent many factors including density, void ratio, degree of saturation, mineral composition, stress history and strain rate. In other words, these factors have been combined and expressed mathematically in the strength parameters.

The effects of changes in net normal stress on mechanical behaviour of an unsaturated soil differ from those caused by changes in matric suction (Jennings and Burland, 1962). The increase in shear strength caused by an increase in net normal stress is characterised by the friction angle, φ'. On the other hand, the increase in shear strength caused by an increase in matric suction is described by the angle, φ^b. The value of φ^b is consistently equal to or less than φ'.

The failure envelope intersects the shear stress against the matric suction plane along a line of intercepts, as illustrated in Figure 4.1. The line of intercepts indicates an increase in strength as matric suction increases. In other words, the increase in shear strength with respect to an increase in matric suction is defined by the angle, φ^b. The equation for the line of intercepts is as follows:

$$c = c' + \left(u_a - u_w\right)_f \tan \varphi^b \tag{4.3}$$

where

> c' = intercept of the extended Mohr–Coulomb failure envelope with the shear stress axis at a specific matric suction, $\left(u_a - u_w\right)_f$, and zero net normal stress; it can be referred to as the "total cohesion intercept".

Relationships between φ^b and χ

Bishop (1959) proposed a shear strength equation for unsaturated soils which had the following form:

$$\tau_f = c' + \left[\left(\sigma - u_a\right)_f - \chi\left(u_a - u_w\right)_f\right] \tan \varphi' \tag{4.4}$$

where

> χ = a parameter related to the degree of saturation of the soil.

Let us assume that the shear strength computed using Equation (4.1) can be made to be equal to the shear strength given by Equation (4.4). Then it is possible to illustrate the relationship between $\tan \varphi^b$ and χ as:

$$(u_a - u_w)_f \tan \varphi^b = \chi (u_a - u_w)_f \tan \varphi' \tag{4.5}$$

It is then possible to solve for the parameter, χ as follows:

$$\chi = \frac{\tan \varphi^b}{\tan \varphi'} \tag{4.6}$$

A graphical comparison between the φ^b representation of strength and the χ representation of strength is shown in Figure 4.2. Using the φ^b method, the increase in shear strength due to matric suction is represented as an upward translation from the saturated failure envelope. The magnitude of the upward translation is equal to $\left[(u_a - u_w)_f \tan \varphi^b\right]_f$ (i.e., point A in the figure). In this case, the failure envelope for the unsaturated soil is viewed as a third-dimensional extension of the failure envelope for the saturated soil. On the other hand, the χ parameter method uses the same failure envelope for both saturated and unsaturated conditions. Matric suction is assumed to produce an increase in the net normal stress. This increase is a fraction of the matric suction at failure (i.e., $\chi(u_a - u_w)_f$). The shear strength at point A using the φ^b method is equivalent to the shear strength at point A′ in the χ parameter method, as depicted in the figure.

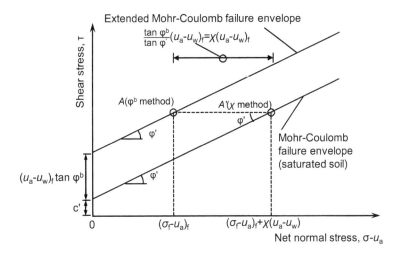

Figure 4.2 Comparisons of the ϕ_b and χ methods of designating shear strength.

Source: After Fredlund and Rahardjo, 1993.

150   Advanced Unsaturated Soil Mechanics

Theoretically, only one χ value is obtained from Equation (4.6) for a particular soil when the failure envelope is planar. A planar failure envelope uses one value of φ' and one value of φ^b. If the failure envelope is bilinear with respect to the φ^b, there will be two values for χ. A χ value equal to 1.0 corresponds to the condition where φ^b is equal to φ'. A χ value less than 1.0 corresponds to the condition where φ^b is less than φ'. For envelopes which are highly curved with respect to matric suction, there will be various χ values corresponding to different matric suctions.

The χ parameter has commonly been correlated with the degree of saturation of the soil. Unfortunately, the χ value has sometimes been obtained from shear strength tests on soil specimens compacted with different water contents. The different initial compacted water content may have resulted in varying initial matric suctions (Fredlund and Rahardjo, 1993). However, soil specimens compacted with different water contents do not represent an "identical" soil. As a result, the χ values obtained from specimens compacted with different water contents are essentially obtained from different soils. The φ^b and χ relationship in Equation (4.6) applies only to initially "identical" soils. These may be soils compacted at the same water content to the same dry density. The χ parameter need not be unique for both shear strength and volume change problems. The χ relationship given in Equation (4.6) only applies to the evaluation of the shear strength of the soil. From a practical engineering standpoint, it would appear to be better to use the $(\sigma - u_a)$ and $(u_a - u_w)$ stress state variables independently to designate the shear strength of an unsaturated soil.

COMPARISONS OF AXIS-TRANSLATION AND OSMOTIC TECHNIQUES FOR SHEAR TESTING OF UNSATURATED SOILS (NG ET AL., 2007)

Introduction

Different techniques have been developed to control the suction of unsaturated soil during a test. To investigate differences between the widely used axis-translation and osmotic control techniques, collaborative research was carried out at the Hong Kong University of Science and Technology (HKUST) and the Ecole Nationale des Ponts et Chaussees (ENPC) in Paris, France. In this research, three series of unsaturated tri-axial shear tests were conducted on a recompacted expansive soil with an initial degree of saturation of 82%. Two of these series were performed at HKUST using the axis-translation technique and the osmotic technique. Another series was performed at ENPC using the osmotic technique and the results were first presented by Mao et al. (2002). In the series using the axis-translation technique at HKUST, single-stage testing procedures and multi-stage testing procedures were compared to investigate the influence of these two testing methods on shear testing. In both series using the osmotic control technique at HKUST and ENPC, test results were compared to study the reliability of test data. The test results using the axis-translation technique and the osmotic control technique were compared to investigate any difference between the techniques for shear testing of the expansive soil at a high degree of saturation. It should be noted that the test results were interpreted using the extended Mohr-Coulomb failure criterion (see Equation (4.1)).

Shear strength and behaviour

Testing equipment

At HKUST, two test systems, one to apply the axis-translation technique and one to apply the osmotic technique, were used to conduct triaxial shear tests on the unsaturated expansive soil sample. In the system using the axis-translation technique, a triaxial apparatus equipped with two water pressure controllers and two pneumatic controllers was used (see Plate 4.1a). Suction was applied to a specimen through one water pressure controller and one air pressure controller. Pore-water pressure was applied to the base of the specimen through a ceramic disc with an air-entry value of 500 kPa. Pore-air pressure was applied at the top of the specimen through a sintered copper filter. A double-cell measurement system invented and developed by Ng et al. (2002), was used to measure the total volume change of the specimen. Details of this testing system are described in Ng et al. (2002). A commercial product of the total volume measuring system is shown in Plate 4.1b. Another type of triaxial set-up to measure the total volume changes of a specimen is to use an internal Hall-effects transducer (see Plate 4.2a).

To shorten the duration of moisture equalisation and testing time, Ng (2015) developed a full suction range direct shear box using the principle of axis-translation, osmotic suction control and humidity control techniques at HKUST (refer to Plate 4.3).

The triaxial system at HKUST using the osmotic control technique (refer to Plate 4.4) is similar to that used at ENPC except that there was no direct measurement of total volume changes for specimens at HKUST during the tests in 2005. However, the dimensions of the specimens tested at HKUST were measured after shearing and then the volume changes were back-calculated. In the osmotic technique at HKUST and ENPC, suction

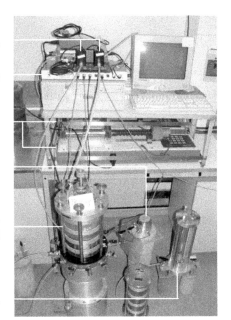

Plate 4.1a A typical set-up equipped with a double-cell total volume measurement system for triaxial testing using the principle of axis-translation technique at HKUST.

152 Advanced Unsaturated Soil Mechanics

Plate 4.1b The double-cell total volume measurement system (with courtesy from Geo-Experts).

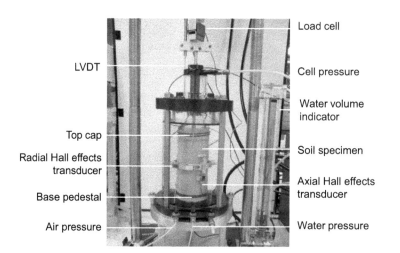

Plate 4.2a A typical set-up equipped with Hall-effects transducers for cyclic triaxial testing using the principle of axis-translation technique at HKUST.

Shear strength and behaviour 153

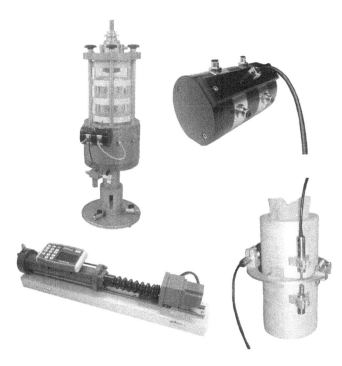

Plate 4.2b A commercial triaxial set-up equipped with Hall effects transducers (with courtesy from Geo-Experts).

was applied to specimens through two pieces of semi-permeable membrane, which were kept in contact with the top and bottom surfaces of the specimens. Details of the triaxial testing system using the osmotic control technique are described in Cui and Delage (1996).

The osmotic control technique in this study involved a Spectra/Por@ 2 Regenerated Cellulose dialysis membrane with a value of 12,000–14,000 Daltons MWCO (Molecular Weight Cut Off). The corresponding PEG (Polyethylene Glycol Solution) has a value of 20,000 Daltons MWCO.

Testing material and specimen preparation

An expansive soil from Zao-Yang (ZY), about 400 km away from Wuhan in China, was used in this study. It was composed of 49% clay, 44% silt and 7% sand, with a liquid limit of 68% and a plastic limit of 29%. The clay fraction is 60% kaolinite, 20–30% montmorillonite and 10–20% illite. According to the Unified Soil Classification System (USCS), the soil can be described as a clay of high plasticity (Wang, 2000).

A drying SWCC of this expansive clay is shown in Figure 4.3, which was obtained from four identical specimens under zero vertical stress in a volumetric pressure plate extractor (Wang, 2000). This estimated air-entry value is 60 kPa. The gentle gradient of the SWCC

154 Advanced Unsaturated Soil Mechanics

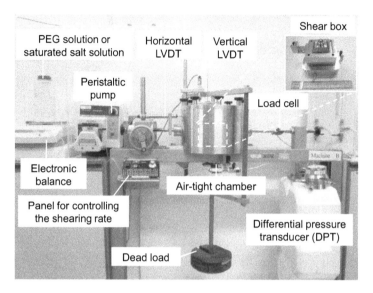

Plate 4.3 A full suction range direct shear box using the principle of axis-translation, osmotic suction control and humidity control techniques at HKUST.

Source: Ng, 2014.

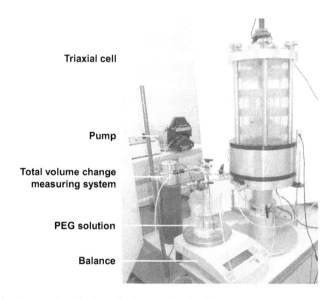

Plate 4.4 A typical set-up equipped with a double-cell total volume measurement system for triaxial testing using the osmotic suction control technique at HKUST.

implies a high-water storage potential for this expansive soil, which enables the soil to maintain a high degree of saturation over a large range of suction.

For triaxial tests, the soil was statically recompacted at a water content of 30.8% in a cylindrical mould 38 mm in diameter and 76 mm in height. Compaction was performed at a rate of 0.3 mm/min in three 25.3 mm high layers to ensure good homogeneity (Cui and Delage, 1996). The final dry density was about 1.36 g/cm³. The initial conditions of the specimens are given in Tables 4.1 and 4.2 for the triaxial tests using the axis-translation and osmotic techniques, respectively.

Testing programme

All triaxial shear tests were the consolidated type under a constant applied suction. The test series using the axis-translation technique includes 5 single-stage tests and 2 multi-stage tests (refer to Table 4.1). In the specimen names used in the table, the letter "A" denotes the axis-translation technique, "S" denotes the single-stage test and "M" denotes the multi-stage test. The applied suction ranged between 0 and 100 kPa and net confining pressure ranged between 25 and 100 kPa. The multi-stage tests maximised the shear strength information from one specimen and assisted in eliminating the effect of soil variability between specimens (Ho and Fredlund, 1982). Both series using the osmotic technique at HKUST and ENPC were single-stage tests and the testing conditions are presented in Table 4.2. In the specimen names used in the table, the letter "O" denotes osmotic technique, "H" denotes tests performed at HKUST and "E" denotes tests performed at ENPC. The applied suction ranged between 0 and 165 kPa and net confining pressure ranged between 25 and 100 kPa.

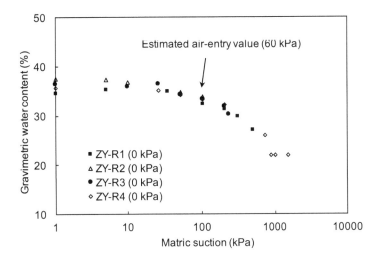

Figure 4.3 SWCC of recompacted ZY expansive soil.

Source: After Wang, 2000; Ng et al., 2007.

156 Advanced Unsaturated Soil Mechanics

Table 4.1 The testing conditions for the soil test specimens using the AS technique

Type	Specimen	Suction: kPa	Mean net stress: kPa
Multi-stage	AM1	50	25
		50	50
		50	100
	AM2	100	25
		100	50
		100	100
Single-stage	AS1	50	100
	AS2	100	50
	AS3	100	100

Source: Ng et al., 2007.

Notes:
A = axis-translation; M = multi-stage; S = single-stage.

Table 4.2 The testing conditions for the soil test specimens using the OS technique

Type	Specimen	Suction: kPa	Mean net stress: kPa
ENPC	OE1	0	50
	OE2	0	100
	OE3	100	100
	OE4	200	25
	OE5	200	50
	OE6	200	100
HKUST	OH1	50	50
	OH2	50	100
	OH3	100	25
	OH4	100	100
	OH5	200	100

Source: Ng et al., 2007.

Notes:
O = Osmotic; E = ENPC; H = HKUST.

Testing procedures

In the test series using the axis-translation technique, a specimen was first equalised at the desired suction under a mean net stress of 10 kPa. Suction equalisation was terminated when the variation in water content was less than 0.05%/day. The duration of this suction equilibrium procedure ranged between 4 and 9 days while the applied suction ranged between 0 and 100 kPa. Thereafter, the specimen was subjected to isotropic consolidation at constant suction. The consolidation stage was terminated when the rate of change of water content was less than 0.05%/day. This stage lasted from 3 to 9 days while the net confining pressure ranged between 25 and 100 kPa. After consolidation, the specimen was sheared at a constant rate under constant net confining pressure and suction. Two types of shearing procedures, i.e., single-stage shearing and multi-stage shearing, were carried out. In the single-stage shearing procedure, the specimen was sheared directly to failure after initial consolidation. The multi-stage shearing procedure consisted of 3 stages. The net confining pressure was varied from one stage to another while the suction was kept constant. At each stage of the multi-stage shearing procedure, each loading stage starts from zero deviator stress, which is called a cyclic loading procedure (Ho and Fredlund, 1982). To ensure the soil was drained, shearing was performed at a constant rate of 3.07×10^{-5} %/s. This rate was lower than the shearing rate of 4.39×10^{-5} %/s used at ENPC (Mao et al., 2002), considering that a longer drainage path (i.e., single drainage) was used when applying the axis-translation technique. Single-stage shearing took about 8 days and multi-stage shearing took about 16 days.

In the series using the osmotic technique at HKUST, suction was equalised by water interchange between the specimen and a PEG solution with the desired concentration under a mean net stress of 10 kPa. After suction equalisation, the specimen was consolidated under a constant net confining pressure. The criteria for ending suction equalisation and consolidation were the same as those adopted in the series using the axis-translation technique. The length of time required to reach suction equilibrium ranged from 5 to 11 days for suction ranging between 0 and 165 kPa, whereas the time required for consolidation ranged from 3 to 7 days for net confining pressure ranging between 25 and 100 kPa. When the consolidation stage was completed, the specimen was sheared to failure at a constant rate of 3.07×10^{-5} %/s under the same constant net confining pressure and suction. The shearing rate was the same as that adopted in the tests using the axis-translation technique, but it was slower than that adopted in the series using the osmotic technique in ENPC (Mao et al., 2002). When applying the osmotic technique, the pore-air pressure was maintained at atmospheric pressure during all the triaxial testing.

In the test series using the osmotic technique at ENPC, specimens were first wrapped in a semi-permeable membrane and kept in a PEG solution for 5–6 days to reach the desired suctions. They were then put in a triaxial cell for 3 days to reach equilibrium under the desired suction and confining pressure. Shearing was performed at a constant rate of 4.39×10^{-5} %/s. Details of the test procedures are described by Mao et al. (2002).

Test results from using the axis-translation technique

Although multi-stage tests can maximise the shear strength information obtained from one specimen, it may be useful to verify this shearing method with a single-stage test result. Figure 4.4 shows the relationships between deviator stress, $q = (\sigma_1 - \sigma_3)$, and axial strain, ε_a, during shearing for one multi-stage test (AM1) and two single-stage tests (AS2 and AS3). These three tests were performed under a suction of 100 kPa. For the multi-stage test AM1 in 3 stages, net confining pressures, p', were 25 kPa, 50 kPa and 100 kPa. For single-stage tests AS2 and AS3, the net confining pressures were 50 kPa and 100 kPa, respectively.

As shown in Figure 4.4, the first stage of test AM1 shows stiffer behaviour than tests AS2 and AS3 at the start of shearing. However, the stress-strain relationship at the second stage of test AMI is consistent with that of the single-stage test AS2, which is under the same net confining pressure of 50 kPa. The stress-strain relationship in the third stage of the multi-stage test is almost identical to that of the single-stage test AS3, which is under the same net confining pressure of 100 kPa. This comparison illustrates that the

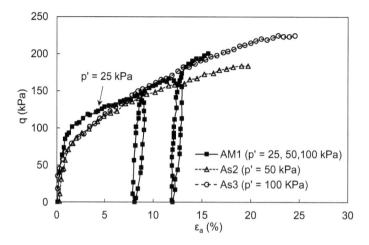

Figure 4.4 Stress-strain relationships of multi-stage and single-stage triaxial tests using the axis-translation technique with s = 100 kPa.

Source: Ng et al., 2007.

stress-strain relationships for multi-stage tests at different stages coincide with those of the corresponding single-stage tests, which are under the same suction and net confining pressure.

The stress points at failure are plotted in Figure 4.5 where $s' = \left(\dfrac{\sigma_1 + \sigma_3}{2}\right) - u_a$ and $t = \left(\dfrac{\sigma_1 - \sigma_3}{2}\right)$.

This figure shows that the shear strengths obtained from single-stage tests are consistent with those obtained from multi-stage tests under the same suction and net confining pressure (e.g., the variations of shear strength for the specimens AS1, AS2 and AS3 are 2.6%, 4% and 2.2% respectively, as compared with the corresponding multi-stage tests).

For this triaxial test series, the internal friction angle, ϕ', ranges from 19.3° to 20.8° for suction range from 0 to 100 kPa, and the average value is 20.0°. By assuming a constant friction angle of 20°, the failure lines for this series fit the test results well (see Figure 4.5). The increasing intercepts on the t axis indicate that apparent cohesion increases with suction.

At saturated state (i.e., zero suction), there appears to be a small intercept of 19.6 kPa on the t axis when the failure envelop is extended linearly to a zero value of s. To demonstrate whether this apparent intercept is attributable to a true cohesion, a soaking test without any applied stress was performed on a specimen compacted to the same initial state as the specimens used in the triaxial tests. It was found that the specimen collapsed completely while accessing to water. This indicates that no true cohesion is present in the recompacted specimen. Hence, the apparent intercept is probably caused either by experimental error or the failure envelope being curved at small stresses (see Figure 4.5).

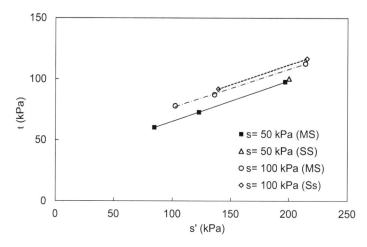

Figure 4.5 Comparison of strength between multi-stage and single-stage triaxial tests using the axis-translation technique.

Source: Ng et al., 2007.

Test results from using the osmotic control technique

The stress points at failure measured using the osmotic technique at HKUST and ENPC are shown in Figure 4.6. Under the same net confining pressure and suction, tests OH4 and OE3 exhibit similar shear strength. Comparing tests OH5 and OE6 gives the same conclusion. The consistency of the results from HKUST and ENPC confirms the reliability of the osmotic technique. The tests at HKUST and ENPC using the osmotic technique are regarded as one test series henceforth.

For this test series, the internal friction angle, φ', ranges from 19.5° to 20.8° for suction ranging from 0 to 165 kPa, and the average value is 20.2°. Again, by assuming a constant friction angle of 20°, the failure lines for this series fit the test results well, as shown in Figure 4.6. The increasing intercepts on the t axis indicate that apparent cohesion increases with suction. As with the results from the axis-translation technique, at saturated state there also appears to be a small intercept of 19.4 kPa on the t axis when the failure envelop is extended linearly to a zero value of s. Previous analysis demonstrated that this apparent intercept is probably due to experimental error or the failure envelope being curved at small stress.

Comparison between test results from the axis-translation and osmotic control techniques

The results from both test series using the axis-translation and osmotic techniques can be approximated by an internal friction angle of 20°. This indicates that there is no difference in the internal friction angle determined when using either the axis-translation technique or the osmotic technique for testing this material.

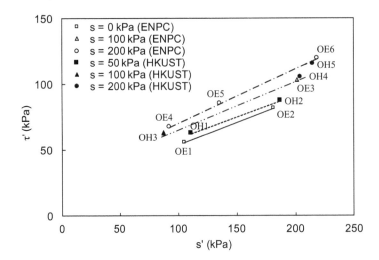

Figure 4.6 Comparison of strength of the triaxial tests performed at ENPC and HKUST using the osmotic technique.

Source: Ng et al., 2007.

In view of the results of the soaking test, it is believed that no true cohesion is present in the recompacted specimens. Thus, the apparent cohesion caused by suction can be calculated from intersects on the t axis at different suctions and subtracting the intersect in the saturated state. The data for apparent cohesion at different suctions are plotted in Figure 4.7a. For the test series using the axis-translation technique, the relationship shows a linear increase in shear strength with suction within the suction range from 0 to 100 kPa. For the test series using the osmotic technique, a nonlinear increase in shear strength with suction is observed within the suction range from 0 to 165 kPa. In the test series using the axis-translation technique, the apparent cohesion due to suction is consistently larger than that in the test series using the osmotic technique. The value of φ^b at different suctions can be calculated from the cohesion versus suction curve in Figure 4.7a and its variation with suction is shown in Figure 4.7b. At zero suction, φ^b may be assumed to be equal to the saturated friction angle, φ' (i.e., 20°) (Gan and Fredlund, 1996). As shown in Figure 4.7b, in the test series using the axis-translation technique, φ^b is slightly larger than that in the test series using the osmotic technique by 2°–3° when the suction is above zero. The larger φ^b indicates that the tests using the axis-translation technique exhibit a larger effect of suction on shear strength as compared with those using the osmotic technique. Inspecting specific volume at failure in the tests under a net confining pressure of 100 kPa (refer to Figure 4.8), it was found that the specimens in the test series using the axis-translation technique have lower values than those using the osmotic technique at different suctions. Perhaps the smaller specific volume at failure induced a larger shear strength and then the tests using the axis-translation technique exhibited a larger value of φ^b. However, the reason for the smaller specific volume in the tests using the axis-translation technique is not yet clear.

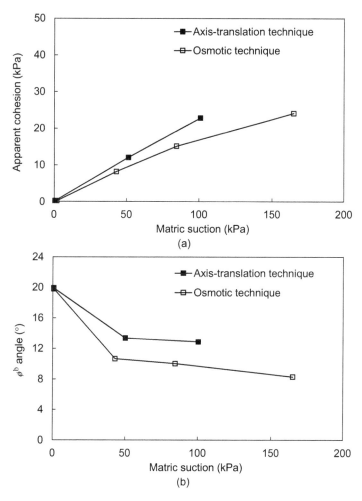

Figure 4.7 Suction versus shear strength: (a) apparent cohesion due to suction; (b) variation in ϕ^b angle with suction.

Source: Ng et al., 2007.

Summary

When applying the osmotic technique, the relationship between osmotic pressure and the concentration of PEG solution is of importance. Different calibration methods lead to different relationships between osmotic pressure and the concentration of PEG solution. The main difference arises from whether a semi-permeable membrane is used in calibration.

For the recompacted expansive clay used in this research, comparison of the results of single-stage and multi-stage tests indicated that these two testing methods provide consistent shear strength information. The consistency of the results using the osmotic technique between HKUST and ENPC showed the reliability of the osmotic technique.

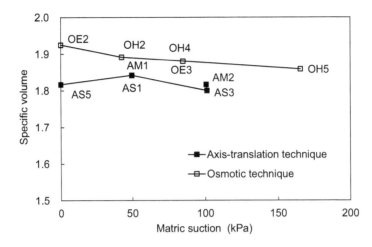

Figure 4.8 Specific volume at failure in the tests with a net confining pressure of 100 kPa.
Source: Ng et al., 2007.

Based on the extended Mohr–Coulomb shear strength formulation, comparison of the results using the axis-translation technique and the osmotic technique showed that there is no difference in determining the internal friction angle φ'. However, the values of φ^b in the test series using the axis-translation technique are slightly larger than those from the test series using the osmotic technique by 2°–3°. Perhaps the larger values of φ^b are due to smaller specific volumes at failure in the tests using the axis-translation technique. However, the reason for the smaller specific volume in the tests using the axis-translation technique is not yet clear.

EFFECTS OF SOIL SUCTION ON THE DILATANCY OF AN UNSATURATED SOIL (NG AND ZHOU, 2005)

Introduction

Soil behaviour is strongly influenced by dilatancy and so it is an important and essential component in the elasto-plastic modelling of soils (refer to Chapter 7). Dilatancy in saturated soils has been studied for many years and is fairly well understood. On the other hand, because of the complexity and long duration of testing unsaturated soils, fundamental understanding of the role of dilatancy in unsaturated soils is limited. Most existing unsaturated constitutive models treat dilatancy using an assumption or deriving it from assumed yield surfaces with associated or non-associated flow rules (Alonso et al., 1990; Wheeler and Sivakumar, 1995). Although Chiu and Ng (2003) have introduced state-dependent dilatancy in their elasto-plastic model, only very limited reliable laboratory data are available to verify these assumptions and calibrate constitutive relationships for unsaturated soils.

Cui and Delage (1996) performed three series of unsaturated triaxial tests on compacted silt using an osmotically controlled suction technique. Their test results obtained at mean constant stress ratio $\left(\eta = \dfrac{q}{p'} = 1 \right)$ and at constant cell pressure under various suctions are shown in Figure 4.9, respectively. Based on the results shown in Figure 4.9a, they drew two conclusions: first, the ratio between incremental plastic volumetric strain to incremental plastic shear strain $d\varepsilon_v^p/d\varepsilon_s^p$ (also called the "flow rule") is independent of suction; second, the final value of $d\varepsilon_v^p/d\varepsilon_s^p$ is 1. For constant cell pressure cases (Figure 4.9b), they interpreted their data by separating these curves into two segments, i.e., one segment with small slope gradients for low η values and the other with large slope gradients for higher η values. Based on a volumetric criterion, they claimed that the η values at the intersection points of these two segments corresponded to yield points. They also claimed that in the plastic zone beyond the intersection points, the gradient of these slopes is independent of suction, thus a linear and suction-independent stress-dilatancy relationship is derived.

Chiu (2001) and Chiu and Ng (2003) have investigated the stress-dilatancy of a decomposed volcanic soil and a decomposed granitic soil using a computer-controlled triaxial apparatus. Figure 4.10 shows the experimental relationships between dilatancy $(d\varepsilon_v^p/d\varepsilon_s^p)$ and normalised stress ratio (η/M) obtained by shearing decomposed volcanic soil specimens using constant water content stress paths where M is the gradient of the critical state line.

Following the approach proposed by Cui and Delage (1996), no distinct yield point can be identified. At high stress ratios, the experimental data formed two groups according to their suction values. If an average line is drawn through these two groups of data at $\eta/M = 0.8$ or higher, the gradients of these two fitted lines may be assumed to be parallel. This implies that the dilatancy rate with respect to η/M is independent of suction.

Here, a series of five laboratory tests on a compacted completely decomposed granite (CDG) were carried out in a suction-controlled direct shear box. Five suctions varying from 0 to 400 kPa were considered. The effects of soil suction on the evolution of dilatancy, stress-dilatancy relationship and maximum dilatancy were investigated.

Soil type and test procedures

The soil used in the experiment was sieved CDG from Beacon Hill (BH), Hong Kong. Based on visual inspection, the BH soil was a coarse-grained sand with 22% silt and 2% clay. The soil was sieved into its constituent particle sizes. Dry sieving was performed and particles larger than 2 mm were discarded. The soil's index properties were determined in accordance with BS1337 (BSI, 1990). The material has a liquid limit, $w_L = 44\%$, and a plastic limit, $w_p = 16\%$. According to the Proctor compaction test results, the maximum dry density was 1845 kg/m³ and the optimum water content was 14.2%. The drying soil-water characteristic curve (SWCC) for the CDG is shown in Figure 4.11. Also included in this figure is a drying SWCC for a silt from Nishimura and Fredlund (2000). The air-entry and residual suction values were determined using an equation proposed by Fredlund and Xing (1994), which are summarised in Table 4.3. The major difference between the two SWCCs is that the silt has a steeper SWCC than CDG does. This indicates that the silt has

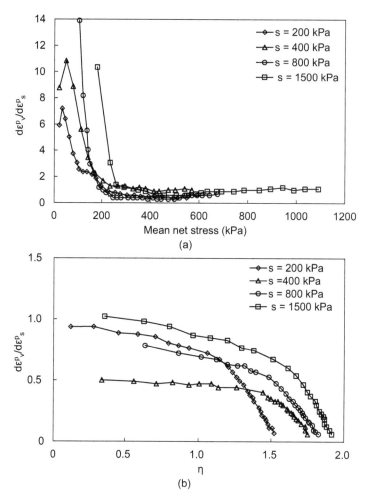

Figure 4.9 Effect of suction on dilatancy of silt under (a) constant stress ratio ($\eta=1$) and (b) constant cell pressure ($\sigma_3=200$ kPa).

Source: **After Cui and Delage, 1996; Ng and Zhou, 2005.**

a more uniform particle size distribution than that of CDG. The relationship between the SWCCs and shear test results is discussed later.

An unsaturated direct shear apparatus was used to carry out a series of unsaturated soil tests. The apparatus was originally developed by Gan et al. (1988) and modified by Zhan (2003) to facilitate measurement of changes in water volume automatically and accurately. The axis-translation technique was used to control the pore-air and pore-water pressures applied to a soil specimen.

A series of five unsaturated direct shear box tests on recompacted CDG were conducted at various controlled suctions. $50 \times 50 \times 21$ mm³ (length × width × height) specimens were used in the tests. Each specimen was prepared using the static compaction technique and

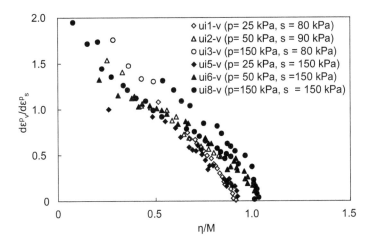

Figure 4.10 Stress-dilatancy relationship for specimens of decomposed volcanic soil under constant water content stress paths.

Source: After Chiu, 2001; Ng and Zhou, 2005.

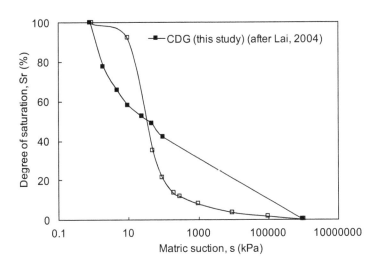

Figure 4.11 SWCCs of two soils.

Source: After Ng and Zhou, 2005.

compacted at the optimum water content of 14.2%. The compacted initial dry density was 1.53 g/cm³, corresponding to 82.7% of the maximum dry density.

After static compaction, each specimen was placed in the chamber of the unsaturated shear box equipped with a ceramic disk having a high air-entry value of 500 kPa, which was saturated according to the procedure proposed by Fredlund and Rahardjo (1993). Then the specimen was soaked under zero total vertical stress over 24 hours as illustrated

Table 4.3 Air-entry values and residual suction values of two soils

Soil	Air-entry value: kPa	Residual suction value: kPa
CDG	1	1,258
Silt	10	200

Source: After Ng and Zhou, 2005.

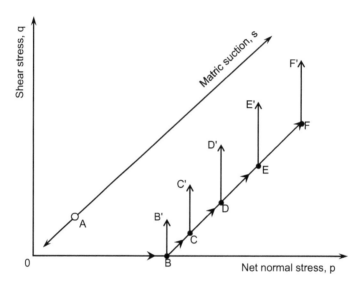

Figure 4.12 Stress paths in direct shear box tests.

Source: **After Ng and Zhou, 2005.**

in the path A→O in Figure 4.12. Then the specimen was loaded along path O→B to a constant vertical normal stress ($\sigma_v - u_a$) of 50 kPa, where σ_v is total vertical stress and u_a is pore-air pressure. Subsequently, pore-air pressure was applied at the top of the specimen step by step while maintaining zero pore-water pressure at the base of the specimen until the target air pressure was reached. Once the target air pressure was reached, suction equalisation was monitored by recording changes in water volume. When the change of water volume was less than 0.01% of the initial water content, the suction equalisation stage was terminated. Then shearing was carried out at a constant suction (i.e., along paths (B→B′, C→C′, D→D′, E→E′ and F→F′).

Evolution of dilatancy during shear

Figure 4.13 shows the relationships between the stress ratio ($t/(\sigma_v - u_a)$) and dilatancy against horizontal displacement (δx) respectively. In the figure, dilatancy is defined as the ratio ($\delta y/\delta x$) of the incremental vertical displacement (δy) to incremental horizontal displacement (δx). A negative sign (or negative dilatancy) means expansive behaviour. In

Shear strength and behaviour 167

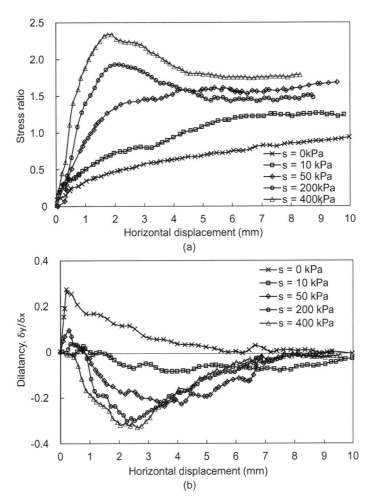

Figure 4.13 Evolution of stress ratio and dilatancy of CDG subjected to shear under different controlled suctions.

Source: After Ng and Zhou, 2005.

Figure 4.13a, it can be seen that at zero suction and suctions of 10 kPa and 50 kPa, the stress ratio-displacement curve displays strain hardening behaviour. With increasing suction, strain softening behaviour was observed at suctions of 200 kPa and 400 kPa. Generally, the measured peak and ultimate stress ratios increased with suction, except for the ultimate stress ratio measured at suction of 200 kPa. These test results are consistent with existing elasto-plastic models (e.g., Alonso et al., 1990; Wheeler and Sivakumar, 1995; Chiu and Ng, 2003; Ng et al., 2020a) for unsaturated soils.

Figure 4.13b shows the effects of suction on the dilatancy of CDG in direct shear box tests. Under saturated conditions, the soil specimen showed contractive behaviour (i.e., positive dilatancy). On the other hand, under unsaturated conditions, all soil specimens displayed contractive behaviour initially but then dilative behaviour as the horizontal displacement continued to increase. The measured maximum negative dilatancy was enhanced by an increase in suction. At the end of each test, the dilatancy of all specimens approached zero, indicating the attainment of a critical state.

An interesting phenomenon observed in Figure 4.13 was that strain hardening was recorded at suctions of 10 and 50 kPa as the soil dilated. A maximum negative dilatancy in the latter test did not lead to strain softening. Another phenomenon observed was that when the controlled suctions were 200 and 400 kPa, strain softening was observed (see Figure 4.13a). However, the measured peak stress ratio in each test did not correspond with the maximum negative dilatancy (see Figure 4.13b). This feature was not consistent with a common assumption in constitutive modelling that the point of peak strength is usually associated with the peak negative dilatancy (Bolton, 1986).

Stress-dilatancy relationship

Figure 4.14 shows the measured stress-dilatancy relationships in the five tests. At low stress ratios (i.e., 0.2 or smaller), positive dilatancy was observed for all specimens. The soil contracted initially during shear. As the stress ratios increased, negative dilatancy was measured in all unsaturated soil specimens. Obviously, there was a phase transformation from positive to negative dilatancy as the stress ratio increased. It is evident that the stress ratio corresponding to the maximum negative dilatancy increased with soil suction. As shearing continued, all the unsaturated soil specimens reached or approached zero dilation (i.e., critical state) at the end of each test. In the critical state, the measured stress ratio increased with suction, except at a suction of 200 kPa. This measured trend is consistent with the test results published by Cui and Delage (1996) and Chiu (2001) as shown in Figures 4.9 and 4.10, respectively. It should be pointed out in Figure 4.14 that a distinct loop was observed in tests with suctions equal to 50 kPa or higher. This implies that a state-dependent dilatancy soil model (Chiu and Ng, 2003) is necessary to capture this type of soil behaviour.

Maximum dilatancy

Figure 4.15 shows the relationships between maximum dilatancy and controlled suction for the CDG specimens. For comparison, experimental data from unsaturated shear box tests on a silt taken from Nishimura (2000) are re-interpreted and included in the figure. Also included in this figure are the air-entry values and the residual suction values of these two soils. Their soil-water characteristic curves (SWCCs) are given in Figure 4.11.

As illustrated in Figure 4.15, the maximum dilatancy of CDG decreases (i.e., more negative or dilative) with an increase in suction, i.e., the soil dilates more at a higher suction. The relationship between maximum dilatancy and suction is highly nonlinear. On the contrary, the maximum dilatancy of the silt published by Nishimura (2000) increases (i.e., is more contractive) linearly with suction. The major difference in the observed maximum dilatancy-suction relationships of these two soils is probably attributable to the difference in their initial soil densities (i.e., void ratios). The CDG and the silty soil have an initial dry

Shear strength and behaviour 169

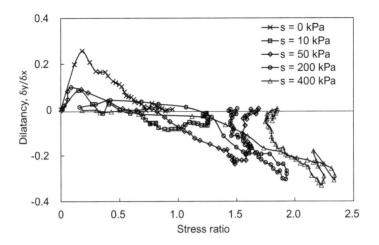

Figure 4.14 Experimental stress-dilatancy relationship.

Source: After Ng and Zhou, 2005.

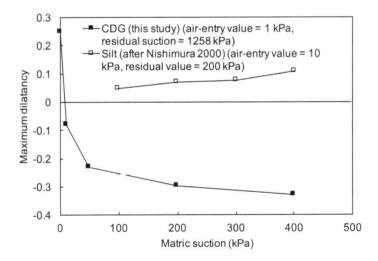

Figure 4.15 Relationship between maximum dilatancy and suction.

Source: After Ng and Zhou, 2005.

density of 1.53 g/cm^3 and 1.27 g/cm^3, respectively. Therefore, it is not surprising to see that CDG is more dilative than the silt for a given suction. By comparing the absolute values of the maximum dilatancy of these two soils, variations in the magnitude of the maximum dilatancy are relatively larger in CDG than in the silt for a given change of soil suctions. In addition to the difference in the initial soil density, the compressibility of CDG is large relative to other silty soils because of the presence of crushable feldspar (Ng et al., 2004a). Therefore, for a given increase in soil suction, it is believed that CDG has undergone a

170　Advanced Unsaturated Soil Mechanics

larger reduction in void ratio than the silt, resulting in the larger absolute maximum dilatancy that is measured.

Summary

A series of consolidated drained shear tests were carried out on compacted CDG to investigate the effects of suction on shear behaviour and dilatancy. The measured stress ratio corresponding to the maximum negative dilatancy increased with soil suction. The measured peak stress ratio in each test did not correspond with the maximum negative dilatancy. The maximum dilatancy of CDG was strongly dependent on suction and soil density. The maximum dilatancy decreased (i.e., became more negative or dilative) with an increase in suction, i.e., the soil dilated more at a higher suction. The relationship between maximum dilatancy and suction was highly nonlinear. State-dependent flow rules are essential and vital for modelling unsaturated soils correctly.

SHEAR STRENGTH CHARACTERISTICS OF COMPACTED LOESS AT HIGH SUCTIONS (NG ET AL., 2017A)

Introduction

Experimental results for the shear strength characteristics of geomaterials under high suction are very limited. The results of unconfined compression tests on a compacted silty soil by Nishimura and Fredlund (2001) indicated that shear strength remains fairly constant beyond the residual suction. The results of drained triaxial compression tests on a compacted silt reported by Alsherif and McCartney (2015) revealed a brittle failure mechanism at high suctions. Merchán et al. (2011) reported a noticeable increase in the residual friction angle of Boom clay when suction increased from zero to 140 MPa. As far as the authors are aware, however, the effects of suction on dilatancy have not been reported in literature. Hence, a series of tests described and reported below examine the significance of suction on the shear behaviour of compacted loess up to 230 MPa.

Test apparatus and measuring devices

A conventional direct shear box was modified to control suction at high magnitudes during a test (Plate 4.5). In the new design (Ng, 2014), the suction range is extended to high magnitudes on the order of tens of MPa by incorporating the vapour equilibrium technique (Blatz et al., 2008). Therefore, suction can be controlled by the vapour transfer mechanism inside the isolated stainless-steel chamber surrounding the whole shear box.

To provide an isolated environment having a particular relative humidity (RH), 8 Plexiglas cylinders were attached to the internal wall of the shear box chamber. The cylinders serve as containers for salt solutions. A total volume of 520 ml is provided by all containers, which is considered sufficient to provide an environment with a stable RH. Moreover, the whole shear box apparatus is housed in a temperature-controlled room with a daily temperature fluctuation of $\pm 0.5°C$. The error in pressure caused by temperature fluctuations was estimated to be 0.01 and 0.43 MPa for the lowest and highest suctions considered (i.e., 7 and 231 MPa), respectively. The error in suction caused by temperature

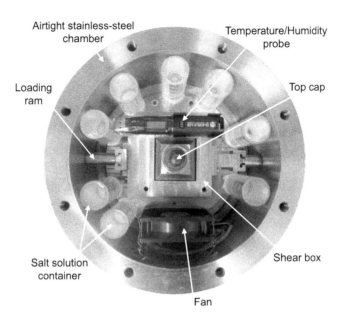

Plate 4.5 Plan view of the modified direct shear box device.

fluctuations is less than 0.2% in all cases. Four types of saturated salt solutions were used to generate target RH values and hence suction. Potassium nitrate, Sodium chloride, Magnesium chloride hexahydrate, and Lithium chloride were used to produce nominal suction potentials of 8 MPa, 40 MPa, 125 MPa, and 230 MPa, respectively.

In addition to the conventional sensors in a direct shear device, a temperature/humidity data logger was used to monitor both relative humidity and temperature during the whole test. The recorded data were used afterwards to calculate the total suction. To accelerate the homogenisation of RH throughout the chamber, an agitation fan was incorporated into the isolated shear box chamber. This modified shear box can measure stress-dependent SWCC (SDSWCC), shear strength and consolidation behaviour of soils (Ng, 2014).

Soil type and its water retention and compression behaviour

Block loess samples were retrieved from a pit in Xi'an, Shaanxi Province of China. Soil samples were taken at four different depths from 1.5 m to 7.5 m at 2.0 m intervals. The sand, silt, and clay contents of loess are 0.1%, 71.9%, and 28.0%, respectively. The soil is therefore classified as a clay of low plasticity or CL according to the USCS (ASTM, 2011). Figure 4.16 shows the results of a compaction test on loess with standard Proctor effort. The results clarify the state of intact loess to be on the dry side of optimum with a relative compaction of 76%.

Recompacted specimens were prepared with the same dry density and water content as *in situ* soil samples. Test specimens were prepared with three layers following the static

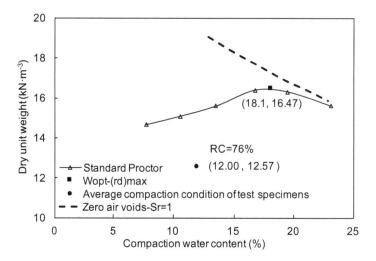

Figure 4.16 Compaction curve of loess.

compaction procedure with a displacement rate of 1 mm/min. A three-piece split mould facilitating the extrusion of specimens was used.

The water retention properties of recompacted loess measured using the modified shear box shown in Plate 4.5, are given in Figure 4.17a. Data in solid black relate to the equalised hydraulic state of the test specimens in the current study. Measured data points from Ng et al. (2016a) for a similar material with an average void ratio of 1.162 were also included in this figure, covering a wider range of suction. Note that the recompacted loess specimen of Ng et al. (2016a) had a slightly higher void ratio than the average void ratio of test specimens in this study. This implies that the test specimens in this study will tend to retain less water at identical suctions. However, the data from both studies are in good agreement at high suctions where adsorption governs water retention behaviour, not capillary forces. The measured water retention curves of recompacted loess suggest a weak bimodal behaviour when the suction is less than 4 MPa. Although a bimodal water retention model (e.g., Zhang and Chen, 2005) may provide a better match than a unimodal curve in the low suction range, this may not be true for the high suction range (> 4 MPa). For simplicity, the unimodal curve fitting proposed by Fredlund and Xing (1994) is used and shown in Figure 4.17a.

Figure 4.17b shows the corresponding volume changes for both sets of data presented in Figure 4.17a. The recompacted specimen that was subjected to wetting and then drying at high suctions (Ng et al., 2016a) experienced a larger volume change than specimens subjected to drying only (this study). Measurements of suction using a small tip tensiometer determined the initial suction to be 62 kPa for the recompacted specimens. The results of this study show that specimens subjected to high suctions from the as-compacted state show limited shrinkage strain of the order of 1.6% which is almost nine times less than the corresponding wetting-induced collapse strain. This is probably because of the low as-compacted water content ($S_r = 0.29$ according to Figure 4.17a), which is dryer than the optimum water content. A smaller amount of shrinkage is expected for a specimen having a moisture content less than the saturated value (Peng and Horn, 2007). In addition, the

Shear strength and behaviour 173

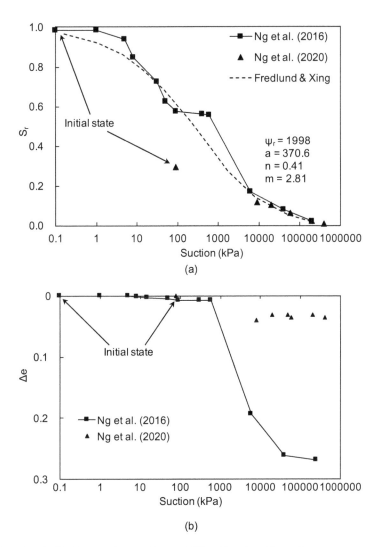

Figure 4.17 Variations in (a) degree of saturation, and (b) void ratio with suction.

negligible variations in void ratio with suction imply the independence of shrinkage from variations in suction over the range of 8 to 230 MPa. This observation suggests that suction values above 8 MPa are within the residual shrinkage zone with insignificant influence on volume changes.

Figure 4.18 shows compression curves from the suction-controlled tests in this study up to a net vertical stress of 200 kPa (solid symbols). In addition, the results of oedometer tests on saturated recompacted loess (Ng et al., 2016a) and on an air-dried specimen under laboratory conditions are included for comparison. The yield stress of the saturated specimen appears to be less than 10 kPa but is about 490 kPa for the air-dried specimen (open square symbol). There is a significant increase in yield stress with suction as the suction

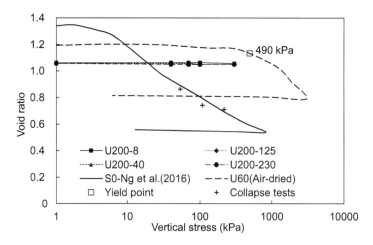

Figure 4.18 Compression curves of saturated and unsaturated recompacted loess.

increases from zero to 60 MPa, which is consistent with the loading collapse concept of the Barcelona Basic Model (Alonso et al., 1990).

The results of the wetting-induced collapse tests are also included in Figure 4.18 for comparison. There is a good agreement between the void ratio of the collapse tests and the consolidation curve of the recompacted loess specimen reported by Ng et al. (2016a). All these results reveal that all the consolidation data points converge on the same virgin compression curve. According to these results, a maximum volumetric collapse strain of up to 13.8% is achieved under 100 kPa vertical stress. There is no indication of yield stress up to 200 kPa net stress for the four suction levels considered. Moreover, the results of suction-controlled tests confirm the similarity between the compression behaviour measured in these tests and the unloading branch obtained from oedometer tests (i.e., S0 and U60). These observations imply that all the unsaturated tests conducted in this study in the high suction range are within the elastic region. The increase in yield stress caused by an increase in suction is well-known.

Shear strength and shear-induced volume change

Two series of saturated and unsaturated suction-controlled tests were conducted to study the shear strength characteristics of loess. Details of the saturated tests are given in Table 4.4. Eight unsaturated tests were run in total, the details of which are summarised in Table 4.5.

The results of tests conducted under a vertical stress of 50 kPa are shown in Figure 4.19a. The saturated test shows a continuous increase in shear strength with a reduced rate as a critical state value is approached at horizontal displacements exceeding 6 mm. The observed strain hardening behaviour, being typical of normally consolidated soils, comes from the post-yield state of the effective consolidation stress. It should be noted that the saturated specimen was consolidated to an effective vertical stress of 50 kPa and the yield stress of the compacted loess was determined to be less than 10 kPa. Constant-suction

Shear strength and behaviour   175

Table 4.4 Summary of test program and conditions for saturated tests on loess

Test ID	Initial void ratio	Initial degree of saturation	Vertical effective stress: kPa
S50	1.111	0.29	50.4
S100	1.114	0.29	100.3
S200	1.085	0.30	199.9

Table 4.5 Summary of test program and conditions for unsaturated tests on loess

Test ID	Initial void ratio	Initial degree of saturation	Vertical net stress: kPa	Suction: MPa
U50-8	1.090	0.29	50.4	13.0
U50-40	1.091	0.29	50.4	35.6
U50-125	1.103	0.30	50.4	128.9
U50-230	1.092	0.29	50.4	231.1
U200-8	1.108	0.31	199.9	6.4
U200-40	1.095	0.29	199.9	37.1
U200-125	1.094	0.29	199.9	121.9
U200-230	1.092	0.29	199.9	228.9

unsaturated tests, however, show a different mode of behaviour. The initial phase (up to 0.3 mm) corresponds to the gap between the cell and the specimen that was created by the application of suction. The behaviour in this phase cannot be attributed to the material as there was almost no change in vertical displacement. Next, a moderate increase in strength (between 0.3 to 1 mm horizontal displacements) was observed followed by a sharp rise up to a peak for all the tests, representing a higher stiffness by comparison with the saturated test. The shear strength decreases thereafter towards a fairly constant ultimate state. The peak strength is reached at horizontal displacements less than 1.5 mm for all the tests. There is a significant increase in shear strength in unsaturated tests by comparison with saturated tests. More importantly, the results of unsaturated tests indicate an increase in both peak and post-peak shear strength with suction, even in the high suction range considered in this study.

Figure 4.19b illustrates the variation of vertical displacement with horizontal displacement for tests carried out under 50 kPa vertical stress. It is noted that contraction is assumed to be positive. As with the strength behaviour, saturated and unsaturated soil specimens show opposite vertical displacement trends. Saturated specimens continuously contract as shearing proceeds until the critical state is reached at large displacements, which is typical behaviour of normally consolidated soils. Unlike saturated tests, unsaturated specimens initially show a very small contraction (after closing up the gap), followed by noticeable dilation. An accelerated rate of dilation can be seen as the suction increases. This agrees with the observations of Hossain and Yin (2015) for a low range of suctions limited to 300 kPa. The critical state is assumed to be reached once no further changes are observed in both volume and shear strength as shearing proceeds. It should be noted that this condition is only partly satisfied on the shear plane. However, this simplified assumption for the critical state approach should not affect the major conclusions drawn from this study.

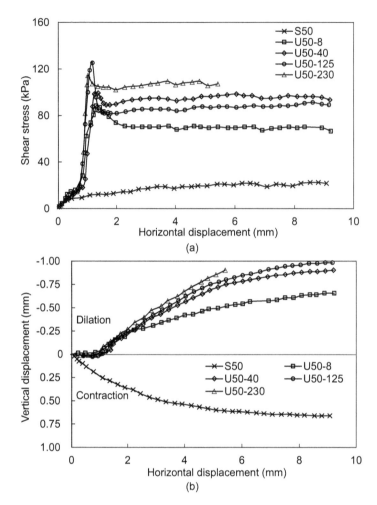

Figure 4.19 Variations in (a) shear stress, and (b) vertical displacement with horizontal displacement for vertical stress of 50 kPa.

The results of tests conducted under a vertical stress of 200 kPa are shown in Figure 4.20a in terms of shear stress versus horizontal displacement. The trends in the change in shear strength are qualitatively similar to those under a lower vertical stress of 50 kPa (Figure 4.19a). However, one of the clearest differences is the higher shear strength achieved under elevated vertical stress. As the vertical stress increases from 50 to 200 kPa, the peak shear strength of both saturated and unsaturated tests roughly quadruples and doubles, respectively. Moreover, the drop in strength of unsaturated specimens after the peak becomes less significant by comparison with the case with 50 kPa vertical stress. The observed behaviour confirms the overwhelming influence of net stress on dilation rate (Hossain and Yin, 2010). Figure 4.20b illustrates the variation of vertical displacement with horizontal displacement below 200 kPa vertical stress. As in the case of low stress,

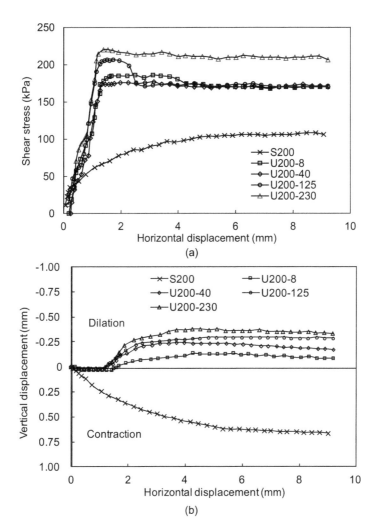

Figure 4.20 Variations in (a) shear stress, and (b) vertical displacement with horizontal displacement for vertical stress of 200 kPa.

the saturated specimen contracts continuously while all the unsaturated specimens dilate at high suctions. In addition, the rate of dilation increases as suction increases from 8 to 230 MPa during the whole shearing process. By comparison with the results of 50 kPa net stress, Figure 4.20b makes it clear that the rise in net stress from 50 to 200 kPa suppresses the tendency of unsaturated specimens to dilate at all suction levels considered. In other words, soil dilation is noticeably lower under 200 kPa net stress than under 50 kPa. Moreover, no further tendency for dilation can be seen at large displacements (> 6 mm) and there is no further dilation.

The results of shear strength measurements are explained and discussed in terms of peak shear strength. Figure 4.21 depicts the maximum shear strength in the conducted tests as

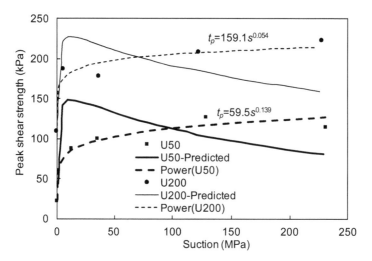

Figure 4.21 Variations in peak shear strength with suction.

a function of suction. A trend of increasing peak strength with suction can be observed in most cases under both net stresses of 50 and 200 kPa. The results of regression analysis demonstrate that the measured data can be fitted well ($R^2 > 0.96$) to power laws. This implies a continuous increase of peak shear strength with suction but at a reduced rate. By contrast with the coarse-grained soils, the contribution of suction to the shear strength of an unsaturated fine-grained soil like loess will not diminish beyond the residual suction. A nonlinear increase of residual shear strength with suctions up to 100 kPa was observed by Hoyos et al. (2014) for silty clayey sand. In spite of numerous studies of the strength characteristics of unsaturated soils over the low to medium range of suction (less than 1 MPa), test results at high suction values are fairly rare. In the current study, one of the recent modeling approaches proposed by Alonso et al. (2010) was used to simulate the observed behaviour. In order to improve the contribution of suction to the effective stress, the original χ parameter in Bishop's effective stress equation is replaced by the effective degree of saturation in this model, to extend the prediction capability of the shear strength model to a higher range of suction in a more rational way. Therefore, the shear strength equation can be expressed as:

$$\tau = c' + \left[(\sigma_v - u_a) + S_r^e \cdot s \right] \tan \phi' \qquad (4.7)$$

where τ is shear strength, c' and ϕ' are the effective saturated strength parameters, σ_v is the total vertical stress, u_a is pore-air pressure, S_r^e is the effective degree of saturation, and s is suction. Based on the results of three saturated tests (Table 4.4), c' and ϕ' were determined to be zero and 27.7°, respectively. In order to find the effective degree of saturation, the second proposal (Alonso et al., 2010) as shown in Equation (4.8) is adopted here as the first approach is based on a piece-wise function which is more suitable for coarse-grained soils.

$$S_r^e = \left(S_r\right)^{\alpha e} \tag{4.8}$$

where S_r is the degree of saturation and αe is a material parameter. The best-fit curve to the measured water retention data (Figure 4.17a) is used as the $S_r - s$ function. Afterwards, the material parameter, αe, was back-calculated from a regression analysis to give the best prediction of shear strength (Equation (4.7)). By analysing the results of 10 saturated and unsaturated tests, the best value of αe was determined to be 1.76. The model predictions were compared with the measured shear strength data and the coefficient of determination R^2 was found to be 0.57. As shown in Figure 4.18, the shear strength increases with increasing suction, but at a reduced rate. This implies that recompacted loess stiffens as suction increases, which in turn results in an increase of shear strength. However, the model predicts a peak shear strength at the optimum value of $(S_r \times s)$ followed by a gradual decrease in strength. Theoretically, the shear strength would continue to decrease and eventually reach the saturated shear strength at the maximum theoretical suction of 1000 MPa (corresponding to zero degrees of saturation). Although this kind of trend is well suited to the real behaviour of coarse-grained soils, it may not represent the strength of fine-grained soils perfectly. The observed trend of a peak shear strength with high suction supports the idea that levelling off the effective degree of saturation (or Bishop effective stress parameter) towards the end of the water retention curve is not physically valid (Oh and Lu, 2014).

Dilatancy

Figure 4.22 shows the evolution of dilatancy ($\delta y/\delta x$) with shear displacement. Contractive behaviour is observed in saturated tests during the whole shearing stage (Hossain and Yin, 2010). On the other hand, the dominant behaviour of unsaturated specimens is dilative throughout shearing except at the initiation of shearing and towards the end. Unsaturated specimens initially show negligible contractive behaviour followed by a sharp rise to a peak dilation rate. Afterwards, dilatancy decreases gradually and eventually vanishes in most cases, being consistent with displacement behaviour. Comparisons of measured dilation rates with three classical stress-dilatancy expressions in the literature are shown as follows:

$$\left(\tau/\sigma_n'\right)_{max} = \tan\varphi_{cs}' + \tan\psi_{max} \tag{4.9a}$$

$$\left(\tau/\sigma_n'\right)_{max} = \tan\left(\varphi_{cs}' + \psi_{max}\right) \tag{4.9b}$$

$$\left(\tau/\sigma_n'\right)_{max} = \tan\left(\varphi_{cs}' + 0.8\,\psi_{max}\right) \tag{4.9c}$$

$$\psi = \tan^{-1}\left(d\right) \tag{4.9d}$$

where φ_{cs}' is the effective critical state friction angle, ψ and ψ_{max} are the angle of dilation and the maximum angle of dilation, respectively. Equations (4.9a) to (4.9c) are known as the Taylor (Taylor, 1948), Coulomb (Budhu, 2008), and Bolton (Bolton, 1986) failure criteria,

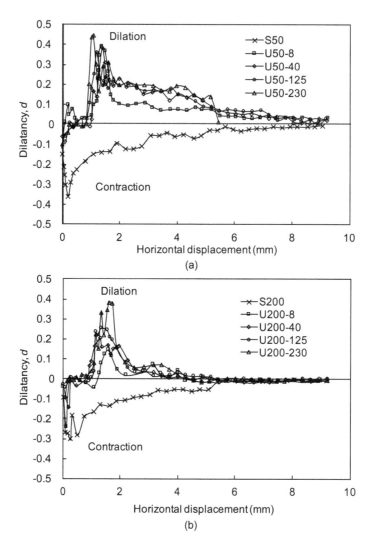

Figure 4.22 Variations in dilatancy with horizontal displacement under vertical stress of (a) 50 kPa, and (b) 200 kPa.

respectively. As pointed out by previous researchers, suction must be taken as a variable in the formulation of stress-dilatancy (e.g., Chiu and Ng, 2003; Alonso et al., 2007). Although the definition of stress ratio is straightforward for saturated soils because of the validity of the effective stress concept, difficulties arise in defining an appropriate effective stress for unsaturated soils. In order to define a rational stress ratio for the unsaturated tests conducted in this study, shear strength is divided by the equivalent vertical effective stress defined as:

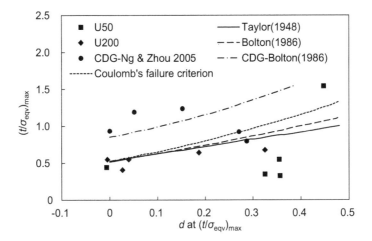

Figure 4.23 Dilatancy at the equivalent peak stress ratio.

$$\sigma_{eqv} = (\sigma_v - u_a) + S_r^e \cdot s \qquad (4.10)$$

Figure 4.23 shows the rate of dilation at the maximum equivalent effective stress. Note that the maximum equivalent effective stress may not be constant during a test. This is because contraction/dilation induced by shearing can result in a change in the degree of saturation and hence suction. This in turn can lead to a change of $(\sigma_{eqv})_{max}$ according to Equation (4.10). Since there was no suction sensor embedded in the specimen, the constant-suction condition was assumed in the high range (> 8 MPa). For tests conducted below 200 kPa net stress, most of the data roughly follow the Taylor criterion. On the other hand, the results of tests with 50 kPa net stress clearly show scattering of the data, mainly to the right of classical models. This observation implies that the three stress-dilatancy equations examined underestimate the dilation rate of unsaturated specimens at low net stress levels. Wang and Leung (2008) also reported that the original expression by Cam Clay and Rowe (Rowe, 1969) overestimates the stress ratio of cemented sand under low confinement if a cohesion term is not incorporated into these models. Similar observations were reported by Cresswell and Powrie (2004) for an uncemented natural sand characterised by a locked fabric having a relative density of 136%. To study the applicability of saturated models further, a different set of unsaturated shear test results from Ng and Zhou (2005) were also examined. The data were re-interpreted following the same methodology used for compacted loess in this study. In other words, the best-fit water retention curve of CDG was used to calibrate the material parameter, ae, in the framework of microstructural-based effective stress by curve fitting the measured shear strength data at different suction levels. The equivalent effective stress was then calculated using Equation (4.10). The results are presented in Figure 4.23 together with the empirical stress-dilatancy equation of Bolton (1986) for comparison. The deviation of measurements from predicted values can be also found for unsaturated silty sand.

Significance of suction

Figure 4.24 shows variations in the maximum dilatancy with suction. The maximum dilatancy was obtained from the results in Figure 4.22. The results are also represented in terms of the maximum dilation angle calculated by the following expression:

$$\psi_{max} = \tan^{-1}(d_{max}) \quad (4.11)$$

In order to distinguish between the effects of suction and void ratio on dilatancy, the void ratio of the test specimens is also depicted in the figure. According to the results of Figure 4.24, there is a linear increasing trend in dilatancy as well as the angle of dilation with suction for both net stresses considered to be within the high suction range. The trend of variations in dilatancy with suction is like those shown in Figure 4.22 for shear strength. In other words, the dilatancy shows a rapid increase from zero suction (the maximum dilatancy is zero for saturated tests) to 8 MPa and is then followed by a linear relationship with a gentler slope approaching a suction value of 230 MPa. In spite of the negligible influence of suction on the dilatancy of compacted limestone gravels (Alonso et al., 2007), the suction effect has quite a significant effect on the dilation rate of compacted loess. The unsaturated specimens compressed to 50 kPa net stress give a consistently higher dilatancy than those compressed to 200 kPa. The observed trends demonstrate that for the suction range considered in this study, the average rate of increase in dilation with suction is 0.02°/MPa for tests carried out under 50 kPa and 0.06°/MPa for tests carried out at 200 kPa. In addition, measurements of volume change after the suction equalisation stage demonstrate that suction greater than 8 MPa has effects changes in void ratio insignificantly (see Figure 4.17b). The numbers indicated in Figure 4.24 are the corresponding void ratio of the soil specimens before the shearing stage. The maximum difference in void ratio is less than 0.03 and 0.01 for tests conducted below 50 kPa and at 200 kPa net

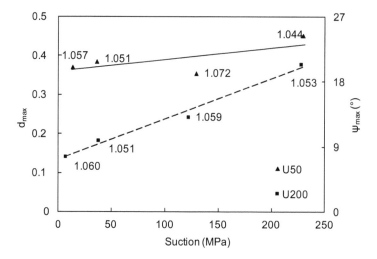

Figure 4.24 The influence of suction on maximum dilatancy of recompacted loess.

stress, respectively. However, the increase of dilatancy with suction is considerable and cannot be neglected. Increasing suction beyond the maximum previous value would result in a considerable increase in the initial stiffness (Vassallo et al., 2007). Further, enhanced suction can induce an aggregated microstructure (Merchán et al., 2011). This implies that as specimens become dryer, larger aggregates will be created. A higher dilatancy is thus expected for a soil with larger and stiffer particles/aggregates. Therefore, an increase in suction will enhance soil dilation even at high suctions.

Summary

The strain softening mode of failure associated with dilation is observed for unsaturated specimens at a high suction range between 8 and 230 MPa. The peak strength is mostly achieved at a horizontal displacement of 1-2 mm and continuously increases with suction but at a reduced rate. The trend of increasing peak strength cannot be predicted by the microstructural-based effective stress framework. One possible reason is that the model diminishes the contribution of high suction to strength towards the tail of the water retention curve at which the degree of saturation is close to zero.

The increase in suction enhances the dilation rate at both net stresses of 50 and 200 kPa. This observation reveals the significance of suction in inducing dilation even at a high range which cannot be captured by existing theoretical models. In most cases, the dilatancy at the equivalent peak stress ratio is greater than the prediction of stress-dilatancy equations for saturated granular soils. This confirms the need to consider suction contributions in dilatancy formulations.

Within the suction range of 8 to 230 MPa, dilatancy increases with suction linearly. Results show that dilatancy increases with suction at rates of 0.02°/MPa and 0.06°/MPa for specimens compressed to 50 and 200 kPa net stress, respectively. It should be also noted that differences in the void ratio of unsaturated specimens before shearing is less than 0.03, implying the independent role of suction in enhancing dilation.

EFFECTS OF MICROSTRUCTURE ON DILATANCY (NG ET AL., 2020B)

Introduction

To investigate the influence of microstructure on the shear strength and dilatancy behaviour of unsaturated soil, two series of direct shear tests were carried out on compacted and intact loess. Figures 4.25a and 4.25b show the measured soil-water retention curve and void ratio versus suction, respectively. Results are presented for both recompacted and intact loess at high suctions measured using the vapour equilibrium technique. In spite of the significant influence of suction on variations in the degree of saturation, the void ratio is not influenced significantly by changes in suction at high values. According to Figure 4.25c, no significant difference can be seen between the two sets of measurements at high values of suction in spite of previous observations for low values. Differences in the water retention characteristics of recompacted and intact loess were presented and discussed by Ng et al. (2016a). The results shown in Figure 4.25c are used to infer the macrovoid (e_M) and microvoid ratios (e_m) for intact and recompacted loess. In order to achieve this, the proposal put forward by Alonso et al. (2013) is used to infer the microvoid

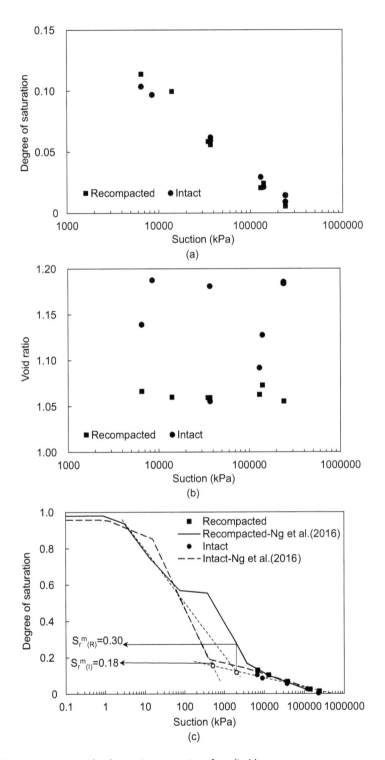

Figure 4.25 Water retention and volumetric properties of studied loess.

Shear strength and behaviour 185

ratio and hence macrovoid ratio from the microstructural parameter, ξ_m, according to the following equation:

$$e_m = \xi_m e \qquad\qquad (4.12a)$$

$$e_M = e - e_m \qquad\qquad (4.12b)$$

It should be also noted that ξ_m is equal to the microstructural degree of saturation and S_r^m, corresponds to the residual suction. The microstructural degree of saturation corresponds to water filling all the micro-voids while the macrovoids remain completely dry. In order to obtain S_r^m, the residual suction was first derived by intersecting two lines tangent to the residual branch and main desorption path after the air-entry value according to Figure 4.25c. The corresponding degree of saturation was then obtained following the procedure shown in the figure. The microstructural properties of recompacted and intact loess are given in Table 4.6. The results for loess microstructure are discussed later.

Pore-size distributions of recompacted and natural loess specimens

A total of six mercury intrusion porosimetry (MIP) tests on freeze-dried specimens were conducted to explore the microstructure of loess at different suction values. Three suction levels of 0.1 MPa (corresponding to *in situ*/as-compacted state), 8 MPa, and 230 MPa were chosen. Figure 4.26 illustrates the pore-size density (PSD) functions of recompacted specimens. The three curves are quantitatively similar although the MR-8 and MR-230 specimens were subjected to severe drying processes. Results indicate a clear peak near 11 μm corresponding to the dominant size of macropores (d_M). However, the peak corresponding to the dominant size of micropores (d_m) in the range between 0.04 and 0.1 μm was much less pronounced. The delimiting diameter of microporosity and macroporosity was derived from the PSD functions, where the derivative with respect to the abscissa becomes zero.

As shown in Figure 4.26b, the as-compacted specimen exhibited a higher maximum value of MIP void ratio, e_{MIP}, than the other two dried specimens because of the shrinkage induced by suction, but the volumetric contraction was negligible. If all the pores within the test specimens can be infiltrated by mercury and hence detected by the MIP apparatus, e_{MIP} should be equal to e. But MIP cannot detect pores below a certain diameter as well as pores larger than a certain diameter, the diameters being dictated by the maximum and minimum mercury pressure, respectively. Results reported by Ng et al. (2016a) confirm

Table 4.6 Microstructural properties inferred from SWRCs of recompacted and intact loess

Specimen type	Microstructural degree of saturation, S_r^m	Microstructural parameter, ξ_m	Average void ratio, e_{ave}	Microvoid ratio, e_m	Macrovoid ratio, e_M
Recompacted	0.30	0.30	1.06	0.32	0.74
Intact	0.18	0.18	1.14	0.21	0.93

that all pores within the intact loess were larger than the minimum detectable diameter of 5 nm. Therefore, MIP test results can be used to evaluate the microporosity of loess reliably. The PSD functions of intact loess are shown in Figure 4.26c. As in the results from recompacted loess, the changes in suction from the *in situ* value to severe drying do not affect the dominant sizes of micropores as well as macropores. Figure 4.26d shows the cumulative void ratio as a function of pore diameter. By comparing Figures 4.26c and 4.26d, it can be inferred that a smaller number of voids can be detected in intact loess than in recompacted loess. Therefore, intact loess contains a higher proportion of undetected voids, e_{nd} (Table 4.7), as a result of the existence of extra-large pores in its natural form (Ng et al., 2016a; Haeri et al., 2016).

The microvoid ratio, e_m was determined from the cumulative void ratio in conjunction with the delimiting diameter. Because of a significant volume of undetected voids

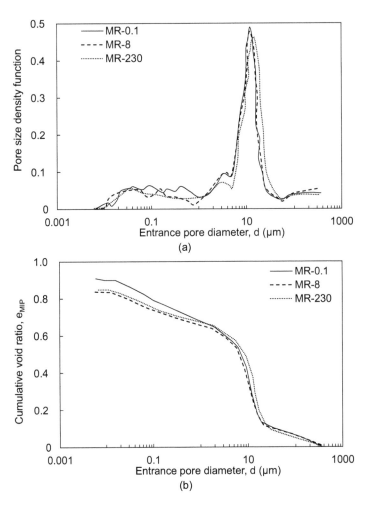

Figure 4.26 MIP test results of recompacted specimens (a) and (b) and intact specimens (c) and (d), in terms of pore-size density functions and cumulative void ratios.

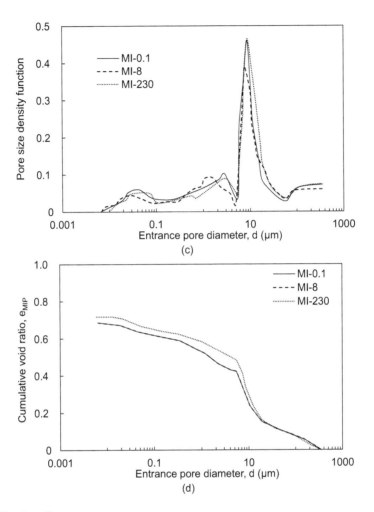

Figure 4.26 (Continued)

(e_{nd}), information on the void ratio is needed to calculate the macrovoid ratio, e_M (Alonso et al., 2013). Following this methodology, the corresponding values of e_m and e_M were determined and are given in Table 4.7. The results obtained from soil-water retention data are also consistent with MIP measurements in terms of trends but not of the exact magnitude. Both methodologies reveal that recompacted loess contains a larger portion of e_m compared to intact loess (Table 4.7). One possible explanation for the difference between the two approaches could be the derivation of S_r^m from water retention data in a region where limited data were obtained using axis-translation and vapour-equilibrium techniques. A complementary approach such as the osmotic control of suction is hence required to fill in the gap. On the other hand, the MIP equipment covers the corresponding range well for precise evaluation of delimiting pore size and hence MIP test results are employed to interpret the shearing behaviour.

188 Advanced Unsaturated Soil Mechanics

Table 4.7 Summary of MIP test results on recompacted and intact loess

Test ID	Specimen type	Suction: MPa	Delimiting pore diameter: um	d_M: um	d_m: um	e	e_{MIP}	e_{nd}	e_m	e_M
MR-0.1	R	0.06	1.059	11.338	0.096	1.094	0.906	0.188	0.231	0.863
MR-8		8	0.682	11.335	0.050	1.064	0.841	0.223	0.171	0.892
MR-230		230	0.841	13.947	0.040	1.055	0.849	0.206	0.168	0.887
MI-0.1	I	0.08	0.151	7.248	0.040	1.156	0.706	0.450	0.093	1.063
MI-8		8	0.121	7.248	0.026	1.135	0.692	0.444	0.076	1.060
MI-230		230	0.183	7.248	0.040	1.126	0.723	0.403	0.086	1.040

Notes:
R = Recompacted, I = Intact.
d_M = dominant macropore diameter; d_m = dominant micropore diameter; e = void ratio; e_{MIP} = cumulative MIP void ratio; e_{nd} = $e - e_{MIP}$: non-detected void ratio; e_m = microvoid ratio; e_M = $e - e_m$: macrovoid ratio.

Significance of macrovoid ratio

The microvoid and macrovoid ratios inferred from MIP test results are used to explain differences in the dilatancy and hence shear behaviour of recompacted and intact loess. Figure 4.27 shows the distribution of void ratios (i.e., e_m and e_M) for recompacted and intact loess specimens. Comparisons are made for three different suction values. The results clearly indicate that the distribution of void ratios is not subject to changes in suction for the range and path considered in this study. More importantly, differences between the microvoid and macrovoid ratios are more significant in the case of intact loess than in recompacted loess. Although both show a non-uniform distribution of voids, the non-uniformity is more pronounced for intact loess specimens. A similar observation was made by Haeri et al. (2016) after comparing SEM photos of undisturbed and recompacted loess. The results in Figure 4.27 confirm that intact loess contains a higher proportion of macrovoids than recompacted loess because of its smaller microporosity and larger population of large pores (Table 4.7). In addition, variations in the macrovoid ratio with suction confirm the independence of e_M from suction.

Figure 4.28 shows the influence of the macrovoid ratio on the peak dilatancy of unsaturated loess at different suctions and net stress levels. According to Figure 4.27a, the maximum dilatancy decreases as the macrovoid ratio increases under a net normal stress of 50 kPa. Although the maximum dilatancy is suppressed by increasing the net stress from 50 to 200 kPa (Figure 4.27b), it still decreases with an increase in macrovoid ratio. In other words, there is a clear reduction in the maximum dilatancy with an increase in the macrovoid ratio for all the tests conducted. It is evident that the macrovoid ratio can describe dilatancy, and hence the shear strength of unsaturated loess with a special aeolian microstructure unlike sedimentary and residual soils, is better than the void ratio.

Although recompacted and intact specimens exhibited similar void ratios, the differences in dilatancy and shear behaviour were very significant at all levels of applied net stress and suction. The microstructural quantifications reveal that differences in the distribution of voids between recompacted and intact loess were large even for the same target void ratio. Therefore, the void ratio may not be a suitable state parameter to describe shear behaviour. In fine-grained soils, silt-clay assemblages also known as aggregates are analogous to individual soil particles in a coarse-grained geomaterial. It is the voids between aggregates

Shear strength and behaviour 189

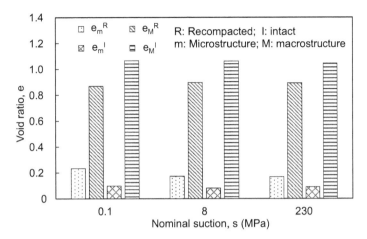

Figure 4.27 Microvoid and macrovoid ratios of recompacted and intact loess at different suctions.

(macrovoid ratio) that govern shear behaviour rather than the voids within individual aggregates (microvoid ratio). According to this hypothesis, silt-clay aggregates in unsaturated fine-grained soils are considered to be individual particles. The inter-aggregate space represents the macrovoid ratio while the intra-aggregate pores are representative of the microvoid ratio. The existence of this pore population may not affect the global behaviour provided that soil aggregates interact as individual rigid bodies during shearing. However, the volume of micropores in a soil with a particular void ratio will affect the volume of the associated macropores. Microstructural measurements revealed that the macrovoid ratio of intact loess is approximately 0.2 higher than that of recompacted loess (Figure 4.27). This notable difference in e_M may be responsible for the more dilative shear behaviour of the latter than the former under different combinations of net stress and suction. Therefore, the macrovoid ratio, e_M, is a more reliable state parameter than the void ratio, e, conventionally used to describe the behaviour of unsaturated fine-grained soil.

Summary

Macroscopic and microstructural evaluations of void ratios confirmed that the observed increase in dilatancy with increased suction is not governed by a change in void ratio but depends on the microstructure and the relationship between micro and macroporosity. Under similar test conditions, recompacted loess consistently showed higher dilatancy than that of intact loess.

At a given void ratio, recompacted and intact loess showed different distributions of micropores and macropores because they had different initial microstructures resulting from sample preparation and natural occurrence in the field, respectively. Analysis of the MIP test results confirmed that the distribution of voids was more uniform in recompacted loess than that in intact loess, because of a relatively larger number of large pores being present in the intact loess. It was therefore proposed that the macrovoid ratio, e_M, can be used to quantify the difference in their microstructures. The observed e_M of intact loess was about

190 Advanced Unsaturated Soil Mechanics

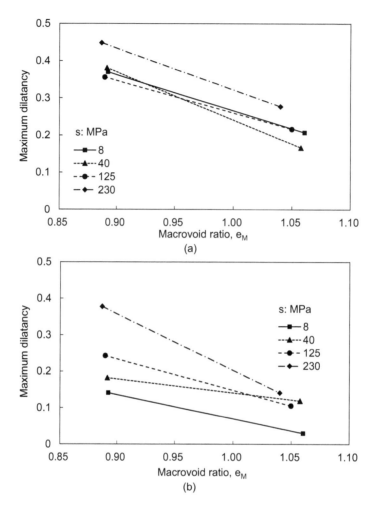

Figure 4.28 Effects of macrovoid ratio on maximum dilatancy at net normal stress of (a) 50 kPa, and (b) 200 kPa.

0.2 higher than that of recompacted loess, which is consistent with the measured lower dilatancy of intact rather than recompacted loess under given stress and suction conditions.

COMPARISONS OF THE SHEAR STRENGTH OF WEATHERED LATERITIC, GRANITIC AND VOLCANIC SOILS: COMPRESSIBILITY AND SHEAR STRENGTH (NG ET AL., 2019C)

Introduction

Lateritic soils are chemically weathered material, commonly formed and found in hot and wet tropical areas. A good understanding of their mechanical behaviour, such as compressibility and shear strength, can improve the design of civil engineering structures built on lateritic soils. By comparison with other tropical soils, lateritic soils are rich in secondary oxides of iron and aluminium (sesquioxides) that are formed during chemical weathering (Alexander and Cady, 1962).

In this section, the compressibility and shear behaviour of a lateritic soil are described and reported within the critical state framework. By comparison, the unique features of lateritic soil with very high sesquioxide contents are highlighted.

Soil types and grain size distribution

Three weathered soils were tested: a lateritic (LAT) soil, a completely decomposed volcanic (CDV) soil and a CDG soil. Their physical properties were determined following the ASTM standard (ASTM, 2011 and are summarised in Table 4.8. According to the Unified Soil Classification System (ASTM, 2011), LAT, CDV and CDG are classified as sandy clay of low plasticity (CL), silty sand of low plasticity (ML) and well-graded gravely sand with little fines, respectively.

Figure 4.29 shows their grain size distributions (GSDs) determined following the ASTM D1140 procedures (ASTM, 2014a). For each soil, GSDs were measured using both the dry sieving (sieving air-dried soil) and wet sieving methods (sieving soil in water, with dispersant added to separate soil particles from aggregates). For each soil, there is a huge difference between the GSDs measured using the two methods. This difference is widely used to quantify the degree of aggregation using the following equation (e.g., Otalvaro et al., 2015):

$$D_a = \frac{P_d - P_w}{P_w} \tag{4.13}$$

where D_a is the degree of aggregation, falling in the range of 0 to 1; P_d is the area below the GSD measured using the dry sieving method; P_w is the area below the GSD measured using the wet sieving method.

Using Equation (4.13), the values of D_a were 52%, 38%, and 21% for LAT, CDV and CDG, respectively. This observation suggests that in LAT, many of the fine particles are strongly attached to form large aggregates, which may not be destroyed by ordinary mechanical remoulding during specimen preparation. Otalvaro et al. (2015) also reported a significant degree of aggregation for a lateritic soil using a similar approach. The highest degree of aggregation of LAT is induced by some of its minerals which are not found in CDG and CDV. Details of this will be provided in a later section.

Mineralogy, chemical and microstructure analysis

The mineralogy, chemical compositions and microstructure of all soils were investigated using X-ray diffraction (XRD), X-ray fluorescence (XRF) and scanning electron microscopy (SEM) tests, respectively. The results were used to interpret the experimental results from compression and shear tests.

Figure 4.30 shows the types of minerals identified by the X-ray diffractometer in LAT, CDV and CDG. Some minerals such as quartz, haematite, and kaolinite are observed

Table 4.8 Physical properties of LAT, CDV and CDG

	LAT	CDV	CDG
Maximum dry density: kg/m³	1696	1540	1670
Optimum water content: %	20	21	20
Percentage of sand: %	42	25	53
Percentage of silt: %	16	65	12
Percentage of clay: %	42	10	14
Specific gravity	2.67	2.66	2.61
Liquid limit: %	44	48	N.A.
Plastic limit: %	24	35	N.A.
Plasticity index: %	20	13	N.A.
Soil classification based on USCS (ASTM, 2010)	Sandy clay (CL)	Silty sand (ML)	Gravelly sand (SW)

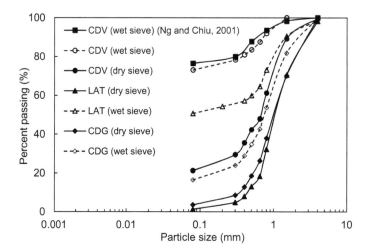

Figure 4.29 Grain size distribution of the studied soils.

in all soils, while goethite and hematite are only found in LAT. Both goethite and hematite are secondary minerals formed during the chemical decomposition of parent rocks on relatively flat ground (Gidigasu, 1976). Goethite is a hydroxide of iron and contains pairs of contiguous singly coordinated OH groups, and hematite is an oxide of iron. These two minerals enhance the formation and stability of soil aggregates which may influence soil behaviour (Schwertmann and Fitzpatrick, 1992; Larrahondo et al., 2011). The influence of these two minerals led to more significant aggregation (see Figure 4.29).

Chemical oxides in the three soils are determined through X-ray fluorescence (XRF) and are summarised in Table 4.9. It is clear that the oxides of iron, aluminium and silicon dominate the three soils. On a quantitative level, LAT contains more iron oxide and aluminium oxide than CDV and CDG. This difference is because the formation process of LAT involves the leaching of silica from the soil, leading to the relative accumulation of

Shear strength and behaviour   193

iron and aluminium oxides (Alexander and Cady, 1962). The oxides of iron and aluminium will provide cementation effects, alternatively described as cladding (Airey et al., 2012; Zhang et al., 2014). Because of cementation effects, the fine particles in LAT are aggregated.

The result from XRD and XRF tests can be used to explain the degree of aggregation reported in Figure 4.30. The highest degree of aggregation in LAT can be attributed to the higher quantity of sesquioxides and the presence of goethite. More discussion of the minerals and chemical oxides is given later when comparing the compressibility and shear strength of LAT, CDV and CDG.

Plate 4.6a shows a SEM photomicrograph of LAT. Fine clay particles are almost invisible, even though the clay content of LAT is up to 42%. This is mainly because clay particles have formed much larger aggregates. Moreover, the surface of the aggregates appears very rough and interlocking between aggregates is very significant.

A SEM photomicrograph of CDV is shown in Plate 4.6b, where many fine clay particles can be identified. Compared with LAT, there are many fewer aggregates in CDV. This difference is mainly because the amount of sesquioxides is much lower in CDV than in LAT, as shown by the results of XRF tests. With fewer sesquioxides and less significant cementation effects, fewer fine clay particles tend to form aggregates.

In Plate 4.6c for CDG, no sign of aggregation is observed. This is because of the low quantity of sesquioxides in CDG. The above observations in Plate 4.6 suggest that the microstructure of soils is closely related to mineral type and quantity of the sesquioxide.

Isotropic compression behaviour

Figure 4.31 shows the responses of LAT, CDV and CDG under isotropic compression. In the stress range considered, LAT and CDG exhibit a distinct yield point and a post-yield normal compression line. The compressibility increases substantially after yielding. For CDV, however, the compressibility does not show any obvious change during the loading process. The difference is probably because the initial void ratio of CDV is much larger than that of LAT and CDG, resulting in a much smaller yield stress (less than 20 kPa).

The slope of their normal consolidation line (NCL) (usually denoted as λ) is 0.07, 0.09 and 0.11 for LAT, CDV and CDG, respectively. This implies that LAT is less compressible than CDV and CDG. It should be noted that by comparison with CDV and CDG, the clay content of LAT is much larger (see Table 4.8). The compressibility of these three soils seems to contradict the conventional understanding that a finer-grained soil is generally more compressible than a coarser soil. The low compressibility of LAT is mainly attributable to its high content of sesquioxides (i.e., Iron and aluminium oxides). Iron and aluminium oxides enhance the formation and stability of soil aggregates, which control soil behaviour (Schwertmann and Fitzpatrick, 1992). In the current study, the LAT, CDV and CDG specimens have almost the same void ratio. The LAT specimen has more intra-aggregate pores but less inter-aggregate pores. Consequently, the compressibility associated with the rearrangement of soil aggregates is smaller. Another reason for the low compressibility of LAT may be goethite content (as evident from XRD tests), which is a hard material with a value of 5.5 on the Mohs scale of hardness (Mukherjee, 2012). Its existence in LAT probably stiffens the soil skeleton and reduces soil compressibility.

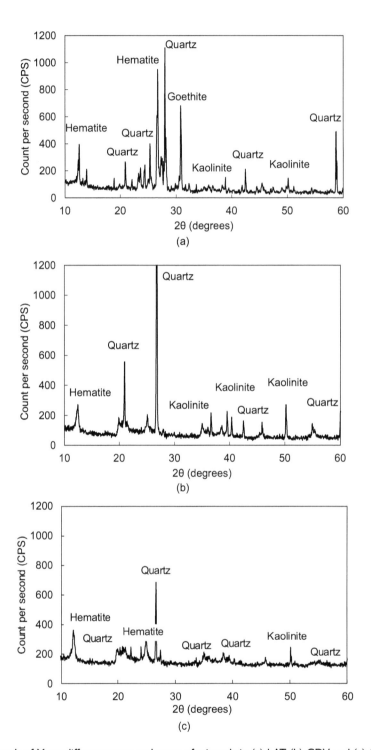

Figure 4.30 Result of X-ray diffractometer and types of minerals in (a) LAT (b) CDV and (c) CDG.

Table 4.9 Major chemical oxides in LAT, CDG and CDV determined by X-ray fluorescence tests

Oxide	LAT	CDV	CDG
Iron II oxide (Fe_2O_3)	10%	4%	2%
Aluminium oxide (Al_2O_3)	28%	23%	17%
Silicon oxide (SiO_2)	60%	70%	72%

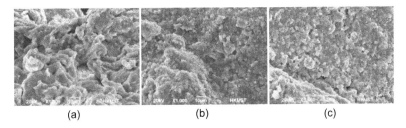

Plate 4.6 Scanning electron micrographs of (a) LAT; (b) CDV and (c) CDG.

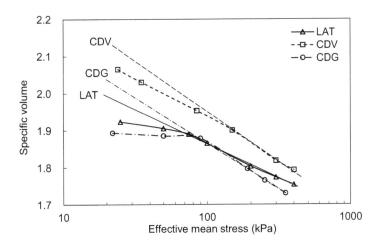

Figure 4.31 Isotropic compression behaviour of LAT, CDV and CDG.

Moreover, the few inter-aggregate pores and significant interlocking observed in the SEM micrograph of LAT also contribute to its low compressibility, as shown in Plate 4.6.

The compressibility of some other chemically weathered soils is summarised in Table 4.10 for comparison, including a lateritic gravel and a sandy clay from Singapore (Toll and Ong, 2003). It is clear that the LAT studied here is less compressible than the other two lateritic soils. The difference between the two lateritic soils can be explained by their parent rock types. The LAT is decomposed from a granitic rock (intrusive igneous rock) consisting of a high proportion of quartz (SiO_2), which is resistant to chemical weathering. On the contrary, the lateritic gravel from a Basalt (extrusive igneous rock) consists of little quartz but mainly of olivine and other mafic minerals, all of which readily degrade

196 Advanced Unsaturated Soil Mechanics

to clay minerals. By comparison with extrusive rocks, intrusive rocks have larger crystals resulting from the slower cooling of magma beneath the earth's surface which encourages the growth of larger crystals and causes a reduced pore size, thereby affecting compressibility (Loughnan, 1969).

Stress-strain relationship during shearing

For each soil, two series of tests were designed and carried out using a triaxial apparatus. The first series was isotropic compression tests to investigate soil compressibility. The other series were consolidated undrained (CU) tests to study shear strength. In CU tests on each soil, three different confining stresses (50, 100 and 200 kPa) were considered and used.

Figure 4.32 shows the stress-strain relationship for the specimen sheared in the undrained condition. For LAT and CDV sheared at 50 kPa, the deviator stress increases monotonically with the axial strain and no sign of softening was observed until the end of the test. This observation for LAT and CDV indicates that both soils exhibit strain hardening behaviour. For the CDG specimens sheared at all confining stresses, the deviator stress reached a peak value at an axial strain of about 5% and then decreases towards a steady state. The contrast between the behaviour of CDG and both LAT and CDV is explained below using the stress path. The behaviour of other specimens sheared at 100 and 200 kPa effective confining stresses is similar to the response described earlier at 50 kPa for each soil type.

Soil dilatancy

Figure 4.33 shows the stress paths of LAT, CDV and CDG under undrained shearing in the $q - p'$ plane. For LAT, the effective mean stress initially reduced (showing a tendency to contract) since the soil state lies on the "wet" side of the critical state line (CSL) before shearing. After reaching a turning point, a phase transformation occurred. Under subsequent shearing, the effective mean stress increased (showing a tendency to dilate, accompanied by an increase in deviator stress) and the soil state finally reached the CSL.

Table 4.10 Comparison of compressibility and friction angles of LAT, CDG and CDV

Soil types	λ	ϕ'_{cs}
LAT	0.07	42
CDV (Ng and Chiu, 2001)	0.09	33
CDG (Ng and Chiu, 2003)	0.11	38
Residual sandy clay (Toll and Ong, 2003)	0.08	31
Lateritic gravel (Toll, 1990)	0.11	38

Note:
All three soils are decomposed materials with similar degrees of weathering.

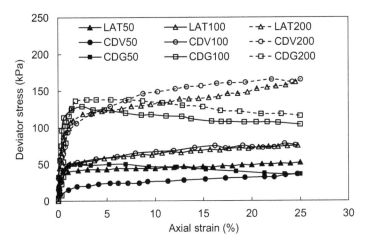

Figure 4.32 Triaxial stress-strain responses from undrained tests at confining stress of 50, 100 and 200 kPa for all specimens.

These results differ from those reported by Toll (1990), where a lateritic gravel exhibits a continuous dilation tendency during shearing. Futai et al. (2004) reported continuous contraction for a lateritic soil without observing any phase transformation.

CDV specimens also showed an initial tendency to contract, followed by a phase transformation to dilative behaviour. As for CDG, all specimens continued to show a tendency to contract until reaching the CSL. Because of the significant contraction, a reduction in mean confining stress and thus deviator stress was observed over the strain range between 5% and 25% (see Figure 4.32). Similar results for CDG were reported by Wang and Yan (2006) and Junaideen et al. (2010). On the other hand, the difference between CDG and LAT/CDV may be because there are large aggregates in LAT and CDV, but not in CDG, as illustrated in Plate 4.6. The initial contraction of the LAT and the CDV is probably attributable to the collapse of large inter-aggregate pores. The subsequent dilative behaviour is caused by rearrangement and interlocking of the large-sized aggregates.

Critical state shear strength

For each soil, the critical state stress state in the $q - p'$ plane is determined as shown in Figure 4.33. The stress ratio of the critical state line M is 1.73, 1.32 and 1.55 for LAT, CDV and CDG, respectively. The corresponding critical state angle of internal friction φ' is 42°, 33°, and 38° for LAT, CDV and CDG respectively. The differences in shear strength of LAT, CDV and CDG probably result from the content of large particles and other factors such as mineralogy and sesquioxide composition. As shown in Figure 4.29, the particle size of LAT is larger than that of CDV and CDG using the dry sieving method. The SEM images (see Plate 4.6) demonstrate that the large particles in LAT are embedded in a matrix of smaller particles. Moreover, the influence of large particles on shear strength was investigated in previous studies by testing soil-rock mixtures (Vallejo and Zhou, 1994; Verma et al., 2016). They found that the presence of large particles increased the shear

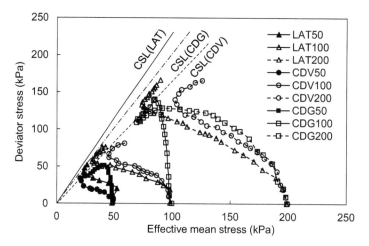

Figure 4.33 Stress paths of LAT, CDV and CDG in $q - p'$ space.

strength, mainly because the large particles increased the number of particle contacts and therefore enhanced particle interlocking. The influence of large particles on shear strength does not occur so strongly in the CDV and CDG samples. Consequently, the large particles in LAT probably result in a higher friction angle. In addition, it should be noted that if the wet sieve method is used, the content of large particles in LAT is reduced significantly. This is mainly because large aggregates are broken down by the addition of Sodium Hexametaphosphate in the wet sieve method (ASTM, 2014a). On the other hand, the high friction angle of LAT is also probably caused by the presence of goethite (iron hydroxide) which affects the surface texture of particles through the configuration of hydroxyl groups on its surface (Barron and Torrent, 1996) and hematite (iron oxide), which increases the particle sizes through aggregation of the fines. Goethite was also described by Airey et al. (2012) as a needle-like material capable of increasing the interlocking between soil particles.

In order to show how the goethite and sesquioxide compositions of lateritic soils influence their friction angle, the comparisons with other chemically weathered soils in the tropics are shown in Table 4.10. It is found that the lateritic soil studied has the highest friction angle. The difference between LAT and lateritic gravel (Toll, 1990) may be attributed to contributions from their parent rock. The surface of soil particles from intrusive granitic rock (LAT) is rough and harder compared to those derived from extrusive rocks already exposed to the earth's surface Loughnan (1969).

Summary

The compressibility and shear behaviour of three weathered soils (LAT (sandy clay), CDV (silty sand) and CDG (gravelly sand)) were compared. Of the three soil types, LAT has about 40% higher content of iron and aluminium oxides than CDG and CDV. These two oxides cause much greater aggregation in LAT, as evidenced by its dry and wet grain size distribution and SEM microphotographs. On the other hand, of these three soils, only LAT contains goethite (a hydroxide of iron), which is formed during the decomposition of the parent rocks. The goethite contains pairs of contiguous singly coordinated OH groups, which are involved in surface adsorption and can enhance inter-aggregate interlocking. It is found that the critical state friction angles of LAT, CDV and CDG are 42°, 33° and 38°, respectively. The highest friction angle of LAT is probably caused by the content of large particles and the presence of goethite. Because of the large particles, the number of particle contacts is increased, enhancing particle interlocking. The goethite mineral found in LAT could also enhance inter-aggregate interlocking through its contiguous singly coordinated OH groups.

SOURCES

Extended Mohr– Coulomb failure criterion (Fredlund and Rahardjo, 1993)
  Comparisons of axis- translation and osmotic techniques for shear testing
  (Ng et al., 2007)
Effects of soil suction on dilatancy (Ng and Zhou, 2005)
Effect of microstructure on shear strength and dilatancy of loess at high suction
  (Ng et al., 2016a; Ng et al., 2020b)
Comparisons of shear strength of weathered lateritic, granitic and volcanic soils:
  compressibility and shear strength (Ng et al., 2019c)

Chapter 5

Shear stiffness and damping ratio

OVERVIEW

The most distinctive features of soils are their strain- and stress path-dependent deformation characteristics and stiffness properties (Ng, 1993). Even at small strains (i.e., 0.001% to 1%), soils exhibit nonlinear behaviour. The nonlinear small-strain deformation characteristics and properties of dry and saturated soils are well documented in the geotechnical engineering and soil testing literature (Jardine et al., 1984; Burland, 1989; Mair, 1993; Ng and Lings, 1995; Atkinson, 2000; Clayton, 2011). It is now widely accepted that the stress-strain relationship of unsaturated and saturated soils is highly nonlinear (Atkinson and Sällfors, 1991; Simpson, 1992; Tatsuoka and Kohata, 1995; Ng and Menzies, 2007). Constructing retaining walls, foundations and tunnels in saturated stiff soils often results in a range of strains over which there is a large variation in soil stiffness (Mair, 1993; Ng and Lings, 1995). A summary of typical mobilised strain ranges for saturated stiff soils under working load conditions is given in Figure 5.1. An understanding of soil's response at low levels of strain is clearly very important, not only for earthquake problems but also because it has practical significance and applications for design and analysis in geotechnical engineering. For ease of identification, Atkinson and Sällfors (1991) define three ranges of strain: very small strain (0.001% or less), small strain (between 0.001% and 1%) and large strains (greater than 1%). These definitions are adopted in this section.

Although substantial research into soil stiffness has been carried out using saturated soils (e.g., Atkinson, 2000; Clayton, 2011), only a relatively limited number of studies have been conducted to determine the soil stiffness of unsaturated soils. In this chapter, laboratory and theoretical studies on the behaviour of unsaturated soils under small strains are reported.

ANISOTROPIC SHEAR STIFFNESS AT VERY SMALL STRAINS (NG AND YUNG, 2008)

Introduction

Pennington et al. (1997) introduced an additional horizontally mounted bender element probe consisting of two pairs of orthogonally oriented bender elements in a triaxial apparatus. The probe can measure the velocities of horizontally propagated shear waves with horizontal and vertical polarisations and also conventional vertically propagated shear

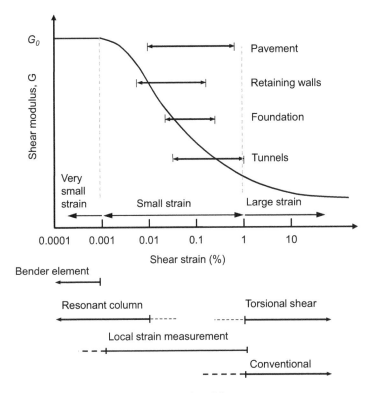

Figure 5.1 Approximate strain limits for soils surrounding different structures.

Source: After Mair, 1993.

waves with horizontal polarisation, within a single soil specimen. The development of this additional horizontally mounted bender element probe enables the evaluation of anisotropic shear moduli in different planes within soils (Ng et al., 2004b).

Theoretical considerations

In dry and saturated soils, the shear wave velocity is known to depend on the effective stresses in the directions of wave propagation and particle motion. The shear wave velocity is known to be almost independent of any effective stress normal to the plane of shear (Roesler, 1979; Knox et al., 1982; Jamiolkowski et al., 1995; Rampello et al., 1997). The shear wave velocity and the small-strain shear modulus are also known to depend on the void ratio. Different void ratio functions have been proposed to describe the influence of the void ratio on shear wave velocity and shear modulus in saturated soils (Hardin and Drnevich, 1972; Iwasaki et al., 1978; Lo Presti, 1989). The velocity $v_{s(ij)}$, of shear waves propagating in dry and saturated soils in the direction of wave propagation, i, and in the direction of particle motion, j, polarised along the principal stress directions ij, can be expressed as

202 Advanced Unsaturated Soil Mechanics

$$v_{s(ij)} = C_{ij} f(e) \left(\frac{\sigma_i - u_w}{p_r} \times \frac{\sigma_j - u_w}{p_r} \right)^{\frac{ns}{2}} \qquad (5.1)$$

where C_{ij} is a material constant (with the dimension of velocity) related to the $i - j$ plane. It is a function of the size, shape and texture of the soil particles, as well as of particle arrangement and inter-particle bonding (if any). $f(e)$ is a void ratio function describing the dependence of shear wave velocity on void ratio, p_r is a reference stress, $\sigma_i - u_w$ and $\sigma_j - u_w$ are the principal effective stresses in the plane in which $v_{s(ij)}$ is measured, and ns is an empirically derived constant.

Applying the two stress state variables, $\sigma - u_a$, and $u_a - u_w$, for unsaturated soils, it is intuitively reasonable to assume the following velocity expression for shear waves travelling in unsaturated soils:

$$v_{s(ij)} = C_{ij} f(e) \left[\frac{(\sigma_i - u_a)}{p_r} \times \frac{(\sigma_j - u_a)}{p_r} \right]^{\frac{ns}{2}} \left[1 + \frac{(u_a - u_w)}{p_r} \right]^{bs} \qquad (5.2)$$

where bs is an empirical exponent reflecting the influence of matric suction on shear wave velocity. When the soil is saturated (i.e., $u_a - u_w = 0$), Equation (5.2) reduces to Equation (5.1). Equation (5.2) allows a smooth transition between an unsaturated soil and a saturated soil. Assuming that the reference pressure, p_r, is 1 kPa, and the void ratio function, $f(e)$, is in the form of an exponential function, e^{ae}, where ae is an empirical void ratio exponent, Equation (5.2) can be rewritten as

$$\frac{v_{s(ij)}}{f(e)} = \frac{v_{s(ij)}}{e^{ae}} = C_{ij} \left[\frac{\sigma_i - u_a}{p_r} \times \frac{\sigma_j - u_a}{p_r} \right]^{\frac{ns}{2}} \left[1 + \frac{u_a - u_w}{p_r} \right]^{bs} \qquad (5.3)$$

Assuming that the strains induced by bender elements in an elastic soil are very small (Dyvik and Madshus, 1985) and knowing the current tip-to-tip distance, L_i, between the transmitting and receiving bender elements (Viggiani and Atkinson, 1995a, 1995b), the velocity of shear waves can be calculated from:

$$v_{s(ij)} = \frac{L_i}{t_{ij}} \qquad (5.4)$$

where t_{ij} is the measured travel time of a shear wave propagating in the polarisation plane, ij, over a distance L_i. Hence, the elastic shear modulus at very small strains, $G_{0(ij)}$, may be determined from:

$$G_{0(ij)} = \rho v_{s(ij)}^2 \qquad (5.5)$$

where ρ is the bulk density of soil, which can be expressed in terms of void ratio, e, and degree of saturation, S_r, as follows:

$$\rho = \left(\frac{G_s + S_r e}{1 + e}\right)\rho_w \qquad (5.6)$$

where G_s is the specific gravity of the soil and ρ_w is the density of water. By measuring the change in water content and volume of a soil specimen, the soil density can be obtained using Equation (5.6) and hence $G_{0(ij)}$ can be determined.

Combining Equations (5.2), (5.5) and (5.6), the elastic shear modulus, $G_{0(ij)}$, can be expressed as

$$G_{0(ij)} = C_{ij}F(e)\left[\frac{\sigma_i - u_a}{p_r} \times \frac{\sigma_j - u_a}{p_r}\right]^{ns}\left[1 + \frac{u_a - u_w}{p_r}\right]^{2bs} \qquad (5.7)$$

where

$$F(e) = \frac{\left[f(e)\right]^2 (G_s + S_r e)}{1 + e}\rho_w \qquad (5.8)$$

F(e) can be considered to be a function of void ratio relating shear modulus to void ratio.

If the degree of anisotropy of unsaturated soils is expressed in terms of the ratio of shear stiffness between the shear modulus in the horizontal plane $(G_{0(hh)})$ and that in the vertical plane $(G_{0(hv)})$ in a triaxial test, then by making use of Equation (5.7), the shear stiffness ratio can be derived as follows:

$$\frac{G_{0(hh)}}{G_{0(hv)}} = \left(\frac{v_{s(hh)}}{v_{s(hv)}}\right)^2 = \frac{C_{hh}^2}{C_{hh}^2}\left(\frac{\sigma_h - u_a}{\sigma_v - u_a}\right)^{ns} \qquad (5.9)$$

where $v_{s(hh)}$ is the velocity corresponding to a shear wave propagating horizontally with horizontal polarisation, and $v_{s(hv)}$ is the velocity corresponding to a shear wave propagating horizontally with vertical polarisation. It can be seen from Equation (5.9) that the ratio of shear modulus equals the square of the ratio of shear wave velocity, $(v_{s(hh)}/v_{s(hv)})^2$. Thus, the degree of stiffness anisotropy of unsaturated soils depends on the soil's inherent material properties, C_{hh}^2/C_{hv}^2, and on the ratio of the net normal stresses in the horizontal and vertical directions, $(\sigma_h - u_a)/(\sigma_v - u_a)$. The degree of anisotropy in the stiffness is predicted to be independent of matric suction by Equation (5.9). For saturated soils (i.e., $u_a - u_w = 0$) with an isotropic fabric (i.e., $C_{hh}^2/C_{hv}^2 = 1$), the degree of stiffness anisotropy should, according to Equation (5.9), depend solely on the stress ratio, $(\sigma_h - u_a)/(\sigma_v - u_a)$, which is commonly recognised as stress-induced anisotropy.

204 Advanced Unsaturated Soil Mechanics

On the other hand, under isotropic stress conditions (i.e., $(\sigma_h - u_a)/(\sigma_v - u_a) = 1$), Equation (5.9) can be simplified as:

$$\frac{G_{0(hh)}}{G_{0(hv)}} = \left(\frac{v_{s(hh)}}{v_{s(hv)}}\right)^2 = \frac{C_{hh}^2}{C_{hv}^2} \tag{5.10}$$

This suggests that the degree of stiffness anisotropy expressed in terms of the ratio of shear moduli is independent of any net normal stress and matric suction in this isotropic stress state. It depends only on the anisotropic inherent material properties.

Testing apparatus and measuring devices

To measure the velocity of shear waves propagating in two planes with different polarisations and hence to determine the anisotropic shear moduli of a soil specimen, a pair of bender elements is fixed to the two end platens and two pairs of bender elements are mounted at the mid-height of the specimen (refer to Plate 4.2a). The arrangement of the bender elements is shown schematically in Figure 5.2. Excited by an external voltage, the transmitting bender element deflects and acts as a source to release energy into the soil specimen and generate shear waves propagating through the soil. On the opposite end of the soil specimen, the shear waves are received by a receiver element, which is aligned so as to detect the transverse motion of the shear waves. The signals of shear waves transmitted and received after propagating through a soil specimen in two different directions are captured by an oscilloscope (HP 3563A system control analyser). For each measurement during a test, the input wave is a single sinusoidal pulse with a frequency of 4 to 10 kHz. The frequency range is selected to obtain as clear a signal as possible and to minimise the near-field effect (Sanchez-Salinero et al., 1986). The arrival time of each wave was taken as the peak-to-peak time, which Viggiani & Atkinson (1995b) have shown to be close to the arrival time derived by numerical analyses of the transmitted and received signals.

A vertically transmitted shear wave with horizontal polarisation ($v_{s(vh)}$) is generated and received by a pair of bender elements embedded in the end platens of the triaxial cell. The bender elements are mounted as described by Dyvik and Madshus (1985). The bender element is cut to a length of 12 mm and a width of 8 mm, with half of its length embedded in the base platen and the remaining half protruding into the soil specimen. It is coated with epoxy resin for insulation (refer to Plate 5.1).

Any horizontally transmitted shear waves, $v_{s(hv)}$ and $v_{s(hh)}$, are generated and received by a pair of bender element probes (see Figure 5.2). Each bender element probe consists of two pieces of bender element with a length of 9 mm and a width of 6 mm potted orthogonally into a short length of plastic tube. The bender elements protrude about 3 mm into the soil specimen. The probe is fitted at the mid-height of the soil specimen by means of a tailor-made silicon grommet (as shown in Plate 5.1), holding the probe in position and preventing the specimen from leaking. The slot cut in the calliper of a Hall-effect radial belt allows the cables of the bender elements to pass through the apparatus.

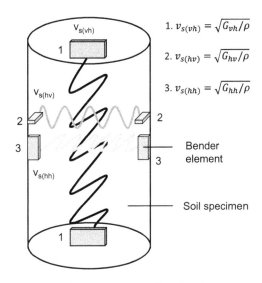

Figure 5.2 Schematic diagram showing the arrangement of bender elements.
Source: After Ng and Yung, 2008.

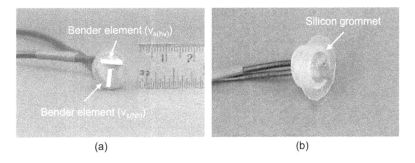

Plate 5.1 Overview of the bender element probe.
Source: Ng et al., 2004b.

In addition to the conventional external measurements of axial strain using a Linear Variable Differential Transformer (LVDT), the triaxial apparatus is equipped with three Hall-effect transducers (Clayton et al., 1989) to measure local axial and radial displacements. These results are used to determine how the volume of a soil specimen changes throughout a test. As a result, the current tip-to-tip travel distance (Equation 5.4) and the current density of the soil specimen (Equation 5.6) could be determined accurately during each stage of the test.

Table 5.1 Index properties of CDT

Index test	Value
Standard compaction tests	
Maximum dry density: kg/m^3	1760
Optimum water content: %	16.3
Grain size distribution	
Percentage of sand: %	24
Percentage of silt: %	72
Percentage of clay: %	4
D_{10} (mm)	0.003
D_{30} (mm)	0.006
D_{60} (mm)	0.015
Coefficient of uniformity D_{60}/D_{10}	4.55
Coefficient of curvature $(D_{30})^2/(D_{60} \times D_{10})$	0.61
Specific gravity	2.73
Atterberg limits (grain size < 425μm)	
Liquid limit: %	43
Plastic limit: %	29
Plasticity index: %	14

Source: After Ng and Yung, 2008.

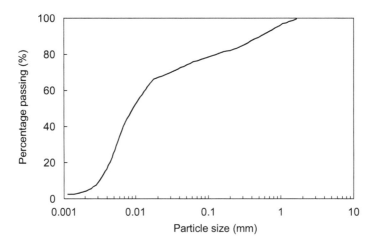

Figure 5.3 Particle size distribution of completely decomposed tuff (CDT).

Source: **After Ng and Yung, 2008.**

Soil types and test specimen preparation

The material tested was a completely decomposed tuff (CDT) from Hong Kong. The material can be described as clayey silt (ML) according to the Unified Soil Classification System. Figure 5.3 shows the particle size distribution of the CDT as determined by sieve and hydrometer analyses (British Standards Institution, 1990). The material was yellowish-brown, slightly plastic, with a very small percentage of fine and coarse sand. The liquid limit and plasticity index of the fines portion (finer than 425 μm) were

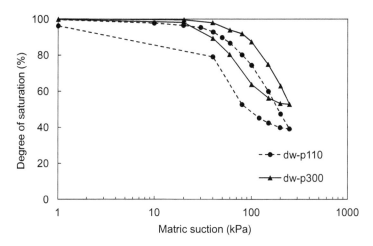

Figure 5.4 Soil-water characteristic curve of recompacted specimens of CDT.
Source: After Ng et al., 2009.

43% and 14%, respectively. The physical properties of the material are summarised in Table 5.1.

Figure 5.4 shows the stress-dependent SWCC (SDSWCCs) of CDT at net stresses of 110 and 300 kPa. The soil specimen subjected to a higher net mean stress tends to have a higher air-entry value. The air-entry values of CDT are estimated to be 55 and 85 kPa at net mean stresses of 110 and 300 kPa, respectively. Further, the specimen subjected to 300 kPa net mean stress shows a lower desorption rate than the one under 110 kPa net mean stress. There is a marked hysteresis between the drying and wetting curves for both specimens. The size of the hysteresis loop appears to be smaller at higher net mean stress. Insights into the effects of stress on SWCC can be found in Chapter 3.

Test programme and procedures

To determine the effects of the inherent material properties induced during sample preparation (refer to Equation (5.10)), a series of isotropic compression tests were conducted. Figure 5.5 shows the initial preparation paths (dotted lines) and test paths (solid lines) for four suction-controlled isotropic compression tests at constant values of matric suction ranging from 0-200 kPa (i.e., i-0, i-50, i-100 and i-200). In order to show the greatest possible effect of suction on soil stiffness, the selection of suction range was based on consideration of the measured air-entry value and the shape of the SWCC shown in Figure 5.4. Details of the testing conditions are summarised in Table 5.2. As illustrated in Figure 5.5, measurements of shear wave velocity (and hence shear stiffness) were carried out at preselected stress states as each isotropic loading proceeded.

Each isotropic compression test consisted of two stages: suction equalisation and compression. The equalisation stage ensured that water moisture reached its equilibrium state

208 Advanced Unsaturated Soil Mechanics

Table 5.2 Summary of testing conditions for isotropic compression tests on CDT

Test ID	Suction: kPa	Stress range: kPa	Initial water content: %	Initial dry density: kg/m³	Relative compaction: %	Void ratio e_0	e_r	e_f
1–0	0	110–400	16.1	1672	95.0	0.632	0.574	0.483
1–50	50	110–500	16.6	1649	93.7	0.656	0.604	0.535
1–100	100	110–500	16.3	1679	95.4	0.626	0.579	0.522
1–200	200	110–500	16.4	1677	95.3	0.628	0.585	0.538

Source: After Ng and Yung, 2008.

Note:
e_0 = initial void ratio; e_r = void ratio at net mean stress of 110 kPa; e_f = void ratio at the maximum net mean stress.

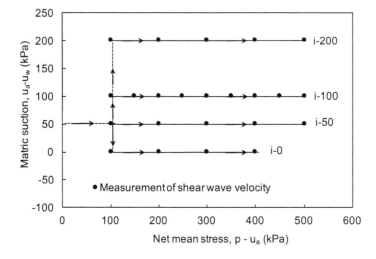

Figure 5.5 Stress paths of triaxial tests conducted on recompacted specimens of CDT.

Source: After Ng and Yung, 2008.

throughout a soil specimen, i.e., pore-water and pore-air pressure were equalised inside the voids of a specimen at each pre-selected stress state. Each equalisation stage generally required 4 to 7 days to complete. Once a specimen was equalised at its specified initial suction, it was then isotropically compressed to a required net mean stress, $(p - u_a)$, at a constant suction, where p is total mean stress. While maintaining constant pore-water and pore-air pressures, the cell pressure was ramped up at 3 kPa per hour to the required target value. This rate of compression was selected to minimise any excess pore-water pressure generated in a soil specimen. This compression rate was reported by Ng and Chiu (2001) to be slow enough to ensure fully drained conditions by monitoring both changes of water flow and volume change of a soil specimen at the end of each compression stage.

Interpretations of experimental results

Effects of changes in soil suction and net mean stress on void ratio

By measuring the local axial and radial strains using Hall-effect transducers, the volume changes of a soil specimen and thus the void ratio were evaluated during the suction-controlled isotropic compression tests. Since the void ratio, e_r, of each specimen at the start of compression at 110 kPa is variable (see Table 5.2), the measured void ratio along each compression path is normalised by its corresponding e_r and is plotted against the logarithm of net mean stress $(p - u_a)$ in Figure 5.6 for comparison. As expected, the normalised void ratio, e/e_r, decreased with an increase in $p - u_a$ at all suctions but at different rates. No distinct yield stress can be identified, especially in the tests with higher suction. For a given change of net mean stresses, soil specimens compressed at a higher constant suction showed a smaller change in normalised void ratios, i.e., a smaller gradient. This demonstrates that matric suction stiffens the soil skeleton of the unsaturated soil during isotropic compression tests over the suction range considered. As shown in Equation (5.7), the void ratio is related to the shear wave velocity. By determining the void ratio, the void ratio function, $f(e)$, can be obtained with known values of matric suction and net normal stress at different states.

Relationships between shear wave velocities ($v_{s(vh)}$, $v_{s(hv)}$, $v_{s(hh)}$) and matric suction

As shown in Figure 5.7, measurements of shear wave velocity were carried out during isotropic compression at constant suctions. Relationships between shear wave velocity and matric suction can be obtained by plotting the measured shear wave velocities in three different polarisation planes against matric suction at different net mean stresses.

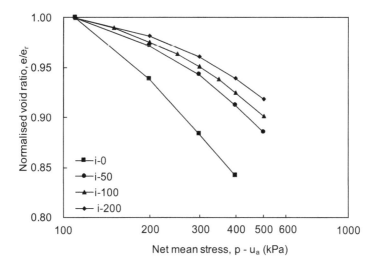

Figure 5.6 Variation of normalised void ratio, e/e_r, with net mean stress.

Source: After Ng and Yung, 2008.

210 Advanced Unsaturated Soil Mechanics

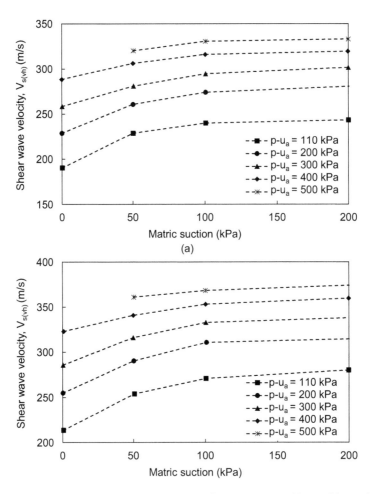

Figure 5.7 Relationships between shear wave velocities and matric suction: (a) $v_{s(vh)}$; (b) $v_{s(hv)}$; (c) $v_{s(hh)}$.

Source: After Ng and Yung, 2008.

The measured shear wave velocities, $v_{s(vh)}$, $v_{s(hv)}$ and $v_{s(hh)}$, increased nonlinearly with increasing matric suction. As suction increased from 0 to 50 kPa, which was close to the estimated air-entry value of the soil (see Figure 5.4), for any given value of $p - u_a$, the measured shear wave velocities in all polarisation planes increased substantially. Beyond a suction of 50 kPa, the measured shear wave velocities continued to increase but at a reduced rate up to a suction of 200 kPa, i.e., the gradients of the curves decreased with an increase in matric suction. The observed significant increase in shear wave velocity for suctions up to the air-entry value was because each specimen remained essentially saturated and bulk-water effects dominated soil stiffness (Mancuso et al., 2002). Any increase in suction was practically equivalent to an increase in mean effective stress resulting in much stiffer soil. On the other hand, as the suction increased beyond the air-entry value, each soil specimen started to desaturate, resulting in the formation of air-water menisci (or contractile

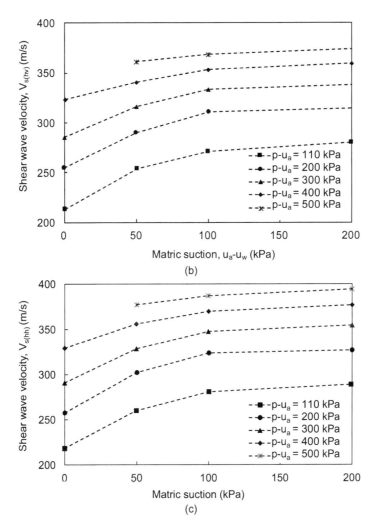

Figure 5.7 (Continued)

skins) at contact points between soil particles. Menisci water caused an increase in the normal force holding the soil particles together, leading to stiffer soil and hence a higher shear wave velocity. However, this beneficial effect on shear wave velocity did not increase indefinitely (Mancuso et al., 2002) but tends towards a limiting value, as the radius of the menisci progressively becomes smaller as the suction increases beyond 100 kPa as shown in Figure 5.7.

Not only does a change of shear wave velocity depend on a change in matric suction but it also depends on net mean stress. At a net mean stress of 110 kPa, the measured shear wave velocities, $v_{s(vh)}$, $v_{s(vh)}$ and $v_{s(hh)}$ increased by 28.5%, 31.0% and 33.8% as the suction increased from 0 to 200 kPa, respectively. On the other hand, at net mean stress equal to 400 kPa, the measured shear wave velocities, $v_{s(vh)}$, $v_{s(vh)}$ and $v_{s(hh)}$ increased by only 10.8%, 12.2% and 15.0% respectively for the same suction change. This clearly indicates that the

212 Advanced Unsaturated Soil Mechanics

behaviour or property (i.e., stiffness) of unsaturated soil depends on two stress state variables, matric suction and net mean stress. As the net mean stress increases, the influence of a given change in suction on shear wave velocity becomes less effective as the soil is stiffer at higher stress.

Influence of suction on the degree of fabric anisotropy

To eliminate differences in frequency and boundary effects, $v_{s(hh)}$ and $v_{s(hv)}$ were chosen for comparison because they were generated by the same bender element probe and had the same travelling length and boundary conditions (Pennington et al., 2001; Ng et al., 2004b). The degree of stiffness anisotropy is therefore expressed in terms of the ratio of the shear modulus in the horizontal plane ($G_{0(hh)}$) to that in the vertical plane ($G_{0(hv)}$) in

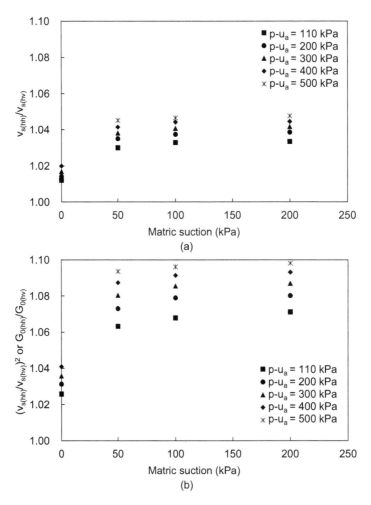

Figure 5.8 Influence of matric suction on: (a) $v_{s(hh)}/v_{s(hv)}$; (b) $(v_{s(hh)}/v_{s(hv)})^2$ or $G_{0(hh)}/G_{0(hv)}$.

Source: After Ng and Yung, 2008.

Shear stiffness and damping ratio   213

this study. Figure 5.8 shows the ratios, $v_{s(hh)}/v_{s(hv)}$ and $\left(v_{s(hh)}/v_{s(hv)}\right)^2$ respectively, plotted

against matric suction. It should be noted that the ratio, $\left(v_{s(hh)}/v_{s(hv)}\right)h2$ is equal to the

shear stiffness ratio, $G_{0(hh)}/G_{0(hv)}$, as demonstrated by Equation (5.9). Under isotropic stress

conditions, the velocity of the horizontally polarised shear wave $v_{s(hh)}$ was higher than that of the vertically polarised shear wave $v_{s(hv)}$ by about 1% to 2% at zero suction as shown in Figure 5.8a. The higher the applied net mean stress, the larger the ratio. Correspondingly, the deduced shear modulus $G_{0(hh)}$, was about 2.5% to 4% higher than $G_{0(hv)}$ when matric suction was zero. Equation (5.10) suggests that this observed stiffness anisotropy can be attributed to the inherent (fabric) anisotropy (i.e., C_{hh}^2/C_{hv}^2) of the tested material during specimen preparation. The higher the net mean stress applied, the greater the anisotropic stiffness induced.

As the suction increased to 50 kPa, the ratio of $v_{s(hh)}/v_{s(hv)}$ and $\left(v_{s(hh)}/v_{s(hv)}\right)^2$ increased

by about 3.5% and 7.5% on average, respectively. There was no significant increase in the velocity and stiffness ratios when the suction was 100 kPa or higher. The observed

suction-induced increase in $\left(v_{s(hh)}/v_{s(hv)}\right)^2$, which is equal to $G_{0(hh)}/G_{0(hv)}$, is consistent with

Equation (5.10), which suggests that matric suction will induce stiffness anisotropy (i.e., anisotropic shear modulus).

Influence of net mean stress on the degree of anisotropy

Figures 5.9a and 5.9b show the measured ratios, $v_{s(hh)}/v_{s(hv)}$ and $\left(v_{s(hh)}/v_{s(hv)}\right)^2$, plotted

against net mean stress, respectively. The ratios, $v_{s(hh)}/v_{s(hv)}$ and $\left(v_{s(hh)}/v_{s(hv)}\right)^2$, appeared

to increase gently with an increase in net mean stress at almost the same rate at different suctions (i.e., the rate of increase is independent of suction). There was about a 1% to 3%

increase in the stiffness ratio, $G_{0(hh)}/G_{0(hv)}$, which is equal to $\left(v_{s(hh)}/v_{s(hv)}\right)^2$, when the net

mean stress increased from 110 to 500 kPa. Although it is generally recognised that shear stiffness will increase with an increase in the mean effective stress for saturated soils, it is somewhat surprising that the stiffness ratio also increases with net mean stress. This observed behaviour may be attributed to coupling effects between the hydraulic and mechanical characteristics of unsaturated soils (Wheeler et al., 2003).

Summary

During suction-controlled isotropic compression tests, the three shear wave velocities increased with increasing suction for a given stress state in a nonlinear fashion with a reduction in the rate of increase when the applied suction was higher than the air-entry value of the soil.

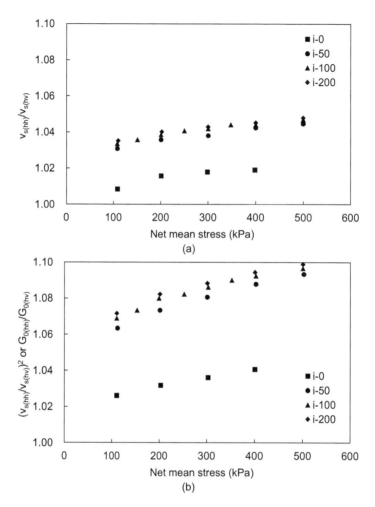

Figure 5.9 Influence of net mean stress on: (a) $v_{s(hh)}/v_{s(hv)}$; (b) $(v_{s(hh)}/v_{s(hv)})^2$ or $G_{0(hh)}/G_{0(hv)}$.

Source: After Ng and Yung, 2008.

By using the theoretical equations derived in this section, it was possible to identify that there is an inherent stiffness anisotropy of the soil with measured $G_{0(hh)}$ at about 2.5% to 4% higher than $G_{0(hv)}$, caused by the soil fabric formed during sample preparation by moist tamping. Moreover, there was a further 7.5% anisotropy in terms of shear stiffness ratio, $G_{0(hh)}/G_{0(hv)}$, induced by suction when the suction was increased from 0 kPa to its air-entry value of 50 kPa. No significant suction-induced stiffness anisotropy was observed at a suction higher than the air-entry value. Although it was intended to eliminate stress-induced anisotropy by carrying isotropic tests, a small increase in the shear stiffness ratio,

$G_{0(hh)}/G_{0(hv)}$, of 1% to 3% was observed when the net mean stress increased from 110 to 500 kPa. This observed behaviour may be attributed to the coupling effects between the hydraulic and mechanical characteristics of the unsaturated soil.

EFFECTS OF CURRENT SUCTION RATIO AND RECENT SUCTION HISTORY ON SHEAR STIFFNESS (NG AND XU, 2012)

Introduction

For a given soil, the shear modulus is affected by the current stress magnitude, stress paths and overall stress history described by the over-consolidation ratio (OCR) (Hardin and Drnevich, 1972, Atkinson et al., 1990). Another factor influencing the small-strain shear modulus is the recent stress history, which is defined as the penultimate stress path before the current stress path (Atkinson et al., 1990). It has been reported that the shear modulus of London clay may be increased by up to an order of magnitude following a change in the direction of the stress path (Atkinson et al., 1990).

Due to changing weather and other environmental conditions, compacted soils in the field always undergo various wetting-drying cycles and hence have various suction histories. The effects of suction magnitude and suction history on the shear modulus reduction curve have rarely been reported in the literature. In this section, suction magnitude refers to the magnitude of the current suction, as illustrated in Figure 5.10. Suction history refers to the current suction ratio (CSR) and recent suction history. The current suction ratio (CSR) is defined as the maximum historical suction experienced by a soil divided by the current suction. CSR= $s_{max}/s_{current}$, where $s = (u_a - u_w)$ is the matric suction. Recent suction history refers to the influence of the penultimate suction path on soil behaviour along the current stress path. The direction of the recent suction path is described by the

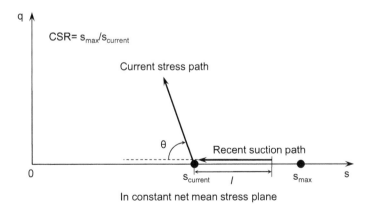

Figure 5.10 Illustration for CSR and recent suction history.

Source: After Ng and Xu, 2012.

216 Advanced Unsaturated Soil Mechanics

angle of rotation from the direction of the penultimate suction path to the direction of the current stress path, denoted by θ in the figure. The magnitude of the recent suction path is quantified by the amount of change in suction in the penultimate suction path, denoted by l.

Theoretical considerations

The elastic stress-strain relationship for a cross-anisotropic saturated soil under axisymmetric stress conditions can be described as follows (Graham and Houlsby, 1983, Atkinson et al., 1990):

$$\begin{bmatrix} \delta\varepsilon_v \\ \delta\varepsilon_s \end{bmatrix} = \begin{bmatrix} \dfrac{1}{K} & \dfrac{1}{J_{qp}} \\ \dfrac{1}{J_{pq}} & \dfrac{1}{3G} \end{bmatrix} \begin{bmatrix} \delta p' \\ \delta p \end{bmatrix} \tag{5.11}$$

where ε_v is the volumetric strain, ε_s is the shear strain (or deviatoric strain); K is the bulk modulus, G is the shear modulus, J_{pq} and J_{qp} are moduli cross-coupling shearing and volumetric effects; p' is the mean effective stress and q is the deviatoric stress.

By assuming that the behaviour of unsaturated soil is governed by the net normal stress and matric suction only, Equation (5.11) may be modified for a cross-anisotropic unsaturated soil in triaxial space:

$$\begin{bmatrix} \delta\varepsilon_v \\ \delta\varepsilon_s \end{bmatrix} = \begin{bmatrix} \dfrac{1}{K} & \dfrac{1}{J_{qp}} \\ \dfrac{1}{J_{pq}} & \dfrac{1}{3G} \end{bmatrix} \begin{bmatrix} \delta p \\ \delta q \end{bmatrix} + \begin{bmatrix} \dfrac{1}{\bar{K}} \\ 0 \end{bmatrix} \delta s \tag{5.12}$$

where

$$\begin{cases} \begin{bmatrix} p \\ q \end{bmatrix} = \begin{bmatrix} \dfrac{1}{3} & \dfrac{2}{3} \\ 1 & -1 \end{bmatrix} \begin{bmatrix} \sigma_a - u_a \\ \sigma_r - u_a \end{bmatrix} \\ s = u_a - u_w \end{cases} \tag{5.13}$$

where p is the net mean stress and s is the matric suction; $\bar{K}$ is the suction modulus, which is defined as the modulus of elasticity with respect to a change in matric suction. In a constant net mean stress test under constant suction conditions, $\delta p = 0$ and $\delta_s = 0$ in Equation (5.12). Then, the shear modulus of an unsaturated soil can be determined using the following equation:

$$G = dq/3d\varepsilon_s \tag{5.14}$$

Therefore, constant net mean stress compression tests under constant suction are carried out to determine the shear modulus at small strains in this study.

Experimental study

In order to investigate the effects of suction magnitude, CSR and recent suction history on the small-strain shear modulus of an unsaturated compacted soil, two series of constant net mean stress (p) compression tests with different suction histories were performed on recompacted CDT using the triaxial apparatus shown in Plate 4.2.

Series 1 tests were carried out to study the effects of suction magnitude and CSR on the small-strain stiffness of CDT. The stress paths in (p: q: s) space for series 1 tests are shown in Figure 5.11. Each specimen was set up in the triaxial testing system after compaction and the initial state is indicated by point A in the figure. An isotropic net stress of 100 kPa was applied to each specimen immediately after setting up (A→B). In order to investigate the effect of the magnitude of suction on the small-strain shear modulus, two specimens S150C1 and S300C1 were brought to two different suctions of 150 and 300 kPa, respectively. The cell pressure, pore-air pressure and pore-water pressure were all ramped at a rate of 1.5 kPa/min, until the matric suction reached the desired value while net mean stress was maintained at 100 kPa (B→C and B→D). After that, the specimen was left to equalise the suction. After suction equalisation at C and D, constant p compression (C→M and D→N) was carried out. According to the definition in Figure 5.10, the CSR equals 1 for both tests S150C1 and S300C1. To study the effect of CSR on small-strain shear modulus, test S150C2 (CSR = 2) was carried out by comparison with test S150C1 (CSR = 1). After an isotropic net stress of 100 kPa was applied (A→B), the specimen S150C2

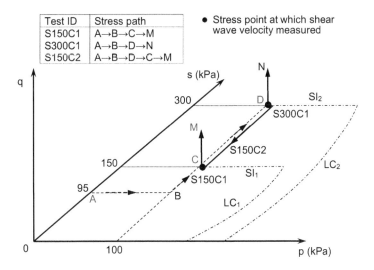

Figure 5.11 Stress paths for series 1 tests.

Source: After Ng and Xu, 2012.

Table 5.3 Summary of testing conditions for investigating effects of suction path on stiffness: series 1

Series No.	Test ID	Suction path before shearing*: kPa	Suction magnitude at shearing*, $s_{current}$: kPa	Current suction ratio, CSR
1	S150C1	95 → 150	150	1
	S300C1	95 → 300	300	1
	S150C2	95 → 300 → 150	150	2

Note:
* Constant net mean stress compression at p = 100 kPa.

Test ID	Stress path	Recent suction history
S150C2I30	A→B→D→C→E→C→M	E→C, θ=90°, l=30 kPa
S150C2I100	A→B→D→C→E→C→M	E→C, θ=90°, l=100 kPa
S150C2	A→B→D→C→M	D→C, θ=90°, l=150 kPa
S150C2D30	A→B→D→C→F→C→M	F→C, θ=-90°, l=30 kPa
S150C2D60	A→B→D→C→F→C→M	F→C, θ=-90°, l=60 kPa

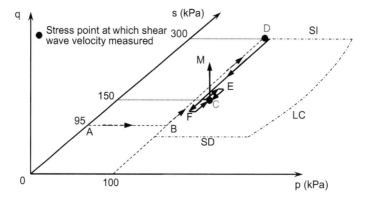

Figure 5.12 Stress paths for series 2 tests.

Source: After Ng and Xu, 2012.

was brought to a suction of 300 kPa for equalisation (B→D). After equalisation at D, the suction was reduced to 150 kPa for equalisation again (D→C). In this case, the specimen has a CSR of 2. Finally, constant p compression (C→M) was conducted. During constant p compression, the deviatoric stress increased at a rate of 3 kPa/hour while the net mean stress and suction were kept constant. The shearing rate of 3 kPa/hour, measured using a high-capacity suction probe and confirmed to be slow enough that the excess pore-water pressure generated during constant p compression, was negligible (less than ±2 kPa). The detailed test conditions for series 1 are summarised in Table 5.3.

Series 2 tests were performed to investigate the effect of recent suction history on the small-strain stiffness of CDT. Both the effect of the direction of the recent suction path (θ = ±90°) and the effect of the magnitude of the recent suction path (l = 30, 60, 100 and 150 kPa) were studied. In order to eliminate the effects of net mean stress, suction

Shear stiffness and damping ratio   219

Table 5.4 Summary of testing conditions for investigating effects of suction path on stiffness: series 2

			Recent suction history	
Series No.	Test ID	Suction path before shearing*: kPa	Direction, θ	Magnitude, l: kPa
2	S150C2I30	95→300→150→180→150	90°	30
	S150C2I100	95→300→150→250→150	90°	100
	S150C2	95→300→150	90°	150
	S150C2D30	95→300→150→120→150	−90°	30
	S150C2D60	95→300→150→90→150	−90°	60

Note:
* Constant net mean stress compression at p = 100 kPa, s = 150 kPa and CSR = 2.

magnitude and CSR, constant p compression for series 2 tests were all performed at p = 100 kPa, s = 150 kPa and CSR = 2. Figure 5.12 shows the stress paths in (p: q: s) space for the series 2 tests. The stress path A→B→D→C was first carried out in all the tests, as described in the series 1 tests. After suction equalisation at point C, the suction path C→E→C was performed for tests S150C2I30 and S150C2I100, while the suction path C→F→C was conducted for tests S150C2D30 and S150C2D60. Finally, constant p compression (C→M) was carried out. Based on the definitions in Figure 5.10, the recent suction path for tests S150C2I30 and S150C2I100 is E→C and θ = 90°. On the other hand, the recent suction path for tests S150C2D30 and S150C2D60 is F→C and θ = −90°. Test S150C2 from series 1 tests is also included in series 2 for comparison. The recent suction path for tests S150C2 is D→C and θ = 90°. The magnitude of the recent suction path (l) for each test is listed in Figure 5.12. l varies from 30 to 150 kPa, to study the effect of l on small-strain stiffness. The detailed test conditions for series 2 are summarised in Table 5.4.

Interpretations of test results

Effects of suction magnitude on small-strain shear behaviour

Figure 5.13a shows the stress-strain relationship during two constant p compression tests at different suctions (s = 150 and 300 kPa). The two sets of test results follow a similar trend. When deviatoric stress increases, the shear strain of both specimens increases nonlinearly at an increasing rate. However, the specimen tested at a suction of 300 kPa (S300C1) produces a consistently lower shear strain (i.e., stiffer) than the one at a suction of 150 kPa (S150C1). It is evident that suction increases the shear stiffness of unsaturated soil. As discussed in the previous section, the specimen at higher suction has a smaller void ratio and lower water content, and thus it has higher shear stiffness.

As illustrated in Figure 5.13a, the secant shear modulus (G_{sec}) of each specimen at small strains can be obtained from the slope of the deviatoric stress-shear strain curve obtained from a constant p compression test according to Equation (5.14). Theoretically, G_0 may be determined from the initial slope of the stress-strain curve. However, because of the limited accuracy of Hall-effect transducers, G_0 cannot be determined reliably from the stress-strain curve. Therefore, G_0 as measured by bender elements is adopted. Since the measured $G_{0(hh)}/G_{0(hv)}$ of CDT varies from 1.01 to 1.03, the CDT is slightly stiffer

220 Advanced Unsaturated Soil Mechanics

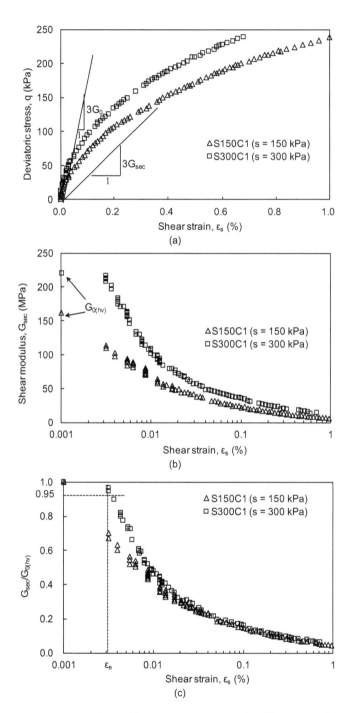

Figure 5.13 Effect of suction magnitude on: (a) stress-strain relationship; (b) shear modulus reduction curve; (c) normalised shear modulus reduction curve.

Source: After Ng and Xu, 2012.

in the horizontal direction. Yimsiri and Soga (2002) proposed a micromechanical model and demonstrated that the shear modulus evaluated by triaxial tests is closer to the shear modulus in the vertical plane ($G_{0(vh)}$ or $G_{0(hv)}$) when the soil is stiffer in the horizontal direction. In addition, $G_{0(hv)}$ is believed to be more accurate than $G_{0(vh)}$ as discussed previously. Therefore, the measured $G_{0(hv)}$ is selected for comparison with G_{sec} at different shear strains obtained from constant p compression tests.

Figure 5.13b shows the variation in G_{sec} with shear strain on a semi-logarithmic scale obtained from two constant p compression tests at suctions of 150 and 300 kPa. $G_{0(hv)}$, measured by bender elements before constant p compression, is also shown in the figure for comparison. Since the maximum shear strain near the bender element is approximately 0.001% (Dyvik and Madshus, 1985), the corresponding shear strain for G_0 is assumed to be 0.001% in this study. The measured $G_{0(hv)}$ at suctions of 150 and 300 kPa are 164 and 220 MPa, respectively. $G_{0(hv)}$ increases by 34% when the suction increases from 150 to 300 kPa. This observation has been explained in the previous section. When the shear strain increases from about 0.003% to 1%, the G_{sec} of both specimens reduces significantly but at different rates. The G_{sec} of the specimen at a suction of 300 kPa is consistently higher than that of the specimen at a suction of 150 kPa.

Figure 5.13c shows the secant shear modulus reduction curve normalised by $G_{0(hv)}$. The stress-strain relationship is elastic, and the shear modulus is constant in the very small-strain range. The stiffness could be roughly constant if a decreasing proportion of the soil behaves elastically as straining proceeds (Simpson, 1992). The shear strain corresponding to the limit of the elastic portion is generally described as the elastic threshold shear strain (ε_e). Stokoe et al. (1995) reported that the ε_e of sand is about 0.001% while the ε_e of clay is much larger (on the order of 0.005%). To avoid subjective assessment, ε_e in this study is quantified as the shear strain corresponding to $G_{sec}/G_{0(hv)} = 0.95$ (as in Clayton and Heymann, 2001). As illustrated in Figure 5.13c, the ε_e of the specimen tested at a suction of 300 kPa is about 0.003%. For the specimen tested at 150 kPa suction, although there is no reliable data shown in the strain ranging from 0.001% to 0.003%, from the trend ε_e is probably smaller than 0.003%. Beyond ε_e, the normalised shear modulus reduces significantly with shear strain, as shown in Figure 5.13c. The rate of reduction is higher for the specimen at a higher suction. When the shear strain is greater than about 0.02%, the two normalised shear modulus reduction curves overlap. This means that the effect of suction magnitude on the normalised shear modulus reduction curve becomes negligible when the shear strain is larger than 0.02%.

The elastic "proportion" of the soil might represent the proportion of contacts between soil particles which are still intact and have not started to slide (Simpson, 1992). For soil specimens at a higher suction and lower degree of saturation, there is more meniscus water in the specimen. Meniscus water causes an increase in the normal force that holds the soil particles together and resists slippage between soil particles. Therefore, the specimen at a higher suction magnitude has a larger ε_e.

Effects of current suction ratio on small-strain shear behaviour

Figure 5.14a shows the stress-strain behaviour of two specimens sheared at the same suction of 150 kPa but at different CSR values (tests S150C1 and S150C2). The shear strain of both specimens increases with deviatoric stress nonlinearly at an increasing rate. However, the stress-strain curve of specimen S150C2 (sheared at CSR = 2) is consistently

222 Advanced Unsaturated Soil Mechanics

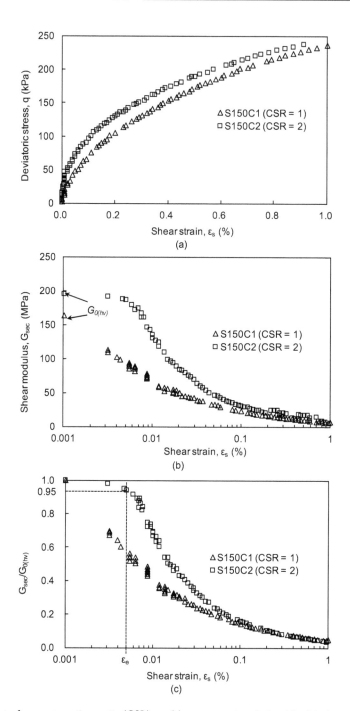

Figure 5.14 Effect of current suction ratio (CSR) on: (a) stress-strain relationship; (b) shear modulus reduction curve; (c) normalised shear modulus reduction curve.

Source: After Ng and Xu, 2012.

Shear stiffness and damping ratio 223

above that of specimen S150C1 (sheared at CSR=1). This means that the shear modulus of the specimen at a higher CSR is consistently higher. Figure 5.14b shows the effect of CSR on the shear modulus reduction curve of the two specimens. The $G_{0(hv)}$ of the two specimens with CSR = 1 and 2 are about 164 and 195 MPa, respectively. Such an increase in $G_{0(hv)}$ due to CSR has been discussed previously. When the shear strain increases from about 0.003% to 1%, the G_{sec} of both specimens reduces significantly but at different rates. The G_{sec} of specimen S150C2 is consistently higher than that of specimen S150C1 and reduces at a higher rate. Figure 5.14c shows the normalised shear modulus reduction curves of the two specimens tested at CSR = 1 and 2. The ε_e of the specimen with CSR= 2 is about 0.005% while the ε_e of the specimen with CSR = 1 is expected to be less than 0.003%. When the CSR is higher in this series, the degree of saturation and water content are lower. As discussed in the previous paragraph, a lower degree of saturation may result in more meniscus water and thus a higher ε_e. The normalised shear modulus decreases significantly with shear strain beyond the ε_e. The rate of reduction is higher for the specimen with higher CSR. The effect of CSR on the normalised shear modulus reduction curve becomes negligible when the shear strain is greater than about 0.2%.

Effect of direction of recent suction path on small-strain shear behaviour

Figure 5.15a shows the effects of the direction of the recent suction path (described by θ) on the stress-strain behaviour of compacted CDT. The direction of the recent suction path for tests S150C2I30 and S150C2D30 is θ = 90° and –90°, respectively. The magnitude of the recent suction path for the two tests is the same and l = 30 kPa. As shown in the figure, tests S150C2I30 (θ = 90°) and S150C2D30 (θ = –90°) show similar stress-strain relationships during compression at constant p. When the deviatoric stress is larger than 100 kPa, the specimen with θ = –90° induces a slightly larger shear strain, indicating a slightly lower shear modulus. Figure 5.15b shows the effect of θ on the shear modulus reduction curve. The $G_{0(hv)}$ of specimens S150C2I30 and S150C2D30 is 194 and 191 MPa, respectively. According to Table 5.5, the void ratio, degree of saturation and water content of the specimen S150C2I30 are slightly less than those of specimen S150C2D30. Therefore, the $G_{0(hv)}$ of specimen S150C2I30 is slightly greater than that of specimen S150C2D30. The shear modulus reduction curves of the two specimens are also quite similar. However, a slight difference is observed at the shear strain ranging from about 0.007% to 0.04%. Figure 5.15c shows the normalised shear modulus reduction curves for the two tests. The difference in ε_e of the two specimens is about 0.001%, which is below the accuracy of the Hall-effect measurement. The rates of stiffness reduction in the two specimens are quite similar.

The similarity in the small-strain behaviour of the two tests illustrated in Figure 5.15 may be explained by small changes in the total volume and water content of both specimens along suction path C→E→C. As shown in Figure 5.12, the stress path A→B→D→C is first carried out for both specimens S150C2I30 (θ = 90°) and S150C2D30 (θ = –90°). Then, the suction of the specimen S150C2I30 is increased from 150 to 180 kPa and then lowered back to 150 kPa (C→E→C). During the suction path C→E, about 0.02% compressive volumetric strain is measured by the Hall-effect transducers. This portion of the deformation is recovered when the suction is lowered from 180 back to 150 kPa (E→C). Therefore, the soil behaviour can be regarded as elastic during the suction path C→E→C. On the other hand,

224 Advanced Unsaturated Soil Mechanics

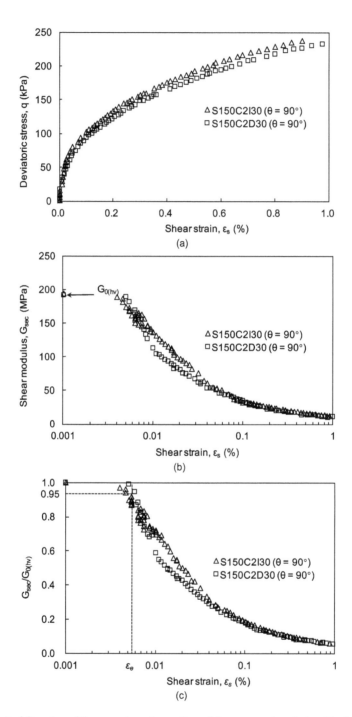

Figure 5.15 Effect of direction of the recent suction path on: (a) stress-strain relationship; (b) shear modulus reduction curve; (c) normalised shear modulus reduction curve.

Source: After Ng and Xu, 2012.

when the suction of the specimen S150C2D30 is lowered from 150 to 120 kPa (C→F), about 0.06% expansive volumetric strain is measured. When the suction is increased from 120 back to 150 kPa (F→C), only about 0.02% compressive volumetric strain is measured. This net expansive volumetric strain of about 0.04% for specimen S150C2D30 ($\theta = -90°$) results in a slightly higher void ratio and thus slightly lower shear modulus. The differences in void ratio, degree of saturation and water content of the two specimens are 1%, 9% and 10%, respectively. Since the differences are all small, the differences between the stiffness of specimens S150C2D30 and S150C2I30 are not significant as shown in Figure 5.15. Therefore, the effect of θ on small-strain shear modulus is not significant in this case.

Effect of magnitude of recent suction path on small-strain shear behaviour

Figure 5.16a shows the stress-strain curves of three specimens during constant p compression with different magnitudes of recent suction path (l). The three specimens have the same direction of recent suction path, $\theta = 90°$. The l for the tests S150C2I30, S150C2I100 and S150C2 is 30, 100 and 150 kPa, respectively. The three stress-strain curves are almost identical. This means that the magnitude of the recent suction path does not affect the stress-strain behaviour at small strains when $\theta = 90°$. The shear modulus reduction curves of the three tests are shown in Figure 5.16b. The magnitude of $G_{0(hv)}$ in the three specimens is very similar. As summarised in Table 5.5, the $G_{0(hv)}$ of the specimens S150C2I30, S150C2I100 and S150C2 is 194, 191 and 195, respectively. The stiffness reduction curves are also similar. Only minor differences are observed in the shear strain range between about 0.005% to 0.03%. Figure 5.16c shows the normalised shear modulus reduction curves in the three tests. The ε_e of the three specimens are almost the same, in the strain range between 0.004% and 0.005%. The rates of reduction of stiffness are also very similar for all three specimens. All the above observations in Figure 5.16 demonstrate that the effect of l on the small-strain shear behaviour is not significant when $\theta = 90°$.

The similar small-strain behaviour of the three tests observed in Figure 5.16 may also be explained by small variations in volume change, degree of saturation and water content along the suction path C→E→C. For specimen S150C2I30, the measured variation in volumetric strain is zero after following the suction path C→E→C. For specimen S150C2I100, only 0.02% net compressive volumetric strain is measured. Therefore, the void ratios of the three specimens are similar (about 1% maximum difference). Furthermore, the

Table 5.5 Measured very small-strain shear moduli before constant p compression

Series No.	Test ID	$G_{0(vh)}$: MPa	$G_{0(hv)}$: MPa	$G_{0(hh)}$: MPa	$G_{0(hv)}/G_{0(vh)}$	$G_{0(hh)}/G_{0(hv)}$	e	S_r: %	w: %
1	S150C1	161	164	167	1.02	1.02	0.575	73.85	15.35
	S300C1	217	220	225	1.01	1.02	0.572	46.73	9.67
	S150C2	195	195	198	1.00	1.02	0.572	53.43	11.05
2	S150C2I30	192	194	200	1.01	1.03	0.565	53.18	10.87
	S150C2I100	193	191	192	0.99	1.01	0.568	56.31	11.57
	S150C2	195	195	198	1.00	1.02	0.572	53.43	11.05
	S150C2D30	186	191	194	1.03	1.02	0.570	58.19	11.99
	S150C2D60	186	189	193	1.02	1.02	0.576	62.53	13.03

226 Advanced Unsaturated Soil Mechanics

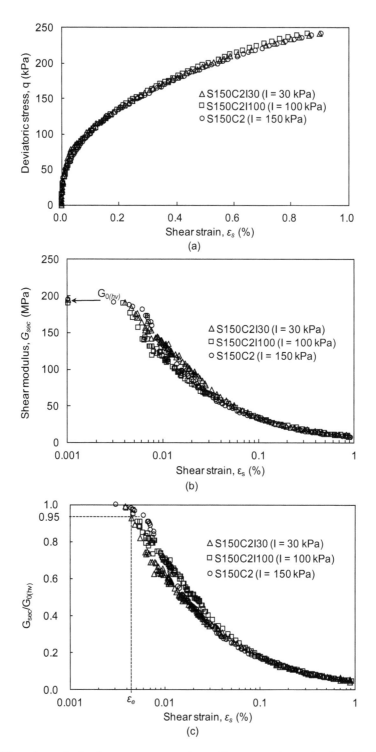

Figure 5.16 Effect of magnitude of recent suction path on: (a) stress-strain relationship; (b) shear modulus reduction curve; (c) normalised shear modulus reduction curve, when $\theta = 90°$.

Source: After Ng and Xu, 2012.

Shear stiffness and damping ratio 227

maximum differences in the degree of saturation and water content of the three specimens are both about 6%, which is quite small. Thus, G_0, ε_e and rate of shear modulus reduction are all similar for the three specimens. In other words, the effect of l on the small-strain shear behaviour is not significant when $\theta = 90°$.

Figure 5.17a shows the effect of l on the stress-strain behaviour of compacted CDT when $\theta = -90°$. The l for tests S150C2D30 and S150C2D60 is 30 and 60 kPa, respectively. The stress-strain curves of both specimens follow a similar trend. The shear modulus reduction curves of both tests are shown in Figure 5.17b. The $G_{0(hv)}$ of specimens S150C2D30 and S150C2D60 is 191 and 189 MPa, respectively (refer to Table 5.5). The shear modulus reduction curves of the two specimens are distinctly different. The shear modulus of specimen S150C2D60 is significantly lower than that of specimen S150C2D30 in the shear strain range between 0.003% and 0.04%. Figure 5.17c shows the normalised shear modulus reduction curves of both tests. The ε_e of specimen S150C2D30 is between 0.005% and 0.006%. However, the ε_e of specimen S150C2D60 is expected to be less than 0.003%. The rate of reduction of normalised shear modulus of specimen S150C2D60 is also much smaller than that of specimen S150C2D30.

In order to explain the above observations, a suction decrease (SD) yield surface was added to the Barcelona basic model (BBM) (Alonso et al., 1990). The suction path C→F→C for specimen S150C2D60 (l = 60 kPa) may touch the SD yield surface, while the suction path C→F→C for specimen S150C2D30 (l = 30 kPa) may still be between the SD and SI yield surfaces (see Figure 5.12). Following the suction path C→F→C, the measured net expansive volumetric strain for specimen S150C2D60 is about 0.09%, while the measured net expansive volumetric strain for specimen S150C2D30 is only about 0.04%. Specimen S150C2D60 has a slightly larger void ratio (see Table 5.5). Further, the degree of saturation and water content of specimen S150C2D60 are also higher. Thus, it has a smaller shear modulus.

Referring to Table 5.5, the maximum differences in $G_{0(vh)}$, $G_{0(hv)}$ and $G_{0(hv)}$ caused by the recent suction history are 5%, 3% and 4%, respectively. Therefore, the effect of recent suction history on G_0 is not significant. From Figure 5.15, it can be seen that the effect of θ on the shear modulus reduction curve is not important when l = 30 kPa. Based on the results shown in Figures 5.18 and 5.19, it is evident that the effect of l on the shear modulus reduction curve is negligible when $\theta = 90°$, but is significant when $\theta = -90°$.

Summary

The measured magnitudes of $G_{0(vh)}$, $G_{0(hv)}$ and $G_{0(hh)}$ are similar and follow a similar trend. G_0 increases by about 35% when the magnitude of suction increases from 150 to 300 kPa. G_0 increases by about 20% when CSR increases from 1 to 2.

When CSR = 1, the shear modulus of the specimen at higher magnitudes of suction is consistently higher in the shear strain range between 0.001% and 1%. However, the effect of suction magnitude on the normalised shear modulus reduction curve becomes negligible when the shear strain is greater than 0.02%. G_0, elastic threshold strain (ε_e) and the rate of stiffness reduction all increase with the magnitude of suction. This may be caused by the dominant effect of meniscus water when the void ratio, degree of saturation and water content decrease.

At a given suction, the small-strain shear modulus of a specimen with a higher CSR is significantly higher. G_0, ε_e and the rate of shear modulus reduction all increase with

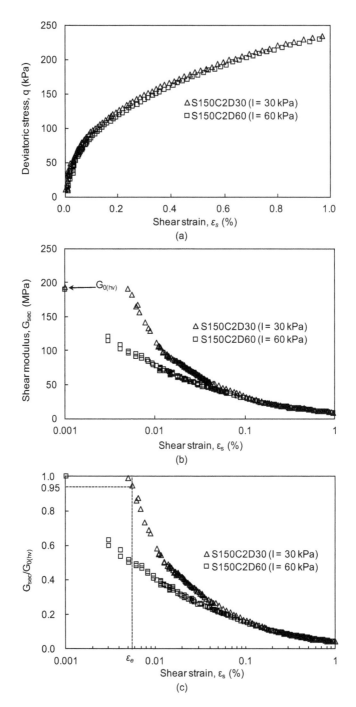

Figure 5.17 Effect of magnitude of the recent suction path on: (a) stress-strain relationship; (b) shear modulus reduction curve; (c) normalised shear modulus reduction curve, when $\theta=-90°$.

Source: After Ng and Xu, 2012.

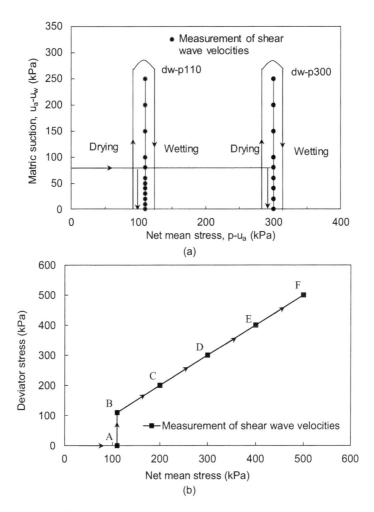

Figure 5.18 Stress paths of (a) the drying and wetting tests under constant isotropic net stresses and (b) the shearing tests at constant suction.

Source: After Ng et al., 2009.

CSR. This may also be interpreted by the dominant effect of meniscus water when void ratio, degree of saturation and water content decrease. When the shear strain is greater than 0.2%, the effect of CSR on the normalised shear modulus reduction curve is insignificant.

The effect of recent suction history on G_0 is not significant. The maximum differences in $G_{0(vh)}$, $G_{0(hv)}$ and $G_{0(hv)}$ induced by the effect of recent suction history are 5%, 3% and 4%, respectively. The effect of direction of recent suction path (θ) on the shear modulus reduction curve is not distinguishable when the magnitude of the recent suction path l = 30 kPa. This observation may be explained by the small changes in void ratio, degree of saturation and water content. The effect of l on the shear modulus reduction curve is

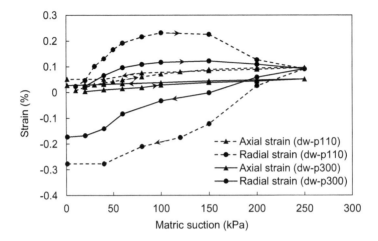

Figure 5.19 Variations in the axial and radial strains with matric suction during the drying and wetting tests.

Source: After Ng et al., 2009.

negligible when $\theta = 90°$. However, l affects the shear modulus reduction curve significantly when $\theta = -90°$, probably because of variations in void ratio, degree of saturation and water content.

EFFECTS OF WETTING-DRYING AND STRESS RATIO ON ANISOTROPIC STIFFNESS (NG ET AL., 2009)

Introduction

To study the effects of wetting-drying and stress ratio on anisotropic shear stiffness at very small strains, two series of tests were carried out on recompacted CDT (refer to Figure 5.3) using the triaxial apparatus shown in Plate 4.2. The stress paths of the two series of tests are shown in Figure 5.18. Two drying and wetting tests were conducted at constant isotropic net stresses of 110 and 300 kPa to study the effects of wetting-drying on the anisotropic shear stiffness at very small strains. Figure 5.18a shows the stress paths of the two tests. After the soil specimen and all measuring devices were set up in the triaxial apparatus, the desired net mean stress was first applied to the soil specimen, which was then brought to zero suction for equalisation. Shear wave velocities were measured at the end of the equalisation stage, and then the matric suction was increased to the next desired value. Upon completion of the drying phase up to a matric suction of 250 kPa, the wetting phase began. The maximum suction applied was limited by the air-entry value of the ceramic disk, which was 300 kPa. When the matric suction decreased to zero, the wetting phase was completed. The drying and wetting tests under constant isotropic net stress states consisted of a sequence of equalisation stages. Measurements of shear wave velocities were performed at the end of each equalisation stage, as shown by the dots in Figure 5.18a.

Table 5.6 Summary of testing conditions for constant-stress drying-wetting tests and constant-stress ratio compression tests

Test ID	Suction: kPa	Net mean stress: kPa	Deviatoric stress: kPa	Initial water content: %	Initial dry density: kg/m³	Relative compaction: %
dw-p110	0–250	110	0	16.1	1667	94.7
dw-p300	0–250	300	0	16.2	1657	94.1
sh-s0	0	110–500	0–500	16.5	1677	95.3
sh-s50	50	110–500	0–500	16.3	1675	95.2

Shearing tests at constant suctions were performed to study the effect of stress ratio on the anisotropic shear moduli at very small strains. Two shearing tests were conducted at constant suctions of 0 and 50 kPa. The stress paths of the two tests are shown in Figure 5.18b. At constant suction, soil specimens were first isotropically loaded to the initial state of the tests (stage A). The desired initial values of net mean stress and deviatoric stress were 110 and 0 kPa, respectively. After equalisation, the deviatoric stress was increased at 5 and 3 kPa per hour for shearing tests conducted at suctions of 0 and 50 kPa, respectively, until the stress ratio $(q/(p - u_a))$ reached 1 (stage B). Finally, the soil specimens were sheared at a constant-stress ratio of 1 following the stress path B-C-D-E-F. The shearing rates during loading under the stress ratio of 1 were also 5 and 3 kPa per hour for tests conducted at suctions of 0 and 50 kPa, respectively. Measurements of shear wave velocities were carried out at stages A, B, C, D, E and F. Measurements of shear wave velocities (and hence the shear moduli) were carried out at pre-selected stress states during each test, as indicated by the dots in the figure. The test conditions are summarised in Table 5.6.

Interpretations of experimental results

Variations in axial and radial strains during drying and wetting tests

During the drying and wetting tests, axial and radial strains were measured by Hall effect local strain transducers. Figure 5.19 shows the variation in the axial and radial strains with matric suction under constant isotropic net stresses of 110 and 300 kPa. The axial strain changes slightly with matric suction, while the radial strain varies significantly. During the drying process at either net stress, the axial strain increased slightly with an increase in matric suction. The radial strain increases significantly as the suction increases from 0 to 100 kPa and then becomes steady or decreases when the suction increases from 100 to 250 kPa. The magnitudes of the strains of the specimen at 300 kPa net mean stress are consistently lower than those at a net mean stress of 110 kPa. This may be because a specimen that is subjected to a higher net mean stress will be stiffer and hence have a higher resistance to volume change due to drying. During the wetting process, the axial strain reduces at either net mean stress with a decrease in matric suction almost along the drying path. The radial strain reduces significantly until the matric suction becomes zero. Marked hysteresis between the drying and wetting curves of radial strain was observed. Besides, as with the SDSWCCs (see Figure 5.4), the size of the hysteresis loop is smaller for the specimen at the higher net mean stress. After the drying and wetting cycle, there was a net volume increase in each specimen (swelling).

However, the net volume increase was not significant. This is because the magnitudes of both the axial and radial strains of each specimen were relatively small, less than 0.3%. The significant difference between the axial and radial strains demonstrates that variation in the suction on soil specimens under constant isotropic stress state can produce anisotropic strains.

Variations in shear moduli during drying and wetting tests

Measurements of shear wave velocities were carried out at the end of each pre-selected equalisation stage during the drying and wetting tests (see Figure 5.18a). The very small-strain shear moduli in these stages can then be determined. Figure 5.20 shows the variation in the very small-strain shear moduli in three different polarisation planes ($G_{0(vh)}$, $G_{0(hv)}$ and $G_{0(hh)}$) with matric suction during the two drying and wetting tests. At a given suction, the values of $G_{0(hv)}$ and $G_{0(hh)}$ are similar, but $G_{0(vh)}$ is slightly smaller. However, in a homogeneous cross-anisotropic elastic continuum, $G_{0(vh)}$ should be equal to $G_{0(hv)}$. The discrepancy between the measured $G_{0(vh)}$ and $G_{0(hv)}$ may have several causes. First, the soil specimen was not actually a homogeneous cross-anisotropic elastic continuum. Because of the preferred orientation of particles in the horizontal direction, a horizontally transmitted shear wave travels through fewer particle contacts per unit distance than does a vertically transmitted shear wave and this results in $v_{s(hv)}$ being higher than $v_{s(vh)}$ (Jardine et al., 1999). Second, the transmitting frequencies used by the horizontally mounted bender element probes ($v_{s(hh)}$ and $v_{s(hv)}$) were normally higher than those used by the platen-mounted bender elements ($v_{s(vh)}$) (Pennington et al., 2001). Third, the bender element embedded in the base pedestal probably deflected more than the horizontally mounted bender elements within the probe did and hence resulted in a lower $v_{s(vh)}$ (Pennington et al., 2001). Fourth, the effect of the platen end on stiffness measurements (Germine and Ladd, 1988; Lacasse and Berre, 1988) could be another cause of the discrepancy.

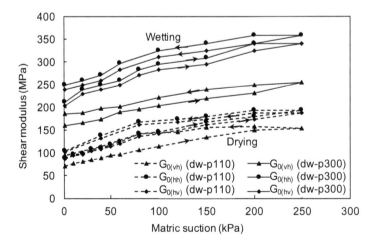

Figure 5.20 Variations in the shear modulus with matric suction during the drying and wetting tests.

Source: After Ng et al., 2009.

Figure 5.20 shows that the variations in $G_{0(vh)}$, $G_{0(hv)}$ and $G_{0(hh)}$ with matric suction follow a similar trend. The shear moduli increase nonlinearly with an increase in matric suction. In the early stages of the drying process, the shear moduli increase significantly, changing more gradually as drying continues. For instance, at a net mean stress of 110 kPa, when matric suction increases from 0 to 100 kPa, $G_{0(vh)}$, $G_{0(hv)}$ and $G_{0(hh)}$ increase by 63.1%, 62.4% and 63.5%, respectively. $G_{0(vh)}$, $G_{0(hv)}$ and $G_{0(hh)}$ increase by only 13.8%, 15.4% and 15.7%, respectively, for suction increases from 150 to 250 kPa. The significant increase in shear moduli in the early stages of the drying process is probably because bulk-water effects dominate the soil stiffness (Mancuso et al., 2002; Ng and Yung, 2008). Any increase in suction is equivalent to an increase in the mean effective stress, resulting in a much stiffer soil. When drying continues, the meniscus water effect dominates increasingly. The meniscus water causes an increase in the normal force holding the soil particles together, leading to a stiffer soil specimen and hence a higher shear wave velocity. However, the beneficial effects on the shear wave velocity cannot increase infinitely, because the meniscus radius reduces progressively as suction increases (Mancuso et al., 2002; Ng and Yung, 2008). These observed variations in shear moduli with matric suction can also be explained by the term, $[1+(u_a - u_w)/p_r]^{2bs}$, in Equation (5.7), where bs is regarded as a positive constant. In comparisons of variations in the shear moduli with matric suction at different net mean stresses, we note that the shear moduli of soil at a net mean stress of 300 kPa are consistently higher than those at a net mean stress of 110 kPa. This is because when the applied net stress is higher, the volumetric deformation is larger. Then, the average pore size of the specimen is smaller, and the soil is stiffer. In turn, specimens with higher stiffness will have a higher resistance to volume change caused by drying. Therefore, at higher net mean stress, the rate of change of the shear modulus with respect to matric suction is lower. At a net mean stress of 110 kPa, the shear moduli, $G_{0(vh)}$, $G_{0(hv)}$ and $G_{0(hh)}$, increase by 118.4%, 114.2% and 115.7%, respectively, as the suction increases from 0 to 250 kPa. At a net mean stress of 300 kPa, the shear moduli, $G_{0(vh)}$, $G_{0(hv)}$ and $G_{0(hh)}$, increase by only 59.6%, 67.1% and 69.0%, respectively.

After drying to the maximum suction of 250 kPa, the matric suction is reduced to wet the soil specimen. As with the SDSWCCs (see Figure 5.4), there is hysteresis between the drying and wetting curves showing the variation in the shear moduli with matric suction. At the same suction, the shear moduli measured during wetting are consistently higher than those obtained during drying. Since the induced axial and radial strains of each soil specimen are relatively small (less than ±0.3%), the total volume change is insignificant. The shear modulus hysteresis may well result from the water content on the adsorption curve being higher than that on the desorption curve at the same suction.

Comparison between shear moduli determined in drying and wetting tests and isotropic compression tests

The shear moduli, $G_{0(vh)}$, $G_{0(hv)}$ and $G_{0(hh)}$, obtained from drying and wetting tests under constant isotropic net stress states and those obtained from isotropic compression tests (Ng and Yung, 2008) on the same material are shown together in Figures 5.21a, 5.21b and 5.21c for comparison. At a net mean stress of 110 kPa, the shear moduli $G_{0(vh)}$, $G_{0(hv)}$ and $G_{0(hh)}$ obtained from the isotropic compression tests are fairly close to those obtained from the drying and wetting tests. However, at a net mean stress of 300 kPa, the shear moduli

234 Advanced Unsaturated Soil Mechanics

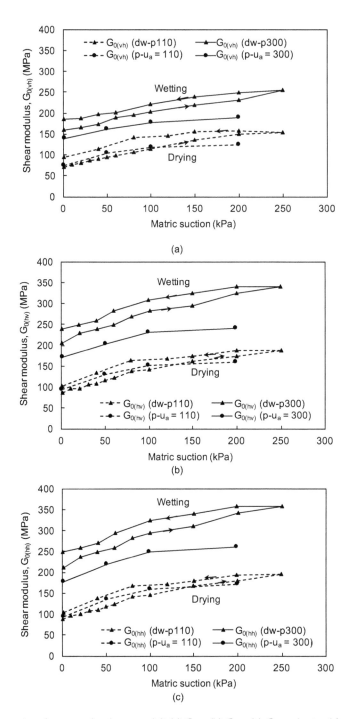

Figure 5.21 Comparison between the shear moduli: (a) $G_{0(vh)}$; (b) $G_{0(hv)}$; (c) $G_{0(hh)}$, obtained from the drying and wetting tests and the isotropic compression tests.

Source: After Ng et al., 2009.

obtained from the isotropic compression tests are obviously lower than those obtained from the drying and wetting tests. Further, the difference increases with higher matric suction. Assuming that all specimens have an identical fabric, which is generally achieved by using identical preparation procedures on the same soil, the difference between the shear moduli obtained from the drying and wetting tests and those extracted from the isotropic compression tests (Ng and Yung, 2008) may be attributed to different stress paths or loading histories.

Variations in the degree of stiffness anisotropy during drying and wetting tests

Figure 5.22 shows the variation in the degree of stiffness anisotropy, $G_{0(hh)}/G_{0(hv)}$, with matric suction during the drying and wetting tests. At different net mean stresses, the variations in the degree of stiffness anisotropy with matric suction follow a similar trend. Initially at zero suction, $G_{0(hh)}$ is about 3% and 4% higher than $G_{0(hv)}$ at net mean stresses of 110 and 300 kPa, respectively. According to Equation (5.10), this observed stiffness anisotropy should be attributed to the inherent anisotropy (i.e., C_{hh}^2/C_{hv}^2) induced in the tested material during specimen preparation (Ng and Yung, 2008). However, this observed stiffness anisotropy at zero suction may not be caused by the inherent anisotropy induced during specimen preparation alone, because the specimens have undergone isotropic compression and wetting from the initial suction to zero suction. Therefore, the $G_{0(hh)}/G_{0(hv)}$ of a specimen at a net mean stress of 300 kPa is higher than at a net mean stress of 110 kPa. When matric suction increases, $G_{0(hh)}/G_{0(hv)}$ increases at a gradually reducing rate, but the increase is less than 1%. Though the magnitudes of the increase in $G_{0(hh)}/G_{0(hv)}$ with matric suction are very small, there is a clear trend showing that the degree of stiffness anisotropy increases with matric suction. As with the SDSWCCs (see Figure 5.4), there is hysteresis between the drying and wetting curves. Although the size of the hysteresis loop is very

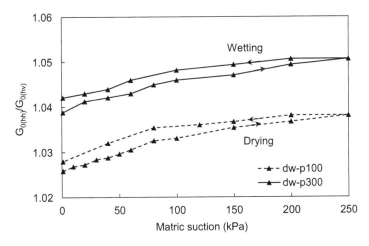

Figure 5.22 Variations in the degree of stiffness anisotropy with matric suction during the drying and wetting tests.

Source: After Ng et al., 2009.

236 Advanced Unsaturated Soil Mechanics

small, the trend is clear. The changing rates of $G_{0(hh)}/G_{0(hv)}$ and the sizes of the hysteresis loops at two different net mean stresses are almost the same. However, the $G_{0(hh)}/G_{0(hv)}$ of the specimen at a net mean stress of 300 kPa is consistently higher than that at a net mean stress of 110 kPa. The degree of stiffness anisotropy does not appear to be independent of the net stresses under isotropic stress conditions. Equation (5.10) cannot predict the variations in the degree of stiffness anisotropy during drying and wetting tests. This may be because the equation does not consider the complex behaviour of an unsaturated soil, such as the coupling effect between the hydraulic and mechanical characteristics, i.e., the possible re-distribution of bulk water within water-filled voids and meniscus water at the inter-particle contacts around air-filled voids in the soil (Wheeler et al., 2003; Ng and Yung, 2008). Therefore, Equation (5.10) should be used with caution.

Summary

During the drying and wetting tests, the measured shear moduli, $G_{0(vh)}$, $G_{0(hv)}$ and $G_{0(hh)}$, increased nonlinearly with increasing matric suction, but at a reduced rate. The differences between the shear moduli obtained from the drying and wetting tests at constant net mean stresses and those extracted from the isotropic compression tests at constant suctions reveal that the stress path or loading history has a significant effect on the stiffness at very small strains.

Anisotropic strains and stiffness anisotropy were observed during the drying and wetting tests. The variations in the radial strains were much more significant than the axial strains throughout the tests. The degree of stiffness anisotropy expressed in terms of $G_{0(hh)}/G_{0(hv)}$ at zero suction was about 1.03 to 1.04. According to Equation (5.10), such stiffness anisotropy is mainly caused by the inherent anisotropic material properties induced during specimen preparation, i.e., C_{hh}^2/C_{hv}^2. The ratio of $G_{0(hh)}/G_{0(hv)}$ increased slightly with the matric suction at a reduced rate. Though the increase was less than 1%, the trend is quite clear. Therefore, it is evident that soil suction can induce anisotropic strains and shear stiffness even under isotropic stress conditions.

The variations in the strains, very small-strain shear stiffness and degree of stiffness anisotropy all showed hysteresis between the drying and wetting curves. These hysteretic variations, which mimic the hysteretic behaviour commonly observed in SDSWCCs, may well be due to the water content difference between the drying and wetting curves.

The variations in $G_{0(vh)}$, $G_{0(hv)}$ and $G_{0(hh)}$ during shearing tests at constant suctions were consistent with Equation (5.7), which predicts that the shear modulus ($G_{0(ij)}$) is proportional to $[(\sigma_i - u_a) \times (\sigma_j - u_a)]^{ns}$. During the shearing tests, significant stress-induced stiffness anisotropy was observed. When the deviatoric stress increased from 0 to 110 kPa, the degree of stiffness anisotropy, $G_{0(hh)}/G_{0(hv)}$, dropped significantly from about 1.03 to 0.87. This significant reduction in $G_{0(hh)}/G_{0(hv)}$ may be explained by the term $[(\sigma_i - u_a)/(\sigma_j - u_a)]^{ns}$ in Equation (5.9). While shearing at a constant-stress ratio ($\eta = q/(p - u_a)$) of 1, $G_{0(hh)}/G_{0(hv)}$ appeared to be constant even though the net mean stress and deviatoric stress increased from 110 to 500 kPa. In addition, it should be noted that the variations in $G_{0(hh)}/G_{0(hv)}$ in the two shearing tests were fairly similar even though the applied suctions were different. Therefore, the influence of suction on the stress-induced stiffness anisotropy was not significant.

RESILIENT MODULUS UNDER CYCLIC SHEARING (NG ET AL., 2013)

Introduction

Fatigue cracking and rutting in the asphalt and concrete layer of a pavement is of great concern to pavement designers and users. Its incidence may be caused by a number of factors such as an increase in traffic volume, deterioration of the asphalt and concrete and differential settlement of subgrade soils (Brown, 1997). Previous researchers found that any non-uniform deformation of subgrade soils under cyclic traffic loads plays an important role in crack generation and thus the performance of a pavement (Hveem, 1955; Seed et al., 1962; Brown, 1996). The resilient modulus (M_R), defined by Seed et al. (1962) as the ratio of repeated deviator stress to axial recoverable strain in cyclic triaxial test, is widely used as a stiffness parameter in pavement engineering to determine soil deformation under cyclic traffic loads (Li and Selig, 1994; Brown, 1996; Kim and Kim, 2007).

Subgrade soil is often unsaturated and subject to seasonal variations in moisture content (i.e., suction) in the field (Khoury and Zaman, 2004; Yang et al., 2008). Although matric suction is very important for understanding the resilient modulus of subgrade soil, it is rarely controlled or measured in most conventional resilient modulus tests in engineering practice with an exception in the design standard "In-situ solidification technical guide for design and construction of highway embankment" published by the Department of Transportation, Zhejiang Province of China in 2018.

Theoretical considerations

The resilient modulus is equivalent to secant Young's modulus following its definition and it is sometimes called resilient Young's modulus (Brown et al., 1986; Li and Selig, 1994). For an isotropic elastic material, Young's modulus is linked to shear modulus G by $E = 2G(1 + \nu)$, where ν is Poisson's ratio. Thus, M_R can be determined from G and ν. For cyclic tests, Equation (5.7) may be modified to predict M_R.

The M_R of a subgrade soil is dependent on various factors, including grain size distribution, density, stress level, stress history, loading frequency and cyclic number. Experimental results obtained by other researchers have demonstrated that stress level is the most important factor (see Lekarp et al. (2000)). Numerous attempts have been made to establish the relationship between M_R and stress level. One of the most widely used formulations was proposed by Uzan (1985) as follows:

$$M_R = k_{u1}\left(\frac{\sigma_1 + \sigma_2 + \sigma_3}{p_{atm}}\right)^{k_{u2}} \left(\frac{q_{cyc}}{p_r}\right)^{k_{u3}} \tag{5.15}$$

where σ_1, σ_2, σ_3 are principal stresses. q_{cyc} and p_{atm} are cyclic stress (i.e., the amplitude of change in deviator stress during cyclic loading-unloading) and atmospheric pressure, respectively. Parameters k_{u1}, k_{u2}, k_{u3} are regression coefficients. Although this equation does not consider seasonal variations of M_R since it ignores any change of moisture content (i.e., soil suction), it can be modified by adopting two stress state variables.

238 Advanced Unsaturated Soil Mechanics

Equation (5.7) incorporates net confining pressure and matric suction, whereas Equation (5.15) considers net confining pressure and cyclic deviator stress. Both equations can be modified to represent the M_R of unsaturated soil. To completely describe the M_R of unsaturated soil, a new equation is proposed as follows:

$$M_R = M_0 \left(\frac{p}{p_r}\right)^{k_1} \left(\frac{q_{cyc}}{p_r}\right)^{k_2} \left(1 + \frac{s}{p}\right)^{k_3} \tag{5.16}$$

where net mean stress p is defined as $((\sigma_1 + \sigma_2 + \sigma_3)/3 - u_a)$. Parameters k_1, k_2 and k_3 are regression exponents. The first term on the right-hand side stands for the resilient modulus at the reference stress state where $p = p_r$, $q_{cyc} = p_r$, $s = 0$. The second term quantifies the influence of net mean stress on M_R. Numerous experimental studies have demonstrated that soil stiffness including M_R increases with confinement (Houlsby and Wroth, 1991; Viggiani and Atkinson, 1995b; Lekarp et al., 2000). Thus, the empirical exponent k_1 should be positive. The third term reflects the variation of M_R with cyclic stress. For a linearly elastic material, k_2 should be 0. For a soil specimen characterised by nonlinear stress-strain behaviour, the power k_2 is negative since soil stiffness decreases with increasing strain (Viggiani and Atkinson, 1995b). The fourth term accounts for the influence of suction on M_R. Experimental results such as those reported by Ng and Yung (2008) have shown that the G_0 of an unsaturated soil increases significantly with matric suction. Similarly, M_R of an unsaturated soil is expected to increase with matric suction. Therefore, parameter k_3 should be positive. Equation (5.16) allows for a smooth transition between an unsaturated soil and a saturated soil. When matric suction is zero, the fourth term reduces to 1.0. Then this equation can be used to determine the M_R of a saturated soil from effective confining pressure and cyclic stress.

When q_{cyc} approaches 0 (i.e., at very small strains), the third term on the right hand side approaches infinity because the power k_2 is negative. Therefore, the M_R predicted by Equation (5.16) approaches infinity when q_{cyc} approaches 0. This prediction contradicts the experimental observation that soil stiffness at very small-strain levels is finite (Mancuso et al., 2002; Ng and Yung, 2008). The limitation of this equation can be simply overcome by replacing the term (q_{cyc}/p_r) by $(1 + q_{cyc}/p_r)$. Then, this equation can be rewritten as:

$$M_R = M_0 \left(\frac{p}{p_r}\right)^{k_1} \left(\frac{q_{cyc}}{p_r}\right)^{k_2} \left(1 + \frac{s}{p}\right)^{k_3} \tag{5.17}$$

This above equation is proposed to represent the influence of net stress and suction on M_R of unsaturated soil. Its validity is verified by the experiments described in the next section.

Experimental study

Using the triaxial apparatus shown in Plate 4.2, three series of cyclic triaxial tests were performed on CDT, which can be described as silt (ML) by AASHTO (2017), to investigate the effects on the M_R of a subgrade soil of (1) two stress state variables (i.e., net stress and matric suction) and (2) the wetting and drying history. Figure 5.23 shows the stress

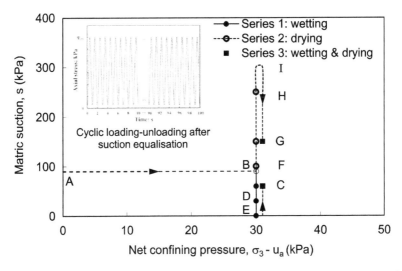

Figure 5.23 Stress path of each soil specimen during suction equalisation stage and variations of axial stress during cyclic loading-unloading.

Source: After Ng et al., 2013.

Table 5.7 Summary of cyclic shear tests

Test ID	Matric suction: kPa	Monotonic shear strength: kPa	Equalisation time: day
W0	95→0	81	12
W30	95→30	113	7
W60	95→60	146	4
D100	95→100	189	4
D150	95→150	243	7
D250	95→250	351	13
WD60	95→0→60	146	14
DW150	95→300→150	243	17

paths of the three series of tests. After compaction, each specimen was set up in the triaxial system. The initial stress state of each specimen is indicated by point A in the figure. First, each specimen was isotropically compressed to a net confining stress of 30 kPa at constant matric suction (A→B). Then, the specimens were subjected to different suction paths at the same net confining pressure of 30 kPa. Finally, cyclic loading-unloading was carried out on each specimen to determine M_R. More details of the testing conditions are summarised in Table 5.7.

In series 1 tests, three CDT specimens, W60, W30 and W0, were wetted by decreasing soil suction from 95 kPa to 60, 30 and 0 kPa (B→C, B→D and B→E), respectively. The measured results were compared to investigate the effects of suction magnitude on M_R along a wetting path. In series 2 tests, three specimens D100, D150 and D250 were dried to suctions of 100, 150 and 250 kPa (B→F, B→G and B→H), respectively. The measured

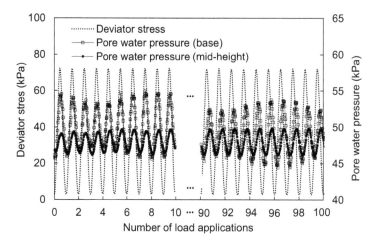

Figure 5.24 Pore-water pressures measured at the base and mid-height of specimen during a cyclic loading-unloading test at net confining pressure of 30 kPa and suction of 30 kPa.

Source: After Ng et al., 2013.

M_R was compared to study the effect of suction on M_R along a drying path. In series 3 tests, the suction of specimen WD60 was decreased from 95 to 0 kPa and then increased to 60 kPa (B→E→C). To investigate the effect of the wetting and drying history, the M_R of this specimen was compared with that of specimen W60 from series 1 (B→C). Similarly, the suction of specimen DW150 in series 3 was increased from 95 to 300 kPa and then decreased to 150 kPa (B→I→G). The measured M_R of this specimen was compared with that of specimen D150 from series 2 (B→G).

Once a specimen had equalised at a given net stress and matric suction, it was subjected to cyclic loading-unloading to determine its M_R. In each cyclic test, the net confining pressure was maintained constant at 30 kPa while the applied axial stress was varied with time following a haversine form. For clarity, variations in axial stress during the first and last 10 cycles are shown in an insert in Figure 5.24. The difference between the maximum and minimum axial stresses is defined as the cyclic stress q_{cyc}. According to the AASHTO (2017) standard for resilient modulus test, four levels of cyclic stress (i.e., 30, 40, 55 and 70 kPa) were considered and applied to each specimen in succession. At each level of q_{cyc}, 100 cycles of loading-unloading at 1 Hz were applied.

During cyclic loading-unloading, a constant water content condition was simulated for subgrade soil because the dissipation rate of excess pore-water pressure was low compared to the rate of repeated traffic loads in the field. The drainage valve for water was closed while pore-water pressure was measured at the base and mid-height of each specimen. Figure 5.24 shows variations in pore-water pressure measured at the base and mid-height during a typical cyclic loading-unloading test. For clarity, only the first and last 10 cycles are shown in the figure. It should be noted that the variations of pore-water pressure during the remaining 80 cycles are like those in these 20 cycles. It can be seen that pore-water pressures measured at the base and mid-height vary similarly with applied deviator stress. The magnitude of variation of measured pore-water pressure is about 10 kPa at the base and 5 kPa at the mid-height.

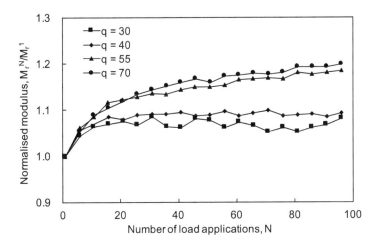

Figure 5.25 Relationship between normalised resilient modulus and number of load applications at zero suction in series I tests.

Interpretations of experimental results

Influence of number of load applications on resilient modulus

During each cyclic stress, there are 100 cycles of loading-unloading. The resilient modulus is determined from each cycle. To investigate the influence of the number of load applications on the resilient modulus, the resilient modulus from the Nth cycle (M_r^N) is normalised by the resilient modulus from the first cycle (M_r^1). Figure 5.25 shows the relationship between the normalised resilient modulus (M_r^N/M_r^1) and the number of load applications (N) after wetting to zero suction (A-B-E in Figure 5.23). M_r^N/M_r^1 increases with N at each cyclic stress (30, 40, 55 and 70 kPa). Dehlen (1969) reported that the application of repeated cyclic stress appears to be a consequence of progressive densification. In this study, volume change was determined from axial and radial strain measured using Hall-effect transducers. At zero suction, the contractive volumetric strains measured after 100 cycles of loading-unloading were 0.03%, 0.04%, 0.09% and 0.25%, corresponding to cyclic stresses of 30, 40, 55 and 70 kPa, respectively. The decreasing volume and hence increasing dry density of each specimen results in an increase in M_r^N/M_r^1 with N.

This figure also reveals that the rate of increase in M_r^N/M_r^1 with N is dependent on cyclic stress. The influence of N on M_r^N/M_r^1 is more significant at higher cyclic stress. When the cyclic stress is 30 and 40 kPa, M_r^N/M_r^1 increases by about 10% during the first 20 cycles of loading-unloading. After the first 20 cycles, there is no obvious variation in M_r^N/M_r^1. When the cyclic stress is 55 and 70 kPa, M_r^N/M_r^1 increases continuously during the 100 cycles of loading-unloading, but at a decreasing rate after the first 20 cycles. This is because the accumulated contractive volumetric strain during cyclic loading-unloading increases with cyclic stress at an increasing rate. The measured contractive volumetric strain is only 0.03% and 0.04% at cyclic stresses of 30 and 40 kPa but increases to 0.09% and 0.25% when the cyclic stress is 55 and 70 kPa. Because of the larger contractive volumetric strain, the densification effect on M_R is more significant at higher cyclic stresses.

Coupled effects of number of load applications and suction on resilient modulus

Figure 5.26 shows the relationship between M_r^N/M_r^1 and N at the same cyclic stress (i.e., 70 kPa) but different suctions (0, 30, 60, 100, 150 and 250 kPa). This figure clearly reveals two types of soil response. At zero suction, M_r^N/M_r^1 increases continuously with N. The total increase during the 100 cycles of loading-unloading is up to 20%. On the other hand, when the suction is equal to or greater than 30 kPa (s = 30, 60, 100, 150 or 250 kPa), M_r^N/M_r^1 varies only slightly with N. One reason is that the contractive volumetric strain during cyclic loading-unloading is much smaller when the suction is equal to or greater than 30 kPa. For example, the contractive volumetric strain under cyclic loading-unloading measured at a cyclic stress of 70 kPa decreases from 0.25% to 0.03% when the matric suction increases from 0 to 30 kPa. Given such a small volumetric strain, the variation of M_r^N/M_r^1 with N becomes insignificant. At a high suction such as 100 kPa, there is even a slight reduction in M_r^N/M_r^1 with N. This is because soil dilation rather than contraction occurs during cyclic loading-unloading. A dilative volumetric strain of -0.03% is measured during cyclic loading-unloading at cyclic stress of 70 kPa and matric suction of 100 kPa. Soil density and hence the resilient modulus decreases with an increase in the number of load applications.

It is revealed in Figures 5.25 and 5.26 that the variation of M_r^N/M_r^1 is negligible when the number of load applications is greater than 20, except when the cyclic stress exceeds 55 kPa at zero matric suction. An unsaturated soil specimen generally achieves a stable resilient modulus within 100 loading-unloading cycles.

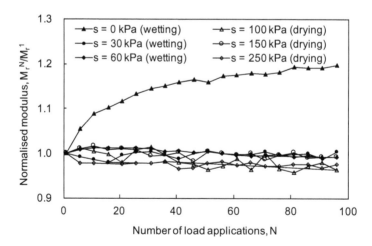

Figure 5.26 Relationship between normalised resilient modulus and number of load applications at a cyclic stress of 70 kPa (s = 0, 30, and 60 kPa in series 1 tests; s = 100, 150, and 250 kPa in series 2 tests).

Source: After Ng et al., 2013.

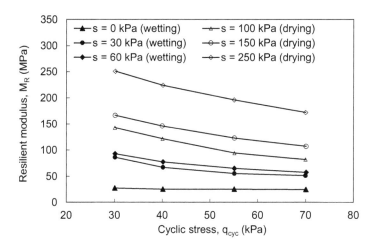

Figure 5.27 Influence of cyclic stress on resilient modulus (s = 0, 30, and 60 kPa in series 1 tests; s = 100, 150, and 250 kPa in series 2 tests).

Source: After Ng et al., 2013.

Influence of cyclic stress on resilient modulus

Figure 5.27 shows the relationship between M_R and q_{cyc} at different suctions (0, 30, 60, 100, 150 and 250 kPa). M_R is the average resilient modulus of the last five cycles at each stress state (i.e., $N = 96$–100). M_R decreases significantly with increasing q_{cyc} at all suctions except $s = 0$. For instance, M_R decreases by about 40% when q_{cyc} increases from 30 kPa to 70 kPa at a suction of 30 kPa. The observed decrease of M_R with q_{cyc} is caused by the nonlinearity of the soil stress-strain relationship, as shown in Figure 5.1. In resilient modulus tests, the strain level increases with q_{cyc}, hence the measured M_R decreases as q_{cyc} increases. The nonlinearity of soils' stress-strain behaviour is captured by the term $(1 + q_{cyc}/p_r)^{k_2}$ in Equation (5.17). Since M_R decreases with q_{cyc}, the parameter k_2 should be negative. For a soil specimen with a greater degree of nonlinearity, the reduction of M_R with q_{cyc} should be more significant and the parameter k_2 should be smaller.

Most experimental data obtained by other researchers have shown that M_R decreases with q_{cyc} (Fredlund et al., 1977; Loach, 1987; Khoury and Zaman, 2004). On the other hand, certain experimental tests have revealed the possibility that M_R increases with increasing q_{cyc} (Seed et al., 1962; Kim and Kim, 2007; Yang et al., 2008). These controversial observations can be explained consistently using Equation (5.17). Based on this equation, M_R increases with net mean stress p but decreases with q_{cyc}. In these resilient modulus tests, cyclic loading is applied under constant confinement. When the axial stress increases, both p and q_{cyc} increase. Thus the measured M_R may either increase or decrease with q_{cyc}, depending on the stress level and parameters k_1, k_2 and k_3.

This figure also reveals that the slope of each curve is generally greater at higher suction. The influence of q_{cyc} on M_R is more significant at higher suction. This is because the degree of nonlinearity of the soil stress-strain relationship is more significant at higher suctions (Xu, 2011). Suction effects on resilient modulus are discussed further in the next subsection.

244 Advanced Unsaturated Soil Mechanics

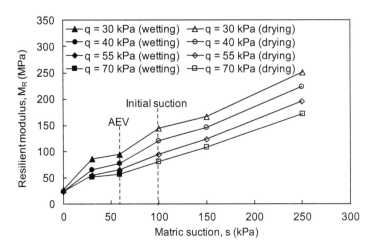

Figure 5.28 Influence of suction on resilient modulus (s = 0, 30, and 60 kPa in series 1 tests; s = 100, 150, and 250 kPa in series 2 tests).

Source: After Ng et al., 2013.

Influence of suction on resilient modulus

Figure 5.28 shows the relationship between M_R and s at different cyclic stresses (30, 40, 55 and 70 kPa). Irrespective of whether it is along a drying or a wetting path, M_R increases significantly with increasing s. At a cyclic stress of 30 kPa, M_R increases eight-fold when s increases from 0 to 250 kPa. The beneficial effects of s on M_R arise for at least two possible reasons. First, when a soil specimen becomes unsaturated, voids are partly filled with water and partly by air, resulting in an air-water interface in each void. When matric suction increases, the radius of the air-water interfaces decreases and induces a larger normal inter-particle contact force (see Figure 1.6). This normal inter-particle contact force provides a stabilising effect on an unsaturated soil by inhibiting slippage at particle contacts and enhancing the shear resistance of the unsaturated soil (Wheeler et al., 2003). Second, an increase in s induces shrinkage of the soil specimen. Because of the stronger inter-normal force between particles and the higher density, M_R measured during cyclic loading-unloading is larger at higher suctions. The influence of s on M_R is captured by the term $(1 + s/p)_3^k$ in Equation (5.17). Since M_R increases with s, the parameter k_3 should be positive. In most state-of-the-art standards for testing the resilient modulus, seasonal variations in soil moisture are not considered. In AASHTO (2017), each soil specimen is prepared and tested for *in situ* soil moisture. The resilient characteristic of an unsaturated subgrade soil is always represented by a single M_R. The use of a single M_R cannot reflect seasonal variation. Observed significant increases in M_R with s from this figure demonstrate that the M_R of a subgrade soil is very likely to be underestimated in a dry season and overestimated in a wet season. To describe the resilient characteristics of an unsaturated subgrade soil appropriately at a given stress state, a suction equalisation stage is necessary before applying cyclic loading-unloading.

Further inspection of this figure reveals that the relationship between M_R and s is highly nonlinear along a wetting path (in series 1 tests). Given the same increase in s, the percentage increase in M_R is much larger in the lower suction range. At a cyclic stress of 30 kPa, M_R doubles when s increases from 0 to 30 kPa. On the other hand, when s increases from 30 to 60 kPa, the percentage increase is only 10%. In series 2 tests, the increase rate of M_R with s is almost constant along a drying path. The different results observed for different suction ranges are probably because bulk-water effects dominate soil behaviour when suction is lower than the air-entry value of the soil (see Figure 5.4) in series 1 tests and meniscus water effects dominate soil behaviour when the suction exceeds the air-entry value in series 2 tests, as discussed above.

Comparisons of the slope of each curve in this figure reveal that the suction effects on M_R are generally more obvious at low cyclic stress levels (i.e., low strain levels). This observation can be explained by experimental results from conventional triaxial compression tests. Nyunt et al. (2010) conducted constant water content triaxial compression tests to investigate the influence of s on the stiffness-strain relationship of an unsaturated soil. They found that the suction effects on secant axial Young's modulus decrease with axial strain. Therefore, the suction effects on M_R decrease slightly with increasing strain level and cyclic stress.

Influence of wetting and drying history on resilient modulus

Figure 5.29 illustrates the influence of the wetting and drying history on measured M_R at two suctions (i.e., 60 and 150 kPa) in test series 3. At a suction of 60 kPa, the M_R measured along a wetting path is greater than that measured along a drying path. Observed differences between these two paths decrease slightly with increasing cyclic stress. Experimental studies have concluded that an unsaturated soil deforms upon suction

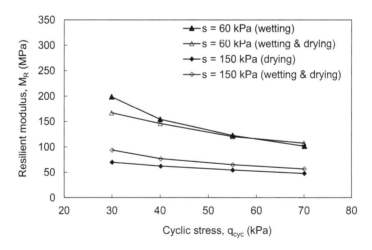

Figure 5.29 Influence of wetting and drying history on resilient modulus (in series 3 tests).

Source: After Ng et al., 2013.

change and irreversible volume changes may occur during cyclic wetting and drying (Ng and Pang, 2000a; Wheeler et al., 2003). The soil specimen along a wetting path may behave like an over-consolidated soil and result in a greater stiffness, at least in the low cyclic stress range.

At a suction of 150 kPa, the M_R measured along a wetting path is greater than that measured along a drying path when the cyclic stress is lower (30 and 40 kPa). When the cyclic stress increases to 50 kPa at suction of 150 kPa, the M_R measured along a drying path becomes even greater than that measured along a wetting path. One possible reason is that the wetting and drying history not only induces effects of over-consolidation but also affects the equilibrium soil water content. Under the same stress and suction conditions, the soil specimen along the drying path has a larger water content as shown in Figure 5.4. At higher water content, the number of air-water interfaces is larger and hence the average skeleton force is higher. Therefore, M_R measured along the drying path could be even higher when the effects of over-consolidation become relatively less important at high cyclic stress levels.

A comparison of Figures 5.25 to 5.29 demonstrates that the stress and suction conditions impose a much more pronounced influence on M_R than the number of load applications and wetting and drying history. This observation is taken into account in the new semi-empirical Equation (5.17). Although this equation does not consider the effects of the number of load applications and the wetting and drying history, it still represents the resilient modulus of an unsaturated subgrade soil with good accuracy.

Verification of the proposed equation

To verify the validity of the newly proposed Equation (5.17) for the resilient modulus, the measured and calculated resilient moduli of four different soils, i.e., CDT, Keuper Marl, Gault clay and London clay, are compared. Resilient moduli measured by CDT tests are first used to fit Equation (5.17) to derive the parameters M_0, k_1, k_2 and k_3. The derived parameters k_1, k_2 and k_3 are then adopted to predict the M_R of Keuper Marl which is in the same category of A-7-6 soils, while M_0 is fitted. This is because M_0 is affected by various factors such as soil density and sampling method. Similarly, experimental data from Gault clay are used to derive parameters, M_0, k_1, k_2 and k_3. The derived parameters k_1, k_2 and k_3 are then adopted to predict the M_R of London Clay which is in the same category of A-7-5 soils. The parameter values are summarised in Table 5.8.

The newly proposed Equation (5.17) is first applied to fit measured the M_R of CDT measured in series 1 and 2 tests. The parameters M_0, k_2 and k_3 are obtained using the least-squares method, while parameter k_1 is assumed 1.0. As this study does not focus on the influence of p on M_R, parameter k_1 is simply determined from previous theoretical and experimental studies. According to the well-known modified Cam-clay model, M_R (i.e., the axial Young's modulus during unloading) is proportional to the effective mean stress p' (Wood, 1990). Viggiani and Atkinson (1995b) also carried out triaxial compression tests to study the relationship between shear modulus and p^n, where n is a regression coefficient. Experimental results have revealed that n is 0.72 at very small-strain levels and increases to 1.0 at a strain level of 0.5%. In this study, the measured total axial strain, including both permanent plastic strain and recoverable resilient strain, is between 0.1% and 1%. Therefore, it may be reasonable to assume that M_R increases linearly with p (i.e., $k_1 = 1$) for simplicity.

Table 5.8 Summary of the soil properties and regression coefficients for Equation (5.17)

	AASHTO classification	Specific gravity	Plastic limit	Liquid limit	Plasticity index	M_0	k_1	k_2	k_3	R^2	S_e/S_y
CDT	A-7-6	2.73	29	43	14	8.32	1.00	−0.65	1.01	0.98	0.14
Keuper Marl	A-7-6	2.69	18	37	19	6.32	1.00	−0.65	1.01	0.66	0.60
Gault clay	A-7-5	2.69	25	61	36	0.61	1.00	−0.36	1.31	0.98	0.14
London clay	A-7-5	2.73	23	71	48	0.53	1.00	−0.36	1.31	0.96	0.21

Figures 5.30a and 5.30b compare the measured and calculated M_R of CDT along a wetting path and a drying path, respectively. In general, they are consistent with a maximum difference of less than 25%. The coefficient of determination (R^2) and S_e/S_y, where S_e and S_y are the residual standard deviation and sample standard deviation, respectively, are determined from measured and calculated resilient moduli under 24 different stress and suction conditions (see Table 5.8). The R^2 and S_e/S_y are found to be 0.98 and 0.14, suggesting a strong correlation between measured and calculated resilient moduli. The strong correlation implies that Equation (5.17) captures the variation of resilient moduli with net stress and matric suction generally. Comparing Figures 5.30a and 5.30b reveals that the results obtained from Equation (5.17) are in slightly better agreement with the corresponding experimental data along a drying path than that along a wetting path. As illustrated in Figure 5.29, the soil specimen tested along a wetting path behaves like an over-consolidated soil because of the wetting and drying history. However, the effects of wetting and drying are not considered by Equation (5.17). As a result, special attention should be taken when Equation (5.17) is used to predict the M_R of a soil specimen along a wetting path, although reasonably accurate predictions can still be made.

To further investigate the validity of Equation (5.17), it was applied to analyse the M_R of three other soils (Keuper Marl, Gault clay and London clay) as reported by Brown et al. (1987). Figure 5.31 compares the measured and calculated M_R of different types of soils. For each soil type, the measured and calculated resilient moduli match very well. It should be pointed out that soil suctions measured by Brown et al. (1986) range from 15 to 75 kPa. More high-quality experimental data are necessary to verify the validity of Equation (5.17) over a wider range of suctions. Close inspection of this figure reveals that the measured and calculated results for A-7-5 soil match even better than those for A-7-6 soil. One possible reason is that all the data obtained for A-7-5 are from the same study. Each test was performed using the same apparatus and test standard. Therefore, any experimental uncertainty is expected to be less for A-7-5 soil.

It can be seen from Figure 5.31 and Table 5.8 that the same values of k_1, k_2 and k_3 can be used to estimate M_R for the same type of soil. Parameter k_2 of each soil is negative as expected, considering that M_R decreases with q_{cyc} due to the nonlinearity of the soil stress-strain relationship. The result shows that A-7-6 has a smaller k_2 than A-7-5, implying that the M_R of A-7-6 soil decreases more rapidly with q_{cyc}. It could be explained by A-7-6 soil having a smaller plasticity index than A-7-5 soil. Vucetic and Dobry (1991) found that the degradation of soil stiffness with strain level is more significant for soils with smaller plasticity indexes. Parameter k_3 is positive for each soil, implying that M_R generally increases

248 Advanced Unsaturated Soil Mechanics

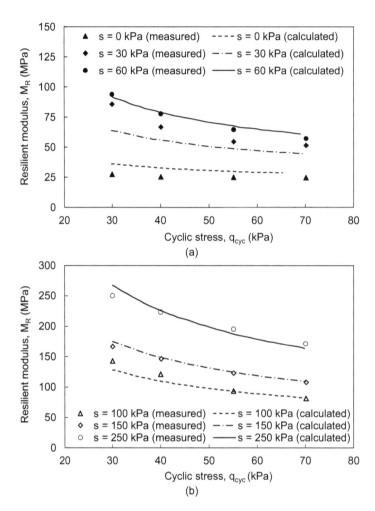

Figure 5.30 Comparisons between measured and calculated resilient modulus using the proposed semi-empirical equation: (a) wetting series; (b) drying series.

Source: After Ng et al., 2013.

with s because of the beneficial effect of air-water interfaces. Its value is also larger for A-7-5 soils, which consist of smaller particles, implying that the suction effect on M_R is more significant for soils containing more fine particles.

Summary

It is found that the M_R of unsaturated CDT increases with the number of load applications when a soil contracts under cyclic loads. On the other hand, M_R decreases slightly with increasing load applications when a soil dilates. An unsaturated soil specimen generally achieves a stable resilient response within 100 cycles of loading-unloading.

Shear stiffness and damping ratio 249

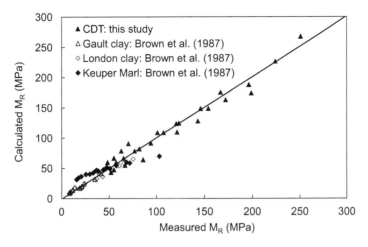

Figure 5.31 Comparison between measured and calculated resilient modulus using the proposed semi-empirical equation of different soils.

Source: After Ng et al., 2013.

M_R is found to be dependent on stress and suction level. It decreases with cyclic stress because soil's stress-strain behaviour under cyclic loads is highly nonlinear. On the other hand, M_R increases significantly with suction up to 250 kPa. When suction increases from 0 to 250 kPa, M_R increases by up to an order of magnitude. This is because an increase in suction induces an additional inter-particle normal force and hence stiffens the soil specimen.

Given the same stress and suction level, the M_R measured along a wetting path is greater than that measured along a drying path at low cyclic stress. The observed difference between the two paths becomes less significant with an increase in cyclic stress.

A new semi-empirical equation describing the stress-dependency of M_R for both saturated and unsaturated soils is proposed. This new equation captures the variation of M_R with both net stress and suction well.

SHEAR MODULUS AND DAMPING RATIO AT HIGH SUCTION OF 40 MPA OF SPECIMENS WITH DIFFERENT INITIAL MICROSTRUCTURES (NG ET AL., 2017B)

Introduction

Intact block samples (cubes 250 mm by 250 mm by 250 mm) of loess were retrieved manually from 7.5 m below the ground surface at an excavation site in Xi'an, Shaanxi Province, China. The sand, silt, and clay contents of loess are 0.1%, 71.9%, and 28.0%, respectively. Some other basic properties of loess are given in Chapter 4. Two series of tests were conducted. In the first series, the loess specimens were at the initial suction conditions. In the second series, the specimens were prepared and equalised at about 40 MPa suction induced by a saturated NaCl solution in desiccators using the vapour equilibrium technique

Table 5.9 Summary of testing conditions for intact, recompacted and reconstituted loess

Test ID	Suction: kPa	Dry density: kg/m³	Void ratio	Water content :%	Degree of saturation (%)
C-1.2-i	70	1183	1.271	11.8	25.1
C-1.3-i	62	1280	1.098	11.6	28.5
R-i	5	1493	0.801	27.3	92.0
C-1.2-40	40000	1190	1.256	2.8	6.0
C-1.3-40	40000	1283	1.095	2.8	6.9
R-40	40000	1621	0.657	2.7	10.8
I-40	40000	1297	1.072	2.8	7.1

Notes:

I = Intact (sampling depth: 7.5 m); C = Recompacted; R = Reconstituted; w_{opt} = 18.1%; $\rho_{d(max)}$ = 1680 kg/m³; Strain amplitude: (0.0003–0.1) %; Confining pressure: 74 kPa.

"i" and "40" represent the initial state and soil suction after wetting (unit: MPa), respectively. Furthermore, the numeric value "1.2" and "1.3" denotes dry densities of recompacted specimens with the unit of g/cm³.

proposed by Blatz et al. (2008). The targeted suction of 40 MPa was similar to that of the intact specimens determined using the filter paper technique (Ng and Menzies, 2007).

Once equilibrium was reached in desiccators, soil specimens were weighed and returned to the desiccators until just before tests were carried out in a novel fixed-free energy-injecting virtual mass resonant column device (Li et al., 1998). In this system, the shear modulus is measured continuously, and the number of vibration cycles and disturbance of the soil specimen are both minimised. The system also allows continuous measurement of shear modulus and damping ratio as shear strain increases. All tests were conducted under an isotropic confining pressure of 74 kPa, corresponding to the total mean stress at the sampling depth of the intact specimens. During the application of cell pressure, the axial strain of each specimen was monitored by an LVDT. It was found that the axial strain was almost negligible (less than 0.07%) for all specimens tested in this study. Details of the test programme are presented in Table 5.9. In addition to the footnote of Table 5.9, "i" and "40" represent the initial state and soil suction after wetting (unit: MPa), respectively. Further, the numeric values "1.2" and "1.3" denote dry densities of recompacted specimens in units of g/cm³.

Interpretations of experimental results

Shear modulus degradation curves of unsaturated loess at initial suctions

Figure 5.32 shows shear modulus as a function of shear strain amplitude for the reconstituted and recompacted test specimens at their initial suctions. The reconstituted specimen shows the lowest shear modulus amongst the three specimens compared for all strain levels. This specimen had the least suction and the lowest void ratio (see Table 5.9). Therefore, we inferred that the effect of initial suction dominates the effect of initial void ratio in enhancing the shear modulus of loess in this hydraulic zone. As discussed above, at high degrees of saturation, bulk-water effects determine the soil stiffness. Once the degree of saturation is reduced, meniscus water dominates soil stiffness and increases in suction will increase the small-strain shear stiffness of fine-grained soils (Mancuso et al., 2002;

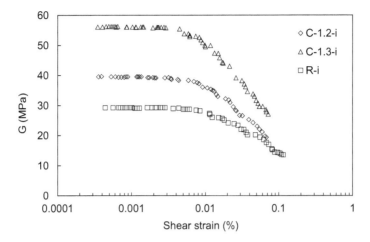

Figure 5.32 Shear modulus degradation curves at initial state.
Source: After Ng et al., 2017b.

Vassallo et al., 2007; Ng and Yung, 2008; Hoyos et al., 2015). The effect of the number of interfaces is captured by the specific air-water interface (the total area of the air-water interface per volume of a representative element volume (REV)). In the drying path, the specific air-water interface increases with a decreasing degree of saturation, until it reaches a maximum after which it decreases (Likos, 2014). On the other hand, the effects of void ratio dominate suction effects if the results of two recompacted specimens are compared. Test C-1.2-*i* with a higher void ratio had a consistently lower shear modulus than C-1.3-*i*. Even though suction in C-1.2-*i* was marginally higher than that in C-1.3-*i* (70 > 62 kPa), the effect of the void ratio remained significant. One possible reason is that both specimens were within the same hydraulic zone (transition) with a limited suction difference of 8 kPa while their void ratios were considerably different.

Shear modulus degradation curves after suction-induced desiccation

Variations in shear modulus with shear strain for four different types of specimen are shown in Figure 5.33. All specimens were tested after being equalised to 40 MPa of suction. By comparing the results in Figure 5.33 with those in Figure 5.32, a significant increase in shear modulus is observed as suction increases to 40 MPa for all specimens at different initial states. This observation is consistent with the results shown in Figure 5.7. However, a suction value as high as 40 MPa would not increase the meniscus water because the hydraulic state of the soil is within the residual zone. The water phase in the residual zone is not continuous and the amount of air-water interface reaches its minimum or vanishes (Lu and Likos, 2004). Therefore, the considerable increase in the shear modulus of desiccated specimens was caused by a different mechanism. As pointed out by Romero et al. (1999), water mainly fills the micro-pores of fine-grained soils at low degrees of saturation corresponding to high suction values. As a result, the microscopic degree of saturation is higher than the macroscopic one, which will result in a stiffer

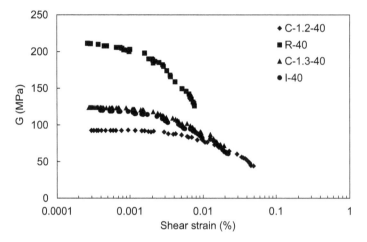

Figure 5.33 Shear modulus degradation curves after suction-induced to 40 MPa in desiccators.
Source: After Ng et al., 2017b.

aggregated structure and hence a higher shear modulus is expected. This will be discussed in the next section.

According to the results in Figure 5.33, the initial shear moduli were 92, 124 and 211 MPa for C-1.2-40, C-1.3-40 and R-40, respectively. The reconstituted specimen exhibited the highest modulus. This is because the shrinkage of the reconstituted specimen was more significant than those of the other two recompacted specimens under the application of high suction resulting in the lowest void ratio. According to the results of Table 5.9, a decrease in void ratio of approximately 0.15 corresponding to 8% of volumetric strain was observed for the reconstituted specimen. However, the maximum volumetric strain caused by desiccation in both recompacted specimens was less than 0.7%. The initial shear moduli of C-1.2-40, C-1.3-40 and R-40 increased respectively by 135%, 121% and 627% as suction was increased to 40 MPa. As a result, the effect of suction in increasing shear modulus is more pronounced for the reconstituted specimen than for the two recompacted specimens because of the significant microstructural modification to the reconstituted specimen.

The recompacted specimen (C-1.3-40) has a similar void ratio to that of the intact specimen (I-40) after desiccation-induced suction of 40 MPa (Table 5.9). It is interesting to note that the shear moduli of these two specimens are very close with a difference of less than 4% (Figure 5.33). Although intact and recompacted soils are considered to have different microstructures (Burton et al., 2015), their dynamic soil properties were found to be similar under the same suction and net stress. This observation suggests that even though the two intact and recompacted specimens experienced different stress and suction histories, the final void ratio and suction seems to control the shear modulus at high suctions. At the suction value considered, the influence of microstructure on the shear modulus of these specimens is very small with a percentage difference less than 4%.

The repeatability of test results was checked and confirmed by conducting tests on two identical specimens of each type before and after desiccation. Two typical repeated test results for two identical specimens before and after desiccation-induced suction of 40 MPa are shown in Figure 5.34. The results of repeated tests match those of the identical

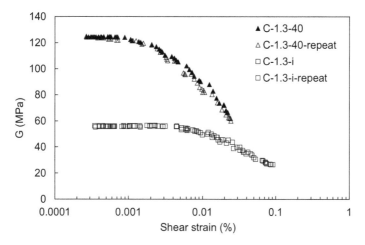

Figure 5.34 Comparison of the shear modulus degradation curves for two identical specimens.
Source: After Ng et al., 2017b.

specimens with the initial shear modulus differing by an average of 0.9%. Further, the maximum deviation of shear modulus for all specimens over the entire shear strain does not exceed 3%. This observation suggests good sampling quality as well as reproducibility of the test results.

Effect of desiccation-induced suction of 40 MPa on normalised shear modulus

Shear modulus degradation curves for all the test specimens were normalised by the initial value and are represented and discussed as normalised curves. Figure 5.35 shows the normalised shear modulus degradation curves for the recompacted and reconstituted specimens in the initial and after desiccation states. Because of the similarity between the results of the intact and recompacted specimens at a lower void ratio (i.e., C-1.3-40), the results for the intact specimen are not shown in this figure for clarity. In the very small-strain range, as the shear strain increased the shear modulus of the soil remained roughly constant and the soil behaved elastically. As the strain approached larger values, the normalised shear modulus was drastically reduced and the strain value separating these two ranges is called the elastic threshold shear strain (Vucetic, 1994). Previous research has quantified the elastic threshold shear strain of soils as the strain value corresponding to $G/G_{max} = 0.95$ (Clayton and Heymann, 2001; Ng and Xu, 2012). In this study, the same criterion was adopted to avoid subjective assessment of the elastic threshold shear strain values. According to Figure 5.35a, the elastic threshold shear strain value for the as-compacted state for the recompacted specimen with a higher void ratio (C-1.2-i) was approximately 0.007%. This value decreased to approximately 0.004% after desiccation and increasing suction to 40 MPa. Figures 5.35b and 5.35c indicate that similar trends in the reduction of threshold strain with increasing suction can be observed for the recompacted specimen with a lower void ratio and for the reconstituted specimen, respectively. However, the significance of suction-induced desiccation for reducing threshold strain is not the same for the three types of specimens studied. Based on the results shown

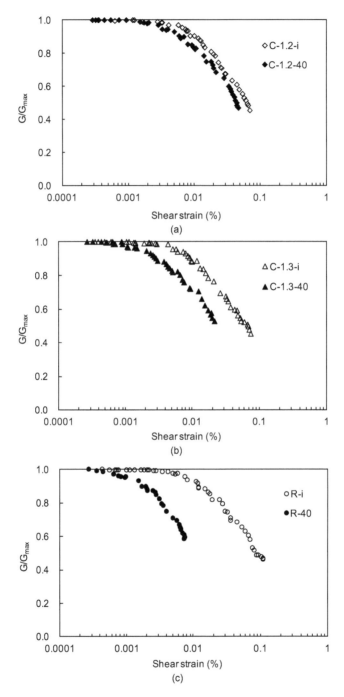

Figure 5.35 Effect of desiccation-induced suction on normalised shear modulus of (a) and (b) the recompacted specimens at two different void ratios, and (c) the reconstituted specimen.

Source: After Ng et al., 2017b.

in Figure 5.35b for the recompacted specimen with a lower void ratio, the elastic threshold shear strain decreased from 0.006% to 0.0021%, highlighting the more pronounced effect of suction on this specimen than on the specimen with a higher void ratio. The most noticeable effect of suction on threshold strain was seen for the reconstituted specimens in Figure 5.35c. The elastic threshold shear strain of the reconstituted specimens in the initial state was 0.0078% which is about one order of magnitude larger than the corresponding value (0.001%) for the specimen desiccated to 40 MPa of suction.

All results show a clear decrease in elastic threshold shear strain after the specimens were desiccated to a suction level beyond the residual suction value. As illustrated above, Ng and Xu (2012) investigated the effect of increasing suction on the elastic threshold shear strain of CDT (see Figure 5.13). An increase in the elastic threshold shear strain was reported caused by the increase in suction and decrease in degree of saturation. The suction range considered in Figure 5.13 was limited to the transition zone. In the transition zone, as suction increases (corresponding to a decrease in the degree of saturation) more air-water interfaces are generated in the soil specimen. The increase in air-water interfaces would increase the normal contact forces holding the particles together and decrease slippage between particles. As a result, the region of elastic behaviour expands, and a larger threshold shear strain is obtained. However, the soil specimens in Figure 5.35 were desiccated to a suction value beyond the transition zone and within the residual zone. How changes in the soil hydraulic state affect the observed mechanical response is discussed later.

Effects of microstructure on cyclic threshold shear strain at different states

Figure 5.36a shows the normalised shear modulus degradation curves for all specimens in their initial state. The two recompacted specimens (with similar as-compacted suctions) exhibited similar stiffness degradation patterns and as stated previously, the elastic threshold shear strain for C-1.2-*i* and C-1.3-*i* were approximately 0.007% and 0.006%, respectively. Of all three types of specimens with different initial microstructures, the reconstituted one having the least suction gave the largest elastic threshold shear strain of 0.0078%. If suction was the only factor affecting the threshold strain, the opposite trend would be expected. However, significant differences in the microstructure of the reconstituted and recompacted specimens will result in a slightly larger threshold strain for the former than the latter. A possible reason could be the existence of more inter-particle repulsion forces and interactions in the dispersed structure of the reconstituted specimen than in the flocculated structure of the recompacted ones.

Figure 5.36b shows the normalised shear modulus degradation curves for all specimens subjected to suction-induced desiccation. The elastic threshold shear strains for C-1.2-40, C-1.3-40 and R-40 were 0.0040%, 0.0021% and 0.0010%, respectively. In spite of the negligible influence of microstructure on threshold strain in the initial state (Figure 5.36a), the effect of microstructure on threshold strain is more significant for desiccated specimens. This is attributed to different evolutions of the microstructure of specimens with different initial states subjected to suction-induced desiccation (Burton et al., 2015). Of the three test results compared, R-40 and C-1.2-40 showed the smallest and the largest threshold shear strain, respectively. Threshold strain exhibited a clear increasing trend with increases in the void ratio. In other words, an increase in void ratio caused an upward shift in the degradation curve associated with enhanced threshold strain. Oztoprak and Bolton (2013), by analysing a database consisting of 454 tests on sands, showed that the elastic threshold

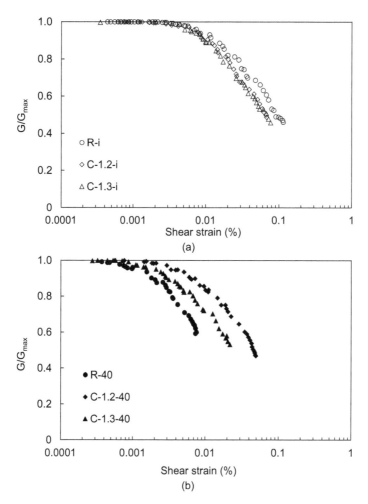

Figure 5.36 Comparison of the normalised shear modulus degradation curves at (a) initial state, and (b) after desiccation-induced suction of 40 MPa.

Source: After Ng et al., 2017b.

shear strain increases consistently with increasing void ratio. Although there was also a clear difference in the void ratio of specimens in their initial state, the influence of void ratio and microstructure on normalised degradation and threshold strain became more pronounced once the specimens were subjected to high suction. This observed behaviour is discussed in further detail later.

Effects of soil microstructure on shear stiffness

As shown in previous sections, suction-induced desiccation increases the initial shear modulus and has a noticeable influence on decreasing the elastic shear strain threshold for all specimens tested. It was also demonstrated that the suction value of 40 MPa considered in this study is within the residual state. According to Romero et al. (2011) and Romero (2013), water is stored inside the aggregates of fine-grained soils in this range of suctions. Therefore, only a small percentage of the soil water is held within the macro-pores and a limited number of menisci would form between the individual aggregates. However, the intra-aggregate water adsorption mechanism dominates in the residual zone and enhances the stiffening of individual aggregates. Previous studies of the microstructure of fine-grained soils by scanning electron microscope (SEM) and mercury intrusion porosimetry (MIP) have demonstrated that clay-silt aggregation occurs during suction-induced drying. The enhanced aggregation process would result in "aggregation stiffening" and the clayey material behaves essentially as a granular material (Merchán et al., 2011; Hoyos et al., 2014). Moreover, the microstructural evolution of specimens due to desiccation is affected by their initial state. According to the results, specimens with a lower void ratio experience more aggregation since the particles are closely bonded in the initial state. The effect of suction on the shear modulus of unsaturated soil is influenced by the amount of all three variables (saturation, the number of interfaces, void ratio/microstructural evolution) and competition between their effects. In the drying path, the contribution of the interfaces increases with decreasing degree of saturation (macro-structural saturation) (Likos, 2014), until it reaches the residual degree of saturation. Although structure formation and aggregation occur when starting from drying, the contribution of interfaces to the shear modulus and elastic threshold is more dominant at suction values below the residual zone (Nikooee et al., 2013). As suction increases above the residual zone, the effect of the evolution of the microstructure overcomes the effect from the increase in interfaces, where the soil is highly deformable, and the elastic threshold is influenced by the desiccation-induced structure (silt-clay aggregation). In order to validate the aggregation phenomenon for loess, SEM micrographs of the reconstituted specimens were taken. Reconstituted specimens were selected since the most significant effects of suction and microstructure on shear modulus and normalised degradation curves were observed for this type of specimen.

Plates 5.2a and b show the microstructure of reconstituted specimens in the initial state and after suction equalisation, respectively. Both images were taken at the same magnification for ease of comparison. By comparing these two figures, a qualitative change in aggregate size can be seen. Although there is a reduction in void ratio due to desiccation, a larger aggregated structure was formed. The enhanced silt-clay aggregation is highlighted in Plate 5.2c, which was taken at 2000 times magnification. The figure shows an aggregate dimension of 40 μm, which is four times larger than the mean particle size of loess obtained from sieve and hydrometer analyses. These observations demonstrate qualitatively that fine-grained soils tend to behave more like coarse-grained soil as high suction develops within the soil specimens. Vucetic and Dobry (1991) reported that the elastic threshold shear strains of sand and clay (mostly saturated) are on the order of 0.001% and 0.005%, respectively. These values suggest that clay has a consistently larger threshold strain than sand. As a result, a smaller threshold shear strain as well as a steeper normalised degradation curve can be expected at high suction (above residual), compared to a much lower range of suction for a fine-grained soil like loess.

Damping ratio

Figure 5.37a shows that the damping ratio exhibits a similar increasing pattern for all specimens in their initial state. The reconstituted specimen, however, had a slightly lower

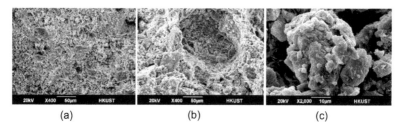

Plate 5.2 SEM micrographs demonstrating the evolution of microstructure in the reconstituted specimen: (a) initial state; (b) after desiccation at 40 MPa, and (c) a large grain formed due to desiccation.

increasing rate than the recompacted specimens. This observation is consistent with the shear modulus degradation curves beyond the elastic shear strain threshold, showing that the degradation rate of the reconstituted specimen was slightly less than that of the recompacted ones. After desiccation-induced suction of 40 MPa (Figure 5.37b), the rate of increase of the damping ratio is consistently lower for the recompacted specimen with a higher void ratio than for the other specimens. In addition, the reconstituted specimen showed the highest rate of increase in the damping ratio with strain. These trends are consistent with the trends previously reported for shear modulus degradation curves. However, the difference between the minimum damping ratios at very small strains for all specimens (in the initial state and at 40 MPa suction) is not significant and the minimum damping ratios fell in the limited range of 3 to 3.7%. Beyond the elastic threshold shear strain, increased interaction between soil particles increases energy dissipation and the damping ratio which will result in a reduction in shear modulus (Senetakis et al., 2013). The typical shear strain of 0.01% is considered to be the threshold between non-linear elastic and nonlinear plastic behaviour by other researchers (Hsu and Vucetic, 2004; Senetakis et al., 2013). In the current study, a comparison between damping ratio and reduced shear modulus was made at a shear strain of 0.007%. The reason was that no measured data are available beyond this level of strain for the reconstituted specimen at a suction of 40 MPa because of the limitations of the resonant column device used. Biglari et al. (2011a) showed that the damping ratio of unsaturated kaolin increases with increasing suction in the medium-to-large strain range. In the current study, beyond the elastic threshold shear strain, the damping ratio increases as suction increases to 40 MPa for all specimens with different initial microstructures. Moreover, for test R-40, the normalised shear modulus (G/G_{max}) and damping ratio were 0.62 and 11%, respectively. The corresponding values were (0.78, 8.5%) for C-1.3-40 and (0.89, 5.5%) for C-1.2-40. The comparisons at high suction imply that the recompacted specimen with a higher void ratio will have less energy dissipation and a lower degradation rate than the other specimens beyond the elastic threshold shear strain. As the void ratio of desiccated specimens decreases, a lower threshold shear strain associated with a steeper degradation curve and an accelerated rate of damping are expected for fine-grained unsaturated loess. The observed pattern for the damping ratio of the reconstituted specimens (with most significant observed effects of suction on microstructure) can be explained by the evolution of the microstructure. Comparison of SEM images before (Plate 5.2a) and after (Plate 5.2b)

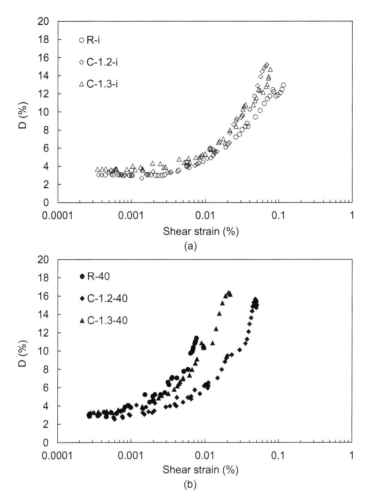

Figure 5.37 Variations in damping ratio with shear strain at (a) initial state, and (b) after desiccation-induced suction of 40 MPa.

Source: After Ng et al., 2017b.

suction-induced desiccation reveals the formation of larger silt-clay aggregates as well as larger pores due to drying. The aggregation process is also evident in Plate 5.2c in which a large silt-clay aggregate can be seen. This aggregate has a dimension of 40 μm, which is four times larger than the mean diameter of the loess particles studied. Microstructural observations suggest aggregation of soil particles caused by severe drying to high suction. Note that the coarse-grained soil shows smaller threshold shear strains as compared to the saturated fine-grained soils. In other words, particles in a coarse-grained soil show relative displacements at a smaller shear strain compared to a fine-grained soil. In this study, the reconstituted specimen in the initial state (nearly saturated) has a damping pattern similar to that of saturated low-plasticity silts. While specimens at high suction (40 MPa) show higher energy dissipation (rate of damping ratio increases) beyond the elastic shear strain threshold, the damping characteristics of the desiccated specimen are similar to that of a coarse-grained soil.

Comparison of test results with design charts

Vardanega and Bolton (2013) proposed design charts to estimate shear modulus reduction curves for fine-grained materials by analysing a database of 67 types of clay and silt (mostly saturated). They found that for fine-grained soil, the plasticity index is the most significant variable influencing the normalised shear modulus reduction curve. As the plasticity index increases, the elastic shear strain threshold value shifts towards higher levels. As stated previously by Ng and Xu (2012) increases in suction below the residual suction zone will increase meniscus water in soil specimens and shift the degradation curve to higher elastic shear strain values. However, as pointed out in the discussion section, the specimens in the current study are in the residual suction zone and increasing suction does not increase air-water interfaces. In order to explain the effect of silt-clay particle aggregation, all data from the current study are summarised in Figure 5.38 together with design curves for plasticity indexes of zero, 100, and 15 (similar to the plasticity index of 17 for loess) for comparison purpose. The curve proposed by Vardanega and Bolton (2013) at PI=15 describes the response of loess at low suction satisfactorily. However, there is a downward shift in the degradation curve towards the lower plasticity index values as suction increases. Vucetic and Dobry (1991) stated that soils with a higher plasticity index are composed of very small particles with a high surface area per unit weight of particles and greater inter-particle contact. On the other hand, soils with a lower plasticity index (e.g., sands) consist of larger particles and have less inter-particle contact. Towhata (2008) attributed this behaviour to the discreteness of soil particles. It implies that soils with lower plasticity (granular) are highly discrete, while soils with higher plasticity and smaller

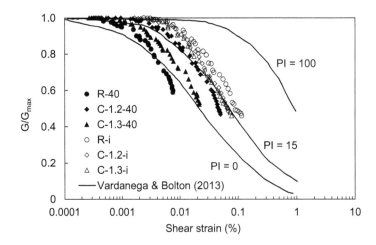

Figure 5.38 Summary of all measured degradation curves and design curves by Vardanega and Bolton (2013).

Source: After Ng et al., 2017b.

particles are less discrete. By the same explanation, soils with small particles can endure larger shear strains before the particles are permanently displaced and degradation effects appear. In the current study, although suction does not affect the plasticity index of loess, it does affect aggregate stiffening and the development of silt-clay particle aggregation. Therefore, it is possible that the accelerated degradation of the normalised modulus with strain in desiccated soil specimens arises from the more aggregated structure associated with less inter-aggregate contact.

Summary

In the initial state, recompacted specimens had a higher shear modulus than the reconstituted one because of higher values of initial (as-compacted) suction. At the desiccation-induced suction of 40 MPa, specimens with lower void ratios showed a higher shear modulus. Moreover, the shear modulus of all specimens increased with suction. The contribution of suction to shear modulus is attributed to stiffening effects caused by aggregation and can be explained within the framework of microstructural-based effective stress.

There was a minor difference in the elastic shear strain threshold of different specimens in the initial state. However, the elastic shear strain threshold decreased significantly (decrease in elastic range) as suction reached values as high as 40 MPa. Furthermore, at higher suctions, both damping ratio and shear modulus degradation reached higher rates. Specimens with lower void ratios attained higher rates of stiffness degradation and lower elastic shear strain thresholds after suction-induced desiccation. This is attributed to different evolutions of the microstructure of specimens with different initial states when subjected to suction-induced desiccation.

The reconstituted specimen at the initial state (nearly saturated) has a damping pattern similar to that of saturated low-plasticity silts. While specimens at high suction of 40 MPa show higher energy dissipation (rate of damping ratio increase) beyond the elastic shear strain threshold. Consequently, the damping characteristics of the desiccated specimen at 40 kPa are similar to that of a coarse-grained soil.

A clear shift of degradation curves towards lower plasticity index design curves was observed as high suction developed in the soil specimens. This was possibly because suction-induced large silt-clay aggregates, which were also demonstrated by SEM observations, would increase the discreteness (i.e., representing how soil particles are separated) of fine-grained soil. Soils with higher discreteness exhibit greater nonlinearity. Therefore, fine-grained materials desiccated to high suctions would behave like coarse-grained sand or gravel.

SOURCES

Extended Mohr– Coulomb failure criterion (Fredlund and Rahardjo, 1993)
Comparisons of axis- translation and osmotic techniques for shear testing (Ng et al., 2007)
Effects of soil suction on dilatancy (Ng and Zhou, 2005) Effect of microstructure on shear strength and dilatancy of loess at high suction (Ng et al., 2016a; Ng et al., 2020b)
Comparisons of shear strength of weathered lateritic, granitic and volcanic soils: compressibility and shear strength (Ng et al., 2019c)

Chapter 6

Thermal effects on soil behaviour and properties

CYCLIC SHEAR BEHAVIOUR OF SILT AT VARIOUS SUCTIONS AND TEMPERATURES (NG AND ZHOU, 2014)

Introduction

As explained in Chapter 5, the deformation of subgrade soils under cyclic traffic loads is crucial to the performance of pavement and earthen infrastructures. In the field, subgrade soils are usually unsaturated and subjected to daily variations in temperature and water content (and hence pore-water pressure (or suction)) (Jin et al., 1994; Ng and Zhou, 2014). Many theoretical and experimental studies have demonstrated that the behaviour of unsaturated soil is dependent not only on suction but also on temperature (Romero et al., 2003; Tang et al., 2008; François and Laloui, 2008; Zhou and Ng, 2013). In this section, a series of cyclic triaxial tests on recompacted CDT at different temperatures are reported. The scope of this study is to investigate the influence of the number of cycles, cyclic stress, suction and temperature on the cyclic behaviour of unsaturated silt, including the accumulation of plastic strain and resilient modulus, M_R, which was defined by Seed et al. (1962) as the ratio of repeated deviator stress to axial recoverable strain. Accumulated plastic strain and resilient modulus are widely used in pavement engineering to predict irreversible and reversible deformations, respectively.

Test apparatus

To test unsaturated soil at various suctions and temperatures, the suction-controlled triaxial apparatus shown in Plate 4.4 and Figure 6.1 was used. The heating system installed in the triaxial apparatus consists of a thermostat, a heater and thermocouple A. Both are connected to the thermostat, forming a closed-loop control and feedback system. During each test, the thermostat adjusts the heater output based on feedback from thermocouple A. When the energy dissipated and the heater output are balanced, the air temperature inside the triaxial cell reaches equilibrium. Depending on the target temperature, 1–3 hours are normally required to achieve thermal equilibrium. To enhance the uniformity of temperature inside the triaxial cell, two small fans and a hollow aluminium cylinder are installed in the triaxial cell. Thermocouples are used to measure the air temperature about 5 mm away from the soil specimen. When thermal equalisation is reached, the readings of the two thermocouples will be almost identical. They remain fairly constant with a maximum fluctuation of ±0.5°C in this apparatus.

Thermal effects on soil behaviour and properties 263

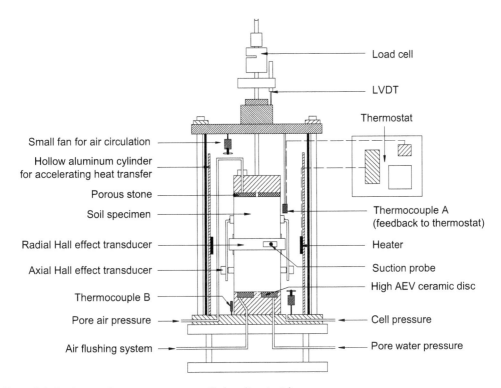

Figure 6.1 Suction- and temperature-controlled cyclic triaxial system.

Test program and procedures

Seven suction- and temperature-controlled cyclic triaxial tests were carried out on recompacted CDT (completely decomposed tuff) (refer to Figure 6.2). Three of them (S0T20, S30T20, S60T20) were carried out at 20°C, but at different suctions (0, 30 and 60 kPa). The other four tests (S0T40, S0T60, S30T40 and S60T40) were conducted at elevated temperatures (40°C and 60°C). Each test consisted of three stages: suction equalisation, thermal equalisation and cyclic loading-unloading. Each test takes about 20 days. Details of the test program are given in Table 6.1.

Figure 6.2 shows the thermo-hydro-mechanical path of each specimen in the first two stages. After compaction, each specimen was set up in the triaxial system. The initial stress state of each specimen was controlled at point A. Each specimen was first isotropically compressed to a net confining stress of 30 kPa at constant water content (A→B). After isotropic compression, each soil specimen was then wetted by decreasing suction from 95 kPa to 0, 30 and 60 kPa (B→E, B→D and B→C). To control soil suction using the axis-translation technique, predefined u_a and u_w were applied at the top and bottom, respectively.

The second stage of each test was thermal equalisation. For the three tests carried out at room temperature, i.e., 20°C (S0T20, S30T20, S60T20), it was not necessary to change the soil temperature. For the four tests carried out at elevated temperatures, however, three

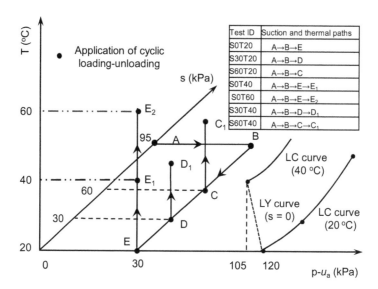

Figure 6.2 Thermo-hydro-mechanical path of each specimen prior to cyclic loading-unloading.

Table 6.1 Summary of testing conditions for temperature-controlled cyclic shear tests

Test ID	Suction: kPa	Temperature: °C	Initial water content: %	Initial void ratio	sS_r after suction and thermal equalisation
S0T20	0	20	16.6	0.57	0
S30T20	30	20	16.6	0.59	26
S60T20	60	20	16.5	0.56	50
S0T40	0	40	16.0	0.56	0
S0T60	0	60	16.4	0.56	0
S30T40	30	40	16.4	0.58	26
S60T40	60	40	16.4	0.58	50

specimens (S0T40, S30T40 and S60T40) were heated to 40°C (E→E$_1$, D→D$_1$, and C→C$_1$), while a fourth (S0T60) was heated to 60°C (E→E$_2$). It should be noted that thermal loading may induce excess pore-water pressure. In this study, three to five days were needed to dissipate the excess pore-water pressure, depending on suction and temperature conditions. This long duration allowed each soil specimen to reach the predefined temperature and suction conditions.

The compression curves of CDT (see Figure 5.6) are used to assist the interpretation of the current data. At suctions of 0, 50 and 100 kPa, the yield stresses of compacted specimens were found to be about 120, 170 and 200 kPa, respectively. Based on these results, the loading collapse (LC) curve at 20°C was determined and is shown in Figure 6.2. It is clear that the yield stress increases with increasing suction (suction hardening; Alonso et al., 1990; Chiu and Ng, 2003; Zhou et al., 2015a). In addition, an isotropic compression test was carried out at zero suction and 40°C in this study. The corresponding yield stress was about 105 kPa. Assuming that the shape of the LC curve is independent of temperature, the LC curve at 40°C was deduced and is also shown in the figure. As expected, yield stress decreases with increasing temperature at a given suction (thermal softening; Hueckel and Borsetto, 1990; Cui et al., 2000; Laloui and Cekerevac, 2003; Zhou and Ng, 2015, 2016). The evolution of the yield surface with temperature is described by the position of the loading yield (LY) curve. It is obvious that the stress paths during the first two stages are all within the LC curves. Consequently, the volume change induced by wetting and heating was small (less than 0.1%).

After equalising both suction and temperature, a cyclic triaxial test was carried out on each specimen to investigate the soil deformation under cyclic loads. The procedures are identical to those reported in Chapter 5 (see Figure 5.23). To investigate soil deformation under cyclic loads quantitatively, two parameters are considered and used. The first parameter is axial plastic strain (ε_a^p) accumulated during 100 cycles of loading-unloading. The other parameter is M_R. As summarised by Brown (1996), ε_a^p and M_R are the most important parameters for analysing cracking and rutting failures of a pavement.

Interpretation of experimental results

Suction effects on axial plastic strain accumulation

Figure 6.3 shows the relationship between ε_a^p and q_{cyc}. It can be seen that ε_a^p increases with an increase in q_{cyc} under all suction and temperature conditions, but at different rates. At zero suction, measured ε_a^p increases exponentially with an increase in q_{cyc}. At suctions of 30 and 60 kPa, the relationship between q_{cyc} and ε_a^p is almost linear. Further, at a given stress and temperature condition, ε_a^p is consistently lower at higher suctions. The decrease in axial plastic strain with increasing suction may be caused by several factors. First, voids in unsaturated soil are occupied by both water and air. The Bishop's stress (σ^*) between soil particles may be expressed as (Gallipoli et al., 2003b):

$$\sigma^* = \sigma_{ij} - u_a + S_r(u_a - u_w) \tag{6.1}$$

where σ_{ij} is the total stress, u_a is the pore air pressure, u_w is the pore-water pressure, and S_r is the degree of saturation. It can be seen that σ^* depends not only on net stress but also on the product of the degree of saturation and suction. As shown in Table 6.1, $S_r(u_a - u_w)$ increases with increasing suction over the suction range considered. An increase in $S_r(u_a - u_w)$ and hence σ^* minimises the possibility of slippage at particle contacts and therefore reduces the accumulation of axial plastic strain.

Second, soil water may be divided into bulk water and meniscus water (Wheeler and Karube, 1996). As for net stress, pore-water pressure in bulk water affects both the normal and tangent forces between soil particles. The effect of bulk water on mechanical behaviour can be captured using Equation (6.1). On the contrary, pore-water pressure in meniscus water only affects the normal forces at particle contacts. Meniscus water provides additional bonding forces at the particle contacts and hence stabilises the soil skeleton (Wheeler and Karube, 1996). This effect may not be captured by Equation (6.1). The normal force induced by meniscus water between two spherical particles can be expressed as:

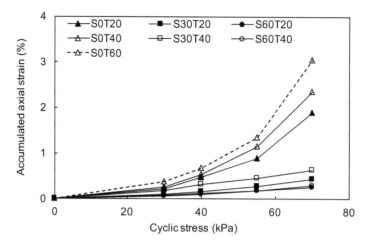

Figure 6.3 Measured plastic axial strain under cyclic loading-unloading.

$$F_n = \frac{\pi T_s^2}{(u_a - u_w)} \frac{\left(\sqrt{9 + 8R_p(u_a - u_w)/T_s} - 3\right)\left(\sqrt{9 + 8R(u_a - u_w)/T_s} + 1\right)}{4} \quad (6.2)$$

where T_s is the coefficient of surface tension, and R_p is the radius of a spherical particle. Based on this equation, the normal force F_n is greater at higher $(u_a - u_w)$. Therefore, as suction increases, the stabilisation effect induced by meniscus water becomes more significant. The accumulation of axial plastic strain is thus reduced.

Thermal effects on axial plastic strain accumulation

The influence of temperature on ε_a^p is clearly illustrated in Figure 6.3. At a given suction and cyclic stress, ε_a^p is consistently higher at 40°C than at 20°C. Cekerevac and Laloui (2010) carried out a series of temperature-controlled cyclic triaxial tests on saturated reconstituted kaolin clay in a normally consolidated state. They observed that cyclic loading-unloading on the heated sample induced a smaller axial strain per cycle than cyclic loading-unloading on the unheated sample. The seeming contradiction between their study and the present one may arise because the soil specimens in the two studies were heated at different over-consolidation ratios (OCRs). According to thermo-hydro-mechanical models (e.g., François and Laloui, 2008), yield stress decreases with an increase in temperature in the p^*-T plane, where p^* is mean Bishop's stress and T is temperature. For over-consolidated soil (CDT in this study), the soil stress state was inside the LY surface during the heating process, as illustrated in Figure 6.2. The apparent over-consolidation ratio decreased as the temperature increased (thermal softening). Given a smaller over-consolidation ratio, a greater axial plastic strain developed in the heated specimen during the subsequent stage of cyclic loading-unloading. For normally consolidated soil, the soil stress state lies on the yield surface. Heating a soil specimen in the drained condition does not alter the pre-consolidation pressure and over-consolidation ratio. This is because the effects of thermal softening could be compensated by the effects of strain hardening. Cekerevac and Laloui (2010) measured contractive volumetric strains of up to 1% when normally consolidated kaolin clay was heated from 22 to 90°C. The thermal loads densified the soil specimen. During the subsequent cyclic loading and unloading stage, a smaller axial plastic strain was induced in the heated specimen than in the unheated specimen.

It is demonstrated in Figure 6.3 that the thermal effects of accumulated axial plastic strain are not obvious at a suction of 60 kPa. The over-consolidation ratio of the soil specimen is up to about 6 and the axial plastic strain induced by the cyclic loads is very small. This is probably because the pre-consolidation pressure increases significantly with suction, as shown in Figure 6.2. The thermal effects on plastic strain accumulation are too small to be measured.

Based on a series of cyclic triaxial tests on unsaturated compacted fine-grained soil, Brown (1996) found that there is a threshold q_{cyc} above which the accumulation of ε_a^p becomes much greater. For the CDT tested in this study, at zero suction, ε_a^p increases non-linearly with an increase in q_{cyc} at an increasing rate. The threshold q_{cyc} is estimated to be between 40 and 55 kPa. On the contrary, the threshold q_{cyc} at suctions of 30 and 60 kPa should be greater than 70 kPa. The increase of threshold q_{cyc} with suction is probably caused by suction hardening. As shown in Figure 6.2, the yield stress increases as suction increases. Given a larger yield stress, a larger q_{cyc} is required to induce a significant accumulation of axial plastic strain. For CDT, the thermal effect is much less significant than the suction effect on threshold q_{cyc}. This is probably because thermal softening is relatively less important than suction hardening within the range of suctions and temperatures considered.

Influence of temperature on resilient modulus

Following the AASHTO (2017) standard, during the last five cycles the resilient modulus M_R was determined for each stress level. Figure 6.4 shows the influence of temperature on M_R at zero suction. As the temperature increases from 20°C to 60°C, M_R increases by about 5% at cyclic stresses of 30 and 40 kPa but decreases by about 5% at cyclic stresses of 55 and 70 kPa. The two different responses may be because temperature can affect M_R through different mechanisms. On the one hand, the electrical force between soil particles becomes relevant for fine-grained soil. Based on the DLVO (Derjaguin, Landau, Vervey, and Overbeek) double layer theory (Israelachvili, 2011), the repulsive electric forces between soil particles increase as temperature increases. This would reduce inter-particle forces and hence cause a decrease in soil stiffness (Santamarina et al., 2001). On the other hand, the effect of soil densification on M_R is more significant at higher temperature (see Figure 6.3). These two mechanisms appear to produce opposing effects.

Figure 6.5 shows the influence of temperature on M_R at different suctions. Two different responses can be identified. At zero suction, the difference in M_R induced by temperature is less than 5 MPa. At suctions of 30 and 60 kPa, M_R measured at 40°C is less than that measured at 20°C by up to 10 MPa. This illustrates that the thermal effects on M_R are greater at suctions of 30 and 60 kPa than at zero suction. Through a series of suction- and temperature-controlled triaxial tests, Uchaipichat and Khalili (2009) observed that for over-consolidated soil, the yield surface shrinks with an increase in temperature. Moreover, the size of the yield surface is more sensitive to temperature at higher matric suctions.

268 Advanced Unsaturated Soil Mechanics

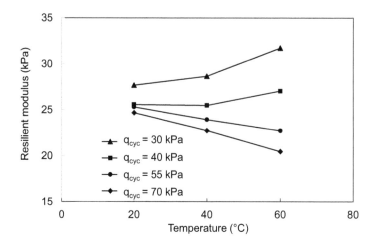

Figure 6.4 Influence of temperature on resilient modulus at zero suction.

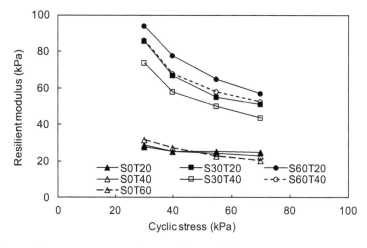

Figure 6.5 Influence of temperature on resilient modulus at different suctions.

The results of their study and the current one suggest that thermal effects on mechanical behaviour are more significant in the unsaturated state than in the saturated state. This may be attributed to the temperature dependence of surface tension. The surface tensions at temperatures of 20, 40 and 60°C are 72.8, 69.6 and 66.2 mN/m, respectively (Ng and Menzies, 2007). A decrease in surface tension affects the behaviour of unsaturated soil in at least two ways. Firstly, as temperature increases, the water retention capacity of unsaturated soil decreases slightly and hence the degree of saturation at a given suction becomes smaller (Romero et al., 2001). A reduction in the degree of saturation would induce a decrease in Bishop's stress, as illustrated by Equation (6.1). Second, Equation

(6.2) suggests that the normal forces induced by meniscus water decrease with decreasing surface tension. For these two reasons, the contribution of suction to soil stiffness may be reduced by thermal loading. Thermal effects on the mechanical behaviour of unsaturated soil would be more significant in the unsaturated state.

The figure reveals that the influence of q_{cyc} on M_R can be classified into two types. As q_{cyc} increases from 30 to 70 kPa, M_R is almost constant at zero suction but decreases by about 40% at suctions of 30 and 60 kPa. This is related to the strain dependency of soil stiffness. Researchers have found that soil stiffness decreases with an increase in strain level for both saturated soil (Atkinson, 2000) and unsaturated soil (Ng and Xu, 2012). The reversible axial strain (ε_a^r) can be calculated using the following equation:

$$\varepsilon_a^r = \frac{q_{cyc}}{M_R} \tag{6.3}$$

Using Equation (6.3), ε_a^r during the last 10 cycles in each test can be estimated from M_R and q_{cyc}. For instance, when cyclic stress increases from 30 to 70 kPa at zero suction and 20°C, ε_a^r increases from 0.1% to 0.3%. An increase in strain induces a decrease in M_R.

The fact that the reduction of M_R is more significant at suctions of 30 and 60 kPa suggests that the degree of non-linearity of soil stress-strain behaviour depends on the suction. As discussed above, for the four levels of q_{cyc}, ε_a^r at zero suction ranges between 0.1% and 0.3%. Similarly, ε_a^r at suctions of 30 and 60 kPa is in the range between 0.03% and 0.15%. On the other hand, Xu (2011) and Ng and Xu (2012) carried out a series of suction-controlled triaxial tests to investigate the small strain stiffness of unsaturated CDT at three suctions (1, 150 and 300 kPa). The soil specimens were prepared using the same method as that used in the current study. Their experimental results showed that the degradation rate of soil stiffness with strain is slightly affected by suction. At suctions up to 300 kPa, the rate of stiffness degradation is consistently larger in the strain range from 0.03% to 0.15% than that in the strain range from 0.1% to 0.3%. Therefore, the rate of decrease of M_R with q_{cyc} is much greater at suctions of 30 and 60 kPa than at zero suction.

Based on the results shown in Figures 6.3 to 6.5, it is clear that the cyclic behaviour of subgrade soil, including the accumulation of axial plastic strain and resilient modulus, is affected by temperature. As far as the authors are aware, current pavement design methods do not take thermal effects on soil behaviour into account since soil specimens are only required to be tested at room temperature (Highways-Agency, 2006; AASHTO, 2008). When the *in situ* temperature is significantly higher than room temperature, ignoring the thermal effects may underestimate the soil deformation induced by cyclic traffic loads. This is because at higher *in situ* temperatures, plastic strain increases (see Figure 6.3) and the resilient modulus decreases (see Figure 6.5) at suctions of 30 and 60 kPa. Pavement design using soil parameters determined at room temperature can therefore fail to be conservative at these suctions.

Influence of suction on resilient modulus at various temperatures

Figure 6.6a shows the influence of S on M_R at 20°C. At each level of q_{cyc}, M_R increases at a decreasing rate non-linearly with an increase in s. For example, at a q_{cyc} of 30 kPa, M_R more than doubles when s increases from 0 to 30 kPa. Then when s increases further from

30 to 60 kPa, M_R increases by only 10%. Figure 6.6b shows the influence of s on M_R at 40°C. The measured M_R also increases at a decreasing rate with suction. These figures illustrate that the suction effects on M_R are more significant in the low suction range. Similarly, Mancuso et al. (2002) carried out suction-controlled resonant column-torsional shear tests to investigate the very small strain stiffness (G_0) of unsaturated soil. They observed that the rate of increase of G_0 is highest when the soil specimen starts to desaturate. A comparison between Figures 6.3 and 6.5 illustrates that the suction effects on both M_R and ε_a^p are greater in the lower suction range (from 0 to 30 kPa). These observations may be related to the water distribution within unsaturated soil. In this study, soil specimens were wetted from 95 kPa to 60, 30 and 0 kPa. The adsorption rate along the wetting path changes abruptly at a suction of about 30 kPa. This may indicate that when the suction is less than 30 kPa, the effects of the air-water interface may start to diminish and hence the stabilising effects described by Equation (6.2) may no longer be relevant. The soil should change from meniscus-water-governed behaviour to bulk-water-regulated behaviour. Thus, the soil stiffness decreases significantly when s is lowered from 30 to 0 kPa. It should be pointed out that due to the limited suction range applied in the water retention test, it is very likely that the residual condition was not reached during drying. On subsequently wetting the soil specimen, it followed a scanning wetting path, not the primary one. If so, it is expected that different scanning wetting paths could result in different estimates of the suction at which the adsorption rate changes abruptly.

A comparison between Figures 6.6a and 6.6b shows that the effect of suction on M_R is slightly more significant at 20°C. This may be attributed to the decrease in surface tension with temperature, as discussed in the preceding section. Some current design guides have considered the effects of soil moisture on soil behaviour using empirical methods. For example, AASHTO (2008) for pavement design suggests that a soil specimen is prepared and tested at a reference moisture condition at or near the optimum water content. Variations in the soil's behaviour with soil moisture are then estimated using empirical models. Since the suction effects on soil behaviour depend on temperature (see Figure 6.6), the regression parameters in these empirical models should be a function of temperature. To incorporate coupled suction and thermal effects on soil behaviour, suction and temperature-controlled cyclic tests are recommended for calibrating the empirical models.

Summary

At a given cyclic stress, the measured axial plastic strain is significantly larger at lower suctions and higher temperatures. This is because the yield stress of unsaturated silt increases with increasing suction (suction-induced hardening) but decreases with increasing temperature (thermal-induced softening). The axial plastic strain induced by cyclic loads increases non-linearly with an increase in q_{cyc}. There is a threshold q_{cyc} above which the accumulation of ε_a^p becomes much more significant. The threshold q_{cyc} increases with suction. This may be because the Bishop's stress and stabilisation effects induced by meniscus water in the unsaturated soil increase with increasing suction.

The influence of temperature on M_R depends on the suction level and cyclic stress. At a given temperature, the resilient modulus increases more than twofold when the suction increases from 0 to 60 kPa. At zero suction, as the temperature increases from 20 to 60°C, the resilient modulus increases by about 5% at low cyclic stress but decreases by about 5%

Thermal effects on soil behaviour and properties 271

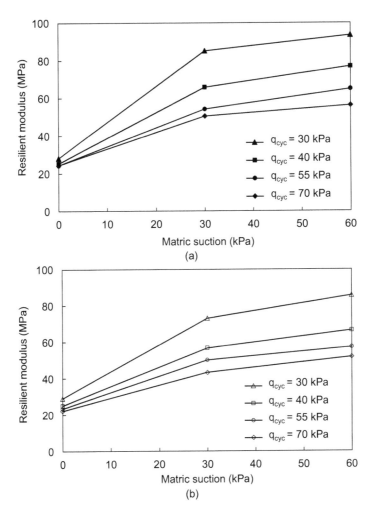

Figure 6.6 Influence of suction on resilient modulus at different temperatures: (a) 20°C; (b) 40°C.

at high cyclic stress. At suctions of 30 and 60 kPa, the resilient modulus at each level of cyclic stress decreases by about 15% as the temperature increases from 20 to 40°C. These observations reveal that thermal effects on resilient modulus are more significant in unsaturated than in saturated states. This may be because when the temperature increases, both the Bishop's stress between soil particles and the stabilisation effect induced by meniscus water decrease.

It is evident that the cyclic behaviour of subgrade soil, including the accumulation of axial plastic strain and resilient modulus, is affected by temperature. Since current pavement design methods do not normally take thermal effects on soil behaviour into consideration, soil deformation induced by cyclic traffic loads may be underestimated when the *in situ* temperature is significantly higher than room temperature.

For empirically based design guides that take into account the effects of soil moisture on soil behaviour and property, the regression parameters in these empirical models should be also affected by temperature because suction effects on soil properties are a function of temperature. To incorporate the coupling effects of suction and temperature on soil behaviour and property, measured results from suction and temperature-controlled cyclic tests should be adopted to calibrate these empirical models.

EFFECTS OF SOIL MICROSTRUCTURE ON THE SHEAR BEHAVIOUR OF LOESS AT DIFFERENT SUCTIONS AND TEMPERATURES (NG ET AL., 2016C)

Introduction

The shear behaviour of unsaturated soil at different suctions and temperatures is important to analyse the ultimate limit state of geothermal structures. Examples of these structures include nuclear waste disposal barriers (Romero et al., 2003; Tang et al., 2008) and energy piles (Cekerevac and Laloui, 2004; Ng et al., 2014a, 2021a, 2022a), which are often constructed in both intact and recompacted soils. Because of the inherent soil structure, intact soil differs in behaviour from recompacted soil in terms of mechanical and hydraulic properties, as expected (Callisto and Rampello, 2004; Muñoz-Castelblanco et al., 2012a; Wen and Yan, 2014; Ng et al., 2016a). The effects of suction (Ng and Zhou, 2005; Zhan and Ng, 2006; Sheng et al., 2011) and temperature (Uchaipichat and Khalili, 2009; Alsherif and Mccartney, 2015) on soil shear behaviour have mainly been studied for recompacted soil.

The objective of this section is to report the effects of soil micro-structure on the shear behaviour of an unsaturated loess at different suctions and temperatures. The basic properties of loess are given in Chapter 4. Two series of direct shear tests at different suctions and temperatures were conducted on intact and recompacted loess specimens. In order to compare measured results consistently between the tests, all specimens were prepared at almost the same initial dry density (with a maximum difference of about 1%), and then subjected to the same wetting path using the same apparatus. The microstructures of intact and recompacted loess in the *in situ* and as-compacted states respectively were measured using a scanning electron microscope (SEM) and a mercury intrusion porosimeter (MIP). The microstructure measurements were interpreted to reveal the difference in shear behaviour between unsaturated intact and recompacted loess.

Test apparatus and program

Figure 6.7 and Plate 4.3 show a recently developed direct shear box with independent control of temperature and suction. To control temperature, a heating system (Ng and Zhou, 2014) was added to the shear box. The system consisted of a thermostat, four heaters and two thermocouples. The thermocouples and heaters were connected to the thermostat to form a closed-loop control and feedback system. During testing, the heat output of the thermostat was adjusted automatically based on feedback from the thermocouple. The soil specimen was heated and maintained at the target temperature. To improve the uniformity of temperature, the four heaters were placed around the chamber and a small fan was installed to accelerate the air circulation. When the temperature reached equilibrium, the two thermocouples, which were set at different locations, showed almost identical measurements (differing by less than 0.1°C). The temperature fluctuated by less than 0.2°C.

Two series of tests (see Table 6.2) were carried out using the new apparatus. For the first series, the shear behaviour of intact and recompacted loess at different suctions and at 20°C was studied. The suction ranged from 0 to 200 kPa. The second series of suction- and temperature-controlled direct shear tests was designed to study thermal effects on the shear behaviour of unsaturated intact and recompacted loess at suctions of 0 and 200 kPa. The temperatures tested were 20 and 60°C. For these two test series, a constant net normal stress of 50 kPa similar to the *in situ* vertical stress was applied. More details of each specimen, such as the water content, void ratio and degree of saturation, are shown in Table 6.2.

Test procedure

Figure 6.8 shows the thermo-hydro-mechanical path of each test before shearing. After specimen preparation, the initial stress state of both recompacted and intact specimens was controlled at point O (suction=223 kPa) and O' (suction=239 kPa) respectively. It should be noted that the paths O-A and O'-A' represent constant water content loading conditions. The soil suction slightly decreased because soil contraction occurred (Muñoz-Castelblanco et al., 2011). A net normal stress of 50 kPa was first applied at constant water

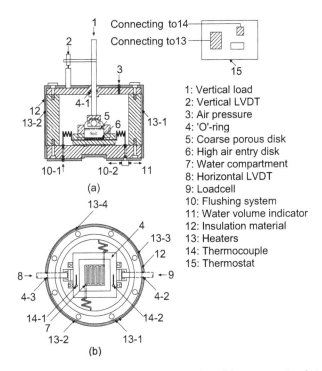

Figure 6.7 Suction- and temperature-controlled direct shear box: (a) cross-sectional view; (b) plan view.

Table 6.2 Summary of testing conditions for suction and temperature-controlled direct shear tests on loess

Series	Test ID	Suction: kPa	Stress path before shearing	w: %	e	S_r: %
I (20°C)	IS0T20	0	O'-A'-B1-E1	9.2	1.08	22.9
	IS50T20	50	O'-A'-B1-D	9.2	1.06	23.6
	IS100T20	100	O'-A'-B1-C	9.2	1.09	22.7
	IS200T20	200	O'-A'-B1	9.2	1.07	23.1
	RS0T20	0	O-A-B1-E1	9.5	1.11	23.0
	RS50T20	50	O-A-B1-D	9.5	1.11	23.0
	RS100T20	100	O-A-B1-C	9.5	1.11	23.0
	RS200T20	200	O-A-B1	9.5	1.11	23.0
II (60°C)	IS0T60	0	O'-A'-B1-E1-E2	9.2	1.09	22.7
	RS0T60	0	O-A-B1-E1-E2	9.5	1.11	23.0
	IS200T60	200	O'-A'-B1-B2	9.2	1.09	22.7
	RS200T60	200	O-A-B1-B2	9.5	1.11	23.0

content for each specimen (i.e., recompacted specimen: O→A; intact specimen: O'→A'). The net normal stress of 50 kPa, which was close to the *in situ* vertical stress, was considered and used to simulate the *in situ* stress state. After applying vertical stress, the soil specimens were wetted by decreasing suction from the initial values to 200, 100, 50 and 0 kPa (i.e., A/A'→B1, A/A'→C, A/A'→D and A/A'→E1) respectively. Then each specimen was subjected to suction equalisation. The second test series, conducted at different suctions and temperatures, shared similar procedures with the first series with the addition of an extra thermal loading process (E1→E2 and B1→B2). For the thermal loading stage, the soil temperature was increased from 20 to 60°C monotonically at intervals of 10°C. The cut-off criteria of suction and thermal equalisation were based on both the water content change (less than 0.04%/day, Sivakumar (1993)) and volumetric strain rate (less than 0.025%/day, Romero et al. (2003)). The duration of suction equalisation varied from 3 to 12 days depending on the target suction values. For the thermal loading stage, it was found that 24 hours for each heating interval was sufficient for both suction and temperature equalisation in this study.

After suction and thermal equalisation, the specimen was sheared at constant suction and temperature. During the shearing stage, the shearing rate should be small enough to allow excess pore-water pressure to dissipate. In this study, two criteria were used to determine the shearing rate: 1) the shear stress remained constant when specimens were sheared at rates below the shearing rate; 2) the water stopped flowing in or out of the specimens after shearing ended at the shearing rate (Tarantino and Tombolato, 2005). Based on these two criteria, a shearing rate of 0.0015 mm/min was found to be appropriate for this study.

Interpretations of experimental results

Microstructures of intact and recompacted specimens

Plate 6.1 shows SEM measurements of intact and recompacted loess in the *in situ* and as-compacted states, respectively. For both intact and recompacted specimens, subangular silt

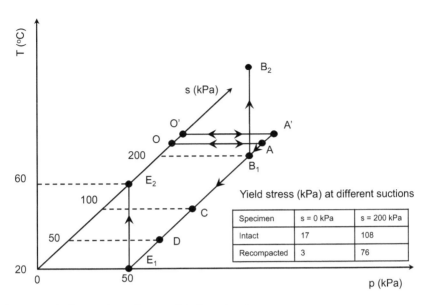

Figure 6.8 Thermo-hydro-mechanical test path before shearing.

grains with diameters of a few tens of micrometres dominate in the SEM observations. This is consistent with the particle size distribution results that show that the test soil consisted of 72% silt-size grains (from 2 to 63 μm). More importantly, the 28% clay fraction is distributed differently in intact and recompacted loess. For intact loess, as shown in Plate 6.1a, the clay particles tend to stick firmly to the silt grain surfaces and also accumulate near the silt grain contacts. The aeolian deposition process is probably responsible for the distribution of clay particles (Kruse et al., 2007). For recompacted loess, as shown in Plate 6.1b, a small amount of clay particles stick to the silt grain surfaces. In addition, the silt grains establish direct contact. Most clay particles in recompacted loess seem to have formed clay aggregates. The distribution of clay particles in recompacted loess is mainly attributed to the specimen preparation method used in this study. The grinding and sieving dislodged the clay particles which were stuck to the grain surfaces and accumulated near the grain contacts. The prepared soil was mixed with water thoroughly to reach a water content of 9.5%. During the subsequent compaction, the clay particles tended to form aggregates because the compaction was conducted dryer than the optimum water content, i.e., 18.2% (Delage et al., 1996).

Figure 6.9 shows MIP measurements of intact and recompacted loess in the *in situ* and as-compacted states, respectively. The cumulative void ratio (CVR) curves and the pore size density (PSD) function are presented. For the CVR curve, the intruded pores correspond to void ratios 0.75 and 0.95 for intact and recompacted specimens, respectively. Whereas the measured void ratios of soil specimens for micro-structure analysis are 1.06 (e_o, intact) and 1.11 (e_o, recompacted), the non-intruded void ratios for intact and recompacted loess are 0.35 and 0.11, respectively. Ng et al. (2016a) illustrated that the non-intruded void ratio is caused by the extra-large pores (i.e., >200 μm) which could not be detected by the MIP because of the minimal pressure. For the PSD function curve, the

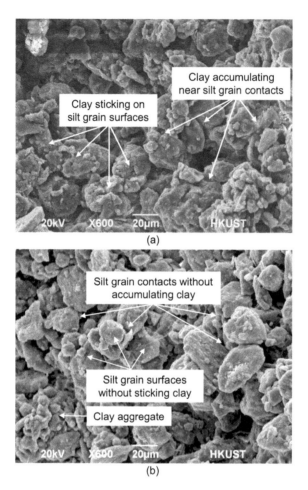

Plate 6.1 Different distribution modes of clay particles: (a) intact specimen; (b) recompacted specimen.

Source: AFTER Ng et al., 2016c.

large pores of both intact and recompacted specimens represented by the peaks around 10 μm are similar. However, the density function of recompacted specimens around the small pores (i.e., 0.007-0.1 μm) is greater than that of intact loess. Previous studies concluded that soil recompacted dry and wet of the optimum water content show bi- and mono-modal pore size density function curves, respectively (Delage et al., 1996). In this study, although the intact and recompacted loess were prepared with a dry densities differing by less than 1% and water contents differing less than 4%, the MIP measurements show that the recompacted loess has a more pronounced bimodal PSD function than the intact loess. The above MIP results are consistent with the SEM observations (see in Plate 6.1b) that the clay particles in the recompacted specimens tend to form clay aggregates. Therefore, more intra-aggregate pores and hence more small pores would be produced. For intact loess, the clay particles tend to be attached to the silt grain surfaces and accumulate

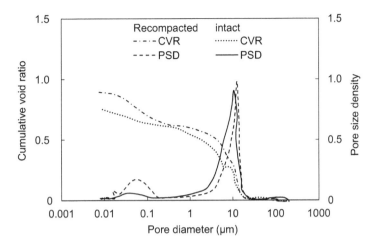

Figure 6.9 Cumulative void ratio (CVR) curve and pore size density (PSD) function of intact and recompacted specimens.

near the silt grain contacts. Within these clay particles, very small pores (smaller than 0.007 μm), which could not be detected by the MIP, are usually formed (Delage et al., 1996). These very small pores are usually identified by the nitrogen sorption technique.

Comparison of the shear behaviour of intact and recompacted unsaturated specimens

Figure 6.10a shows the relationship between shear stress and horizontal displacement of intact and recompacted specimens at different suctions. For the intact specimens at a suction of 0 kPa, the shear stress increases monotonically with horizontal displacement and reaches a plateau at a displacement of approximately 2 mm. At suctions of 50 and 100 kPa, a similar trend is observed but the horizontal displacement is much smaller when the shear stress reaches the plateau (i.e., less than 1 mm). At a suction of 200 kPa, the shear stress reaches a peak before gradually reducing. These observations suggest that as suction increases, the stress-displacement curves of intact specimens change from strain-hardening to strain-softening behaviour. For recompacted specimens at all suctions, the shear stress increases gradually with horizontal displacement. The displacement is approximately 2.5 mm, much larger than that for intact specimens, when the shear stress finally hit the plateau.

Figure 6.10b shows the relationships between the vertical displacement and horizontal displacement of intact and recompacted specimens at different suctions. For both intact and recompacted specimens, the deformation changes from contractive to dilative as the suction increases. This shows that the soil tended to dilate as suction increased. In previous studies (Ng and Zhou, 2005; Hossain and Yin, 2010), the conclusion that the tendency for soil to dilate could be enhanced by an increase in suction was reached by conducting suction-controlled shear tests following a drying path. With an increase in suction, soil specimens would shrink, and the void ratio would then decrease. In this study, the tests follow a wetting path, and the specimens show dilative behaviour at high suction and a large void ratio. The intact specimens show dilative behaviour until the suction decreased to 100 kPa while the recompacted specimens exhibit dilative behaviour only when the suction reaches 200 kPa. This indicates that the soil's dilation is determined not only by the stress state and void ratio but also the specimen preparation method (i.e., intact or recompacted).

It is clear that the intact specimens have a more dilative shear behaviour, probably because of the distribution of clay particles. For intact specimens, the clay particles sticking to the grain surface as well as accumulating near the grain contacts (see Figure 6.10a) result in a resistant structure (Kruse et al., 2007; Muñoz-Castelblanco et al., 2012b). Upon wetting, the soil structure formed by clay particles is destroyed because of the significant plastic axial strain (Muñoz-Castelblanco et al., 2012b). The difference in stress-strain behaviour between intact and recompacted specimens decreases. Constant water content compression tests were conducted on intact and recompacted specimens at a suction of 200 kPa. At a suction of 200 kPa, the gravimetric water content of intact and recompacted specimens are 9.8% and 10.3%, respectively. During compression at such low water contents, the suction is expected to remain fairly constant. This is because compression causes the largest dry pores to collapse, but the smaller hydrated pores that determine suction changes are little affected (Muñoz-Castelblanco et al., 2011). As shown in Figure 6.8, the measured pre-consolidation pressures for intact and recompacted specimens are 108 kPa and 76 kPa respectively. In addition, the pre-consolidation pressures at s suction of 0 kPa are also shown in Figure 6.8. The data were obtained from the compression tests conducted by Ng et al. (2016a). According to the results, the over-consolidation ratio (OCR) of the intact specimens and recompacted specimens at a suction of 200 kPa are 2.0 and 1.5, respectively. Therefore, the larger OCR value of intact specimens results in dilative stress-strain behaviour.

Figure 6.10c shows the changes in water content with horizontal displacement for intact and recompacted specimens at different suctions. In the tests with non-zero suction, both intact and recompacted specimens absorb water, resulting in an increase in water content during shearing. The total amount of absorbed water decreases as suction increases. At s suction of 0 kPa, water continues to drain out of the specimen during the shearing. For suctions of 0, 100 and 200 kPa in intact specimens and 0 and 200 kPa in recompacted specimens, the water content change is consistent with the deformation behaviour (see Figure 6.10b), in the sense that the water content increases (decreases) when the soil dilates (contracts). However, for a suction of 50 kPa in intact specimens and 50 and 100 kPa in recompacted specimens, the water content change seems be opposite to the deformation behaviour (see Figure 6.10b). Although the soil shows contractive behaviour, the water content increases during shearing. Similar test results were obtained in previous studies (Gallipoli et al., 2003a; Zhan and Ng, 2006). The increase in water content is because a number of pores reduce in size below a critical dimension corresponding to flooding at the applied value of suction (Gallipoli et al., 2003a). In addition, water absorption may also be caused by dilation of individual saturated intra-aggregate pores although total volume contraction is observed (Zhan and Ng, 2006).

Figure 6.11a shows the measured relationships between shear stiffness and suction for intact and recompacted specimens. The shear stiffness is determined from the slope of the stress-displacement relationship when the horizontal displacement is less than 0.4 mm. As the suction rises from 0 to 200 kPa, the shear stiffness of intact and recompacted

Thermal effects on soil behaviour and properties 279

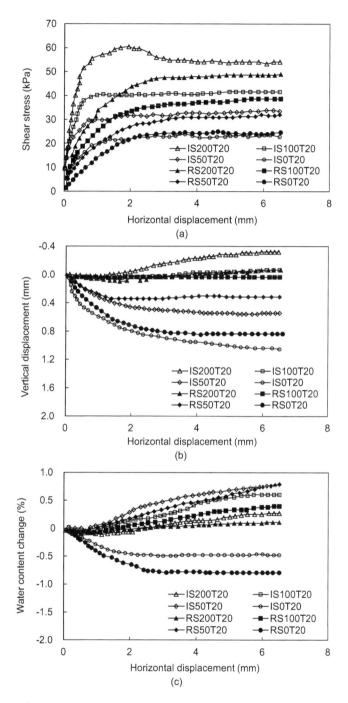

Figure 6.10 Suction effects on the shear behaviour of intact and recompacted specimens at 20°C and net vertical stress of 50 kPa: (a) stress-displacement relationship; (b) deformation behaviour; (c) water content change.

specimens increases by 337% and 229%, respectively. The observed differences in shear stiffness between intact and recompacted specimens can be also explained by Equation (5.7). The shear stiffness of intact specimens is consistently larger than that of recompacted specimens at a given suction, because C_{ij} is larger for intact specimens. Moreover, the difference between intact and recompacted specimens is greater at higher suctions. At zero suction, the difference in shear stiffness between intact and recompacted specimens is only about 40%. As the suction increases to 200 kPa, the shear stiffness of intact specimens is about 2.3 times greater than that of recompacted specimens. In the present study, when a soil specimen is wetted from the initial suction (about 230 kPa) to 200 kPa, both intact and recompacted specimens show negligible volumetric strain (less than 1%). Before shearing, the void ratios of intact and recompacted specimens are very close (see Table 6.2). The shear stiffness of intact specimens is significantly

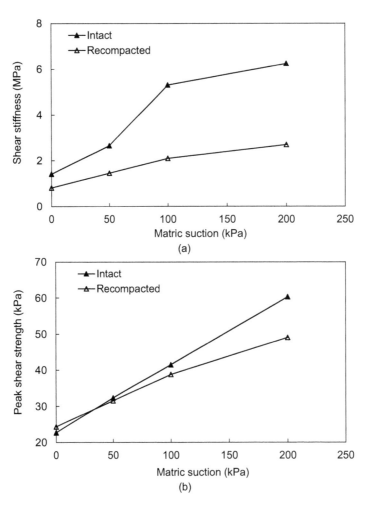

Figure 6.11 Effects of suction on the (a) shear stiffness; (b) shear strength of intact and recompacted specimens at 20°C.

greater than that of recompacted specimens, because intact specimens have a larger C_{ij}. As the suction decreases to 0 kPa, the wetting-induced plastic contraction is 8.1% and 2.3% for recompacted and intact specimens, respectively (see Table 6.2). Consequently, the density and hence $F(e)$ of intact specimens are lower than those of recompacted specimens. Density effects on shear stiffness may negate the influence of inherent structure at zero suction.

Figure 6.11b shows variations in the peak shear strength of intact and recompacted specimens with suction. The peak shear strength is the largest shear stress in the stress-displacement relationship. With an increase of suction from 0 to 200 kPa, the peak shear strength of intact and recompacted specimens increases by 165% and 102%, respectively. The contribution of suction on the peak shear strength can be explained by the extended Mohr-Coulomb failure criteria taking soil dilatancy into consideration (Zhan and Ng, 2006):

$$\tau_f = c' + (\sigma - u_a)\tan(\varphi' + \psi) + (u_a - u_w)\tan\varphi^b \qquad (6.4)$$

where τ_f is the shear strength, σ is the normal stress, u_a and u_w are the air pressure and pore-water pressure, respectively, c' is the true cohesion, ϕ' is the internal friction angle associated with the net normal stress, ϕ^b is the angle indicating the rate of increase in shear strength relative to matric suction. $(u_a - u_w)\tan\phi^b$ is defined as the apparent cohesion, and ψ is the dilation angle, which may be defined as $\tan\psi = -\partial y/\partial x$. According to Equation (6.4), as suction increases, the apparent cohesion increases. Further, the dilatancy ψ increases with an increase in suction (Ng and Zhou, 2005). Because of a larger apparent cohesion and dilatancy, the shear strength of unsaturated soil is therefore larger at a higher suction. At suctions of 0, 50 and 100 kPa, the difference in peak shear strength between intact and recompacted specimens is negligible. On the other hand, as suction increases to 200 kPa, the peak shear strength of intact specimens is much greater than that of recompacted specimens. The obvious difference at 200 kPa is probably because the dilatancy (ψ in Equation (6.4)) of intact specimens is much greater than that of recompacted specimens at 200 kPa, as illustrated in Plate 6.1.

Thermal effects on the shear behaviour of unsaturated intact and recompacted specimens

Figure 6.12 shows the heating-induced volumetric strain for unsaturated intact and recompacted specimens at suctions of 0 and 200 kPa. For recompacted specimens, as the soil temperature increases from 20 to 60°C, contractive volumetric strains are observed at suctions of 0 and 200 kPa. For intact specimens, as the soil temperature increases from 20 to 60°C, contractive volumetric strains also manifest at suctions of 0 and 200 kPa. On the other hand, no strong trend can be found in the change of thermal strain with suction.

Figure 6.13 shows the thermal effects on the shear behaviour of intact specimens at suctions of 0 and 200 kPa. At a suction of 0 kPa, negligible thermal effects are observed in the stress-displacement relationship and deformation behaviour at different temperatures. At a suction of 200 kPa, as the soil temperature increases from 20 to 60°C, the shear stiffness and peak shear strength decrease by 35% and 12%, respectively. The dilation decreases with increasing temperature. The thermal softening illustrated in Figure 6.12 shows a similar trend to that observed by Uchaipichat and Khalili (2009) through a series of triaxial tests on a recompacted silt. These observations imply that thermal effects on the stress-strain behaviour of intact specimens are more significant at higher suctions. It should be noted that at a suction of 200 kPa, a contractive plastic volumetric strain of 0.45% is induced (see Figure 6.12) with an increase in soil temperature. According to previous studies, an increase in shear stiffness and peak shear strength would result in subsequent mechanical loading because of plastic strain hardening effects (Cekerevac and Laloui, 2004; Tang et al., 2008). In this study, the intact specimens tested at a suction of 200 kPa are close to the *in situ* stress state and the soil structure is strongly preserved (see Plate 6.1a). The plastic soil strain of 0.45%, which is mainly attributed to particle rearrangement (Campanella and Mitchell, 1968; Di Donna and Laloui, 2015), is expected to destroy the soil structure partially instead of inducing strain hardening for intact specimens. According to previous studies (Callisto and Rampello, 2004), the degradation of the soil structure is mainly caused by the plastic volumetric strain. Then the shear stiffness and peak shear strength would decrease with an increase in soil temperature at a suction of 200 kPa. In addition, the critical shear strength at suctions of 0 and 200 kPa seems to be independent of temperature for intact specimens.

Figure 6.14 shows the thermal effects on the stress-displacement behaviour of recompacted specimens. As the soil temperature increases from 20 to 60°C, the shear stiffness increases by 47% and 34% at suctions of 0 and 200 kPa, respectively. With an increase in temperature, contractive deformation is reduced at a suction of 0 kPa while dilation is enhanced at a suction of 200 kPa. These observations suggest that the recompacted specimens become stiffer with an increase in soil temperature. This behaviour is unlike that seen in the intact specimens (see Figure 6.12) which show a reduction in shear stiffness and peak shear strength as temperature increases. For the recompacted specimens, as the soil temperature

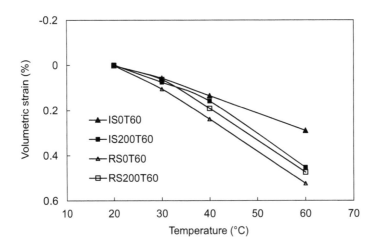

Figure 6.12 Thermally induced volumetric strains for intact and recompacted specimens at different suctions.

Thermal effects on soil behaviour and properties 283

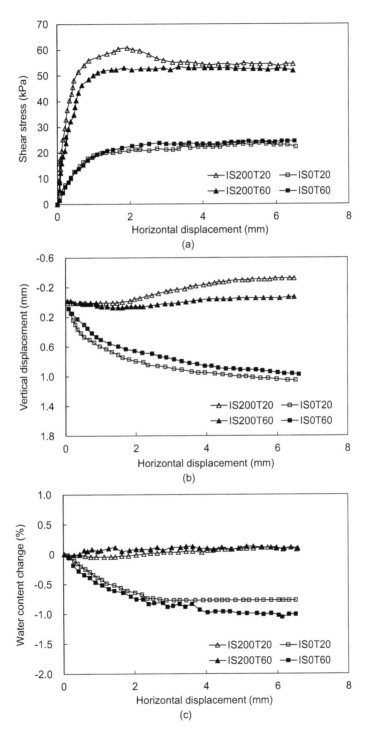

Figure 6.13 Thermal effects on the shear behaviour of intact specimens at net vertical stress of 50 kPa: (a) stress-displacement relationship; (b) deformation behaviour; (c) water content change.

rises from 20 to 60°C, plastic volumetric strains of 0.48% and 0.51% are observed at suctions of 0 and 200 kPa, respectively (see Figure 6.12). The plastic volumetric strain is expected to induce strain hardening for recompacted specimens. Then the shear stiffness increases with an increase in soil temperature at suctions of 0 and 200 kPa. Uchaipichat and Khalili (2009) carried out a series of temperature-controlled triaxial tests on an unsaturated recompacted silt. They found that the shear stiffness and peak strength decrease with an increase in soil temperature. Their observation differs from our observations in the current study, where both the shear stiffness and peak strength of recompacted loess increase with an increase in temperature. The observed differences between these two studies may be explained by differences in the thermal volume change of recompacted silt and loess. In the study of Uchaipichat and Khalili (2009), as temperature rises from 20 to 60°C, the specimens show very small plastic contraction (less than 0.2%) and even expansion. This suggested that thermal softening, i.e., the pre-consolidation pressure and hence over-consolidation ratio decreases with an increase in temperature (Zhou and Ng, 2016), dominates soil behaviour. In the current study, however, plastic contractive volumetric strains of about 0.5% are observed with an increase in soil temperature from 20 to 60°C, inducing significant volumetric strain hardening of soil specimens. In addition, at a given horizontal displacement (i.e., around 1 mm), the mobilised shear stress at 60°C is about 10% larger than that at 20°C. This small difference induced by thermal hardening may be neglected when considering possible experimental artifacts and statistic variability.

Effects of suction and temperature on soil dilatancy

The results of soil dilatancy at different suctions (s = 0 and 200 kPa) and temperatures (T = 20 and 60°C) are shown in Figure 6.15. Dilatancy is defined as $-\partial y/\partial x$, where ∂y and ∂x are increments in the vertical and horizontal displacements, respectively. The positive and negative values of soil dilatancy represent dilation and contraction, respectively. For intact specimens, as the suction increases from 0 to 200 kPa, the maximum dilatancy increases by 433% and 500% at 20 and 60°C, respectively (see IS0T20 and IS200T20; IS0T60 and IS200T60 in Figure 6.15). These observations indicate that soil dilatancy increases significantly with increasing soil suction. For recompacted specimens, the same conclusion can be drawn.

As all the tests conducted at a suction of 0 kPa show contractive behaviour, the effect of temperature on soil dilatancy is evaluated based on the tests carried out at a suction of 200 kPa. For intact specimens, the maximum dilatancy decreases by 68% as the soil temperature rises from 20 to 60°C. The reduction in soil dilatancy may be explained by degradation of the soil structure induced by the thermal plastic strain (see Figure 6.12). On the other hand, for recompacted specimens, the maximum dilatancy increases by 63% with the same increase in temperature. The increase in soil dilatancy is probably caused by the strain hardening effects induced by thermal plastic strain (see Figure 6.12). Further, the test results show that the value of Ψ (see Equation (6.4)) for unsaturated soil is also affected by soil temperature, especially for the unsaturated intact specimens which show a decrease in soil dilatancy as the soil temperature rises. Note that only the effect of suction on Ψ was considered in previous studies (Zhan and Ng, 2006; Hossain and Yin, 2010). The temperature variation should be considered when predicting shear strength using the extended Mohr-Coulomb yielding criterion.

Thermal effects on soil behaviour and properties 285

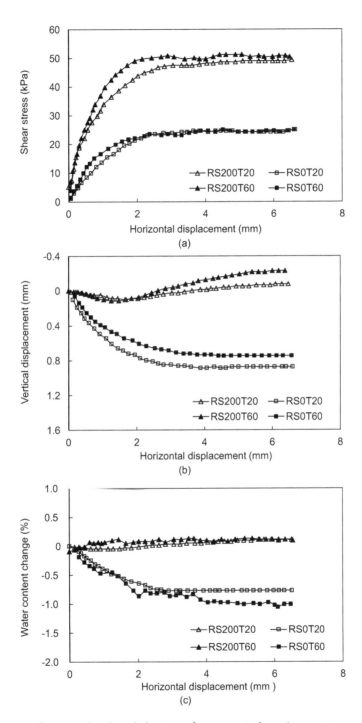

Figure 6.14 Thermal effects on the shear behaviour of recompacted specimens at net vertical stress of 50 kPa: (a) stress-displacement relationship; (b) deformation behaviour; (c) water content change.

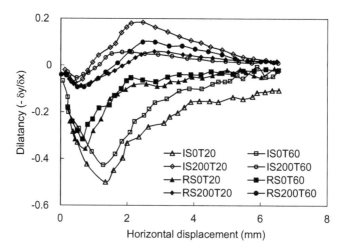

Figure 6.15 Soil dilatancy of intact and recompacted specimens at different suctions (s = 0 and 200 kPa), different temperatures (T = 0 and 60°C) and net vertical stress of 50 kPa.

Summary

For a given suction and temperature, intact loess specimens show higher shear stiffness and more dilative behaviour than recompacted ones at almost the same initial dry density. The differences are mainly attributed to different distributions of clay particles in intact and recompacted loess, as revealed by SEM and MIP tests. In intact loess, more clay particles accumulate near the silt grain contacts and stick to the silt grain surfaces. The clay particles in recompacted loess tend to form clay aggregates. The distribution of clay particles in intact loess results in a more resistant structure and higher yield stress. Further, the differences between intact and recompacted loess decrease with soil suction, mainly because the resistant structure of intact specimens is destroyed during wetting.

Suction has a significant influence on the shear behaviour of unsaturated intact and recompacted loess. When following a wetting path, both intact and recompacted loess show dilative shear behaviour at a high suction as well as a large void ratio. Unlike previous shear tests that followed a drying path, the coupled effects of soil shrinkage and suction on soil dilatancy could be distinguished in this study. The observation further confirms that soil dilatancy is enhanced with an increase in suction. In addition, the critical shear strength increases with suction for both intact and recompacted loess.

For recompacted specimens at suctions of 0 and 200 kPa, the shear stiffness and dilatancy increase with temperature. The shear stiffness and maximum dilatancy increase by up to 47% (s = 0 kPa) and 63% (s = 200 kPa), respectively. These increases in shear stiffness and dilatancy are mainly because of the continuous plastic volumetric contraction during heating, inducing strain hardening effects. On the other hand, thermal effects on the stress-strain behaviour of intact specimens are negligible at zero suction. At a suction of 200 kPa, as the soil temperature rose from 20 to 60°C, the shear stiffness, maximum dilatancy and peak shear strength of intact specimens decreased by 35%, 68% and 12%, respectively.

These observations imply that the thermal effects for intact specimens differ from those for recompacted specimens. The difference is probably because heating-induced plastic strain, in addition to plastic strain hardening effects, also destroys the resistant structure of intact specimens.

EFFECTS OF TEMPERATURE AND SUCTION ON SECANT SHEAR MODULUS (ZHOU ET AL., 2015)

Introduction

To investigate suction and thermal effects on the secant shear modulus of unsaturated soil, five suction- and temperature-controlled triaxial compression tests were carried out on CDT. Three of them (S1T20, S150T20 and S300T20) were carried out at the room temperature of 20°C and at suctions of 1, 150 and 300 kPa. The other two (S1T60 and S150T60) were conducted at 60°C and at suctions of 1 and 150 kPa. Details of the test program are given in Table 6.3.

Each test consisted of four stages: isotropic compression, suction equalisation, thermal equalisation and shearing. During the compression stage, the net confining pressure of each specimen was increased directly from 0 to 100 kPa. Based on the initial and final readings of the Hall-effect transducers, the contractive volumetric strain of soil during this stage was only about 0.1%. Then, soil specimens in tests S1T20 and S1T60 were wetted from initial suction (95 kPa) to 1 kPa. The soil specimens in tests S150T20 and S150T60 were dried from 95 kPa to 150 kPa. Similarly, the soil specimen in test S300T20 was dried to 300 kPa. The third stage was the application of thermal loading. For the three tests carried out at room temperature (20°C) (S1T20, S150T20 and S300T20), it was unnecessary to change the soil temperature. For the tests carried out at a higher temperature (60°C) (S1T60 and S150T60), the soil specimens were heated to 60°C. Then the soil specimens were sheared at constant mean net stress (p) and suction to determine the decay in shear modulus with increasing shear strain. The choice of constant p stress path was made to eliminate the coupling effects of dp and dq on shear modulus (Atkinson and Sallfors, 1991; Ng et al., 2000). The deviator stress was increased at a rate of 3 kPa per hour, slowly enough to generate excess pore-water pressures less than ±2 kPa during shearing.

Table 6.3 Summary of testing conditions for suction and temperature-controlled constant p compression

	Soil state after compaction					Soil state prior to shearing		
Test ID	Dry unit weight: kN/m³	Water content: %	Suction: kPa	Suction path: kPa	Thermal path: °C	S_r	s S_r: kPa	σ^*: kPa
S1T20	17.2	16.4	94	95→1	20	0.91	1	101
S150T20	17.2	16.3	95	95→150	20	0.74	111	211
S300T20	17.2	16.2	96	95→300	20	0.47	140	240
S1T60	17.1	16.2	96	95→1	20→60	0.91	1	101
S150T60	16.9	16.4	93	95→150	20→60	0.73	110	210

Interpretations of experimental results

Suction effects on secant shear modulus

Figure 6.16a shows measured stress-strain relations at 20°C and suctions of 1, 150 and 300 kPa. As expected, the deviator stress of all specimens increases non-linearly with increasing strain. At the same deviator stress, soil specimen at higher suction is affected by consistently smaller strain than that at lower suction. The corresponding increase of shear stiffness with increasing suction may be explained by several reasons. First of all, the average skeleton stress (σ^*) of unsaturated soils is expressed in Equation (6.1). As illustrated in Table 6.3, the value of $S_r(u_a - u_w)$ increases with increasing suction and stiffens the soil skeleton. Another reason for soil stiffness increases with increasing suction stands on the fact that pore water may be classified as bulk water and meniscus water (Wheeler and Karube, 1996). Similar to net stress, u_w in bulk water affects both normal and tangent forces between soil particles, as captured by Equation (6.1). On the contrary, u_w in the meniscus water affects only normal forces at particle contacts, providing additional bonding forces at particle contacts and stabilising the soil skeleton (Wheeler and Karube, 1996; Gallipoli et al., 2003b). This stabilisation effect is not captured by Equation (6.1). The normal force induced by meniscus water between two spherical particles can be expressed in Equation (6.2). According to this equation, ΔN is larger at higher ($u_a - u_w$). Furthermore, Table 6.3 shows that S_r greatly decreases as suction increases. Thus, the number of menisci (related to $1 - S_r$) is expected to increase with an increase in suction. As suction increases, both the stabilisation effect exerted by ΔN and the number of menisci become more significant and therefore soil skeleton behaves stiffer. Finally, it should be pointed out that apart from the above two reasons, the history of suction may affect the shear modulus of unsaturated soil. According to the results of Ng and Zhou (2014), for example, the yield stress of unsaturated CDT at suction of 1 and 150 kPa and 20°C is about 120 and 220 kPa, respectively. The distance between the current stress level (100 kPa) and the yield stress is larger at a suction of 150 kPa than at a suction of 1 kPa. Therefore, a part of the soil stiffening might also be attributed to the change in the OCR value, which was activated by the different tests.

From the slope of the stress-strain curve, secant shear moduli (G) at different strains are determined and presented in Figure 6.16b. As strain increases from 0.003% to 1%, measured G values of all specimens greatly reduce, but they remain always greater at higher suction in the investigated strain range.

Thermal effects on secant shear modulus

Figures 6.17a and 6.17b show the thermal effects on small strain behaviour at different suctions (1 and 150 kPa). It can be seen from the figure that the measured G is consistently smaller at 60°C than that at 20°C under constant suction conditions. Furthermore, when the mobilised shear strain is 0.003%, the thermal effect on G is greater at a suction of 1 kPa (a reduction of about 30% as the soil temperature increases from 20 to 60°C) than that at 150 kPa (a reduction of about 15%). At some other strains (for example, 0.03%), however, the thermal effect on G seems less than at a suction of 1 kPa (a reduction of about 15%) than that at 150 kPa (a reduction of about 30%).

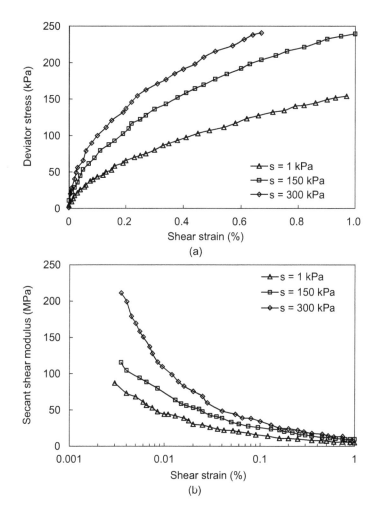

Figure 6.16 Effects of suction on small strain stiffness (20°C).

The observed reduction in G may be attributed to the temperature dependence of T_s, which is 72.8 and 66.2 mN/m at 20°C and 60°C respectively (Ng and Menzies, 2007). As already stated, the decrease in T_s with increasing temperature affects both the retention capacity of the soil (Romero et al., 2003), causing a reduction of S_r and σ^*, and the intergranular extra stress induced by the meniscus water (see Equation (6.2)). However, the data in Figure 6.17b suggest that the former effect (i.e., the reduction of S_r with increasing temperature) is quite limited for the soil tested. Thus, the second mechanism is considered to be the main reason for the observed reduction in G.

The observed behaviour can be also explained by thermo-plasticity theory. As a matter of fact, thermo-mechanical models for unsaturated soil (e.g., François and Laloui, 2008) suggest that the yield stress and OCR decreases as an over-consolidated soil is heated at constant stress. Given a smaller OCR, the soil response would be softer during shearing.

Effects of suction and temperature on shear-induced volume change

Figure 6.18 shows measured shear-induced volumetric strains at various suctions and temperatures. It demonstrates that soil becomes more dilative when suction increases under isothermal conditions. A similar observation is also reported in Ng et al. (2013), who carried out a series of suction-controlled cyclic triaxial tests on unsaturated CDT at room temperature. The increase in dilative tendency with increasing suction is mainly attributable to the higher yield stress at a higher suction over the suction range considered. At higher yield stress and OCR, the soil is expected and observed to be more dilative during shearing.

As far as thermal effects on shear-induced volume change are concerned, at a suction of 1 kPa, soil is more contractive at 60°C than that at 20°C at a suction of 1 kPa. The observed increase in contractive strain with increasing temperature can be explained by

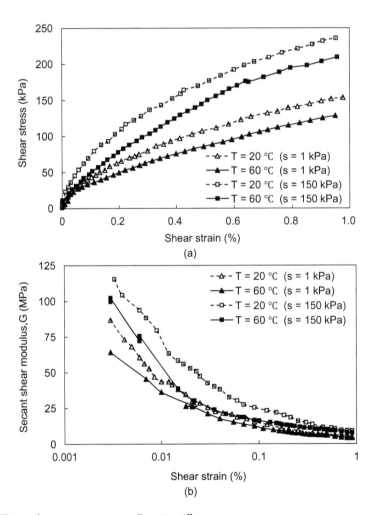

Figure 6.17 Effects of temperature on small strain stiffness.

the thermo-plasticity model proposed by François and Laloui (2008). With an increase of soil temperature, OCR decreases and hence the shear-induced contractive strain becomes larger (François and Laloui, 2008). At a suction of 150 kPa, however, the thermal effects on shear-induced volume change are almost negligible. This is probably because this suction mobilised small volumetric strains. Thus, any difference in the volumetric strain induced by thermal effects is relatively small.

Summary

Over the range between 0.003% and 1%, the secant shear modulus G of CDT is consistently greater at higher suction. This is mainly attributable to an increase in the average skeleton stress and inter-particle normal forces between soil grains with increasing suction. Consequently, an increase of suction will stabilise and stiffen the soil skeleton.

At the same suction, G at a given strain is slightly less at higher temperatures. This is probably caused by a reduction in the air-water surface tension increasing temperature, weakening the average skeleton stress and the stabilisation effects on the soil skeleton. The contribution of suction to shear modulus seems to be less significant.

THERMAL EFFECTS ON YIELDING AND WETTING-INDUCED COLLAPSE OF RECOMPACTED AND INTACT LOESS (NG ET AL., 2018; CHENG ET AL., 2020)

Introduction

In this section, a series of isotropic compression tests on recompacted and intact loess specimens from the Shaanxi Province were conducted at different suctions (0 and 100 kPa) and temperatures (5, 23 and 50°C). Moreover, a series of wetting tests were carried out on recompacted and intact loess specimens at different confining stresses (50 and 110 kPa)

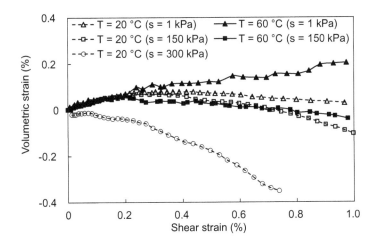

Figure 6.18 Effects of suction and temperature on volume change during shearing.

and temperatures (5, 23 and 50°C). The results of the wetting tests were analysed using the measured yield stress.

Test programme

The principal objective is to investigate thermal effects on yielding and wetting-induced collapse of recompacted and intact loess. To achieve this, two series of suction and temperature-controlled isotropic tests were conducted. In the first series, eight compression tests were carried out to determine the suction and temperature dependent yield stress and yielding behaviour. Three tests (RS0T5, RS0T23 and RS0T50) were conducted on saturated recompacted loess at temperatures of 5°C, 23°C and 50°C. Similarly, another three tests (RS100T5, RS100T23 and RS100T50) were carried out on recompacted loess at a suction of 100 kPa at different temperatures (5°C, 23°C and 50°C). The other two tests (IS0T23 and IS100T23) were carried out on intact loess at room temperature (i.e., 23°C), but at suctions of 0 and 100 kPa. Details of the suction and temperature-controlled compression tests are summarised in Table 6.4.

The second series of tests included six wetting tests with stress and temperature control. Three tests (R50T5, R50T23 and R50T50) were carried out on recompacted loess at a confining stress of 50 kPa, but at different temperatures of 5°C, 23°C and 50°C. To study stress effects on the wetting-induced collapse of recompacted loess, a test (R110T23) was carried out at confining stress of 110 kPa at room temperature. Similarly, two tests (I50T23 and I110T23) were designed and conducted to investigate stress effects on the wetting-induced collapse of intact loess. Details of the suction and temperature-controlled wetting tests are summarised in Table 6.5.

Test apparatus

A suction- and temperature-controlled double-cell triaxial apparatus, as shown in Figure 6.19, was used in this study. The principle of the conventional double-cell triaxial apparatus is described in Chapter 4. To control soil temperature, a heating/cooling bath connected with a spiral copper tube installed between the inner cell and outer cell was installed. The heating/cooling bath consists of a thermostat, a heating/cooling unit, a water bath, an inbuilt pump and a thermocouple. The thermocouple is installed in the water

Table 6.4 Summary of suction and temperature-controlled compression tests

Test ID	Specimen type	Suction: kPa	Temperature: °C	Stress path (see Figure 6.20)
RS0T5	Recompacted	0	5	A'→B2→C2→D3
RS0T23	Recompacted	0	23	A'→B2→D2
RS0T50	Recompacted	0	50	A'→B2→C1→D1
RS100T5	Recompacted	100	5	A'→B1→C4→D6
RS100T23	Recompacted	100	23	A'→B1→D5
RS100T50	Recompacted	100	50	A'→B1→C3→D4
IS0T23	Intact	0	23	A→B2→D2
IS100T23	Intact	100	23	A→B1→D5

Table 6.5 Summary of stress and temperature-controlled wetting tests

Test ID	p: kPa	T: °C	e (initial)	Stress path (see Figure 6.21)	ε_v calculated using Eq. (6.6) (%)	Sr: % Before	Sr: % After
R50T5	50	5	1.17	A'→B'→D$_1$→E$_1$→F$_3$	2.6	46.9	95.2
R50T23	50	23	1.18	A'→B'→D$_1$→F$_1$	2.4	44.9	92.4
R50T50	50	50	1.17	A'→B'→D$_1$→E$_2$→F$_4$	2.3	41.9	90.1
R110T23	110	23	1.17	A'→C'→D$_2$→F$_2$	0.9	49.5	93.6
I50T23	50	23	1.12	A→B→D$_1$→F$_1$	2.0	30.4	89.1
I110T23	110	23	1.13	A→C→D$_2$→F$_2$	2.8	60.0	95.7

Note:
I and R means intact and recompacted specimens, respectively.

of the inner cell. The thermostat is used to adjust the output of the heating/cooling unit according to the current and target temperatures. Both the thermocouple and the heating/cooling bath are connected to the thermostat, forming an automatic control and feedback system. In this study, 48 hours were allowed to reach the target temperature and to achieve thermal equilibrium. After reaching thermal equilibrium, the temperature fluctuation is less than 0.2°C. More details of the temperature control system were reported by Ng et al. (2016e).

Test procedures

Figure 6.20 shows the thermo-hydro-mechanical path of each compression test. Each compression test consists of a suction equalisation stage, a thermal equalisation stage and an isotropic compression stage. After setting up in the double-cell triaxial apparatus, the initial state of intact specimens was fixed at point A (initial suction of 200 kPa) and that of recompacted specimens was controlled at point A' (initial suction of 180 kPa). The suction equalisation stage ensures that the water content reaches its equilibrium state throughout the soil specimen at the target suction value. For the tests RS100T5, RS100T23, RS100T50 and IS100T23, the soil specimens were wetted to 100 kPa (A→B1, A' →B1). Similarly, for the tests RS0T5, RS0T23, RS0T50 and IS0T23, the soil specimens were wetted to 0 kPa (A→B2, A' →B2). When the water flow rate is less than 0.1 ml/day, the equilibrium state is considered to have been reached (Ng et al., 2012). This stage usually takes 7–10 days to achieve suction equilibrium. Following suction equilibrium, the second stage is thermal equalisation. Soil specimens in tests RS100T50 and RS0T50 were heated to 50°C (B1→C3, B2→C1). Similarly, soil specimens in tests RS100T5 and RS0T5 were cooled to 5°C (B1→C4, B2→C2). For the two tests RS100T23 and RS0T23, the temperatures were maintained at room temperature. This stage lasted for 2 days, which was found sufficient to achieve thermal equilibrium based on the readings of thermocouples. The last stage was to compress all soil specimens isotropically to 300 kPa step by step (5-10-20-40-80-150-300 kPa) in the drained condition (C1→D1, C2→D3, C3→D4, C4→D6, B1→D5, B2→D2). The detailed stress paths of each compression test are summarised in Table 6.5.

Figure 6.21 shows the thermo-hydro-mechanical path of each wetting test. Each test consists of four stages: isotropic compression, wetting from the initial suction to 100 kPa,

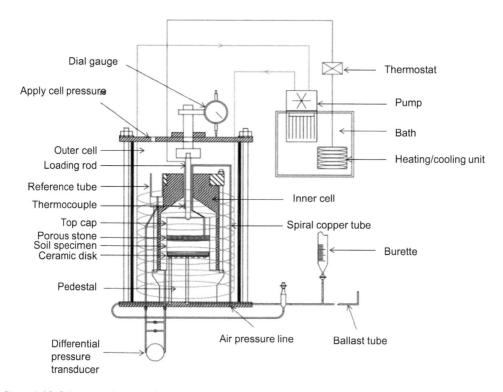

Figure 6.19 Schematic diagram of a suction- and temperature-controlled double-cell triaxial apparatus.

Source: After Ng et al., 2016e.

thermal equalisation and wetting from 100 to 0 kPa. Each specimen was first isotropically compressed to a target confining stress in the drained condition. The confining stress was 50 kPa for tests R50T5, R50T23 and R50T50 and 110 kPa for tests R110T23 and I110R23. The second stage was to wet all specimens to 100 kPa (B→D1, B'→D1, C→D2, C'→D2). This stage required 7–10 days to satisfy the equilibrium criterion that the water flow rate is less than 0.1 ml/day. The soil temperature was changed in the following stage. The soil temperatures in tests R50T5 and R50T50 were changed to 5 and 50°C, respectively (D1→E1, D1→E2). For the remaining six tests conducted at room temperature, the temperatures were kept constant. As in compression tests, two days were allowed for each specimen to achieve thermal equilibrium. The final stage is wetting from a suction of 100 to 0 kPa (E1→F3, E2→F4, D1→F1, D2→F2) under constant stress and temperature conditions. During the wetting process, the soil suction was decreased step by step (100-50-10-1-0 kPa). At each suction, the equilibrium state was reached. Details of all stress paths of are summarised in Table 6.5.

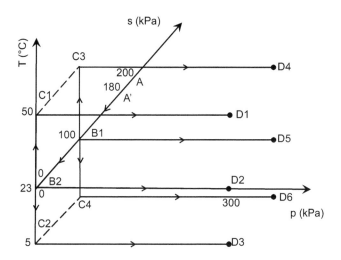

Figure 6.20 Thermo-hydro-mechanical path of compression tests.

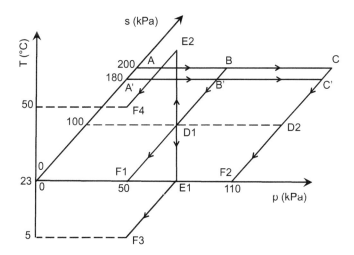

Figure 6.21 Thermo-hydro-mechanical path of wetting tests.

Interpretations of experimental results

Thermal effects on yielding behaviour of recompacted loess

Figure 6.22 shows the thermal effects on the isotropic compression behaviour of recompacted loess. As expected, with an increasing in confining stress, the void ratio of each specimen decreases. The initial yield stresses of all specimens were determined using the method proposed by Sridharan et al. (1991). Each isotropic compression curves clearly consists of two approximately linear segments. The stress at the intersection of these two linear segments is defined as the yield stress. The yield stresses determined at different suctions and temperatures are also shown in the figure. At a suction of 0 kPa, the yield stresses are about 13, 10 and 7 kPa at 5°C, 23°C and 50°C, respectively. At a suction of 100 kPa, the yield stresses are about 42, 40 and 38 kPa at 5°C, 23°C and 50°C, respectively.

In addition, the soil compressibility at various suctions and temperatures can be deduced from the compression behaviour shown in Figure 6.22. The values of the plastic compressibility index λ, which is the slope of the second linear segment of each curve, are 0.12 and 0.15 at suctions of 0 and 100 kPa, respectively. At a higher suction, the plastic compressibility index λ becomes slightly larger. On the other hand, thermal effects on λ are not obvious in each suction condition. The observed thermal effects on compressibility in this study remain consistent with previous experimental results (Cekerevac and Laloui, 2004; Abuel-Naga et al., 2007a; Tang et al., 2008; Di Donna and Laloui, 2015).

Figure 6.23 shows the thermal effects on the loading collapse characteristics, which represents the relationship between suction and yield stress, of recompacted loess. The yield stresses at various suctions and temperatures are obtained from Figure 6.22. It can be seen that, at a given temperature, the yield stress increases significantly with increasing suction. This is because an increase in suction induces more meniscus water, which provides additional stabilising effects on the soil skeleton and minimises particle rearrangements (Wheeler et al., 2003). The stabilising effects can be described by a suction-dependent inter-particle normal force ΔN. Assuming that the contact angle is zero, ΔN between two spherical particles with the same radius can be expressed by Equation (6.2). ΔN increases with increasing suction, suggesting that the stabilising effects of suction become more significant with increasing suction (Gallipoli et al., 2003b; Wheeler et al., 2003; Zhou et al., 2015b). Consequently, the yield stress of unsaturated soil increases with increasing suction. The observed reduction in yield stress with decreasing suction is applicable at relatively low suction. At extremely high suctions with a degree of saturation below about 30%, the loss of water meniscus during drying would cause a reduction of suction-induced stabilising effects (Wan et al., 2014).

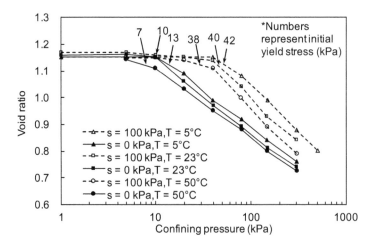

Figure 6.22 Thermal effects on compression behaviour of recompacted loess at suctions of 0 and 100 kPa.

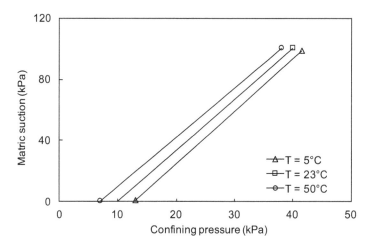

Figure 6.23 Thermal effects on LC characteristics (i.e., relationship between yield stress and suction) of recompacted loess.

It can be also seen from the figure that the yield stress at a given suction decreases with increasing temperature (thermal softening). This may be explained by thermal effects on T_s and R_p in Equation (6.2). According to Gittens (1969), the surface tension T_s decreases with increasing temperature. For example, when the temperature increases from 23°C to 50°C, T_s decreases from 72.3 to 67.9 mN/m. At 5°C, T_s is 74.9 mN/m. The variation in R_p can be estimated using the thermal expansion coefficient of soil particles. According to Horseman and McEwen (1996), for clay particles, the thermal expansion coefficient is about $2.9 \times 10^{-3}\%/°C$. Taking into account the thermal effects on T_s and R_p, when temperature increases from 23°C to 50°C, ΔN decreases by 6.0% and 5.6% at suctions of 0 and 100 kPa, respectively. Similarly, when the temperature decreases from 23°C to 5°C, ΔN increases by 3.4% and 2.1% at suctions of 0 and 100 kPa, respectively. ΔN decreases with increasing temperature, suggesting that the stabilising effects of suction become less significant with increasing temperature (Ng and Zhou, 2014). Thus, with increasing temperature, the yield stress at a given suction decreases.

Comparison between the yielding characteristics of recompacted and intact loess

Figure 6.24 shows the isotropic compression behaviour of recompacted and intact loess at suctions of 0 and 100 kPa and at 23°C. At zero suction, the yield stress of the recompacted specimen is around 10 kPa. When the suction increases to 100 kPa, the yield stress of the recompacted specimen is 4 times larger than that at zero suction (40 kPa). For intact loess, the yield stress is about 25 kPa at a suction of 0 kPa. When the suction becomes 100 kPa, the yield stress of an intact specimen is 90 kPa, which is 2.6 times greater than that at zero suction.

As expected, at zero suction, the yield stress of an intact specimen is 1.5 times larger than that of a recompacted specimen. At a suction of 100 kPa, the yield stress of an intact specimen is 1.3 times larger than that of a recompacted one. The difference between recompacted and intact specimens is mainly attributed to the more resistant soil fabric of intact specimens than recompacted ones (Burland, 1990). As shown by microstructural observations (Ng et al., 2017c), more clay aggregates accumulate at inter-particle contacts in intact loess than in recompacted loess. The enlarged contact area minimises slippage at silt contacts and stiffens the soil skeleton.

The plastic compressibility indices λ of recompacted soil specimens at suctions of 0 and 100 kPa are 0.12 and 0.15, respectively. For intact soil specimens, the plastic compressibility indices λ are 0.13 and 0.17 at suctions of 0 and 100 kPa. At a given suction, the λ of the intact specimen is greater than that of the recompacted one. The larger λ of intact specimens may be because unlike recompacted loess, there are some extra-large pores with a diameter of over 200 μm in intact loess (Bai et al., 2014; Ng et al., 2016d). These extra-large pores collapse easily under compression, inducing a larger compressibility for intact loess.

Figure 6.25 compares the loading collapse characteristics of recompacted and intact loess at 23°C. The data points in this figure are the yield stresses obtained from Figure 6.24. As expected, wetting-induced softening is observed in both recompacted and intact loess specimens. Wetting-induced softening in intact loess is more significant than in recompacted loess. By comparison with recompacted loess, more clay aggregates accumulate at inter-particle contacts in intact loess, as shown by microstructural observations (Ng et al., 2017c). These clay aggregates in the intact loess enlarges the contact area at silt contacts. With a larger contact area, the contribution of suction to the inter-particle normal force in intact specimens will be larger than in recompacted specimens. As a result, the wetting-induced softening of intact loess specimens is greater than that of recompacted specimens.

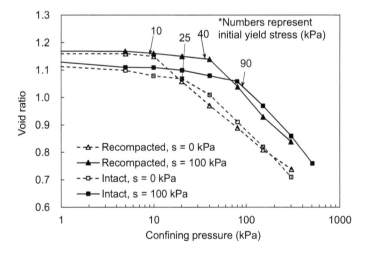

Figure 6.24 Compression behaviour of recompacted and intact loess at suctions of 0 and 100 kPa (T = 23°C).

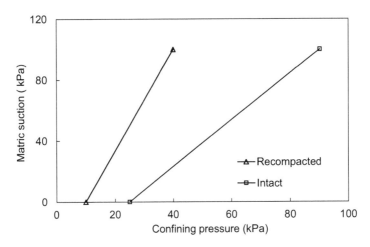

Figure 6.25 Loading collapse characteristics of recompacted and intact loess (T = 23°C).

Thermal effects on wetting-induced collapse of recompacted loess

Figure 6.26 shows the thermal effects on the wetting-induced collapse of recompacted loess during wetting from a suction of to 100 kPa to 0 kPa. The volumetric strain ε_v during this process is calculated using the following equation:

$$\varepsilon_v = -\frac{e_{100} - e_i}{1 + e_{100}} \tag{6.5}$$

where e_{100} is the void ratio at a suction of 100 kPa; e_i is the void ratio at a given suction value during the wetting process. Before wetting from 100 kPa to 0 kPa, contractive volumetric strains of about 9% were observed in the three specimens during compression to 50 kPa and wetting from the initial suction to 100 kPa. It is reasonable to assume that all three specimens are normally consolidated in the stress state with a suction of 100 kPa and confining stress of 50 kPa.

It can be seen from the figure that with decreasing suction, the wetting-induced volumetric strain of recompacted loess increases, at an increasing rate. Moreover, with increasing temperature, the cumulative collapse strain increases. When wetting to 0 kPa, the cumulative collapse volumetric strain at 5°C is about 4%, which is 20% less than that at room temperature of 23°C (about 5%). The collapse volumetric strain at 50°C is around 12%, which is three times of that at 5°C. The thermal effects on wetting-induced collapse can be explained using the yield characteristics reported in Figure 6.23. As shown by Figure 6.23, the wetting-induced softening becomes greater with increasing temperature. The coupled effects of suction and temperature on yield stress can be explained using Equation (6.2). When the suction decreases from 100 kPa to 0 kPa, ΔN decreases by about 8.4%, 9.9% and 10.3% at 5°C, 23°C and 50°C, respectively. Because of the larger reduction of the stabilising inter-particle normal

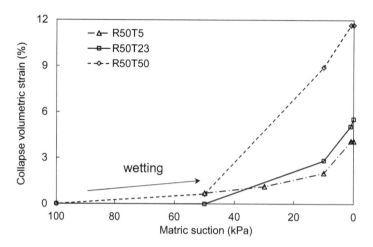

Figure 6.26 Thermal effects on wetting-induced collapse of recompacted loess during wetting from 100 to 0 kPa.

force at higher temperatures, the wetting-induced softening becomes greater at 50°C than at 5°C.

According to the elasto-plastic model for unsaturated soil (e.g., Alonso et al., 1990), there is a positive relationship between incremental volumetric strain $d\varepsilon_v$ and incremental yield stress:

$$d\varepsilon_v = \frac{\lambda(0)}{1+e} \frac{dp_c^0}{p_c^0} \qquad (6.6)$$

where $\lambda(0)$ is the plastic compressibility index at zero suction, e is the void ratio, p_c^0 is the initial yield stress at zero suction and dp_c^0 is the incremental yield stress at zero suction. To determine the values of p_c^0 and dp_c^0 in Equation (6.6), the yield stress of recompacted loess before and after wetting are shown in Figure 6.27. As discussed above, the soil specimens at different temperatures were all normally consolidated before wetting from 100 to 0 kPa. The stress states of each specimen should be located on the corresponding loading collapse (LC) curves. In the current study, we assume that the shape of the loading collapse curve remains the same before and after wetting. This assumption is supported by the experimental results of Nowamooz and Masrouri (2008) for a bentonite/silt mixture. It was found that the loading collapse curves are almost parallel over a similar suction range (0 to 100 kPa) to that in the current study. Under this assumption, the LC curves before wetting from 100 kPa to 0 kPa are obtained by parallel shifting of those shown in Figure 6.23. During the subsequent wetting process, all three LC curves at different temperatures shift to the right. After wetting to 0 kPa, the stress state in each stress and temperature condition is on the corresponding LC curve. The ratio dp_c^0/p_c^0 can thus be obtained. As can be seen from Figure 6.27, the dp_c^0/p_c^0 induced by the wetting process increases with increasing temperature. Consequently, based on Equation (6.6), the wetting-induced yielding and

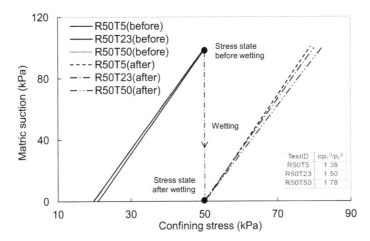

Figure 6.27 Thermal effects on the evolution of LC characteristics of recompacted loess during wetting from 100 to 0 kPa.

volumetric contraction are greater at higher temperatures, as shown in Figure 6.26. This observation implies that at a higher temperature, the wetting-induced soil contraction and ground movement will be much larger. The experimental results of the current study are interpreted within the BBM framework (Alonso et al., 1990), which uses net stress and suction as constitutive variables. Alternatively, these results may be analysed within the effective stress framework for unsaturated soils. Through a series of stress-controlled wetting tests at room temperature, Sun et al. (2007) found that wetting-induced collapse strain is more sensitive to variations in the degree of saturation than variations in the suction. The degrees of saturation of each specimen before and after wetting from a suction of 100 kPa to 0 kPa are listed in Table 6.5. It can be seen that the incremental degree of saturation and collapse volumetric strains of the tested loess are consistently larger at higher temperatures. This result suggests that the finding of Sun et al. is applicable at different temperatures.

Comparisons between wetting-induced collapse of recompacted and intact loess

Figure 6.28 shows the volumetric collapse of recompacted and intact loess at 23°C during wetting from 100 kPa to 0 kPa at various stresses. The volumetric strain ε_v during this wetting process is also calculated using Equation (6.5). As illustrated by the compression curves shown in Figure 6.22, the yield stress of recompacted specimen at a suction of 100 kPa is 40 kPa. Thus, the recompacted specimens at confining stresses of 50 kPa and 110 kPa are normally consolidated. For the intact specimen, the yield stress is 90 kPa at a suction of 100 kPa. The intact specimen is normally consolidated at a confining stress of 50 kPa, but over consolidated at a confining stress of 110 kPa.

It can be seen from the figure that for recompacted specimens, when the confining stress increases from 50 kPa to 110 kPa, the volumetric strain induced by wetting from 100 kPa to 0 kPa decreases from around 5.5% to 2.4%. On the other hand, over the same confining

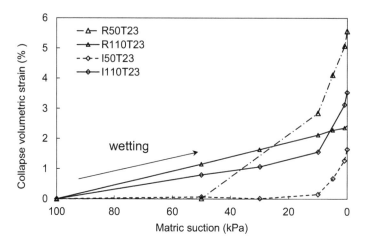

Figure 6.28 Wetting-induced collapse of recompacted and intact loess at various stresses (T = 23°C).

stress range, the wetting-induced volumetric strain of an intact specimen increases from about 1.6% to 3.5%. Furthermore, at a confining stress of 50 kPa, the wetting-induced volumetric strain of a recompacted specimen is about 3.4 times greater than that of an intact specimen. At a confining stress of 110 kPa, the wetting-induced volumetric strain of a recompacted specimen is 30% less than that of an intact specimen.

Figure 6.29a shows the change in yield stress of recompacted specimens before and after wetting from 100 kPa to 0 kPa at different stresses. As discussed above, the recompacted specimens at a confining stress of 50 and 110 kPa are normally consolidated before wetting from 100 kPa to 0 kPa. The stress states are on the corresponding LC curves. Thus, the LC curves before wetting from 100 kPa to 0 kPa are obtained by shifting the initial LC curve of the recompacted specimen shown in Figure 6.25. During the following wetting process, the LC curve continues to shift to the right. After wetting to 0 kPa, the soil stress states are on the corresponding shifted LC curves at both stresses of 50 kPa and 110 kPa. Based on the initial and final LC curves, the value of dp_c^0/p_c^0 is deduced and shown in the figure. It is clear that dp_c^0/p_c^0 in Equation (6.6) is larger at smaller stresses. Hence, the wetting-induced yielding and volumetric contraction decreases with increasing stress over the confining stress range, as shown in Figure 6.28.

Figure 6.29b shows the change of yield stress in intact specimens before and after wetting from 100 kPa to 0 kPa at different stresses. The initial loading collapse curve of the intact specimen was determined by suction-controlled isotropic compression tests and is shown in Figure 6.22. During isotropic compression from 0 kPa to 50 kPa and wetting from the initial suction to 100 kPa, no plastic strain was observed, suggesting that there was no change in the loading collapse curve. Hence, the loading collapse curve at a confining pressure of 50 kPa and suction of 100 kPa was the same as the initial loading collapse curve. During the following wetting process, the LC curve is touched and shifted to the right. The intact specimen at a confining stress of 110 kPa is like the two recompacted specimens, as illustrated in Figure 6.29a. Before and after wetting from 100 kPa to 0 kPa, the stress states are both on the LC curve. As can be seen from the figure, dp_c^0/p_c^0 at a

Thermal effects on soil behaviour and properties 303

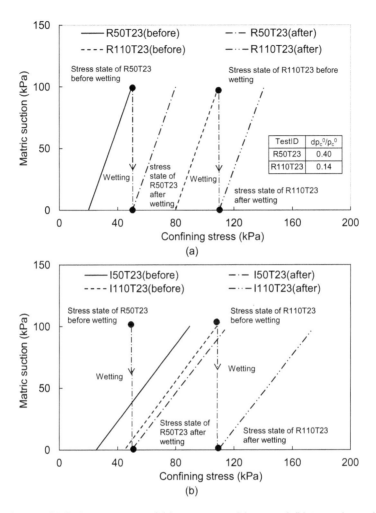

Figure 6.29 Evolution of LC characteristics of (a) recompacted loess and (b) intact loess during wetting from 100 to 0 kPa (T = 23°C).

confining stress of 110 kPa is greater than that at confining stress of 50 kPa. Based on Equation (6.6), we find that the wetting-induced yielding and volumetric contraction of intact specimens increases with increasing confining stress, as shown in Figure 6.28.

The collapse volumetric strain of each specimen, induced by wetting from suction of 100 kPa to 0 kPa, is calculated using Equation (6.6) and summarised in Table 6.5. For recompacted loess, this equation underestimates wetting-induced strain by 57% and 63% at confining stresses of 50 and 110 kPa, respectively. For intact loess, the wetting-induced strain is overestimated by 22% at a confining stress of 50 kPa but underestimated by 20% at a confining stress of 110 kPa. The discrepancy is probably because this equation predicts soil volume changes based on the initial and final conditions of suction and stress only, but ignores the influence of stress, suction history and path, and the degree of saturation (Alonso et al., 1990; Ng and Xu, 2002; Sun et al., 2007).

On the other hand, at a confining stress of 50 kPa, a recompacted loess specimen is normally consolidated and an intact specimen is over consolidated. Hence, recompacted loess has a larger dp_c^0/p_c^0 than intact loess at a stress of 50 kPa, resulting in a greater wetting-induced collapse in recompacted loess. The initial void ratios of the intact and recompacted specimens are about 1.13 ± 0.02 and 1.17, respectively. The maximum difference in the initial void ratio between the intact and recompacted specimens is about 4%. The influence of the initial void ratio on wetting-induced collapse was investigated by previous researchers (Sun et al., 2007; Kholghifard et al., 2012) for pearl clay and silty clay, which are classified as clay of low plasticity (the same as loess). Based on their experimental results, when the void ratio increases by 4%, the wetting-induced collapse volumetric strain increases by 10% to 20% over the stress range of 0 kPa to 100 kPa. A change of 10% to 20% is almost negligible, considering that the collapse volumetric strain of recompacted loess is about 2.5 times greater than that of an intact specimen. Hence, it is reasonable to presume that the difference in initial void ratios will not affect any key conclusion drawn in this study. At a confining stress of 110 kPa, both recompacted and intact loess specimens are normally consolidated. Concerning the greater wetting-induced softening of intact specimens than recompacted ones (see Figure 6.25), the dp_c^0/p_c^0 of intact specimens is greater than that of recompacted one during the wetting process. Consequently, the wetting-induced yielding and volumetric contraction of normally consolidated intact specimens is more significant than that of normally consolidated recompacted ones.

Moreover, according to the degrees of saturation listed in Table 6.5, we found that a greater increment in the degree of saturation does not always induce a higher collapse strain when comparing intact and recompacted loess specimens. It seems that the finding of Sun et al. cannot be generalised for both intact and recompacted loess specimens, probably because the influence of the degree of saturation is much less significant than structural effects.

Influence of soil micro-structure on thermal softening

According to previous studies (Burland, 1990; Liu and Carter, 2002), the behaviour of reconstituted soil can be used as a reference. The isotropic compression behaviour of reconstituted and structured soils (i.e., intact and recompacted specimens) can be linked as follows:

$$e = e^* + \Delta e \qquad (6.7)$$

where e is the void ratio of the structured soil at a given stress state, e^* is the void ratio of the corresponding reconstituted soil in the same stress state, and Δe, the additional void ratio, is the difference in void ratio between the tested soil and the reconstituted soil. Δe at the initial yield stress of a structured soil is defined as Δe_i to quantify the structure of intact, recompacted and reconstituted specimens (Liu and Carter, 2002). The values of Δe_i are 0.27, 0.25 and 0 for intact, recompacted and reconstituted specimens, respectively. A larger Δe_i suggests a more resistant soil structure.

Figure 6.30a shows the relationship between $p_c(T)$ and T for intact, recompacted and reconstituted specimens. In order to determine the thermal softening of yield stress for each type of specimen (intact, recompacted and reconstituted), the yield stress at a given temperature is normalised by that at a reference temperature (5°C). According to the theoretical equation derived by Zhou and Ng (2018), the normalised yield stress $p_c(T)/p_c(T_0)$ can be calculated using the following equation:

$$\frac{p_c(T)}{p_c(T_0)} = \exp\left(\frac{v(T) - v(T_0)}{\lambda - \kappa}\right) \tag{6.8}$$

where $v(T)$ and $v(T_0)$ are the specific volumes of a normally consolidated soil at T and T_0, respectively, under the same stress condition, λ is the plastic compressibility index and κ is the elastic compressibility index. The relationship between normalised yield stress and normalised temperature is shown in Figure 6.30b. According to Laloui and Cekerevac (2003), the relationship between yield stress and temperature can be expressed by the following equation:

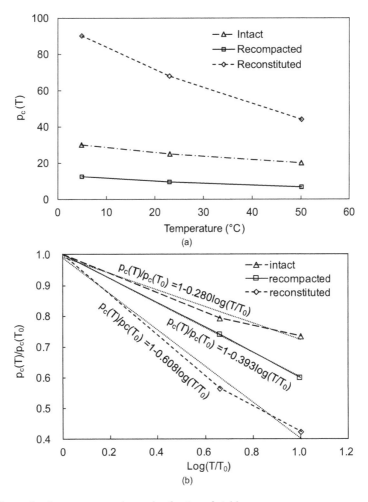

Figure 6.30 Effects of soil structure on thermal softening of yield stress.

$$\frac{p_c(T)}{p_c(T_0)} = 1 - \gamma_T \log\left(\frac{T}{T_0}\right) \tag{6.9}$$

where γ_T is a material parameter describing the evolution of yield stress with temperature. For intact, recompacted and reconstituted loess specimens, the values of γ_T are found to be 0.280, 0.393 and 0.608, respectively. In other words, a smaller material parameter γ_T represents a structure more resistant to thermal softening. It is clear that the behaviour of recompacted loess is more like that of intact loess than to that of reconstituted loess. This is consistent with the SEM images shown in Plate 6.2, which clearly reveal that the microstructures of intact and recompacted loess are similar. Moreover, the thermal softening of yield stress is least significant for intact loess over the temperature range considered. This is because in intact loess, clay particles form aggregates and accumulate at the inter-particle contacts, as shown in Plate 6.2a. These aggregates stabilise the structure of intact loess as they enlarge the inter-particle contact areas and hence stabilise the soil skeleton (Ng et al., 2016c, 2017c). On the other hand, the thermal softening of yield stress of reconstituted specimens is most significant, mainly attributable to its microstructure. For reconstituted loess, silt particles are dominant but clay particles float on the surface of silt particles, as shown in Plate 6.2c. Without the stabilising effects of clay aggregates on inter-particle contacts, the thermal softening of yield stress is most severe in reconstituted loess. The observed phenomenon can be used to explain the experimental

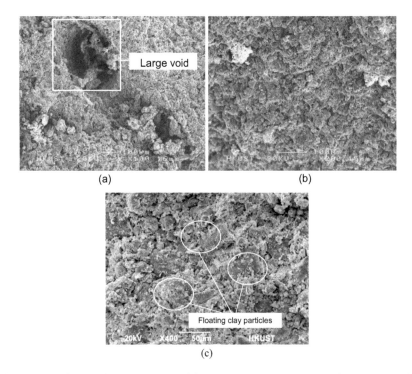

Plate 6.2 SEM images of loess (a) intact specimen; (b) recompacted specimen and (c) reconstituted specimen.

results obtained by Ng et al. (2018). According to existing thermo-mechanical models (Cui et al., 2000; Abuel-Naga et al., 2007a; Laloui and Cekerevac, 2008; Yao and Zhou, 2012; Zhou and Ng, 2018), there is a positive relationship between incremental volumetric strain and incremental yield stress. Therefore, within the same temperature range, the thermally induced plastic volumetric strain of reconstituted loess is greater than those of intact and recompacted loess.

Summary

The yield stress of loess decreases significantly with decreasing suction (wetting-induced softening), because the stabilising inter-particle normal force lessens during wetting. Moreover, the wetting-induced softening of recompacted loess become greater at a higher temperature. The thermal effects are probably because, for a given suction reduction, the reduction in the stabilising inter-particle normal force is greater at a higher temperature. In addition, the wetting-induced softening of intact loess is more significant than that of recompacted loess.

Under wetting from a suction of 100 kPa to 0 kPa at a confining stress of 50 kPa, the wetting-induced volumetric strain of recompacted loess increases from 4.1% at 5°C to 11.7% at 50°C. This is mainly because the wetting-induced softening is greater at higher temperatures, resulting in more significant yielding and volumetric collapse during wetting. This observation implies that at a higher temperature, wetting-induced soil contraction and ground movement would be much larger.

When the confining stress increases from 50 kPa to 110 kPa, the wetting-induced volumetric strain of intact specimens increases from about 1.6% to 3.5%. This is because before wetting, the soil specimen is over consolidated at 50 kPa but normally consolidated at 110 kPa. On the other hand, the wetting-induced volumetric strain of recompacted specimens decreases from around 5.5% to 1.6% when the confining stress increases from 50 kPa to 110 kPa. This is most probably because before wetting from 100 kPa to 0 kPa, recompacted specimens at both stresses are normally consolidated. Soil specimens at a stress of 110 kPa have a higher density and thus a smaller wetting-induced collapse.

At a lower stress of 50 kPa (similar to the *in situ* stress level), the wetting-induced softening of intact soil is much less than that of recompacted soil. This is because, after many drying and wetting cycles in the field, intact soil has a greater yield stress and more resistant structure. At the higher stress of 110 kPa (much higher than the *in situ* stress level), at which both intact and recompacted soils become normally consolidated, wetting-induced volumetric contraction of the intact specimen is about 30% greater than that of the recompacted specimen.

When the temperature increases from 5°C to 50°C, the yield stresses of the intact, recompacted and reconstituted specimens decreased by about 33%, 46% and 51%, respectively. The highly resistant structure to thermal softening of intact specimens results from the clay aggregates accumulating at the inter-particle contacts and stiffening the soil skeleton. Reconstituted specimens have the least resistant structure to thermal softening. This is because the clay particles in reconstituted specimens float on the grain surfaces rather than accumulating at inter-particle contacts.

VOLUME CHANGES OF AN UNSATURATED CLAY DURING HEATING AND COOLING (NG ET AL., 2016D)

Introduction

The thermally induced volumetric behaviour of soils can have a significant influence on many geotechnical engineering problems such as landfill cover systems and energy pile foundations (Gens, 2010). In many previous studies, all the temperature cycles were applied to the drying path and the influence of wetting was neglected. Few of the previous studies considered the effect of soil micro-structure (intact and recompacted) on the thermally induced volumetric behaviour of unsaturated soils.

Test programme and test apparatus

The principal objective of the tests was to reveal and report the volume changes of intact and recompacted low plasticity clay specimens (loess soil) under heating and cooling over a wide temperature range of 5°C to 53°C. Four suction- and temperature-controlled heating and cooling tests were carried out. Two of them (R0 and R100) were carried out on recompacted specimens, but at different suctions (0 kPa and 100 kPa). The other two (I0 and I100) were carried out on intact specimens at suctions of 0 kPa and 100 kPa. Details of the test program are summarised in Table 6.6. In order to interpret the effects of soil structure, a scanning electron microscope (SEM) was used to investigate the microstructure of intact and recompacted specimens, respectively. In order to control suction and temperature independently in unsaturated soils, a double-cell triaxial apparatus (Ng et al., 2012) was modified by adding a temperature control system (see Figure 6.19).

Test procedures

Figure 6.31 shows the thermo-mechanical paths of the loess specimens. Each test consisted of three stages: isotropic compression, wetting and thermal cycle. After set-up in the triaxial apparatus, the initial state of each specimen was fixed at point A (or A'). For intact and recompacted specimens, the initial suction is about 200 kPa and 180 kPa, respectively. In the figure, the initial state of intact and recompacted specimens is denoted by Points A and A', respectively. Each specimen was first compressed isotropically to a net confining stress of 50 kPa (A→B) in the drained condition. The next stage was to apply the

Table 6.6 Summary of compression/wetting tests on recompacted and intact loess

Test ID	Specimen type	Suction: kPa	Temperature: °C	Void ratio Initial	After wetting
R0	Recompacted	0	23→33→43→53→43→33→23→13→5→13→23	1.17	0.86
R100	Recompacted	100		1.17	0.97
I0	Intact	0		1.15	0.93
I100	Intact	100		1.17	1.01

target suction. In tests R100 and I100, the soil specimens were wetted to 100 kPa (B→C1). Similarly, the soil specimens in tests R0 and I0 were wetted to 0 kPa (B→C2). The suction equilibrium was considered to be complete when the rate of water content change was less than 0.09% per day (Chiu and Ng, 2012). Generally, 7 to 10 days were required to reach the equilibrium condition. For all specimens tested in this study, a contractive strain of over 10% was measured during the stages of isotropic compression and wetting, as shown in Table 6.6. It is reasonable to assume that all specimens are normally consolidated before heating. The third stage was to change the temperature step by step (the change of soil temperature is about 10°C in each step). The heating process was from typical room temperature (about 23°C) to 53°C (C1→D1, C2→D2), followed by cooling to 5°C (D1→E1, D2→E2) and re-heating to room temperature (E1→F1, E2→F2). Each step lasted for 24 hours to achieve thermal equilibrium.

Interpretation of experimental results

Typical thermally induced volumetric behaviour

Figure 6.32a shows the volumetric behaviour of recompacted soil during heating and cooling at two different suctions of 0 kPa and 100 kPa. During the heating process, the contractive volumetric strain increases with increasing temperature, at an increasing rate. This is probably because the heating-induced expansion of soil particles triggered particle rearrangement and facilitated plastic contraction. These observations can be explained using

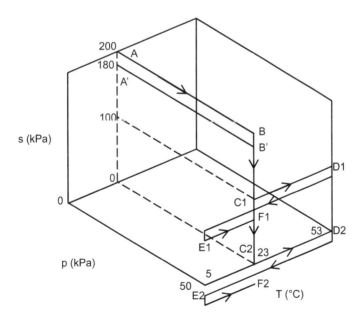

Figure 6.31 Thermo-mechanical path of each specimen.

310 Advanced Unsaturated Soil Mechanics

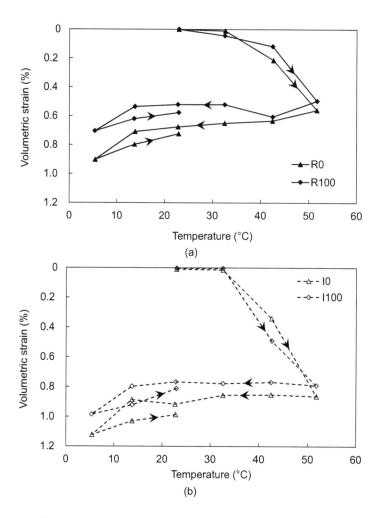

Figure 6.32 Suction effects on volume change of unsaturated soil during heating and cooling: (a) recompacted specimen; (b) intact specimen.

elasto-plastic theory, as shown in Figure 6.33a. The pre-consolidation pressure decreases as temperature increases. When temperature increases until the yield curve is reached, heating-induced yielding and plastic volumetric contractive strain can be observed and the yield curve shifts to the right.

It can be seen from Figure 6.32a that when the temperature decreases from 53°C to 13°C, the contractive volumetric strain continues to increase, but at a much slower rate of around 2×10^{-3}%/°C. This process is within the yield surface, which means that it is an elastic deformation caused by thermal contraction of the soil particles, as shown in Figure 6.33a, and the ratio of thermal strain to temperature change is close to the thermal expansion coefficient of soil particles (for clay particles this is 2.9×10^{-3}%/°C, Horseman and McEwen, 1996). When temperature continues to decrease, from 13°C to 5°C, a

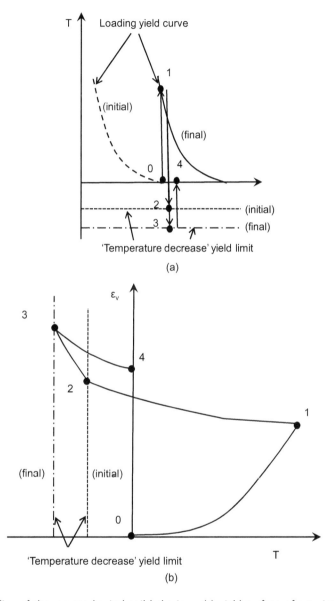

Figure 6.33 Modelling of thermo-mechanical soil behaviour: (a) yield surface of a typical existing model; (b) postulation of the "temperature decrease" yield limit.

plastic contractive volumetric strain at a higher rate of $2.5 \times 10^{-2}\%/°C$ can be observed, as revealed in Figure 6.32a. It should be pointed out that this observation differs from previous studies of saturated soil. Campanella and Mitchell (1968) and Di Donna and Laloui (2015) measured volume changes of remoulded illite and natural silty clay during cooling over the temperature range of 5°C to 60°C. For both soils, only elastic volume changes were observed during cooling. The discrepancy between the current study and the two previous studies (Campanella and Mitchell, 1968; Di Donna and Laloui, 2015) may be because the void ratio of the loess specimens is about 20% and 60% greater than those of the remoulded illite and natural silty clay specimens, respectively. Given a much larger void ratio, some large voids in loess are unstable and can be destroyed by cooling-induced particle contraction. Consequently, cooling may induce particle re-arrangement and plastic contraction in the loess specimen. On the other hand, remoulded illite and natural silty clay specimens are expected to have more stable structures. The cooling-induced responses of the remoulded illite and the natural silty clay specimens would be essentially elastic with minimum particle re-arrangement. Note that through temperature-controlled oedometer tests on saturated soft Bangkok clay, Abuel-Naga et al. (2007a) observed that pre-consolidation pressure and void ratio show negligible effects on the thermal strain of soil specimens at a given OCR. Their tests focused on the heating process, with a variation of void ratio of less than 15% only. Recently, Ng et al. (2016d) investigated the thermal volume changes of saturated loose and dense sand specimens using a temperature-controlled triaxial cell. They found that under thermal cycling, loose specimens with a relative compaction of 20% show plastic strain, but the response of dense specimens with a relative compaction of 90% is almost elastic. It is expected that soil specimens will show different responses when there is a wide variation in void ratio.

Further, the observed contraction during cooling cannot be predicted using existing elasto-plastic theory which predicts elastic contraction during cooling. A new "temperature decrease" yield limit is introduced in existing elasto-plastic thermo-mechanical models (for example, Zhou and Ng (2015)) to simulate the observed elasto-plastic behaviour during cooling, as shown in Figure 6.33b. When the temperature decreases to a critical value, the cooling-induced contraction of soil particles causes irreversible particle re-arrangement. This critical temperature is represented by the TD yield surface, as illustrated in Figure 6.33b. When the soil state reaches the "temperature decrease" yield surface, the soil response under continuous cooling is elasto-plastic. The volumetric contraction during cooling beyond the TD line is substantially less than the thermal volumetric contraction observed during heating beyond the loading yield curve. It is noted that the lowest temperature experienced by each recompacted specimen was the ambient temperature (i.e., 20°C). During the cooling process, each recompacted specimen exhibited elasto-plastic behaviour only when soil temperature was below 13°C. It may be postulated that the yield temperature could be slightly less than the ambient temperature. This observation suggests that the mechanical and thermal behaviours of soils are coupled. In the current study, large volumetric contractions (larger than 0.5%), leading to stiffening of the soil skeleton, were observed before cooling. During subsequent cooling, a lower temperature would be required to induce further particle re-arrangement and plastic strain.

The trend of thermally induced volume change of intact specimens, as shown in Figure 6.32b, is qualitatively similar to that of recompacted specimens. The differences between intact and recompacted specimens are analysed in the section "Effects of soil microstructure on volume changes during heating and cooling". As illustrated in Figure 6.32, the trends of soil behaviour obtained under thermal loads are the same for all four tests. This consistency suggests that the new laboratory data are limited but reliable. However, the above observations and discussion should be treated with caution since they may be specific and only applicable to a particular soil type and temperature range.

Suction effects on volume changes during heating and cooling

As shown in Figure 6.32 for both intact and recompacted specimens, the cumulative thermal volume changes at zero suction are greater than at a suction of 100 kPa. The difference, which varies with temperature from 0% to 20%, is about 5% on average. This value is much less significant by comparison with suction effects on Boom clay (Romero et al., 2003). In their study, the volume change at 60 kPa was 30% larger than that at 200 kPa with the same change of temperature. The differences between the current study and the study of Romero et al. (2003) mainly result from a different response during the wetting before heating. For the Boom clay in their study, the pre-consolidation was greater at a higher suction (suction hardening) and swelling was observed during wetting. For unsaturated loess in the current study, substantial volumetric contraction (about 17%, 25%, 13% and 19% for R100, R0, I100 and I0, respectively) was recorded during wetting. With a larger suction, the void ratio after wetting is larger. The effects of suction-induced hardening are compensated by the effects of strain hardening induced by wetting collapse.

At both suctions (i.e., 0 kPa and 100 kPa) considered in the current study, an increase in soil temperature induces volumetric contraction. At a high suction range (>100 MPa), only heating-induced expansion was observed by previous researchers (Tang et al., 2008; Alsherif and McCartney, 2015). The two different thermal responses of unsaturated soils may be attributed to a significant increase in soil pre-consolidation pressure when a high suction was applied (Alonso et al., 1990). Given a larger pre-consolidation pressure induced, the soil state would be within the yield surface during heating, leading to an essentially elastic thermal expansion of soil particles.

Effects of soil micro-structure on volume changes during heating and cooling

Figure 6.34a shows the volumetric behaviour of intact and recompacted soils at a suction of 0 kPa. The cumulative volumetric strain of intact specimens is about 25% greater than that of recompacted ones. Similar results are found at a suction of 100 kPa, as shown in Figure 6.34b. The differences between the intact and recompacted specimens may be attributed to their different soil structures. For intact specimens, there are a number of large pores (Bai et al., 2014) with a diameter greater than 200 µm, as shown in the SEM image in Plate 6.2a. The pores of the recompacted specimens are relatively uniform and no large inter-aggregate pores are observed (Plate 6.2b). In addition, the wetting-induced collapse (contractive volumetric strain during the wetting process (Alonso et al., 1990)) has caused a larger void ratio in intact specimens before heating and cooling than in recompacted specimens. With more large pores and a larger void ratio, the intact specimen shows a larger volume change under the heating and cooling cycles.

Summary

For both intact and recompacted low plasticity clay specimens (i.e., loess soil), contractive volumetric strain increases as temperature increases incrementally from a room temperature of 23°C to 53°C. This may be because the heating-induced expansion of soil particles triggered particle re-arrangement and facilitated plastic contraction. During the cooling process from 53°C to 13°C, contractive volumetric strain continues to increase at a rate

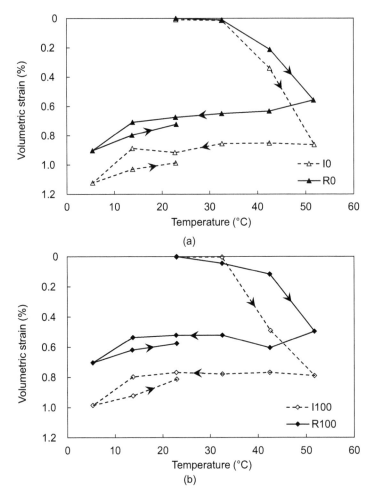

Figure 6.34 Effects of soil micro-structure on volume change of unsaturated soil during heating and cooling: (a) s = 0 kPa; (b) s = 100 kPa.

of around $2 \times 10^{-3}\%/°C$. When the soil temperature further decreases from 13°C to 5°C, a plastic volume contraction at a much higher rate of $2.5 \times 10^{-2}\%/°C$ is observed. This differs from previous studies on remoulded illite and natural silty clay. The change of contraction rate and plastic volume changes during cooling probably result from cooling-induced contraction of soil particles, leading to plastic particle re-arrangement in loess. More importantly, the observed plastic response during cooling cannot be captured by existing thermo-mechanical models, which predicts elastic contraction during cooling. A new "temperature decrease" yield surface may be introduced to existing thermo-mechanical models to simulate the observed elasto-plastic behaviour during cooling.

The cumulative volumetric strain induced in an intact specimen by the applied thermal cycle is about 25% greater than that in recompacted specimens. This is probably because of the presence of larger pores in the intact specimens.

For the given thermal cycle applied, the difference between thermally induced volume changes at suctions of 0 kPa and 100 kPa is 5% on average, which is much less than that reported for other soils such as Boom clay. This is probably because strain hardening induced by wetting collapse of the loess soil compensates for the effects of suction-induced hardening on soil behaviour.

SOURCES

Cyclic shear behaviour of silt at various suctions and temperatures (Ng and Zhou, 2014)

Effects of soil structure on the shear behaviour of loess at different suctions and temperatures (Ng et al., 2016c)

Effects of temperature and suction on secant shear modulus (Zhou et al., 2015)

Thermal effects on yielding and wetting- induced collapse of recompacted and intact loess (Ng et al., 2018; Cheng et al., 2020)

Volume changes during heating and cooling (Ng et al., 2016d)

Chapter 7

Constitutive modelling of state-dependent behaviour of soils

ELASTOPLASTIC MODELLING OF STATE-DEPENDENT SATURATED AND UNSATURATED SOIL BEHAVIOUR (NG ET AL., 2020A)

Introduction

The state and properties of unsaturated soils can change significantly with external loads, weather conditions and groundwater levels. It is obvious that the fully saturated state (i.e., degree of saturation equal to 100%) and completely dry state (i.e., degree of saturation equal to 0%) are just the two limiting conditions for an unsaturated soil. Proper modelling of the state-dependent behaviour of soils at different degrees of saturation is crucial to analysing the performance of almost all earthen civil engineering infrastructure and slopes. During the past three decades, many constitutive models for unsaturated soils have been reported, reviewed and compared in literature (Alonso et al., 1990; Wheeler and Karube, 1996; Chiu and Ng, 2003; Gens et al., 2006; Sheng et al., 2008b; D'Onza et al., 2011; Sheng, 2011; Zhou et al., 2023; Ng et al., 2024a). These reviews have discussed various important aspects of the constitutive modelling of unsaturated soils, such as constitutive variables, wetting-induced collapse, compressibility, yielding, shear strength, failure criteria, water retention behaviour and hydromechanical coupling. However, these various aspects of unsaturated soil behaviour have not been discussed in a unified theoretical framework, resulting in some confusion and potential misinterpretation. Moreover, some important aspects of the behaviour of unsaturated soil have not been addressed and discussed well, including state-dependent small-strain stiffness and large-strain dilation, and stress-dependent water retention behaviour.

In this chapter, a unified and relatively simple constitutive framework is presented for unsaturated soils. To express the framework in terms of clear physical meanings, a three-by-three compliance matrix is used to link volumetric strain, deviator strain and incremental degree of saturation to incremental mean net stress, deviator stress and suction. In this way, all nine variables in the compliance matrix can be shown to have clear physical meanings, which are illustrated and measured using various unsaturated soil tests. Based on this constitutive framework, a systematic approach to the modelling of the state-dependent behaviour of unsaturated soils is discussed in this section.

A unified and simple framework for simulating the state-dependent behaviour of soils

In the constitutive modelling of any soil, one of the key issues is the proper choice of constitutive variables. Many different constitutive stress variables have been proposed to model the mechanical behaviour of unsaturated soils in the literature (e.g., Wheeler et al., 2003; Russell and Khalili, 2006; Duan et al., 2019). Gens et al. (2006) reviewed the various variables adopted in existing elastoplastic models. They stated that "Different constitutive stresses stand on an equal footing and the matter of adopting one or the other must be decided using the criteria of convenience". In the unified and simple framework of this section, net stress and matric suction are used for simplicity.

In any constitutive model, the relationship between strain increment $\{d\hat{\varepsilon}\}$ and stress increment $\{d\hat{\sigma}^*\}$ can be described using a general formulation:

$$\{d\hat{\varepsilon}\} = [C^*]\{d\hat{\sigma}^*\} \tag{7.1}$$

where $[C^*]$ is the compliance matrix and $\{\hat{\sigma}^*\}$ and $\{\hat{\varepsilon}\}$ are the constitutive stress and strain variables, respectively. In the discussion here, the strain variables $\{\hat{\varepsilon}\}$ are defined as $\{\varepsilon_v, \varepsilon_q, S_r\}$, where ε_v is the volumetric strain, ε_q is the deviator strain and S_r is the degree of saturation.

By adopting net stress and suction in this chapter, the constitutive formulations for soils can be expressed in a general incremental form as follows:

$$\begin{bmatrix} d\varepsilon_v \\ d\varepsilon_q \\ dS_r \end{bmatrix} = \begin{bmatrix} I_{11} & I_{12} & I_{13} \\ I_{21} & I_{22} & I_{23} \\ I_{31} & I_{32} & I_{33} \end{bmatrix} \begin{bmatrix} dp \\ dq \\ ds \end{bmatrix} \tag{7.2}$$

where dp is the increment in the mean net stress, dq is the increment in the deviator stress, ds is the increment in suction and I_{ij} (i = 1, 2 and 3; j = 1, 2 and 3) are state-dependent variables for a given soil. According to Equation (7.2), the variables I_{11}, I_{21}, and I_{31} in the compliance matrix describe the behaviour of soils during compression, including the development of volumetric strain, deviator strain and degree of saturation. Similarly, I_{12}, I_{22}, and I_{32} describe the hydromechanical behaviour during the shearing process, while I_{13}, I_{23} and I_{33} capture the behaviour of soil subjected to drying or wetting. All nine variables can be calibrated and determined using suction- and stress-controlled tests, to be discussed later. It can be seen readily that Equation (7.2) is valid when the soil is saturated, which is considered to be a special case of unsaturated soil with S_r = 100%. In this special condition, the net stress can be replaced by the Terzaghi's effective stress and the values of I_{13}, I_{23}, I_{31}, I_{32} and I_{33} are zero.

To obtain the relationship between $[C^*]$ (i.e., the compliance matrix in Equation (7.1)) and $[C]$ (i.e., the compliance matrix defined in Equation (7.2)), $\{d\hat{\sigma}\}$ can be expressed as (Chen, 2007)

$$\{d\hat{\sigma}^*\} = [T_a]\{d\hat{\sigma}\} + [T_b]\{d\hat{\varepsilon}\} \tag{7.3}$$

where $\{d\hat{\sigma}\}$ is the incremental form of the constitutive stress variables $\{dp, dq, ds\}$ used in Equation (7.2); $[T_a]$ and $[T_b]$ are two matrices and their values depend on the constitutive stress variables in the constitutive model under investigation. Substituting Equations (7.2) and (7.3) into Equation (7.1), we obtain:

$$[C] = [I - C^* T_b]^{-1} [C^*][T_a] \tag{7.4}$$

where $[I]$ is a unit matrix. It should be noted that Equations (7.1) and (7.4) are general equations.

Although some unsaturated soil models reported in the literature are based on other constitutive stress variables other than net stress and suction, these models can be also converted to Equation (7.2) by matrix transformation. However, when different models are used, $[T_a]$ and $[T_b]$ take different forms.

For example, the model of Wheeler et al. (2003) uses the following three constitutive stress variables: $\{\hat{\sigma}^*\} = \{p^*, q, s^*\}$, where mean Bishop's stress p^* and modified suction s^* are defined as $(p + sS_r)$ and (ns), respectively. From the incremental form of $\{\hat{\sigma}^*\}$, it can be derived readily that:

$$[T_a] = \begin{bmatrix} 1 & 0 & S_r \\ 0 & 1 & 0 \\ 0 & 0 & n \end{bmatrix} \tag{7.5}$$

$$[T_b] = \begin{bmatrix} 0 & 0 & s \\ 0 & 0 & 0 \\ (1-n)s & 0 & 0 \end{bmatrix} \tag{7.6}$$

Another example is the model of Khalili et al. (2008), which adopts the following constitutive stress variables: $\{\hat{\sigma}^*\} = \{p_k^*, q, s\}$. p_k^* is the mean effective stress proposed by Khalili and Khabbaz (1998): $(p + \chi s)$, where χ is defined as follows:

$$\chi = \begin{cases} 1 & \text{for } s \leq s_e \\ \left(\dfrac{s}{s_e}\right)^{-0.55} & \text{for } s > s_e \end{cases} \tag{7.7}$$

where s_e is the suction value marking the transition between saturated and unsaturated states. $[T_a]$ and $[T_b]$ are calculated using the following two equations:

$$[T_a] = \begin{bmatrix} 1 & 0 & 2(s/s_e)^{-0.55} \\ 0 & 1 & 0 \\ 0 & 0 & 1 \end{bmatrix} \tag{7.8}$$

$$[T_b] = \begin{bmatrix} 0 & 0 & (s/s_e)^{0.45}(\partial s_e/\partial \varepsilon_v) \\ 0 & 0 & 0 \\ 0 & 0 & 0 \end{bmatrix} \quad (7.9)$$

Lu et al. (2010) proposed a new effective stress formulation $(\sigma - \sigma^s)$ based on the concept of suction stress σ^s (Lu and Likos, 2006). For constitutive models based on this effective stress formulation, $[T_a]$ and $[T_b]$ are calculated using the following two equations:

$$[T_a] = \begin{bmatrix} 1 & 0 & (S_r - S_{rr})/(1 - S_{rr}) \\ 0 & 1 & 0 \\ 0 & 0 & 1 \end{bmatrix} \quad (7.10)$$

$$[T_b] = \begin{bmatrix} 0 & 0 & (S_r - S_{rr})/(1 - S_{rr}) \\ 0 & 0 & 0 \\ 0 & 0 & 0 \end{bmatrix} \quad (7.11)$$

where S_{rr} is the residual degree of saturation.

The above three examples clearly show that all constitutive models can be converted to Equation (7.2) by matrix transformation. Within this unified framework (i.e., Equation (7.2)), the constitutive formulations for unsaturated soil behaviour are discussed in terms of their physical meaning, which enables the determination of the variables to be described in the following paragraphs.

Determination of the variable I_{11}

The variable I_{11} in Equation (7.2) can be determined using the following equation:

$$I_{11} = \frac{\partial \varepsilon_v}{\partial p} \quad (7.12)$$

According to Equation (7.12), the variable I_{11} is the ratio between the incremental volumetric strain and incremental mean net stress when q and s are constant. The value of this variable corresponds to the soil's volumetric compressibility, which can be measured using compression tests under constant q and s. For example, Ng and Yung (2008) carried out a series of suction-controlled isotropic compression tests on a compacted completely decomposed tuff (CDT, a silt). Four suction levels, 0, 50, 100 and 200 kPa, were considered and applied. Figure 7.1 shows the measured relationship between volumetric strain and mean net stress. One of the key findings was that the measured compressibility decreased with increasing suction over the stress and suction ranges considered, because

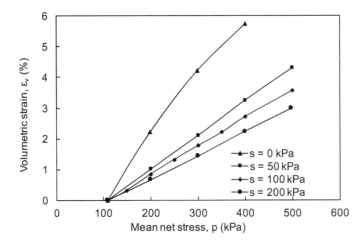

Figure 7.1 Measured relationship between volumetric strain and mean net stress of a compacted silt at various suction conditions.

Source: Data from Ng and Yung (2008).

of the stiffening effects of water menisci. This observation implies that the value of I_{11} is lower at a higher suction. In contrast, some other soils have been found to be more compressible at a higher suction (e.g., Wheeler and Sivakumar (1995)), probably because drying a soil results in a greater number of compressible macro pores (Gallipoli et al., 2003b; Zhou et al., 2012). These differing trends suggest that suction-controlled compression tests should be carried out to obtain the value of I_{11} accurately.

To model the volume change behaviour of unsaturated soils during the loading and unloading process, various formulations have been reported in the literature. They were compared and discussed in detail by Sheng (2011). To illustrate the relationship between volume change behaviour and Equation (7.2), this current study adopts the following equation as an example:

$$de = \frac{\alpha_p(s)dp}{p} \tag{7.13}$$

where e is the void ratio; $\alpha_p(s)$ is the compressibility. It should be noted that although Equation (7.2) does not consider the yield surface explicitly, the values of variables such as I_{11} are affected by it. For unloading or reloading inside the yield surface and loading on the yield surface, the members of $\alpha_p(s)$ are equal to $\kappa(s)$ and $\lambda(s)$ respectively. Equation (7.13) assumes that the compressive behaviour of an unsaturated soil can be described by a straight line in the e-ln p plane. This equation has been used widely in elastoplastic models (e.g., Chiu and Ng (2003)), mainly because it is a simple but effective model of the volume change behaviour of unsaturated soils.

Based on Equations (7.12) and (7.13), the following equation can be derived:

$$I_{11} = \frac{\alpha_p(s)}{(1+e)p} \tag{7.14}$$

Equation (7.14) clearly reveals that the value of I_{11} is affected by net stress, suction and void ratio. Therefore, the state-dependent compressibility is taken into consideration by this equation and hence by Equation (7.2).

Determination of the variable I_{12}

According to Equation (7.2), the variable I_{12} is described by

$$I_{12} = \frac{\partial \varepsilon_v}{\partial q} \tag{7.15}$$

This is the ratio of incremental volumetric strain to incremental deviator stress when p and s are constant. Volumetric strain can be induced by dilation or contraction during the shearing process, which is irreversible (i.e., the elastic volumetric strain is equal to zero). Hence,

$$d\varepsilon_v = \left(d\varepsilon_q - \frac{dq}{G_0}\right) D_q \tag{7.16}$$

where G_0 is the elastic shear modulus; and D_q is the dilatancy associated with the plastic mechanism of shearing. From Equations (7.15) and (7.16), the following equation can be derived:

$$I_{12} = \left(\frac{\partial \varepsilon_q}{\partial q} - \frac{1}{G_0}\right) D_q \tag{7.17}$$

where $\partial \varepsilon_q/\partial q$ is defined as I_{22}. Equation (7.17) suggests that the value of I_{12} is governed by three variables, I_{22}, G_0 and D_q. The variable D_q is discussed here, but detailed discussion of I_{22} and G_0 is given later.

Ng and Chiu (2001, 2003) carried out two series of triaxial tests on compacted decomposed volcanic (CDV, a silty clay) and compacted decomposed granitic (CDG, a sand silt) soils. Triaxial undrained and constant water content shear tests were conducted on saturated and unsaturated specimens, respectively. They found that a higher stress ratio is required to mobilise the same amount of dilatancy when the suction is higher. A similar behaviour was found for the CDG soil. Ng and Zhou (2005) reported a series of suction-controlled direct shear tests on another coarse-grained CDG soil. For the five tested specimens, specimens subjected to suctions of 200 and 400 kPa exhibited brittle stress-strain behaviour, while the other three specimens at suctions of 0, 10 and 50 kPa exhibited ductile behaviour. All four unsaturated specimens exhibited a phase transformation from positive to negative dilatancy with increasing stress ratio. It was also observed that the stress ratio corresponding to a maximum negative dilatancy increased with suction. In addition, the maximum negative dilatancy decreased (i.e., the soil became more dilative) with increasing suction. Through microscopic analysis, Ng et al. (2020b) illustrated that suction-induced dilatancy is not governed by a change in void ratio, but depends on suction effects in the micro and macro pores. Based on the above experimental results, it is evident that the dilatancy of unsaturated soils depends on suction. Therefore, the formulations developed for saturated soils should be modified for unsaturated soils.

A dilatancy equation (or plastic flow rule) is an essential component of a constitutive model for soils. Alonso et al. (1990) presented one of the first elastoplastic models for unsaturated soils. This model is commonly referred to as the Barcelona basic model (BBM) and adopts the modified Cam Clay model (MCCM) as the reference model in the saturated state. Hence, the yield curve is an ellipse at constant suction associated with a pre-consolidation stress p_0 (or yield stress), which increases with increasing suction. The relationship between p_0 and suction is referred to as the loading-collapse (LC) curve in the BBM. It should be noted that the shape of the LC curve depends on the isotropic compression lines at different values of suction. In the three-dimensional stress and suction space, the yield surface is a series of ellipses. As the associated flow rule is used in the MCCM, the plastic potential function is the same as the yield curve. However, the MCCM overpredicts the volumetric deformation in the K_0 condition (Gens and Potts, 1982). Thus, a non-associated flow rule is adopted in the BBM (see Equation (A1) in Table 7.1). A parameter α is used such that no lateral strain is predicted by Equation (A1) under the K_0 condition. p_s is a parameter describing the contribution of suction to the tensile strength of unsaturated soil. In Equation (A1) parameters p_0 and p_s are functions of suction. Thus, the contribution of suction to the dilatancy is incorporated by these two parameters. If the stress states of a soil lie on a yield curve corresponding to a suction s_1 for a given stress ratio η, the same stress states of the soil will lie inside a yield curve corresponding to a higher suction s_2 ($s_2 > s_1$). In other words, soil subjected to suction s_2 is modelled as an over-consolidated soil in the BMM. The elasticity and Equation (A1) are used to predict the shear-induced volume change when the stress states lie inside and on the current yield surface, respectively. Further, a higher stress is required to reach zero dilatancy in the critical state for soil subjected to a higher suction (i.e., a higher p_s).

Because of inherent shortcomings of the MCCM, the BBM and the other models derived from the MCCM also cannot satisfactorily predict the shear-induced volumetric behaviour of unsaturated granular and over-consolidated soils. Two different approaches can be identified to address this limitation. In the first approach, other dilatancy equations for saturated sand have been modified for unsaturated soils (Cui and Delage, 1996; Chávez and Alonso, 2003; Chiu and Ng, 2003; Alonso et al., 2007). In the second approach, new constitutive models have been developed based on the bounding surface plasticity (Morvan et al., 2010) and sub-loading surface plasticity (Zhou and Sheng, 2015; Luo et al., 2020) to model the dilatancy of unsaturated over-consolidated soil. Cui and Delage (1996) used the Nova–Wood Equation (see Equation (A2) in Table 7.1), where η_r is the stress ratio at zero dilatancy. It should be noted that η_r corresponds to not only the stress ratio at the critical state but also the stress ratio in the phase transformation state, i.e., it changes from contractive to dilative behaviour for an over-consolidated soil. η_r is dependent on both

Table 7.1 Dilatancy expressions for unsaturated soils

Dilatancy expression	No.	Reference
$\dfrac{d\varepsilon_p^p}{d\varepsilon_q^p} = \dfrac{M^2(2p + p_s - p_0)}{2\alpha q}$	A1	Alonso et al. (1990)
$\dfrac{d\varepsilon_p^p}{d\varepsilon_q^p} = \dfrac{\eta_r - \eta}{\mu}$	A2	Cui and Delage (1996)
$\dfrac{d\varepsilon_p^p}{d\varepsilon_q^p} = d_1(s)\left(e^{m\psi} - \dfrac{\eta}{M}\right)$	A3	Chiu and Ng (2003)
$\dfrac{d\varepsilon_p^p}{d\varepsilon_q^p} = (1 + k_d \cdot \psi) - \eta$	A4	Russell and Khalili (2006)
$\sin\varphi_m = \dfrac{\sin\phi_m - \left(\dfrac{e}{e_{cr}}\right)^\beta \sin\phi_{cr}}{1 - \left(\dfrac{e}{e_{cr}}\right)^\beta \sin\phi_m \sin\phi_{cr}}$	A5	Chávez and Alonso (2003)
$\dfrac{d\varepsilon_p^p}{d\varepsilon_q^p} = \left(a + \dfrac{b}{\left(\eta w^p / p\right)^2}\right)^2 - b^2$	A6	Alonso et al. (2007)

suction and stress. Two of the key limitations of their equation are that: (1) a finite dilatancy ($d = \eta_r/\mu$) is predicted for isotropic compression ($\eta = 0$); and (2) the effects of density on dilatancy are not considered.

To improve the modelling of unsaturated soil dilatancy, Chiu and Ng (2003) extended the framework of state-dependent dilatancy (Li and Dafalias, 2000) from saturated to unsaturated conditions. In their formulation (i.e., Equation (A3) in Table 7.1), $d_1(s)$ is a model parameter that is a function of suction, and ψ is a state parameter defined as the difference between the current void ratio and the void ratio at the critical state for a given mean stress (Been and Jefferies, 1985). The state parameter describes the density and stress level of soils. Based on experimental evidence (Ng and Chiu, 2001; Ng and Chiu, 2003), Chiu and Ng (2003) revealed that ψ is a function of density, mean net stress and suction for unsaturated soils. They illustrated that by using a single set of model parameters, Equation (A3) can effectively capture the shearing-induced volume changes of unsaturated CDV and CDG soils with different initial densities and confining pressures. Russell and Khalili (2006) also used ψ in the formulation of dilatancy for a boundary surface plasticity model as depicted in Equation (A4) in Table 7.1. The model parameter k_d may vary for different soils and can be assumed to be a material constant if high-precision simulations are not required. When k_d becomes zero, the dilatancy equation of Cam Clay is recovered. Chávez and Alonso (2003) have also adopted a similar state-dependent dilatancy framework (Wan and Guo, 1998) in their constitutive model. The dilatancy angle φ_m is expressed by Equation (A5) in Table 7.1, where φ_m = mobilised friction angle; φ_{cr} = friction angle at the critical state; e and e_{cr} = current void ratio and void ratio at the critical state for a given mean stress and β = model parameter. In the equation, ϕ_{cr} and e_{cr} both depend on suction. A major assumption of the three models proposed in Chiu and Ng (2003), Russell and Khalili (2006) and Chávez and Alonso (2003) is that the tested materials reach the critical state after large shear deformation over the range of suctions studied. The triaxial test results for compacted DV and DG soils support such a hypothesis (Ng and Chiu (2001, 2003)). On the other hand, experimental results for compacted shale and limestone gravels did not reach the critical state after shearing to large deformations (Chávez and Alonso, 2003; Alonso et al., 2007). Thus, Alonso et al. (2007) have proposed an alternative parameter, plastic work input (W^p) to describe the dilatancy. It can be seen that Equation (A6) in Table 7.1 gives a good fit to the measured dilatancy for suction-controlled triaxial tests conducted on the compacted limestone gravel. In the equation, $\eta W^p/p$ is a dimensionless parameter. The effect of confining pressure p on the constraint of dilatancy is expressed by the parameter. Further, η is added so that Equation (A6) can predict an infinite value of dilatancy for isotropic compression, i.e., $\eta = 0$. Parameters a and b are two fitting variables that are functions of the total suction. In the second approach, some recent constitutive models still adopt the yield function and plastic potential of MCCM, but were developed based on the bounding surface plasticity (Morvan et al., 2010) and sub-loading surface plasticity (Zhou and Sheng, 2015) to model the dilatancy of unsaturated over-consolidated soil. Morvan et al. (2010) extended Bardet's boundary surface model for saturated soil (Bardet, 1986) to unsaturated soil. In this series of constitutive models, a limit state line (LSL) is defined which represents an upper bound for the admissible stress domain. The hardening modulus is formulated as a function of the stress ratio of LSL, which influences the amplitude of dilatancy and post-peak softening. Zhou and Sheng (2015) adopted a framework of sub-loading surface plasticity to model the effect of initial density on the mechanical behaviour of unsaturated soil. In this model, a unified hardening (UH) parameter proposed by Yao et al. (2009) was used to model the hardening of the yield surface. The UH parameter depends on the similarity ratio (R) between the sub-loading surface and the reference yield surface. If the soil is normally consolidated or over-consolidated, R will be equal to 1 and less than 1, respectively. The magnitude of the dilatancy and post-peak softening depend on R. Recently, Luo et al. (2020) proposed a function for the UH parameter including a state variable that describes the degree of over-consolidation for the current void ratio with reference to the anisotropic consolidation line. This state variable increases with increasing degree of over-consolidation, which controls the amount of strain softening and shear-induced dilatancy.

Figure 7.2 shows the measured and calculated values of dilatancy during shearing. The measurements were obtained from two triaxial tests on a gravelly sand at suctions of 0 and 40 kPa (Ng and Chiu, 2001). Both specimens were consolidated to the same confining pressure and similar void ratio but different suctions before shearing. The theoretical results were calculated using Equations (A1) to (A4) in Table 7.1. The value of the model parameters is summarised in Table 7.2. It is clear that Equation (A1) (Alonso et al., 1990) overestimates the dilatancy of the gravelly sand, as expected. This is mainly because the model of Alonso et al. (1990) was developed from the modified Cam Clay model. The theoretical results calculated using Equations (A2) to (A4) are generally consistent with the trend of the experimental data. All three equations were modified from Rowe's dilatancy equation. Equation (A2) cannot take into account density effects. Thus, a new set of calibrated parameters must be obtained for different densities. Khalili et al. (2008)

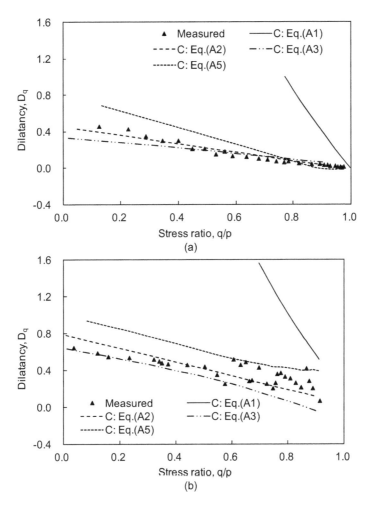

Figure 7.2 Comparisons between dilatancy measured (M) (Ng and Chiu, 2003) during shearing and theoretical results calculated (C) using the equations in Table 7.1: (a) s = 0; (b) s = 40 kPa.

also used Equation (A2) in a coupled flow deformation model but included an effective stress using the χ parameter as presented in Equation (7.7). The merit of using the effective stress approach is that the parameters are independent of suction. On the other hand, both Equations (A3) (Chiu and Ng, 2003) and (A4) (Chávez and Alonso, 2003) take the effects of density into account. To model the effect of state, (A3) and (A4) adopt the variables $(e - e_c)$ and (e/e_c) respectively, where e and e_c are the current void ratio and void ratio in the critical state respectively. It seems that Equation (A3) predicts the experimental results better, particularly in the range of stress ratios below 0.6. On the other hand, none of these equations explicitly consider the influence of the suction path. Some researchers found that at the same suction, the values of shearing-induced dilatancy are obliviously different along the drying and wetting paths (Chen, 2007; Goh et al., 2014; Chen et al., 2018). The

Table 7.2 Model parameters for the simulations in Figure 7.2

Equation	s = 0 kPa	s = 40 kPa
A1	M = 1.55	M = 1.55, p_s = 4.5 kPa
A2	η_r = 1.55, μ = 3.36	η_r = 1.62, μ = 2.05
A3	M = 1.55, d_1 = 0.2, m = 5	M = 1.55, c = 7 kPa, d_1 = 0.6, m = 5
	Γ = 0.824, λ = 0.11	Γ = 0.982, λ = 0.186
A4	sin ϕ_{cr} = 0.616, β = −2.5	sin ϕ_{cr} = 0.638, β = −2.5
	Γ = 0.824, λ = 0.11	Γ = 0.982, λ = 0.186

Note:

Critical state line is modelled using $q = Mp$ and $e = -\lambda \cdot \ln\left(\dfrac{p}{p_{at}}\right)$ where p_{at} = 100 kPa.

effects of the suction path on soil dilatancy need more experimental and theoretical study in the future.

One of the existing formulations for dilatancy can be used to determine I_{12}. Taking Equation (A3) as an example here, based on Equations (7.15) through (7.17), the following equation can be derived:

$$I_{12} = \left(I_{22} - 1/G_0\right) d_1(s) \left(\exp(m\psi) - \frac{\eta}{M} \right) \tag{7.18}$$

According to Equation (7.18), the value of I_{12} is affected by suction. Therefore, the state-dependent dilatancy of unsaturated soils is taken into consideration by this equation and hence by Equation (7.2).

Determination of the variable I_{13}

The variable I_{13} in Equation (7.2) can be calculated by

$$I_{13} = \frac{\partial \varepsilon_v}{\partial s} \tag{7.19}$$

Equation (7.19) denotes that I_{13} is the ratio of incremental volumetric strain to incremental suction when p and q are constant. This variable can describe the shrinkage and swelling of unsaturated soils subjected to drying and wetting. Figure 7.3 shows the experimental results of drying-induced shrinkage reported by Chiu and Ng (2012). The test soil was a compacted CDT, classified as a silt. Three different mean net stresses, 0, 40 and 80 kPa, were applied and maintained at a constant value during the drying process. It is evident that the void ratio lessened nonlinearly with increasing suction. More importantly, the reduction rate was affected by stress, suggesting that I_{13} is a function of stress.

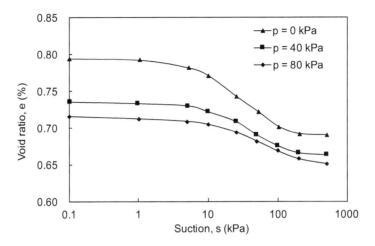

Figure 7.3 Drying-induced shrinkage of a compacted silt at different mean net stresses.
Source: Measured results from Chiu and Ng (2012).

To model the suction-induced volume changes of unsaturated soils, several formulations have been reported in the literature, as reviewed by Sheng (2011). Among these formulations, the equation of Sheng et al. (2008a) considers the influence of mean net stress on the shrinkage or swelling of unsaturated soils. It is used to describe the suction-induced volume change in the unified framework (i.e., Equation (7.2)):

$$de = \frac{\alpha_s ds}{p+s} \qquad (7.20)$$

where α_s is the compressibility of unsaturated soils upon a change in suction. The value of α_s strongly depends on the suction history, which governs the location of the yield surface. Four different cases are considered: (1) When the soil is subjected to drying and the current suction is equal to the maximum suction in the suction history (i.e., on the yield surface), α_s is equal to the shrinkage index λ_s; (2) when the soil is subjected to drying but the current suction is below the maximum suction in the suction history (i.e., within the yield surface), α_s is equal to the swelling index κ_s; (3) when the soil is subjected to wetting and the soil is over-consolidated (i.e., the soil state is within the yield surface), α_s is equal to the swelling index κ_s; and (4) when the soil is subjected to wetting and the soil is normally consolidated (i.e., soil state on the yield surface), wetting collapse occurs and α_s is equal to the rate of accumulation of plastic strain.

Based on Equations (7.19) and (7.20), the following equation can be derived:

$$I_{13} = \frac{\lambda_s}{(1+e)(p+s)} \qquad (7.21)$$

Determination of the variable I_{21}

The variable I_{21} in Equation (7.2) can be determined using the following equation:

$$I_{21} = \frac{\partial \varepsilon_q}{\partial p} \qquad (7.22)$$

According to Equation (7.22), the variable I_{21} is the ratio of the incremental deviator strain to the incremental mean net stress when q and s are constant. Moreover, the incremental deviator strain can be calculated using

$$d\varepsilon_q = \frac{d\varepsilon_v - \kappa(s)dp/p}{D_p} \qquad (7.23)$$

where D_p is the dilatancy associated with the plastic compression mechanism. The D_p of unsaturated soils has been studied previously using compression tests under constant ratio of suction to stress. Figure 7.4 shows experimental results for the D_p of a compacted silt at suctions of 200, 400, 600 and 1500 kPa (Cui and Delage, 1996). During the compression process, the stress ratio was maintained at 1. D_p was clearly affected by soil suction, particularly at mean net stresses below 300 kPa. To model the suction dependent D_p, Chiu and Ng (2003) proposed the following equation:

$$D_p = (\lambda(s) - \kappa(s)) d_2(s) \frac{M}{\eta} \qquad (7.24)$$

Based on Equations (7.22) through (7.24), the following equation can be derived:

$$I_{21} = \frac{I_{11} - \kappa(s)/p}{(\lambda(s) - \kappa(s)) d_2(s) M/\eta} \qquad (7.25)$$

It can be seen that the value of I_{21} is affected by suction. Therefore, the state-dependent dilatancy of unsaturated soils is taken into consideration by Equation (7.25) and hence by Equation (7.2).

Determination of the variable I_{22}

The variable I_{22} in Equation (7.2) can be determined using

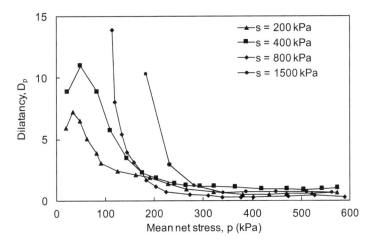

Figure 7.4 Suction effects on the dilation of a compacted silt subjected to compression at a constant stress ratio.

Source: Test data from Cui and Delage (1996).

$$I_{22} = \frac{\partial \varepsilon_q}{\partial q} \tag{7.26}$$

According to Equation (7.26), I_{22} is the ratio of the incremental deviator strain to incremental deviator stress when p and s are constant. This variable is closely related to the tangent shear modulus G (i.e., $I_{22} = 1/G$), which has been studied by many researchers (Atkinson, 2000; Clayton, 2011). It is recognised that the G of soil depends on strain, as shown in Figure 5.1. At very small strains below 0.001%, the shear modulus is almost constant and is denoted by G_0. This value is widely used for different purposes, such as the calculation of ground movement under dynamic loads. At small strains between 0.001 to 0.01%, the shear modulus lowers significantly with increasing strain. Under the working conditions of many civil engineering structures in relatively medium and dense or stiff soils, typical strains encountered fall within the small-strain range (Atkinson and Sallfors, 1991; Mair, 1993; Ng et al., 2020d). In the following section, the formulation of G_0 and the strain dependence of shear modulus are discussed. A formulation for I_{22} is then derived.

Initial shear modulus G_0 at very small strains

Through a series of resonant column tests on unsaturated sand, Wu et al. (1984) found that as the degree of saturation increased, G_0 first increased and then decreased. The maximum value occurred at a degree of saturation of about 20%. A similar relationship between G_0 and the degree of saturation was observed by Qian et al. (1993). In these two studies, soil suction was not controlled or measured and soil specimens were compacted at different water contents, resulting in different soil microstructures and hence these specimens could not be regarded as "identical" (Ng and Menzies, 2007). In more recent studies, soil specimens were generally prepared with the same water content and density to obtain the same microstructure. After preparation, the soil specimens were subjected to drying and wetting in a suction-controlled apparatus before determining G_0. Using a suction-controlled resonant column, Mancuso et al. (2002) investigated the G_0 of a compacted silty sand subjected to isotropic compression at constant suction. They found that G_0 increased with suction. Similar findings from various unsaturated soils have also been reported by many researchers (Vassallo et al., 2007; Ng and Yung, 2008; Sawangsuriya et al., 2009; Hoyos et al., 2015). Apart from suction, the effects of suction history on anisotropic G_0 were identified by Ng et al. (2009), as shown in Figure 5.20. They applied a drying and wetting cycle to unsaturated silt at constant net stress. They found that at the same suction, G_0 was higher along the wetting path than along the drying path. This revealed that apart from the current suction, the suction path also affected the stiffness of unsaturated soils. Their finding was confirmed by Khosravi and McCartney (2012) who tested another compacted silt.

Some semi-empirical equations have been reported in the literature to describe the G_0 of unsaturated soil. In this review, these equations are classified into four types based on their constitutive variables. The first type of G_0 model, such as Equations (B1) through (B3) in Table 7.3, adopts net stress suction as constitutive variables. Equation (B1) proposed by Leong et al. (2006) is an example of this. This equation takes the influence of stress and suction into consideration, but ignores the influence of stress path and suction path. As a consequence, the effects of degree of saturation and density on G_0 cannot be captured. More importantly, this equation is only applicable for isotropic conditions, including isotropic soil fabric and isotropic stress state. Ng and Yung (2008) proposed Equation (B2) to describe the anisotropic G_0 of unsaturated soil. According to Equation (B2), the anisotropic G_0 of unsaturated soils is affected by the soil fabric, void ratio, stress and suction. In addition, Sawangsuriya et al. (2009) proposed Equation (B3) for the G_0 of unsaturated soil. Unlike Equations (B1) and (B2), this equation uses multiplication rather than addition. Hence, the suction-induced variation of G_0 is independent of stress in Equation (B3). In other words, the effects of stress and suction on G_0 are assumed to be independent. On the other hand, we note that the first type of G_0 model predicts a unique relationship between G_0 and suction when the stress and void ratio are constant. The effects of suction history are not captured by the first type of G_0 model. To improve these models, a possible approach is to incorporate the degree of saturation or water content.

The second type of G_0 model, such as those described by Equations (B4) and (B5) in Table 7.3, is based on the effective stress of unsaturated soil. In Equation (B4) proposed by Sawangsuriya et al. (2009), the effective stress of unsaturated soil is taken as $(\sigma - u_a) + S_r^\kappa(u_a - u_w)$. Typical values of parameter κ were found to lie between 1 and 3. At a given suction, the calculated G_0 is therefore less along the wetting path than that along the drying path. This predicted trend is unlike the experimental results reported by many researchers (see Figure 7.5). Recently, Pagano et al. (2018) developed a microscale-based model (i.e., Equation (B5)) for G_0 in unsaturated granular geomaterials. In their formulation, σ_i is the intergranular stress of the unsaturated packing and σ_i^b and σ_i^m are the intergranular stresses in the regions of bulk water and meniscus water, respectively. It is conceptually good to differentiate the intergranular stresses in bulk water and meniscus water, as these two types of

Table 7.3 Stiffness expressions for unsaturated soils

Model type	Stiffness expression	No.	Reference
I	$G_0 = (G_0)_{ref}\left(1+\left[(\sigma_3 - u_a)/p_{atm}\right]^n\right)\left(1+\ln(1+s/p_{atm})\right)^m$	B1	Leong et al. (2006)
	$G_{0(ij)} = C_{ij}^2 f(e)\left[(\sigma_i - u_a)/p_{atm} \cdot (\sigma_j - u_a)/p_r\right]^n (1+s/p_r)^{2m}$ where $p_r = 1\text{kPa}$	B2	Ng and Yung (2008)
II	$G_0 = Af(e)(\sigma - u_a)^n + C(S_r)^K s$	B3	Sawangsuriya et al. (2009)
	$G_0 = Af(e)\left[(\sigma - u_a) + (S_r)^K s\right]^n$	B4	Sawangsuriya et al. (2009)
	$G_0 = \dfrac{2}{7}\dfrac{l^2}{V}\dfrac{3}{2}k_{n_0}^{2/3}\left[\dfrac{\sigma_i}{2l}V\right]^{1/3}$ where $\sigma_i = \sigma + \sigma_i^m\left(\dfrac{S_r - S_{rm}}{1-S_{rm}}\right) + \sigma_i^m\left(1 - \dfrac{S_r - S_{rm}}{1-S_{rm}}\right)$	B5	Pagano et al. (2018)
III	$G_0 = Ap_{atm}\left[\dfrac{p'_{c0}}{p_0}\exp\left(\dfrac{e^p}{\lambda-\kappa}\right)\right]^K\left[\dfrac{p_n}{p'}\exp(b(S_{e0}-S_e))\right]^K\left(\dfrac{p'}{p_{atm}}\right)^n$ where $p' = p_n + \dfrac{S_r - S_{r,res}}{1-S_{r,res}}s$	B6	Sawangsuriya et al. (2009)
	$G_0 = A(p_{atm})^{1-n} f(e) OCR^m (p+sS_r)^n \left[1-a(1-\exp(b\xi))\right]$	B7	Biglari et al. (2011)
	$G_0 = Ap_r f(e)(p'/p_r)^n S_r^{-k/\lambda}$ where $p' = p + (s_e/s)^{0.55} s$	B8	Wong et al. (2014)
	$G_0 = G_0^{sat}(1/s_e)^a (1+\sigma'/p_{atm})^b$ where $\sigma' = (\sigma - u_a) + S_e s$	B9	Dong et al. (2016)
IV	$G_0 = G_0^{dry}(1+H(S_r))$	B10	Wu et al. (1984)
	$G_0 = G_{0(s*)}\left[(1-r)\exp(-\beta(s-s*)) + r\right]$	B11	Mancuso et al. (2002)
	$G_0 = G_0^{sat} + (G_0^{ref} - G_0^{sat})(s/s_{ref})(S_r/S_{r,ref})^\xi$	B12	Han and Vanapalli (2016)

water change the intergranular stresses in very different ways (Wheeler et al., 2003). This microscopic approach, however, leads to great difficulty in calibrating the model.

To consider the effects of suction history on G_0 properly, some recent models have used two constitutive variables, at least one of which is a function of the degree of saturation. Equations (B6) through (B9) in Table 7.3 all belong to this type of G_0 model (Biglari et al., 2011b; Khosravi and McCartney, 2012; Wong et al., 2014). Equation (B6), which was proposed by Sawangsuriya et al. (2009), has six model parameters (A, K, b, n, λ and κ). Many test results are required to calibrate all of these parameters. Similarly, Equation (B7) (Biglari et al., 2011b) requires information about NCLs to compute the OCR of a soil, in addition to five model parameters (A, n, m, a and b). Wong et al. (2014) developed a semi-empirical equation based on the effective stress formulation of Khalili and Khabbaz (1998). By comparison with the model of Biglari et al. (2011b), the model of Wong et al.

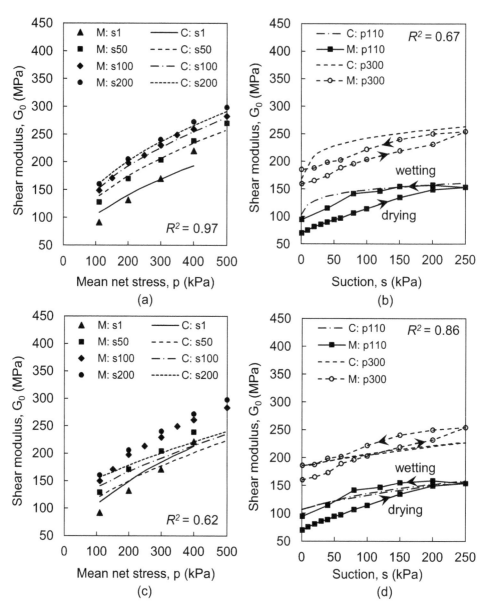

Figure 7.5 Comparisons between measured (M) shear moduli G_{vh} (Ng and Yung, 2008; Ng et al., 2009) and theoretical results calculated (C) using (a)-(b): Equation (B2) of Ng and Yung (2008); (c)-(d): Equation (B4) of Sawangsuriya et al. (2009); (e)-(f): Equation (B9) of Dong et al. (2016); (g)-(h): Equation (B12) of Han and Vanapalli (2016).

Note: the number after the p and s are net mean stress and suction respectively.

Constitutive modelling of state-dependent behaviour of soils 333

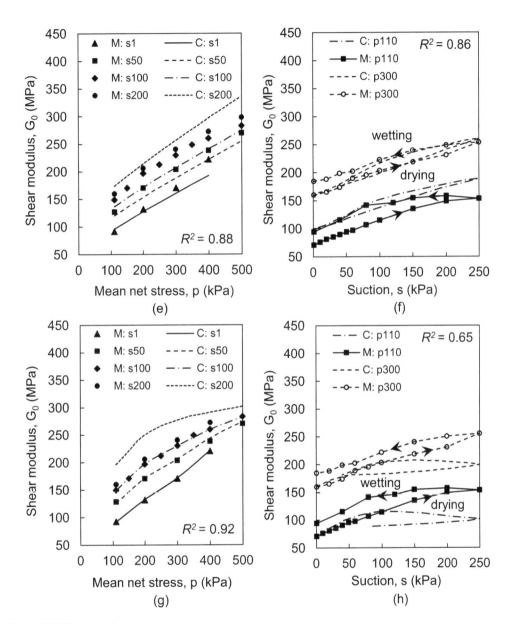

Figure 7.5 (Continued)

(2014) (i.e., Equation (B8)) adopts a void ratio function to take the effects of stress history into account, instead of incorporating both OCR and void ratio. Consequently, fewer model parameters are required in the model of Wong et al. (2014). Wong et al. (2014) applied the above three models to predict the G_0 of different soils along various stress paths. They found that along the drying and isotropic compression paths, the predictions given by these three models are quite consistent with measured data. Further, the equations of Biglari et al. (2011b) and Wong et al. (2014) can capture the variation of G_0 through drying and wetting cycles. Dong et al. (2016) proposed a G_0 model based on effective stress and effective degree of saturation (refer to Equation (B9)). This model is qualitatively similar to that of Wong et al. (2014).

The last type of G_0 model includes Equations (B10) through (B12) in Table 7.3. These models use a reference G_0 at a specific moisture condition (generally the fully saturated or completely dry condition) and calculate the variation of G_0 with soil moisture. Hence, these models do not require explicit consideration of stress state variables. Equation (B10), which was proposed by Wu et al. (1984), uses the function $H(S_r)$ to calculate the variation of G_0 with soil moisture. Mancuso et al. (2002) proposed Equation (B11), in which parameter β, controls the rate of increase of G_0 with increasing suction; r is the ratio of shear modulus at a very high suction and $(G_0)_{s^*}$. Han and Vanapalli (2016) proposed Equation (B12) to calculate the variation of G_0 with increasing suction and degree of saturation. To improve the model's prediction, two reference values of G_0 are used, including the fully saturated state and an unsaturated state. This type of model is simple, but may not be able to capture some important aspects of soil stiffness. For example, by using a scalar (water content/degree of saturation), the effects of suction on stiffness anisotropy (see Figure 5.20) cannot be simulated. In addition, the hysteresis of stiffness during drying and wetting cannot be captured.

As discussed above, theoretical formulations of G_0 in Table 7.3 may be classified into four types based on the constitutive stress variables. One equation from each type, including Equations (B2), (B4), (B9) and (B12), was selected to simulate the very small-strain behaviour of an unsaturated silt. The very small-strain moduli G_0 of this soil were measured by Ng and Yung (2008) and Ng et al. (2009) along two different stress paths, including constant-s compression and constant-p drying and wetting. The measured and calculated results are compared in Figure 7.5 and the parameter values are shown in Table 7.4. To evaluate the performance of each equation, the coefficient of determination (R^2) is calculated and shown in the figure. During constant-s compression, Equation (B2) (Ng and Yung, 2008) gave the best prediction of G_0, with R^2 values of 0.97. This is mainly because Equation (B2) uses the two independent stress state variables. The rate of increase of G_0 with increasing suction and stress can be simulated well using two independent terms. When Equation (B4) (Sawangsuriya et al., 2009) was used, however, there was an obvious discrepancy between measured and calculated results with an R^2 of 0.62. This problem is mainly because Equation (B4) is based on a single constitutive stress variable, which is insufficient to capture the influence of net stress, suction and degree of saturation in a unified formulation. During constant-p drying and wetting, Equation (B9) (Dong et al., 2016) gave the best prediction of experimental results (R^2 = 0.89). This is because this equation considers at least two different effects of the suction path properly: (a) altering the suction stress and hence the effective stress (Lu et al., 2010); (b) affecting suction hardening (Khalili et al., 2004). Equation (B12) (Han and Vanapalli, 2016) gives the lowest value of R^2 (i.e., 0.65). This is because at a given suction, the model predicts a higher G_0 along

Table 7.4 Model parameters for the simulations in Figure 7.6

Equation	Value of model parameters
B2	$C_{vh} = 11$ MPa; $n = 0.17$, $m = 0.045$
B4	$A = 9$ MPa; $n = 0.4$, $\kappa = 0.3$
B9	$G_0 = 53$ MPa; $n = 0.4$, $m = 0.8$
B13	$\xi = 2$; $s_{ref} = 100$ kPa
	Other variables $G_0^{sat}, G_0^{ref}, S_r, S_{r,ref}$ depend on suctions and stresses. The measured results for them are used in the numerical simulation.

Note:
The same void ratio function $1/(0.3 + 0.7e^2)$ is used in all equations, considering that the current study focuses on stress and suction effects on G_0.

the drying path, while the experimental results reveal that G_0 is greater along the wetting path. In addition, Equation (B2) (Ng and Yung, 2008) is cannot capture the variation of G_0 during drying and wetting ($R^2 = 0.67$) well, because this equation does not include S_r. In addition, during constant-p drying and wetting, the values of R^2 are less than 0.9 for all equations. Further studies are therefore required to improve the effect of modelling the suction path on G_0.

Reduction in shear modulus with increasing strain

To determine the stiffness strain relationship of a Singapore residual soil (clayey sand), Leong et al. (2006) carried out a series of undrained triaxial compression tests. Ten specimens with different suctions were sheared at constant water content and constant confining stress. They observed that the shear modulus increased consistently with initial suction and confining stress. It should be noted that suction was not controlled during the isotropic compression and shearing process, and the reported degradation of shear stiffness was a function of deviator strain as well as varying with suction during shearing. Ng and Xu (2012) carried out a series of suction-controlled constant mean net stress shear tests to investigate the effects of suction on the small-strain behaviour of an unsaturated CDT. The suction was controlled using the axis-translation technique. After the specimens were equalised at the target mean net stress and target suction, they were sheared under constant mean net stress and constant suction. As shown in Figure 7.6, the initial shear stiffness and degradation of stiffness are affected by suction.

Various semi-empirical equations have been proposed to model the degradation of shear modulus with strain based on experimental studies of saturated soils (see for example (Zhang et al., 2005; Oztoprak and Bolton, 2013; Vardanega and Bolton, 2013)). One example is the following hyperbolic equation proposed by Vardanega and Bolton (2013):

$$\frac{G}{G_0} = \begin{cases} 1 & \text{for } \varepsilon_q < \varepsilon_{qe} \\ \dfrac{1}{1 + \left(\dfrac{\varepsilon_q - \varepsilon_{qe}}{\varepsilon_{qref} - \varepsilon_{qe}}\right)^a} & \text{for } \varepsilon_q \geq \varepsilon_{qe} \end{cases} \quad (7.27)$$

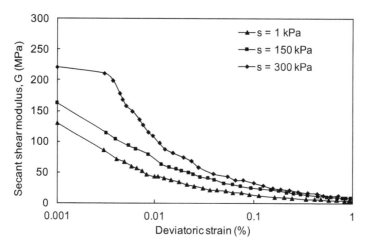

Figure 7.6 Suction effects on the stiffness degradation curve of a compacted silt.
Source: Test data from Ng and Xu (2012).

where G is the secant shear modulus at any deviator strain, ε_q is the deviator strain, α is a curvature parameter that controls the degradation rate of shear modulus with strain and ε_{qe} is the elastic threshold strain beyond which the shear modulus decreases with increasing deviator strain. ε_{qref} is a characteristic reference strain, defined as the deviator strain at which the secant shear modulus is reduced to $0.5G_0$. Parameter α is mainly affected by the soil type, while ε_{qe} and ε_{qref} depend on not only the soil type but also the soil state (for example, the void ratio and stress level) (Vardanega and Bolton, 2013). Equation (7.27) was originally developed by Vardanega and Bolton (2013) for saturated soils. Zhou (2014) illustrated that it can be used for unsaturated soils with a minor modification. In unsaturated soil, meniscus water increases the inter-particle normal force, which stabilises the soil skeleton. Hence, the elastic threshold strain (ε_{qe}) in Equation (7.27) is expected to increase when soil becomes desaturated.

Formulation for I_{22}

According to Equations (7.26) to (7.27), it can be shown that the value of I_{22} depends on the current deviator strain. When it is less than the elastic threshold strain, the shear modulus is constant. Hence,

$$I_{22} = \frac{1}{C^2 f(e) \left(\dfrac{p}{p_r}\right)^{2n} \left(1+\dfrac{s}{p_r}\right)^{2k}} \qquad (7.28)$$

When the current deviator strain is above the elastic threshold value, the strain dependence of the shear modulus should be considered. I_{22} is calculated using the following equation:

$$I_{22} = \frac{1 + \left(\dfrac{\varepsilon_q - \varepsilon_{qe}}{\varepsilon_{qref} - \varepsilon_{qe}}\right)^a}{C^2 f(e)\left(\dfrac{p}{p_r}\right)^{2n}\left(1 + \dfrac{s}{p_r}\right)^{2k}} \quad (7.29)$$

There are four parameters in Equations (7.28) and (7.29): C, n, k and α. The first three parameters can be obtained from stress and suction-controlled bender element and resonant column tests, while the last can be determined from a constant-p shear test.

Determination of the variable I_{23}

According to Equation (7.2), the variable I_{23} is equal to

$$I_{23} = \frac{\partial \varepsilon_q}{\partial s} \quad (7.30)$$

Equation (7.30) suggests that I_{23} is the ratio of incremental or decremental deviator strain to incremental or decremental suction when p and q are constant. Hence, the value of I_{23} can be calibrated using stress-controlled drying/wetting tests. Figure 7.7 shows the development of the deviatoric strain of a compacted gravelly sand subjected to wetting (Ng and Chiu, 2003). This study considered four mean net stresses of 0, 50, 100 and 200 kPa, and applied a stress ratio of 1.4 in all tests. During the wetting process, the deviator strain accumulated at an increasing rate, because yielding occurred when the wetting path reached the yield surface. Assuming that the wetting-induced deviator strain

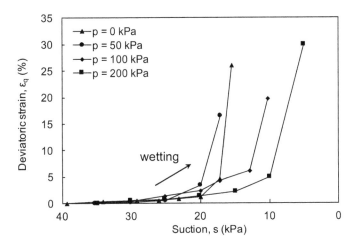

Figure 7.7 Deviatoric strain of a compacted gravelly subjected to wetting.

Source: Test data from Ng and Chiu (2003).

is essentially plastic (i.e., the elastic strain is zero), the slope of the deviator strain-suction relationship equals I_{23}.

It should be pointed out that in most drying/wetting tests reported in the literature, only volumetric strain was measured under an assumption of isotropic soil behaviour. Hence, experimental results for suction-induced deviator strain are very limited. If suction-induced elastic deviator strain is small and can be ignored, Equation (7.30) suggests

$$I_{23} = \frac{1}{D_s} \frac{\partial \left[\varepsilon_v - \frac{\kappa_s ds}{(p+s)(1+e)} \right]}{\partial s} \quad (7.31)$$

where D_s is the dilatancy during drying/wetting. D_s generally take the value of D_q or D_p (see Equations (A3) and (7.15)), when suction-induced yielding occurs for the plastic mechanisms of shearing and compression, respectively. Based on Equations (7.20) and (7.31), the following equation can be derived:

$$I_{23} = \frac{I_{13}}{D_s} - \frac{\alpha_s}{D_s(p+s)(1+e)} \quad (7.32)$$

Equation (7.32) implies that I_{23} is a function of I_{13}, α_s and D_s, which can all be determined based on experimental results, as presented and discussed previously. Hence, the value of I_{23} can be calculated readily using this alternative approach. No extra tests are required to calibrate I_{23}.

Determination of the variable I_{31}

The variable I_{31} in Equation (7.2) can be determined by

$$I_{31} = \frac{\partial S_r}{\partial p} \quad (7.33)$$

According to Equation (7.33), the variable I_{31} is the ratio of the incremental degree of saturation to incremental mean net stress when q and s are constant. Hence, this variable is closely related to the stress-dependent SWRC of unsaturated soils. The following paragraphs first discuss the measurement and modelling of a SWRC. Then, a specific SWRC model is used as an example to derive a formulation for I_{31}.

Stress-dependent SWRC

Various experimental technologies have been developed to determine the SWRC in soil science and agriculture-related disciplines, such as the Tempe pressure cell, the volumetric pressure plate extractor, the pressure membrane extractor and the osmotic desiccator (Fredlund and Rahardjo, 1993). However, these apparatuses do not take into account the deformation and stress of the soil as well as their influence on the SWRC. It is implicitly assumed that the SWRC of a given soil is unique. Therefore, the value of I_{31} is assumed to be zero.

Unlike soil science and agriculture-related disciplines, in geotechnical engineering the soil state, including the density and stress, is an important variable. This is because soil behaviour depends strongly on the soil state. Since about 20 years ago, the influence of soil density and deformation on SWRC has been recognised and actively investigated. Although the net stress of the soil specimens was not controlled in their study, Vanapalli et al. (1999) tested several specimens with different initial void ratios. With a reduction in the void ratio from 0.59 to 0.54, the water retention capacity of the soil improved greatly, particularly at suctions below 100 MPa. A further reduction in the void ratio from 0.54 to 0.51 did not affect the water retention ability substantially. The influence of the initial density on the SWRC has been reported by many researchers (Ng and Pang, 2000a; Tarantino, 2009). The observed effects of soil density on the SWRC are attributable to the fact that the average pore size of a soil specimen decreases as a result of deformation. The influence of pore size distribution on a SWRC cannot be known and modelled explicitly.

Ng and Pang (2000b) developed a new stress-controllable pressure plate apparatus, which can be used to determine the SWRC of unsaturated soils subjected to different stress states. Using the modified apparatus, the researchers measured the SWRC of unsaturated silt at vertical net stresses of 0, 40 and 80 kPa. The results are shown in Figure 3.10. It is clear that the SWRC was greatly affected by stress. The water retention capacity of the soil specimens increased with stress. Similar observations were reported by Lee et al. (2005b) and their results are shown in Figure 3.11. The effects of stress on the SWRC partially resulted from the average density-dependence of the SWRC. As reported by Vanapalli et al. (1999), when stress increases, the density increases and so does the water retention capacity. However, stress effects are not equivalent to density effects. This is because the application of stress affects not only the soil density (or void ratio) but also the pore size distribution, as illustrated by the results of MIP tests shown in Figure 7.8.

Modelling the stress-dependent SWRC

Many SWRC models have been reported in literature, and they may be classified into three types. The major difference comes from their approaches to considering the influence of soil state on the SWRC. The first type of SWRC models adopts a unique relationship between suction and soil moisture. Examples include Equations (C1) through (C3) in Table 7.5 (Gardner, 1956; Van Genuchten, 1980; Fredlund and Xing, 1994). The SWRC calculated using these models is a curve in the S_r–s plane. Even though these models in widespread practical use because of their simplicity, they do not take the coupling effects between the hydraulic and mechanical behaviour of unsaturated soils into consideration. Because of this limitation, I_{31} predicted by these equations is always regarded as zero.

The second type of SWRC models explicitly considers the effects of soil density on the water retention behaviour of unsaturated soil, such as Equations (C4) and (C6) in Table 7.5. Among these models, Gallipoli et al. (2003a) employed a closed-form equation which was modified from the Van Genuchten (1980) model by including the void ratio (e)

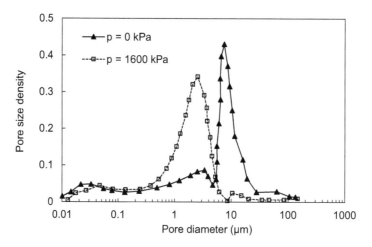

Figure 7.8 Stress effects on the pore-size distribution of a compacted clay with low plasticity.
Source: Data from Ng et al. (2016a).

Table 7.5 SWRC expressions for unsaturated soils

Model type	SWRC expression	Reference	No.
I	$S_r = \dfrac{1}{1+as^n}$	Gardner (1956)	C1
	$S_r = \left(1+\left(\dfrac{s}{a}\right)^{m_2}\right)^{-m_1}$	Van Genuchten (1980)	C2
	$S_r = \left(\ln\left(2.7+\left(\dfrac{s}{a}\right)^n\right)\right)^{-m}$	Fredlund and Xing (1994)	C3
II	$S_r = \left(1+\left(\dfrac{se^{m_4}}{m_3}\right)^{m_2}\right)^{-m_1}$	Gallipoli et al. (2003a)	C4
	$S_r = \left(1+\left(\dfrac{se^{m_4}}{m_3}\right)^{m_2}\right)^{-m_1}$ where $m_4 = 1/m_1/m_2$	Tarantino (2009)	C5
	$dS_r = E - B\dfrac{S_r}{n}(1-S_r)^\xi ds - \dfrac{S_r}{e}(1-S_r)^\xi de$	Sheng and Zhou (2011)	C6
III	$S_r = \left(1+\left(\dfrac{s(e_0 - \alpha_p \ln(1+p/p_{atm}) - \alpha_s \ln(1+s/p_{atm}))^{m_4}}{m_3(1+p/p_{atm})^{m_5}}\right)^{m_2}\right)^{-m_1}$ where $m_4 = 1/m_1/m_2$	Zhou and Ng (2014)	C7

in the formulation. Hence, S_r is a function not only of suction but also of the void ratio. The calculated SWRC becomes as a surface in the S_r–s–e space. Tarantino (2009) modified the model of Gallipoli et al. (2003a) by reducing one parameter based on extensive tests on both fine-grained and coarse-grained soils. Experimental results revealed that the product of S_r and e is almost independent of e at high suction. Based on this experimental evidence, it can be deduced that the product $m_1 m_2 m_4$ should be 1. Sheng and Zhou (2011) proposed an incremental form. The influence of void ratio on the SWRC was considered. This type of model implicitly assumes that stress effects are equivalent to density effects on the SWRC.

The last type of SWRC model (denoted as Equation (C7) in Table 7.5) was proposed by Zhou and Ng (2014). The derivation and verification of this equation are introduced in Chapter 3. It is illustrated that this equation shows a fundamental advantage in modelling the stress-dependence of SWRC. Based on Equations (7.33) and (C7), the following equation can be derived:

$$I_{31} = \frac{\partial S_r}{\partial e}\frac{\partial e}{\partial p} + \frac{\partial S_r}{\partial \xi}\frac{\partial \xi}{\partial p} \tag{7.34}$$

Then,

$$I_{31} = \frac{\partial S_r}{\partial e}(1+e)I_{11} + \frac{\partial S_r}{\partial \xi}\frac{\partial \xi}{\partial p} \tag{7.35}$$

It should be noted that the first type of SWRC model (i.e., Equations (C1) to (C3) in Table 7.5) assume that I_{31} is zero, while the second type of SWRC models (i.e., Equations (C4) to (C6) simply consider the first term on the right-hand of Equation (7.35).

Determination of the variable I_{32}

The variable I_{32} in Equation (7.2) can be determined using the following equation:

$$I_{32} = \frac{\partial S_r}{\partial q} \tag{7.36}$$

According to Equation (7.36), the variable I_{32} is the ratio of the incremental or decremental degree of saturation to the incremental or decremental deviator stress when p and s are constant. Ideally, I_{32} is determined from the relationship between S_r and q based on shear tests under constant-p and s conditions. Such data are very limited in literature because it is uncommon to carry out constant-p and s shear tests. Alternatively, it can be determined using data from SWRCs at various q conditions. Ng et al. (2012) measured the SWRCs of a compacted silt. Three different values of the stress ratio q/p (0, 0.75 and 1.2), corresponding to different deviator stresses, were applied. They found that the influence of deviator stress on the SWRC was much less than the effects of mean net stress. Based on this observation, the mean net stress is directly incorporated in the SWRC model, while the effects of deviator stress on SWRC are described using void ratio for simplicity. Hence, Equation (7.36) can be derived as follows:

$$I_{32} = \frac{\partial S_r}{\partial e} \frac{\partial \varepsilon_v}{\partial q}(1+e) \tag{7.37}$$

Based on Equations (7.16), (7.36) and (7.37), the following equation can be derived:

$$I_{32} = \frac{\partial \left[1 + \left(s \frac{e^{m_4}}{m_3} \left(\frac{\xi_m}{\xi_m^{ref}}\right)^{-m_m}\right)^{m_2}\right]^{-m_1}}{\partial e} - D_q(1+e)I_{22} \tag{7.38}$$

Equation (7.38) suggests that the value of I_{32} is affected by several factors, including density effects on the SWRC, D_q and I_{22}. A lot of data is available in literature to determine each of them, so I_{32} can be determined readily from experimental results.

Determination of the variable I₃₃

The variable I_{33} in Equation (7.2) can be calculated using the following equation:

$$I_{33} = \frac{\partial S_r}{\partial s} \tag{7.39}$$

According to Equation (7.39), the variable I_{33} is the ratio of the incremental degree of saturation to incremental suction when p and q are constant. This variable is related to the desorption/adsorption rates of unsaturated soils. Based on Equations (C7) and (7.39), the following equation can be derived:

$$I_{33} = \frac{\partial \left[1 + \left(s \frac{e^{m_4}}{m_3} \left(\frac{\xi_m}{\xi_m^{ref}}\right)^{-m_m}\right)^{m_2}\right]^{-m_1}}{\partial s} \tag{7.40}$$

It should be noted that the value of I_{33} is not constant, but depends on suction and stress. Moreover, the influence of stress on I_{33} is related to the change in the average void ratio and pore size distribution.

Apart from stress-dependence, hydraulic hysteresis also has a great influence on the water retention behaviour of unsaturated soil. At a given suction, the equilibrium water content along the drying path is higher than or equal to that along the wetting path. So far, some theoretical models have been proposed for the hysteretic water retention behaviour of unsaturated soil. In each model with an assumption of rigid soil, two boundary curves are generally defined, including the main drying curve and main wetting curve. Any state in the $S_r - s$ plane is bounded by these two curves. Similarly, a main drying surface and a main wetting surface are defined in the S_r–s–e space or S_r–s–p space, if density/stress effects on SWRC are considered in the model. When the soil state is on the main drying and wetting curves/surfaces, S_r is simply calculated from s, e and p. When soil state is between the main drying and wetting curves/surfaces, S_r is affected by not only s and e but also by the suction path, because of the existence of hydraulic hysteresis. Different methods have been used in the theoretical model to simulate the scanning curves between the main drying and wetting curves/surfaces. Wheeler et al. (2003) modelled the water retention behaviour in a classic elastoplastic framework. Two processes were defined, including the elastic and elastoplastic processes. During each process, a constant adsorption/desorption rate was assumed. Similar approaches were used in many unsaturated soil models (Nuth and Laloui, 2008; Sheng et al., 2008a; Gallipoli, 2012; Wong and Mašín, 2014). These approaches cannot predict a smooth transition between scanning curves and main curves/surfaces, since the adsorption/desorption rate during the elastoplastic process is much greater than that during the elastic process. To solve this problem, Li (2005) applied bounding surface theory to model the hysteretic water retention behaviour of unsaturated soil. Unlike the model of Wheeler et al. (2003), the adsorption/desorption rate is not constant when the soil state is between the main drying and wetting curves. The adsorption/desorption rate is affected by the suction history, the current suction and degree of saturation. The model of Li (2005) assumes a rigid soil, so it cannot capture the effects of density and stress on SWRC. In recent years, some advanced water retention models have been proposed with that incorporate density and hysteresis effects (Zhou et al., 2012; Gallipoli et al., 2015; Zhou et al., 2015a). Zhou et al. (2015a) developed a bounding surface model that includes the influence of soil deformation on water retention behaviour. The main drying/wetting curves in the model of Li (2005) was extended from the S_r – s plane to S_r – s – e space. Gallipoli et al. (2015) proposed a term (i.e., scaled suction s^*), which is a function of suction and void ratio. By adopting the scaled suction, the main drying/wetting surfaces in the S_r – s^* plane become two curves. It should be pointed out that none of the above models explicitly consider the reasons for hydraulic hysteresis. These models are therefore semi-empirical. Hence, some researchers attempt to model the hysteretic water retention behaviour by explicitly considering specific reasons for hydraulic hysteresis. For instance, Zhou (2013) applied different contact angles for advancing and receding water meniscus, leading to different water retention abilities during the drying and wetting processes. Cheng et al. (2019) incorporated the influence of pore non-uniformity on the water retention behaviour. These models have clear physical meaning, but the calibration of these soil parameters is not very straightforward. More studies at the micro and macro levels are needed in the future.

Discussion on the simple and unified framework

Based on the discussion above, it is illustrated that I_{ij} (i = 1, 2 and 3; j = 1, 2 and 3) in Equation (7.2) can be derived theoretically. More importantly, these nine variables can all be derived from suction- and stress-controlled unsaturated soil tests. Hence, Equation (7.2) provides a simple but effective framework for modelling the state-dependent behaviour of unsaturated soils.

Equation (7.2) does not explicitly differentiate elastic and elastoplastic processes. To consider these two processes in numerical analysis, the variables I_{ij} (i = 1, 2 and 3; j = 1,

2 and 3) should take different values for different processes. For instance, in the formulation for I_{11} (i.e., Equation (7.13)), $\alpha_p(s)$ is equal to $\kappa(s)$ and $\lambda(s)$ for the elastic process and elastoplastic process respectively.

The cross terms in the compliance matrix of Equation (7.2) are not independent. First, I_{12} and I_{21} describe soil dilatancy during constant-p shearing and constant-q compression, respectively. Soil dilatancy during these two processes is modelled using a unified formulation in many constitutive models such as the BBM (Alonso et al., 1990). Second, I_{12} and I_{32} are closely related because shearing-induced volumetric strain (described by I_{12}) affects the water retention behaviour (described by I_{32}) of unsaturated soil. Finally, I_{13} and I_{23} govern the volumetric and deviator strains induced by drying/wetting in conditions of constant p and q. The ratio of I_{13} and I_{23} represents soil dilatancy.

Summary

In this section, the state-dependent hydro-mechanical behaviour of unsaturated soil is discussed based on a unified and relatively simple framework. This framework uses mean net stress, deviator stress and suction as the constitutive stress variables. Theoretical models based on other constitutive stress variables can also be converted to this framework by matrix transformation. The nine variables, which have clear physical meanings, in the compliance matrix are derived. Moreover, the calibration methods for these nine variables are discussed and explained.

The small-strain stiffness of unsaturated soil is greatly affected by many factors, including strain, suction and suction path. So far, extensive formulations for G_0 at the very small strains (less than 0.001%) can be found in the literature. A performance study of these formulations suggests that at least two constitutive stress variables are required to capture well the variation of G_0 along various stress paths, such as compression and drying/wetting. Moreover, most existing models cannot predict the hysteresis of stiffness during drying and wetting because they do not incorporate the effects of the suction path properly. On the other hand, the modelling of the stiffness degradation curve at small strains (between 0.001% and 1%) is relatively less studied. More models of stiffness degradation for unsaturated soil are needed. Some other topics may also need further study, including suction-induced anisotropy effects of recent suction paths.

The dilatancy of unsaturated soil is affected by not only the stress ratio and density but also other factors such as suction and its path. With an increase in suction, soil dilatancy generally increases. It is therefore essential to model state-dependent dilatancy, which has not been incorporated in most existing models. To model state-dependent dilatancy, the choice of void ratio function is important. Based on the test results analysed in this study, the use of $(e-e_c)$ seems better than e/e_c, where e and e_c are the current void ratio and void ratio at the critical state respectively. Further, as far as the authors are aware, none of the existing models can capture the influence of drying and wetting cycles on the dilatancy of unsaturated soil. The effects of suction path on dilatancy needs further experimental and theoretical study.

The effects of stress and density on SWRC are fundamentally different. This is because net stress not only reduces the average void ratio but, more importantly, alters the pore structure of unsaturated soil. The use of the average void ratio to describe stress effects on water retention capability is therefore insufficient. Hence, SWRC models including the effects of stress on pore structures are required. More studies are required to understand better the effects of stress on the pore structure and hence SWRC of unsaturated soil.

SIMULATING THE CYCLIC BEHAVIOUR OF SOILS AT DIFFERENT TEMPERATURES USING BOUNDING SURFACE PLASTICITY APPROACH (ZHOU ET AL., 2015A; ZHOU AND NG, 2016)

Introduction

As discussed in the previous section, many elastoplastic models have been developed for unsaturated soil based on the critical state framework (Alonso et al., 1990; Gens, 2010; D'Onza et al., 2011; Sheng, 2011). Most of these models include a loading collapse yield surface, and assume that the size of yield surface increases with increasing suction. They generally assume that soil's response is essentially elastic inside the yield surface. Consequently, nonlinear soil behaviour at small strains (less than 1%) such as the degradation of shear modulus with strain cannot be captured fully. Further, none of the previous models were developed with the intent of capturing the cyclical behaviour of unsaturated soils at various temperatures.

In this section, a state-of-the-art (SOA) thermo-mechanical model for simulating the behaviour of unsaturated soil is described. This model is intended to capture the small-strain and cyclic behaviour of unsaturated soil at various temperatures. Its unique capability to model suction and thermal effects on small-strain soil behaviour, such as the degradation of soil stiffness with strain, will be demonstrated. Bounding surface plasticity theory is used to simulate the degradation of soil stiffness beyond the elastic range allowing for the influence of suction and its history. Unlike conventional elastoplasticity theory with one yield surface, bounding surface plasticity will predict the elastoplastic response even for stress states inside the bounding surface. The plastic modulus is determined to be dependent on the distance between any soil stress state and the bounding surface and hence highly nonlinear mechanical behaviour can be captured.

Mathematical formulations

Constitutive variables

As unsaturated soil mechanics has developed, various constitutive variables have been proposed for unsaturated soils. Adopting the thermodynamic point of view, Houlsby (1997) demonstrated that a complete description of the mechanical behaviour of unsaturated soil requires at least two constitutive variables. Bishop's stress ($\sigma^* = \sigma^t - u_a + S_r(u_a - u_w)$) and suction ($s$) are adopted in this study with their work-conjugate strain variables of ε and $-ndS_r$, where σ^t, u_a, u_w, S_r, ε and n are total stress, pore air pressure, pore water pressure, degree of saturation, soil skeleton strain and porosity, respectively. This set of constitutive variables ensures a smooth translation between the unsaturated and saturated states because σ^* becomes Terzaghi's effective stress (σ') when the soil specimen becomes saturated (i.e., $S_r = 1$).

The SOA model is formulated in triaxial space. For simplicity, three variables are used to define the stress state of a soil specimen, including mean Bishop's stress (p^*), deviator stress (*q*) and suction (*s*). Two volumetric variables, i.e., specific volume (*v*) and degree of saturation (S_r), are chosen to define the relative proportions of solid, water and air within an unsaturated soil element. Moreover, the temperature (*T*) is used to model thermal effects on soil behaviour. These variables except *T* are defined as follows:

$$p^* = (p^t - u_a) + S_r(u_a - u_w) \tag{7.41}$$

$$q = \sigma_1 - \sigma_3 \tag{7.42}$$

$$s = u_a - u_w \tag{7.43}$$

$$v = 1 + e \tag{7.44}$$

$$S_r = \frac{wG_s}{e} \tag{7.45}$$

where p^t is mean total stress, u_a is pore air pressure, u_w is pore water pressure, σ_1 is major principal stress, σ_3 is minor principal stress, *e* is void ratio, *w* is gravimetric water content, and G_s is specific gravity. In addition, $p^t - u_a$ is defined as the mean net stress (*p*). The work-conjugate strain rates for p^*, *q* and *s* are the volumetric strain increment ($d\varepsilon_v$), shear strain increment ($d\varepsilon_q$) and $-ndS_r$.

State parameters

As well as the above six stress and strain variables, two state parameters (ξ and ψ) are adopted in constructing the current model. The first (ξ) was defined by Gallipoli et al. (2003b) as follows:

$$\xi = f(s)(1 - S_r) \tag{7.46}$$

The first term, $f(s)$, on the right-hand side of Equation (7.46) is related to the stabilising inter-particle normal force F_N, which results from the water meniscus between soil particles. It is defined as F_N at any suction *s* and normalised by the value of ΔN at zero suction. When the contact angle is zero, F_N between two identical spherical particles can be expressed by the analytical solution in Equation (6.2). According to Equation (6.2), F_N is equal to $4/3\pi RT_s$ as *s* approaches zero. Normalising ΔN by $4/3\pi RT_s$, $f(s)$ can readily be obtained as follows:

$$f(s) = \frac{3T_s}{Rs} \frac{\left(\sqrt{9 + 8Rs/T_s} - 3\right)\left(\sqrt{9 + 8Rs/T_s} + 1\right)}{16} \tag{7.47}$$

Figure 7.9 shows the relationship between $f(s)$ and *s*. This relationship is consistent with the curve given by Gallipoli et al. (2003b). For simplicity, a constant *R* value of 1×10^{-6} m and a single value of T_s equal to 72.8 mN/m were assumed to calculate $f(s)$ in this study. These two assumed values should not affect the model prediction significantly since the

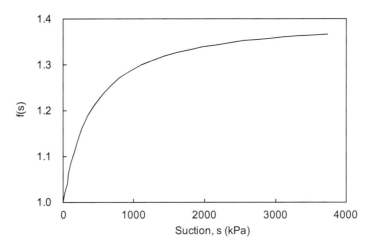

Figure 7.9 Increase of inter-particle normal forces due to meniscus water lens.
Source: Fisher, 1926.

value of $f(s)$ is insensitive to R and T_s. According to Equation (7.47), for all values of R and T_s, the value of $f(s)$ is limited to a narrow range between 1 and 1.5 over the full suction range (Gallipoli et al., 2003a).

Figure 7.10 shows the definition of another state parameter (ψ) that was proposed by Been and Jefferies (1985) for saturated soil. It is defined as the difference between the current void ratio (e) and critical state void ratio (e_c) corresponding to the current stress in the v-lnp^* plane. Following this definition, the value of ψ is positive for soil states on the wet side of CSL, while it is negative for soil states on the dry side of CSL. It is well known that ψ is related to the elastoplastic behaviour of saturated and unsaturated soils (Li, 2002; Chiu and Ng, 2003; Dafalias and Manzari, 2004). Dafalias and Manzari (2004) proposed the following two formulations to describe the influence of ψ on dilatancy and shear strength, respectively:

$$M_d = M\exp(n_d\psi) \tag{7.48}$$

$$M_b = M\exp(-n_b\psi) \tag{7.49}$$

where n_b and n_d are soil parameters, M_d is the dilation stress ratio, at which soil response changes from contractive to dilative, M_b is the peak stress ratio attainable at the current state and M is the stress ratio at the critical state. The influence of M_d and M_b on soil behaviour is discussed further below.

Normal compression line (NCL)

Experimental results in the literature reveal that when temperature increases, the NCL of saturated soil in the v-lnp' plane shifts towards a smaller void ratio. Furthermore, NCLs

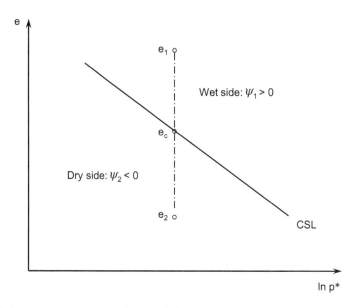

Figure 7.10 Definition of state parameter for soil skeleton.

at different temperatures are always parallel (Campanella and Mitchell, 1968; Cekerevac and Laloui, 2004). Based on these experimental observations, the NCLs of saturated soil are modelled as parallel straight lines in the $v - \ln p'$ plane:

$$v = N(T) - \lambda \ln\left(\frac{p'}{p_{atm}}\right) \qquad (7.50)$$

$$N(T) = N_0 - r_N(T - T_0) \qquad (7.51)$$

where λ and $N(T)$ are the slope and intercept of the NCL at a temperature of T respectively, N_0 is the intercept of NCL at the reference temperature T_0, r_N is a soil constant describing the thermal effects on the location of NCL and the atmospheric pressure p_{atm} (101 kPa) is included as a reference pressure. Zhou and Ng (2014) showed that according to Equations (7.50) and (7.51), thermal effects on the pre-consolidation pressure of soil follow

$$p_0(T) = p_0(T_0)\exp\left[-\beta(T - T_0)\right] \qquad (7.52)$$

$$\beta = \frac{r_N}{\lambda - \kappa} \qquad (7.53)$$

where $p_0(T)$ and $p_0(T_0)$ are the pre-consolidation pressures of soil at temperatures T and T_0 respectively and κ is the slope of the loading and unloading lines in the v-$\ln p^*$ plane. It should be noted that Equation (7.52) was first proposed by Cui et al. (2000). According to Equations (7.52) and (7.53), thermal effects on $p_0(T)$ are more important at a larger β. Experimental data available in literature (e.g., Cekerevac and Laloui, 2004) suggests that the value of β for saturated clay in general ranges from 0.001 to 0.004. As the temperature increases from 20 to 80°C, $p_0(T)$ decreases by about 5% and 20% when parameter β is 0.001 and 0.004 respectively.

For unsaturated soil, the NCL of a soil shifts towards a larger void ratio as soil desaturates, because of the stabilisation effects of meniscus water on the soil skeleton. To link NCLs at saturated and unsaturated states, the semi-empirical equation proposed by Gallipoli et al. (2003a) is adopted:

$$\frac{e}{e_s} = 1 - a(1 - \exp(b\xi)) \tag{7.54}$$

where e and e_s are void ratios in the unsaturated and saturated state at a given p^* (or p'), respectively, and a and b are soil parameters. According to Equations (7.50) through (7.54), the NCL at any given condition of ξ, and T is a straight line in the $v - \ln p^*$ plane. The slope and intercept are expressed as

$$\begin{cases} \lambda(\xi) = \lambda\left[1 - a(1 - \exp(b\xi))\right] \\ N(T,\xi) = \left[N_0 - r_N(T - T_0)\right]\left[1 - a(1 - \exp(b\xi))\right] \end{cases} \tag{7.55}$$

If $p_0(T, \xi)$ and $p_0(T, \xi_0)$ are the pre-consolidation pressures at ξ and ξ_0 respectively, then

$$\ln p_0(T,\xi) = \ln p_0(T,\xi_0)\frac{\lambda(\xi_0) - \kappa}{\lambda(\xi) - \kappa} + \frac{N(T,\xi) - N(T,\xi_0)}{\lambda(\xi) - \kappa} \\ + \ln p_{atm}\left[\lambda(\xi) - \lambda(\xi_0)\right] \tag{7.56}$$

It is clear that thermal and suction effects on the pre-consolidation pressure of unsaturated soil are described by Equations (7.52) and (7.56), respectively. These two equations can be used to calculate $p_0(T, \xi)$ for any suction and temperature.

The critical state line (CSL)

Experimental results reported in literature show that the CSL of saturated soil in the $v - \ln p'$ plane shifts towards a smaller void ratio as temperature increases (Cekerevac and Laloui, 2004). Like NCLs, CSLs of saturated soil are modelled using the following equations:

$$v = \Gamma(T) - \omega \ln\left(\frac{p'}{p_{atm}}\right) \tag{7.57}$$

$$\Gamma(T) = \Gamma_0 - r_\Gamma(T - T_0) \tag{7.58}$$

where ω and $\Gamma(T)$ are the slope and intercept of CSL at temperature of T respectively, Γ_0 is the intercept of the CSL at a reference temperature T_0, and r_Γ is a soil parameter describing thermal effects on the location of the CSL. The spacing between NCL and CSL is temperature dependent if parameters r_N and r_Γ are not identical.

For unsaturated soil, like the NCL, the CSL shifts towards a larger void ratio as soil desaturates. Furthermore, Equation (7.54) is applicable not only for NCL but also for CSL and the same set of parameters (a and b) can be used in both cases. According to Equations (7.54), (7.57) and (7.58), the CSLs in the $v - \ln p^*$ plane are straight lines for any given condition of ξ, and T. The slope and intercept are expressed as follows:

$$\begin{cases} \omega(\xi) = \bar{\omega}\left[1 - a\left(1 - \exp(b\xi)\right)\right] \\ \Gamma(T,\xi) = \left[\Gamma_0 - r_\Gamma(T - T_0)\right]\left[1 - a\left(1 - \exp(b\xi)\right)\right] \end{cases} \tag{7.59}$$

On the other hand, the CSL of saturated and unsaturated soils in the $p^* - q$ plane is described by the following expression:

$$q = Mp^* \tag{7.60}$$

where M is the stress ratio at the critical state. As summarised by Zhou and Ng (2014), experimental results available in the literature suggest that M is generally insensitive to temperature. When the temperature increases from 20 to 90°C, the variation in M is generally less than 5% for most soils (Uchaipichat and Khalili, 2009; Hueckel et al., 2009). For simplicity, thermal effects on M are not incorporated in this study.

Water retention curve (WRC)

In this SOA model, a main drying curve and a main wetting curve are used to model hydraulic hysteresis, as shown in Figure 7.11. Any state in the $S_r - s$ plane is bounded by these two curves. Further, the main drying curve and main wetting curve are described by the following equations:

$$S_r = \left[1 + \left(\frac{se^{m_4}}{m_3^I}\right)^{m_2}\left(\frac{T}{T_0}\right)^{m_2 m_T}\right]^{-m_1} \tag{7.61}$$

$$S_r = \left[1 + \left(\frac{se^{m_4}}{m_3^D}\right)^{m_2}\left(\frac{T}{T_0}\right)^{m_2 m_T}\right]^{-m_1} \tag{7.62}$$

where m_1, m_2, m_3^I, m_3^D, m_4 and m_T are soil parameters. Previous experimental results reveal that the water retention ability of unsaturated soil increases with decreasing e, but

Constitutive modelling of state-dependent behaviour of soils 351

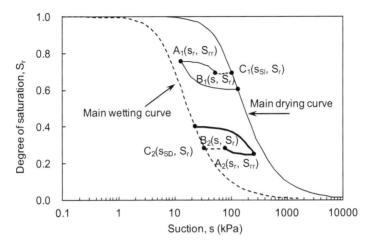

Figure 7.11 Modelling stress-dependent hydraulic hysteresis using bounding surface plasticity theory.

decreases slightly with an increase in T (Romero et al., 2001). Both density and thermal effects are incorporated in Equations (7.61) and (7.62).

The change of S_r with s is modelled as an elastoplastic process. The main drying and wetting curves, which are dependent on density and temperature, serve as bounding surfaces. When the soil state is on the main drying and wetting curves, S_r can be calculated simply from s, T and e using Equations (7.61) and (7.62). If the soil state is on the scanning curves, however, S_r is not only affected by s, T and e, but also depends on the suction history. Then, the hysteretic water retention behaviour is computed using the bounding surface theory. More details are given in the following section.

Elastoplasticity

The bounding surface plasticity theory of Dafalias (1986) is adopted in the current model. For each loading process, both elastic and plastic strains occur. The total strain increment $(d\varepsilon)$ is the sum of the elastic $(d\varepsilon^e)$ and plastic strain increments $(d\varepsilon^p)$. We assume that plastic strain arises from three different mechanisms: shearing, compression and suction change. As a result, $d\varepsilon_v$, $d\varepsilon_q$ and dS_r are expressed as follows:

$$\begin{cases} d\varepsilon_v = d\varepsilon_v^e + d\varepsilon_{v(c)}^p + d\varepsilon_{v(s)}^p + d\varepsilon_{v(h)}^p \\ d\varepsilon_q = d\varepsilon_q^e + d\varepsilon_{q(c)}^p + d\varepsilon_{q(s)}^p + d\varepsilon_{q(h)}^p \\ dS_r = dS_r^e + dS_{r(c)}^p + dS_{r(s)}^p + dS_{r(h)}^p \end{cases} \quad (7.63)$$

where $d\varepsilon_v^e$ is the elastic volumetric strain, $d\varepsilon_v^p(c)$ is the plastic volumetric strain due to compression,

$d\varepsilon_v^p(s)$ is the plastic volumetric strain due to shearing, $d\varepsilon_v^p(h)$ is the plastic volumetric strain due to suction change, $d\varepsilon_q^e$ is the elastic shear strain, $d\varepsilon_q^p(c)$ is the plastic shear strain due to compression, $d\varepsilon_q^p(s)$ is the plastic shear strain due to shearing, $d\varepsilon_q^p(h)$ is the plastic shear strain due to suction change, dS_r^e is the elastic incremental degree of saturation, $dS_r^p(c)$ is the plastic incremental degree of saturation due to compression, $dS_r^p(s)$ is the plastic incremental degree of saturation due to shearing and $dS_r^p(h)$ is the plastic incremental degree of saturation due to suction change.

It should be noted that Equation (7.63) does not explicitly incorporate the thermally induced plastic increments of strain and degree of saturation. However, thermo-plasticity is indirectly included, since the NCL, CSL and WRC in the current model are all temperature dependent (see Equations (7.51), (7.58), (7.61) and (7.62)). A change of temperature would alter the yield stresses for the plastic mechanisms of shearing, compression and suction change.

Given an increment in stress, suction and temperature, $d\varepsilon^e$ and $d\varepsilon^p$ are determined using a decoupled approach. As soil is isotropic, elastic strains can be calculated using the following formulations:

$$\begin{cases} d\varepsilon_v^e = \dfrac{dp^*}{K} - \dfrac{\alpha_{sk}}{1+e} dT \\[2mm] d\varepsilon_q^e = \dfrac{dq}{3G_0} \\[2mm] dS_r^e = -\dfrac{ds}{K_w} \end{cases} \quad (7.64)$$

where K is the elastic bulk modulus for the soil skeleton, G_0 is the very small-strain shear modulus, K_w is the bulk modulus for the water phase, e is the void ratio and α_{sk} is the isotropic thermal expansion coefficient for the soil skeleton.

On the other hand, the plastic strains are expressed as

$$\begin{cases} d\varepsilon_v^p = \Lambda_{(c)} + D_{v(s)}\Lambda_{(s)} + D_{v(h)}\Lambda_{(h)} \\[2mm] d\varepsilon_q^p = D_{q(c)}\Lambda_{(c)} + \Lambda_{(s)} + D_{q(h)}\Lambda_{(h)} \\[2mm] dS_r^p = D_{w(c)}\Lambda_{(c)} + D_{w(s)}\Lambda_{(s)} + \Lambda_{(h)} \end{cases} \quad (7.65)$$

where $\Lambda_{(c)}$, $\Lambda_{(s)}$ and $\Lambda_{(h)}$ are loading indices for the compression, shearing and suction change mechanisms, $D_{q(c)}$ and $D_{w(c)}$ are dilatancy factors, defined as the ratios $d\varepsilon_q^p/d\varepsilon_v^p$ and $dS_r^p/d\varepsilon_v^p$ respectively arising from compression, $D_{v(s)}$ and $D_{w(s)}$ are the ratios $d\varepsilon_v^p/d\varepsilon_q^p$ and $dS_r^p/d\varepsilon_q^p$ respectively arising from shearing, $D_{v(h)}$ and $D_{q(h)}$ are the ratios $d\varepsilon_v^p/dS_r^p$ and $d\varepsilon_q^p/dS_r^p$ respectively arising from suction change. The loading indices are determined through hardening law and condition of consistency, while the dilatancy factors are obtained from the flow rule. Mathematic formulations for determining them are presented later.

Khalili et al. (2010) illustrated that thermo-elastic strain does not induce any change in the void ratio of the soil skeleton, assuming that the thermal expansion coefficient of the soil skeleton is identical to the thermal expansion coefficient of soil particles. As the rearrangement of soil particles during elastic deformation is always negligible, their assumption is adopted in the current study. Therefore (Mašín and Khalili, 2011),

$$de = (1+e)\left(d\varepsilon_v + \frac{\alpha_{sk}}{1+e}dT\right) \quad (7.66)$$

Elastic moduli

According to Equation (7.64), four stiffness parameters (K, G_0, K_w and α_{sk}) are required to compute the elastic increments of strain and degree of saturation. K and G_0 are calculated using the following equations:

$$K = \frac{v\, p^*}{\kappa} \quad (7.67)$$

$$G_0 = C_0(1+e)^{-3}\left(\left(\frac{p^*}{p_{atm}}\right)^{0.5} + C_s \xi^{0.5}\right) \quad (7.68)$$

where κ is the slope of the isotropic unloading-reloading line in the ($v - \ln p^*$) plane and C_0 and C_s are soil parameters describing the shear modulus of unsaturated soils. In these equations, there are several assumptions. First, three exponents in Equation (7.68) may take default values (−3, 0.5 and 0.5). These two suggested values were determined from extensive experimental evidence (Vardanega and Bolton, 2013; Oztoprak and Bolton, 2013). Second, the elastic moduli (K and G_0) are assumed to be independent of temperature. This is because some experimental results reported in literature suggest that the parameter κ of soil seems to be independent of temperature, at least the magnitude of variation is negligible (Campanella and Mitchell, 1968; Cekerevac and Laloui, 2004; Abuel-Naga et al., 2007b). Third, to guarantee energy conservation, the dependence of the elastic shear modulus on the effective mean stress must be accompanied by the dependence of the bulk modulus on deviator stress (Hueckel et al., 1992). That is, the elastic moduli defined by Equations (7.67) and (7.68) may violate the principle of energy conservation. Note that the use of these two equations is for simplicity and their drawbacks do not affect the model performance significantly.

Figure 7.11 reveals that the slope of the water retention curve is very small at low suctions. This suggests that $\partial S_r / \partial s$ is approximately equal to 0 for low suctions. Given an increase in suction, the total incremental degree of saturation is almost zero. Hence, the elastic component dS_r^e should be very small. For simplicity, we assume that dS_r^e is zero, implying that K_w is very large. As dS_r^e is very small, it is expected that this assumption will not affect the fidelity of the model much.

It should be noted that several formulations for α_s have been proposed in some existing thermo-mechanical models. Cui et al. (2000) assumed it to be constant, whereas

Abuel-Naga et al. (2007b) considered it to be temperature dependent. On the other hand, Laloui and François (2009) suggested it was affected by both temperature and OCR. Hong et al. (2013) evaluated these three approaches by comparing measured and computed results for heating and cooling tests on a saturated clay. They found that a constant value of α_s is sufficient to describe the thermo-elastic strain of soil accurately. Thus a constant α_s is used in the current SOA model for simplicity.

Bounding surfaces

Three bounding surfaces (F_c, F_s and F_h) are constructed in the SOA model. They are closely related to the three plastic mechanisms: compression, shearing and suction change. The first, F_c, is defined to describe elastoplastic behaviour during compression. It is well known as the loading collapse (LC) surface in the p^*–q–ξ space and is shown in Figure 7.12, while it is known as the loading yield (LY) surface in the p^*–q–T space and shown in Figure 7.12. It can be described as:

$$F_c = p^* - p_0(T, \xi) \tag{7.69}$$

where $p_0(T, \xi)$ is the pre-consolidation pressure at given T and ξ. It is related to NCLs and isotropic unloading-reloading lines (URLs):

$$\frac{p_0(T, \xi)}{p_{atm}} = \exp\left(\frac{N(T, \xi) - v - \kappa \ln\left(\frac{p^*}{p_{atm}}\right)}{\lambda(\xi) - \kappa}\right) \tag{7.70}$$

According to this equation, $p_0(T, \xi)$ is dependent on T and ξ. In addition, ξ is a function of not only s but also S_r. The effects of hydraulic hysteresis on the size of bounding surface are therefore taken into account. The coupling of hydraulic hysteresis and mechanical behaviour models the effects of drying and wetting on stress-strain relationship more reasonably.

The second bounding surface F_s is relevant to elastoplastic behaviour during shearing. It is also shown in Figure 7.12 and defined as:

$$F_s = q - M_m p^* \tag{7.71}$$

where M_m is the maximum value of stress ratio q/p^* in the stress history. Implicitly, it is assumed that F_s surface is planar.

The third bounding surface F_h, which is used to predict elastoplastic behaviour as suction changes, is defined in the S_r–s plane. It is expressed as:

$$F_h = s - s_{s\alpha} \tag{7.72}$$

Constitutive modelling of state-dependent behaviour of soils 355

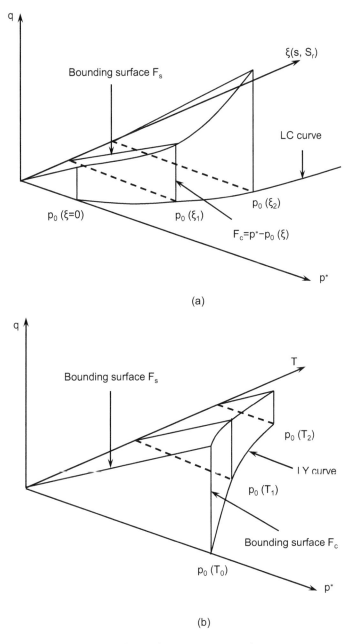

Figure 7.12 Bounding surfaces: (a) in the $p^* - q - \xi$ space; (b) in the $p^* - q - T$ space.

where s_{sa} is the soil suction obtained from main drying/wetting curve at current degree of saturation, as illustrated in Figure 7.11. According to Equation (7.61), s_{sa} is a function of S_r, s, e and T:

$$s_{sa} = \frac{m_3}{e^{1/m_1 m_2}} (T_0/T)^{m_T} \left[(S_r)^{-1/m_1} - 1 \right]^{1/m_2} \tag{7.73}$$

Yield surfaces

Figure 7.13 shows that two yield surfaces are used to define a "wedge" in the p^*–q plane (Dafalias and Manzari, 2004). These yield surfaces are used to model the elastic threshold for shearing. When the soil stress state is inside the "wedge", the soil behaviour is assumed to be purely elastic. When the soil stress state is on the yield surface and the incremental stress is outwards, elastoplastic behaviour occurs. During elastoplastic behaviour, kinematic hardening theory is used to describe the evolution of yield surfaces. The "wedge"

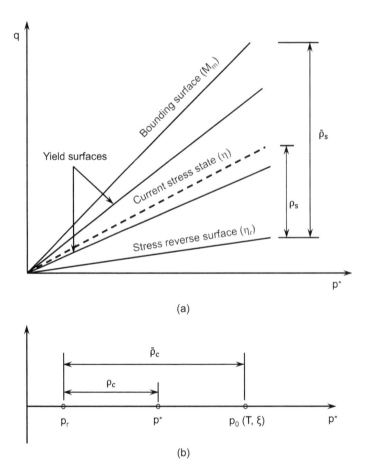

Figure 7.13 Mapping rule for (a) constant p^* shearing mechanism; (b) constant stress ratio compression mechanism.

rotates in the p^*–q plane while maintaining a constant size and shape. At a given mean Bishop's stress, the elastic range is $2mp^*$, where m is a soil parameter controlling the size of the elastic range. The elastic "proportion" of a soil specimen might represent the proportion of the contacts between soil particles which remain intact and have not started to slide (Simpson, 1992). For a soil specimen at a higher matric suction and lower degree of saturation, there is more meniscus water in the specimen. Meniscus water increases the normal force holding the soil particles together and resisting slippage between soil particles. Therefore, a specimen at a higher suction has a wider elastic range (Ng and Xu, 2012). In this study, we assume that the relationship between m and ξ can be described by the following equation:

$$m = m_a \xi^{m_b} \tag{7.74}$$

where m_a and m_b are soil parameters. For simplicity, the variable m is assumed to be independent of temperature.

For compression, it is simply assumed that there is no purely elastic region. In other words, no yield surface is used for compression. As described earlier, the elastic increment of S_r is zero and also no yield surface used for suction change.

Mapping rule

The mapping rule is an important component in bounding surface plasticity theory. When the soil stress state is inside the bounding surface, it is used to project the actual stress state onto the bounding surface and hence define a new image stress state (Dafalias, 1986). As well as the actual stress state, the image stress state may affect many aspects of soil behaviour, including dilatancy and loading indices. In the current model, three mapping rules are defined to map the actual stress state to F_c, F_s and F_h.

The mapping rule for F_s takes a form similar to that proposed by Li (2002) for saturated soil. Some key features of this mapping rule are shown in Figure 7.13a. The stress ratio at which the last loading reversal occured (η_r) serves as the projection centre. The maximum stress ratio in the stress history (M_m) is the image stress ratio. Based on the relative positions of the projection centre, actual stress state and image stress state, two Euclidian "distances" (ρ_s and $\bar{\rho}_s$) are defined. As illustrated in the figure, ρ_s is the "distance" between the projection centre and the current stress ratio (η), while $\bar{\rho}_s$ is the "distance" between the projection centre and the image stress ratio. The ratio of these two "distances" is defined as follows:

$$\frac{\rho_s}{\bar{\rho}_s} = \frac{\eta - \eta_r}{M_m - \eta_r} \tag{7.75}$$

According to this mapping rule, small fluctuations of stress require η_r to be updated and therefore induce numerical errors. To avoid this problem, η_r is not updated immediately upon stress reversal. The update is only made when the soil stress state reaches the yield surface. Consequently, numerical errors induced by small fluctuations of the stress inside yield surfaces can be eliminated effectively in the current model.

Figure 7.13b shows the mapping rule for F_c. The stress p_r at which the last loading reversal occurs serves as the projection centre, while $p_0(T, \xi)$ is the image stress. Like the mapping rule for F_s, two Euclidian "distances" (ρ_c and $\bar{\rho}_c$) are defined. ρ_c is the "distance" between p_r and the current mean stress (p^*) and $\bar{\rho}_c$ is the "distance" between p_r and $p_{0(\xi)}$. The ratio of these two "distances" is defined as follows:

$$\frac{\rho_c}{\bar{\rho}_c} = \frac{p^* - p_r}{p_0(T,\xi) - p_r} \qquad (7.76)$$

The mapping rule for F_h is shown in Figure 7.11. The suction at which the last loading reversal occurred (s_r) serves as the projection centre. $s_{s\alpha}$ (s_{sI} or s_{sD}) on the main drying/wetting curve serves as the image suction. Two Euclidian "distances" (ρ_h and $\bar{\rho}_h$ on the main drying) are defined. Their ratio is expressed as follows:

$$\frac{\rho_h}{\bar{\rho}_h} = \frac{s - s_r}{s_{s\alpha} - s_r} \qquad (7.77)$$

where ρ_h is the "distance" between the projection centre and the current suction (s). $\bar{\rho}_h$ is the "distance" between the projection centre and image suction. The three ratios ($\rho_s/\bar{\rho}_s$, $\rho_c/\bar{\rho}_c$ and $\rho_h/\bar{\rho}_h$) are necessary to determine the loading index and dilatancy parameters. More details are given later.

Flow rule

Flow rules for the three bounding surfaces are defined individually. First, many formulations have been reported describing soil dilatancy during shearing, represented by the parameter $D_{v(s)}$ in Equation (7.65). Li (2002) showed that $D_{v(s)}$ for saturated soil depends not only on stress level but also on the internal state (e.g., void ratio). For unsaturated soil, Chiu and Ng (2003) suggested that $D_{v(s)}$ is also affected by suction. By comparing experimental and computed results, Chiu and Ng (2003) verified that incorporating the void ratio and suction in the flow rule significantly improved model capability. In the current model, the formulation proposed by Chiu and Ng is modified as follows:

$$D_{v(s)} = \frac{d_1}{M}\left(M_d\sqrt{\frac{\bar{\rho}_s}{\rho_s}} - M_m\right) \qquad (7.78)$$

where d_1 is a soil parameter. According to Equation (7.48), M_d is a function of ψ which is affected by suction, degree of saturation and temperatures. Unlike the formulation of Chiu and Ng (2003), Equation (7.78) includes temperature and degree of saturation, modelling the effects of hydraulic hysteresis and temperature soil dilatancy. Further, the term $\rho_s/\bar{\rho}_s$ is used to take the influence of stress history on flow rule into consideration. This feature is important to modelling the cyclic behaviour of soil.

Constitutive modelling of state-dependent behaviour of soils

Second, the ratio between incremental shear strain and incremental volumetric strain is denoted as $D_{q(c)}$ in Equation (7.65). The following formulation is used:

$$D_{q(c)} = d_2 \left(\frac{\lambda}{\lambda - \kappa} \right) \frac{\eta}{M} \tag{7.79}$$

where d_2 is a soil parameter. According to this equation, parameter $D_{q(c)}$ increases with an increase in η. This is consistent with experimental evidence that during constant stress ratio compression, greater shear strain develops at higher stress ratios.

As illustrated in Equation (7.65), apart from $D_{v(s)}$ and $D_{q(c)}$, two more parameters ($D_{w(s)}$ and $D_{w(c)}$) are required to define the flow rule during shearing and compression. For simplicity, both are assumed to be zero in this study, i.e.:

$$D_{w(s)} = 0 \tag{7.80}$$

$$D_{w(c)} = 0 \tag{7.81}$$

It should be noted that $D_{w(c)} = 0$ does not imply constant S_r during compression. This is because compression induces a change of void ratio, leading to a corresponding movement of the bounding surface F_h. The evolution of F_h would induce non-zero $\Lambda_{(h)}$ and hence dS_r. More details are given in the section "hardening law and condition of consistency".

Third, to describe the direction of plastic strain increments related to the surface F_h, the following formulations are used:

$$D_{v(h)} = 0 \tag{7.82}$$

$$D_{s(h)} = 0 \tag{7.83}$$

Similarly, $D_{v(h)} = 0$ does not mean that the current model cannot predict wetting-induced collapse. This is because wetting induces a change of $p_0(\xi, T)$, resulting in a corresponding movement of the bounding surface F_c. The evolution of F_c induces non-zero $\Lambda_{(c)}$ and hence plastic volumetric strain.

Hardening law

The loading indices ($\Lambda_{(s)}$, $\Lambda_{(c)}$ and $\Lambda_{(h)}$) in Equation (7.65) are determined from the hardening law. From the bounding surface F_s, $\Lambda_{(s)}$ is calculated using the following equations:

$$\Lambda_{(s)} = \frac{1}{K_s^p}(dq - M_m dp^*) \tag{7.84}$$

$$K_s^p = \frac{G_0 h}{M_m}\left(M_b\left(\frac{\bar{p}_s}{p_s} \right) - M_m \right) \tag{7.85}$$

It can be seen from Equation (7.85) that the plastic modulus K_s^p decreases as the ratio $p_s/\bar{p}_s$ increases. This is consistent with the experimental observation that more plastic strain is induced as the stress state approaches the bounding surface. Moreover, this is one of the key features of the current SOA model unlike many conventional elastoplastic models, in which the soil specimen is assumed to be purely elastic when the soil state is inside the yield surface.

From bounding surfaces F_c and F_h, $\Lambda_{(c)}$ and $\Lambda_{(h)}$ can be calculated using the following equation:

$$\begin{cases} \dfrac{\partial F_c}{\partial p^*}dp^* + \dfrac{\partial F_c}{\partial q}dq + \dfrac{\partial F_c}{\partial p_0}\dfrac{\partial p_0}{\partial \xi}\dfrac{\partial \xi}{\partial s}ds + \dfrac{\partial F_c}{\partial p_0}\dfrac{\partial p_0}{\partial \varepsilon_v^p}\left(\dfrac{1}{R_p}\Lambda_{(c)} + D_{v(s)}\Lambda_{(s)} + D_{v(h)}\Lambda_{(h)}\right) \\ \quad + \dfrac{\partial F_c}{\partial p_0}\dfrac{\partial p_0}{\partial \xi}\dfrac{\partial \xi}{\partial S_r^p}\left(D_{w(c)}\Lambda_{(c)} + D_{w(s)}\Lambda_{(s)} + \Lambda_{(h)}\right) + \dfrac{\partial F_c}{\partial p_0}\dfrac{\partial p_0}{\partial T}dT = 0 \\ \dfrac{\partial F_h}{\partial s_{s\alpha}}\dfrac{\partial s_{s\alpha}}{\partial p^*}dp^* + \dfrac{\partial F_h}{\partial s}ds + \dfrac{\partial F_h}{\partial s_{s\alpha}}\dfrac{\partial s_{s\alpha}}{\partial \varepsilon_v^p}\left(\Lambda_{(c)} + D_{v(s)}\Lambda_{(s)} + D_{v(h)}\Lambda_{(h)}\right) \\ \quad + \dfrac{\partial F_h}{\partial s_{s\alpha}}\dfrac{\partial s_{s\alpha}}{\partial S_r^p}\left(D_{w(c)}\Lambda_{(c)} + D_{w(s)}\Lambda_{(s)} + \dfrac{1}{R_s}\Lambda_{(h)}\right) + \dfrac{\partial F_h}{\partial s_{s\alpha}}\dfrac{\partial s_{s\alpha}}{\partial T}dT = 0 \end{cases} \quad (7.86)$$

where R_p and R_s are assumed to take the following forms:

$$R_p = \begin{cases} \dfrac{p_c}{\bar{p}_c} & \text{if } \Lambda_{(c)} \geq 0 \\ 0 & \text{if } \Lambda_{(c)} < 0 \end{cases} \quad (7.87)$$

$$R_s = \dfrac{p_h}{\bar{p}_h} \quad (7.88)$$

The factors R_p and R_s are introduced to simulate the reduction in plastic modulus as the distance between the stress state and the bounding surface decreases, using an approach similar to Raveendiraraj (2009). Their function is like that of $p_s/\bar{p}_s$ in Equation (7.85).

Experimental program to verify the SOA model

Experimental results from both monotonic (Xu, 2011) and cyclic shear tests (Ng and Zhou, 2014) are used to verify the model. Xu (2011) studied the effects of net stress and suction as well as drying and wetting cycles on the small-strain behaviour of CDT through suction-controlled triaxial compression tests. Figure 7.14 shows the stress paths of five tests in the p-q-s space. After setting up the soil specimen, the initial state of each specimen was controlled at point A. The desired net mean stresses and matric suctions were first

Constitutive modelling of state-dependent behaviour of soils 361

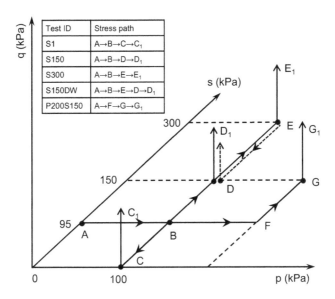

Figure 7.14 Stress paths for suction and stress-controlled shear tests.

applied to the soil specimens for equalisation (A→B→C, A→B→D, A→B→E, A→B→E→D and A→F→G). After the equalisation stage, the deviator stress was increased at a rate of 3 kPa per hour while the net mean stress and matric suction were kept constant. The shearing rate was confirmed by measurement using a suction probe to be slow enough that the excess pore water pressure during shearing stage was negligible (less than ±2 kPa).

The stress paths for the suction and temperature controlled cyclic triaxial tests are shown in Figure 6.2. Each test consisted of four stages: isotropic compression, wetting, heating and cyclic loading-unloading. In the first three stages, the stress, suction and temperature of soil specimen were controlled to predefined values. In the last stage, deviator stress in haversine form was applied, while the net confining pressure was maintained constant at 30 kPa. Four levels of cyclic stress (q_{cyc}) (30, 40, 55 and 70 kPa) were applied to each specimen in succession. As the soil behaviour at the four levels of q_{cyc} is similar, only the last stage is reported here. To take the influence of stress history induced by previous three levels of q_{cyc} into consideration, the initial value of hardening parameter M_m is taken as 55 kPa/p^*. In addition, the soil state after compaction was as follows: $s = 95 \pm 2$ kPa, $S_r = 0.79 \pm 0.01$ and $e = 0.58 \pm 0.02$. The average values are used to define the initial state for model simulation: $p^* = 75$ kPa, $q = 0$, $s = 95$ kPa, $T = 20°C$, $e = 0.58$ and $S_r = 0.79$.

Determination of soil parameters

According to their functions, the parameters defined in the SOA model are grouped into six categories: water retention, elastic, isotropic compression, critical state, flow rule and hardening. The determination of parameters from experimental data of the CDT is described. The calibrated value of each parameter is summarised in Table 7.6.

Parameters for water retention

The water retention behaviour of soil is described by five independent parameters in the model. Figure 3.12 shows drying WRCs of CDT at three different stress levels, measured by Ng et al. (2012) using a modified triaxial apparatus. Because of the application of net stress, these curves correspond to different initial void ratios. Based on experimental results, the parameters for drying water retention curves are obtained by three-dimensional least-square fitting in $S_r - s - e$ space. Parameters m_1, m_2, m_3^l are 0.99, 1.88 and 100, respectively. Similarly, m_3^D is found to be about 1 kPa by fitting wetting the water retention curves reported in Ng et al. (2012).

Parameter m_T describes thermal effects on the water retention ability of unsaturated soil. According to Equation (7.61), m_T is calculated using the following equation:

$$m_T = \log_{(T/T_0)} \frac{a_{T_0}}{a_T} \qquad (7.89)$$

where a_{T_0} and a_T are the value of parameter a, which is approximately equal to AEV and can easily be determined from experimental data, at temperatures of T_0 and T, respectively. As the WRCs of CDT at elevated temperatures are not available, parameter m_T was estimated based on the experimental results for Boom clay as reported by Romero et al. (2001): $m_T = 0.3$. It is expected that this simplified method will not affect the predicted results much, since thermal effects on the WRC of unsaturated soil are not significant for most soils (e.g., Romero et al., 2001; Uchaipichat and Khalili, 2009).

Elastic parameters

To simulate elastic behaviour, six elastic parameters are defined in the model. κ is the slope of isotropic URLs in the $\upsilon - \ln p^*$ plane. Since it is assumed to be independent of suction, it can be obtained from the unloading and reloading paths of an isotropic compression test at zero suction. For CDT, it was found to be 0.01. On the other hand, Figure 7.15 shows the G_0 of CDT measured at different stress and suction conditions (Ng and Yung, 2008). Based on their experimental results, C_0 and n_s are 330 MPa and 1.2, respectively. The variable m controls the size of the elastic range in the $p^* - q$ plane. In constant p^* triaxial compression tests, soil reaches the yield surface and the secant shear modulus starts to decrease when q is equal to mp^*. Thus, m can be calculated using the following equation:

$$m = \frac{\gamma_e G_0}{p^*} \qquad (7.90)$$

where γ_e is the threshold shear strain corresponding to the limiting elastic range (Stokoe et al., 1995). From the measured shear modulus degradation curve of CDT (Ng and Xu, 2012), γ_e at different levels of ξ are determined. The variable m is found to be $0.24\xi^{1.33}$, with parameters m_a and m_b equal to 0.24 and 1.33 respectively.

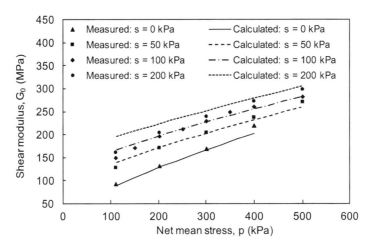

Figure 7.15 Calibration of very small strain shear modulus (G_0).

Parameter α_{sk} describes the elastic deformation of the soil skeleton induced by temperature change. By assuming that soil deformation during cooling is elastic ($d\varepsilon_v = d\varepsilon_v^e$), the following can be obtained from Equation (7.64):

$$\alpha_{sk} = -v \frac{d\varepsilon_v}{dT} \tag{7.91}$$

Experimental data available in literature suggests that α_{sk} for most soils falls into a narrow range between 3×10^{-5} and 4×10^{-5} per °C (Zhou and Ng, 2014). For CDT, it is assumed to be 3.5×10^{-5} per °C.

Parameters for normal compression and the critical state

Five parameters are used to define NCLs. λ and N are obtained from the gradient and intercept of the NCL in the saturated state, while a and b are determined by fitting isotropic compression tests at different ξ. The CSLs in the $e - \ln p^*$ plane can be deduced from NCLs by assuming that ω is equal to λ and that N is larger than Γ by $(\lambda - \kappa)\ln 2$. Figure 7.16a shows the experimental results for a series of suction-controlled isotropic compression tests on CDT (Ng and Yung, 2008). From data measured at zero suction, the values of λ and N are found to be 0.07 and 1.65, respectively. The relationship between e/e_s and ξ are determined and shown in Figure 7.16b. By fitting the relationship using Equation (7.54), a and b are found to be -0.2 and -7, respectively.

Finally, parameter r_N describes the distance between NCLs at different temperatures. According to Equation (7.53), the following equation can be obtained:

$$r_N = \frac{N_0 - N(T)}{(T - T_0)} \tag{7.92}$$

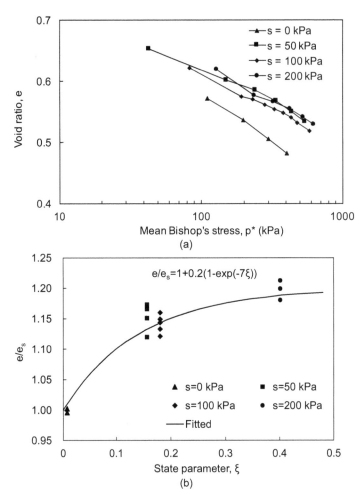

Figure 7.16 Calibration of parameters for NCLs: (a) measured NCLs at various suction; (b) calibration of parameters by fitting the experimental results.

Parameters for dilatancy

Two parameters are used to define dilatancy for shearing mechanisms, while another parameter is adopted to define dilatancy for compression mechanisms. In the saturated state, n_d and $d_1(0)$ are determined from undrained triaxial compression tests. According to Equation (7.78), M_d is equal to η when soil is in a phase transformation state. Then n_d can be calculated using the following equation:

$$n_d = \frac{1}{\psi}\ln\left(\frac{\eta}{M}\right) \tag{7.93}$$

As illustrated in Chiu and Ng (2003), by ignoring the elastic shear strain, the dilatancy during undrained shearing can be expressed as:

$$\frac{d\varepsilon_v^p}{d\varepsilon_q^p} = \frac{-d\varepsilon_v^e}{d\varepsilon_q} = \frac{-dp'/K}{d\varepsilon_q} \qquad (7.94)$$

Substituting Equation (7.78) into Equation (7.94) yields:

$$\frac{d_1}{M}\left(M_d\sqrt{\frac{\overline{p}_s}{p_s}} - \eta\right) = \frac{-dp'/K}{d\varepsilon_q} \qquad (7.95)$$

According to Equation (7.95), parameter $d_1(0)$ at zero suction can be obtained from untrained triaxial shear tests. Figure 7.17 shows the measured response of saturated CDT during undrained shearing, with an initial effective confining pressure of 50 kPa. Based on experimental data, n_d and $d_1(0)$ are found to be 5 and 0.3, respectively.

The other dilatancy parameter $d_2(0)$ can be determined from drained constant stress ratio compression tests. Since there no experimental data from constant stress ratio compression tests are available, the following procedures are used to estimate $d_2(0)$. During K_0 compression, there is no lateral strain. By ignoring small elastic shear and volumetric strain, it can be deduced that

$$\frac{d\varepsilon_{v(c)}^p}{d\varepsilon_{q(c)}^p} = \frac{d\varepsilon_v^p}{d\varepsilon_q^p} = \frac{d\varepsilon_1}{2/3\,d\varepsilon_1} = 1.5 \qquad (7.96)$$

Combining Equations (7.79) and (7.96), it can be derived that

$$d_2(0) = \frac{2}{3}\frac{\lambda-\kappa}{\lambda}\frac{M}{\eta_{K_0}} \qquad (7.97)$$

where η_{K_0} is the stress ratio during K_0 compression. It can be estimated using empirical equations proposed by Jaky (1944) as follows:

$$K_0 = 1 - \sin(\varphi') \qquad (7.98)$$

Based on Equations (7.97) and (7.98), the value of $d_2(0)$ for CDT is about 1.7. In this study, $d_1(\xi)$ and $d_2(\xi)$ are assumed to remain unchanged for simplicity (i.e., $d_1(\xi) = d_1(0)$ and $d_2(\xi) = d_2(0)$).

Parameters for hardening

The hardening law in the model is controlled by two parameters. They can be determined from triaxial shear tests, assuming that M_m and $\overline{p}_s/p_s$ are equal to η' and 1 respectively. When the elastic shear strain is negligible, it can be obtained that

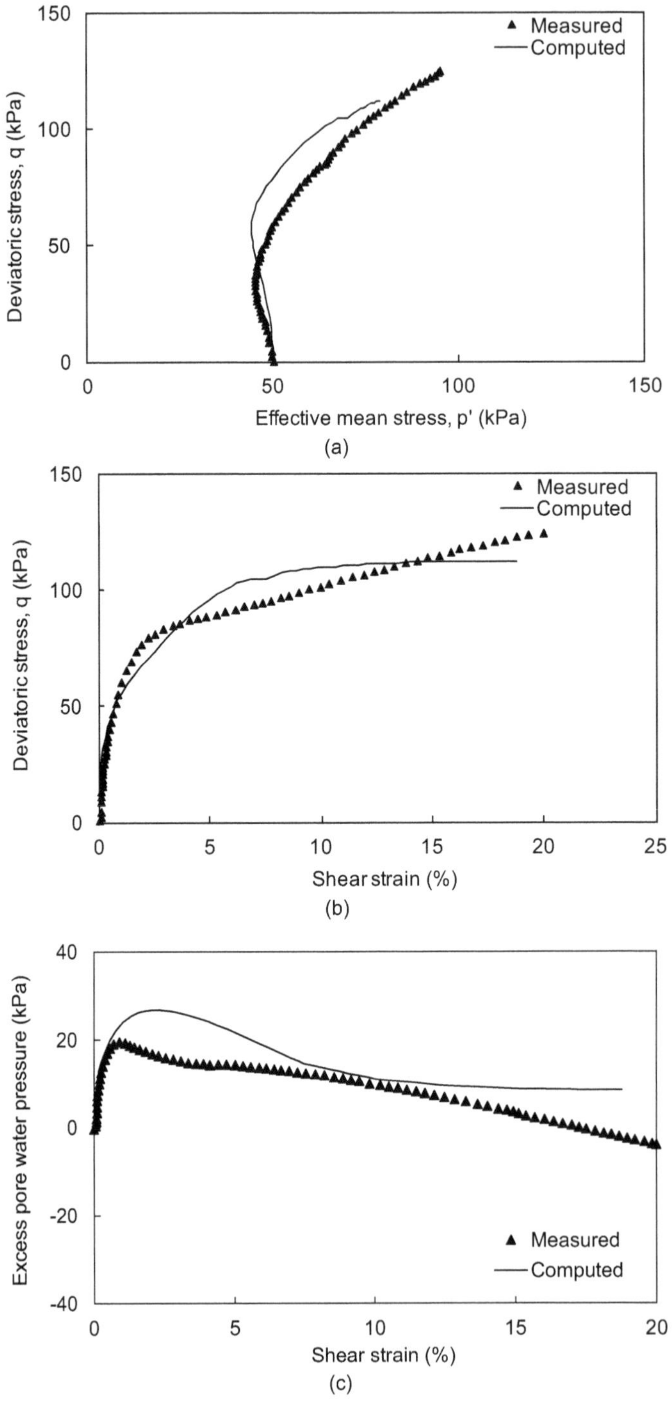

Figure 7.17 Calibration of parameters for flow rule and hardening law.

$$\frac{dq}{d\varepsilon_q} = \frac{dq}{d\varepsilon_q^p} = \frac{dq}{(1/K_p)(dq - \eta'dp')} = \frac{K_p}{1 - \eta'(dp'/dq)} \tag{7.99}$$

Substituting Equations (7.48) and (7.78) into Equation (7.99) yields:

$$\frac{dq}{d\varepsilon_q} = \frac{G_0 h \left[M \exp(-n_b \psi)/\eta' - 1 \right]}{1 - \eta'(dp'/dq)} \tag{7.100}$$

Parameter n_b and h can be obtained by fitting the measured stress-strain relationship using Equation (7.100). The computed stress-strain relationship is shown in Figure 7.17 for comparison. When n_b and h are 5 and 0.07, the computed and measured results match well.

Comparisons between measured and computed results

To verify the validity of the SOA model, the suction-controlled triaxial tests described in Figures 6.2 and 7.14 were simulated using a single set of parameters listed in Table 7.6. The rate constitutive equations were integrated using an explicit sub-stepping stress point algorithm (Potts et al., 2001). To minimise numerical errors, a very small stress increment was used for each step.

Figure 7.18a shows the measured and computed results for the stress-strain behaviour of unsaturated CDT, obtained from three constant p compression tests at suctions of 1, 150 and 300 kPa. The shear stress increases nonlinearly as the shear strain increases, at a decreasing rate. This nonlinearity at small strains (less than 1%) is captured well by the current model at various suctions. This is one of the advantages of adopting bounding surface plasticity theory. Elastoplastic behaviour is predicted by the current SOA model even at small strains, while constitutive models based on classic elastoplastic framework (for example, (Alonso et al., 1990; Chiu and Ng, 2003)) may predict purely elastic behaviour at small strains.

From the slope of the stress-strain relationship, the secant shear modulus (G) is calculated and shown in Figure 7.18b. The very small-strain shear modulus G_0 measured by bender elements is also included for comparison. The corresponding shear strain for G_0 is assumed to be 0.001%. The measured and computed results are fairly consistent over a wide range of strain between 0.001% and 1%. This is because Equation (7.68) represents the suction effects on G_0 accurately, as shown in Figure 7.15. Another important reason is that bounding surface plasticity theory predicts nonlinear stress-strain behaviour even at small strains. These two aspects are very important to the prediction of shear modulus at very small strains (0.001% or less) and to the modelling of shear modulus degradation at small strains (between 0.001% and 1%), respectively.

Figure 7.18c shows the relationship between the ratio G_0/G and shear strain. It illustrates that G_0/G decreases significantly with shear strain, starting from shear strains as low as 0.001%. At shear strains of 0.01% and 0.1%, the ratio G_0/G is only about 0.5 and 0.2, respectively. Further, the rate of stiffness degradation is slightly lower for the specimen at the higher suction. Moreover, at suctions of 1 and 150 kPa, the soil stiffness decreases

368 Advanced Unsaturated Soil Mechanics

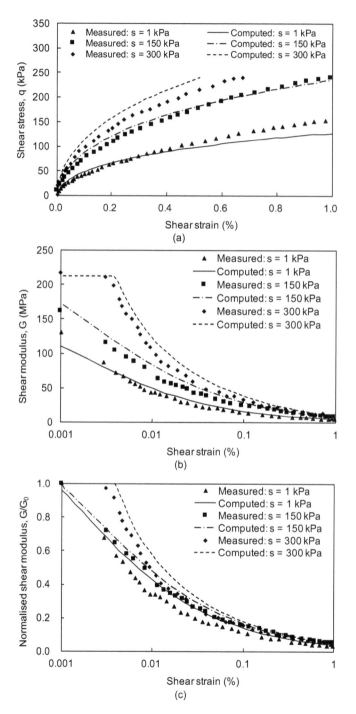

Figure 7.18 Comparisons between measured and computed results: effects of suction magnitude on small-strain behaviour.

Table 7.6 Summary of model parameters for CDT

Soil parameters	Symbol	Values
Water retention	m_1	0.99
	m_2	1.88
	m_3^i: kPa	178
	m_3^D: kPa	1
	m_T	0.3
Elastic	κ	0.01
	C_0: MPa	330
	C_s	1.2
	m_a	0.24
	m_b	1.33
	α_{sk} per °C	3.5×10^{-5}
Isotropic compression	λ_0	0.085
	N_0	1.6
	r_N per °C	1×10^{-3}
	a	−0.2
	b	−7
Critical state	M	1.42
	ω_0	0.085
	Γ_0	1.55
	r_Γ per °C	1×10^{-3}
Flow rule	nd	5
	d_1	0.3
	d_2	1.7
Hardening law	nb	5
	H	0.07

with increasing strain from as low as 0.001%. At a suction of 300 kPa, the threshold shear strain corresponding to the limits of the elastic range is about 0.003%. It is evident that the effects of suction on the small-strain degradation are captured well by the current SOA model. This is attributable to several key features of the model: firstly, the variable m is larger at higher suction, as illustrated in Equation (7.74). Consequently, degradation of yielding and stiffness occur at a larger strain when the suction is higher. Secondly, the void ratios of different specimens are almost the same after suction equalisation. Thus, Equation (7.59) suggests that the CSL of unsaturated soil in the $v - \ln p^*$ plane shifts towards to a larger void ratio as ξ (or suction) increases. When suction increases, the state parameter ψ decreases and thus M_b increases. According to Equation (7.85), the plastic modulus therefore becomes higher. During shearing, a larger plastic modulus leads to a smaller plastic strain and a lower rate of stiffness degradation.

Figure 7.19 shows the influence of drying and wetting on the small-strain behaviour of unsaturated soil. It can be seen from the figure that when the state of soil specimen moves along the wetting path (S150DW), the soil is stiffer than when the state moves along the drying path (S150) at the same suction. The threshold shear strain is also greater for S150DW. At a given strain, the ratio G_0/G is also smaller for a soil specimen along the wetting path. Model predictions are also included in the figure for comparison. The experimental results are represented well by the current model. According to the current model, S150DW has a larger value of ξ than that of S150. Given a larger ξ, the variable m

and hence the elastic range is larger. Although the initial states of both specimens before shearing are within the yield surface, the distance between the stress state and yield surface is larger for S150DW because of the effects of drying and wetting. Yielding and stiffness degradation occur at a higher strain for S150DW. Furthermore, a larger ξ value suggests a higher plastic modulus, a stiffer soil response and a lower rate of stiffness degradation.

Figure 7.20 compares model predictions and experimental data for the stress effects on the small-strain behaviour of unsaturated soil. The measured and computed results reveal consistently that the soil specimen tested at higher net stress behaves much more stiffly. This is because p^* increases with an increase in net mean stress. During shearing, the elastic modulus G_0 (Equation (7.68), the elastic range described by threshold shear strain (mp^*/G_0), the plastic modulus K_s^p (Equation (7.85)) are all higher at a larger value of p^*.

Figure 7.21a compares the measured and computed relationships between accumulated deviator strain and the number of cycles at zero suction but different temperatures (20, 40 and 60°C). At each temperature, the measured and computed results match well. Further, it is clearly revealed that the plastic strain induced by cyclic loads increases with increasing temperature. The observed thermal effects on plastic strain accumulation are captured well, which is attributable to key features of the current model. The most important is Equation (7.59) which suggests that the CSL shifts to a smaller void ratio with increasing temperature. As the CSL shifts to a smaller void ratio, the state parameter ψ increases and hence the plastic modulus K_s^p decreases (see Equation (7.85)). Given a smaller K_s^p, the subsequent shearing load will induce a greater plastic strain.

Figures 7.21b and 7.21c show the measured and computed plastic strain accumulation with number of cycles at suctions of 30 and 60 kPa, respectively. Like the findings in Figure 7.21a, the plastic strain at a given number of cycles is greater at higher temperatures. This thermal effect is captured well by the current model at each suction level. In addition, we see that the model tends to overestimate the accumulated plastic strain slightly at suction of 60 kPa. This deficiency could be minimised by modifying the formulation for plastic modulus (i.e., Equation (7.85)), for example, by expressing the plastic modulus as a function of suction (Chiu and Ng, 2003). Considering that the differences between measured and computed results are small compared to the differences induced by a change of suction and temperature, Equation (7.85) is not modified further to keep the formulation simple.

Figure 7.22 shows the measured and computed relationships between accumulated deviator strain and number of cycles at temperatures of 20, 40 and 60°C. The measured and computed results are reasonably consistent. Furthermore, at each temperature, the plastic strain at a given number of cycles is less at a higher suction. The observed suction effects are captured well. According to Equation (7.59), the CSL shifts to a larger void ratio with increasing suction. At a larger void ratio, the state parameter ψ decreases and hence the plastic modulus K_s^p increases (see Equation (7.85)). Given a larger K_s^p, a subsequent shearing load will induce less plastic strain.

Summary

Using bounding surface plasticity theory, the SOA state-dependent model to simulate the cyclic behaviour of soils is described. This model focuses on the degradation of shear modulus with strain at small strains (between 0.001% and 1%). The model is formulated in terms of mean average skeleton stress, deviator stress, suction, specific volume and

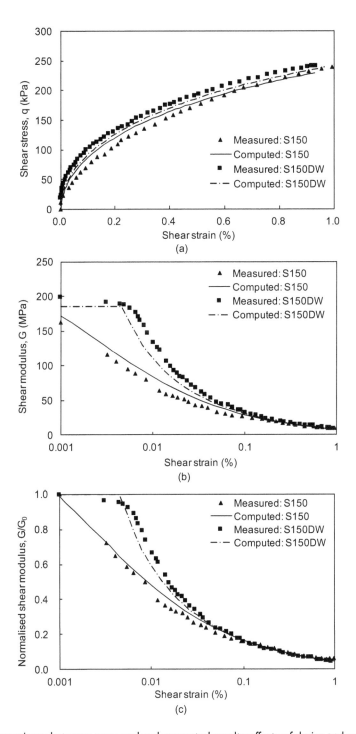

Figure 7.19 Comparisons between measured and computed results: effects of drying and wetting on small-strain behaviour.

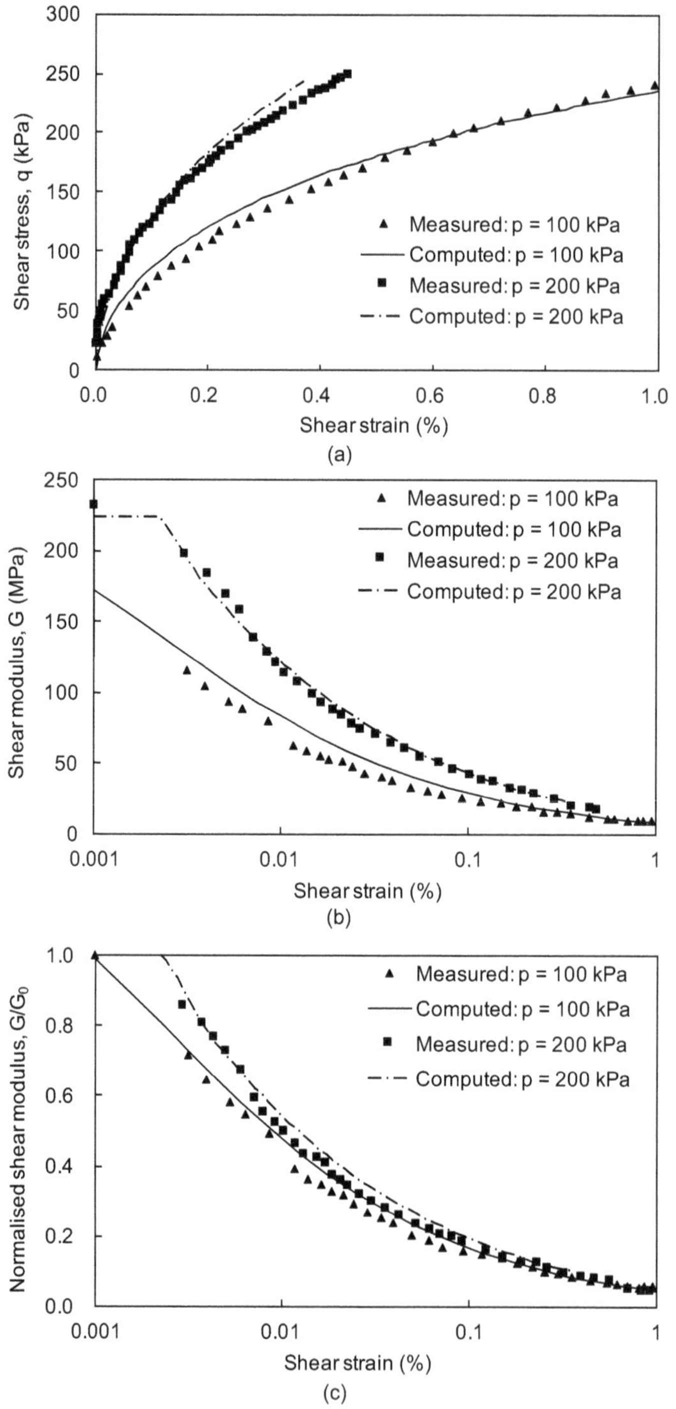

Figure 7.20 Comparisons between measured and computed results: effects of net stress on small-strain behaviour.

Constitutive modelling of state-dependent behaviour of soils 373

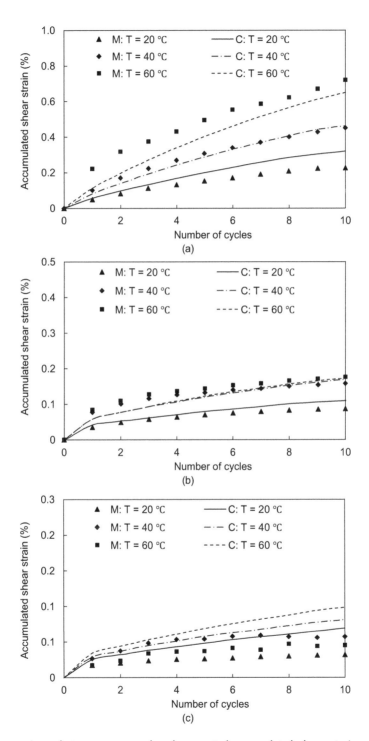

Figure 7.21 Comparisons between measured and computed accumulated shear strains at suctions of (a) 0 kPa; (b) 30 kPa; (c) 60 kPa.

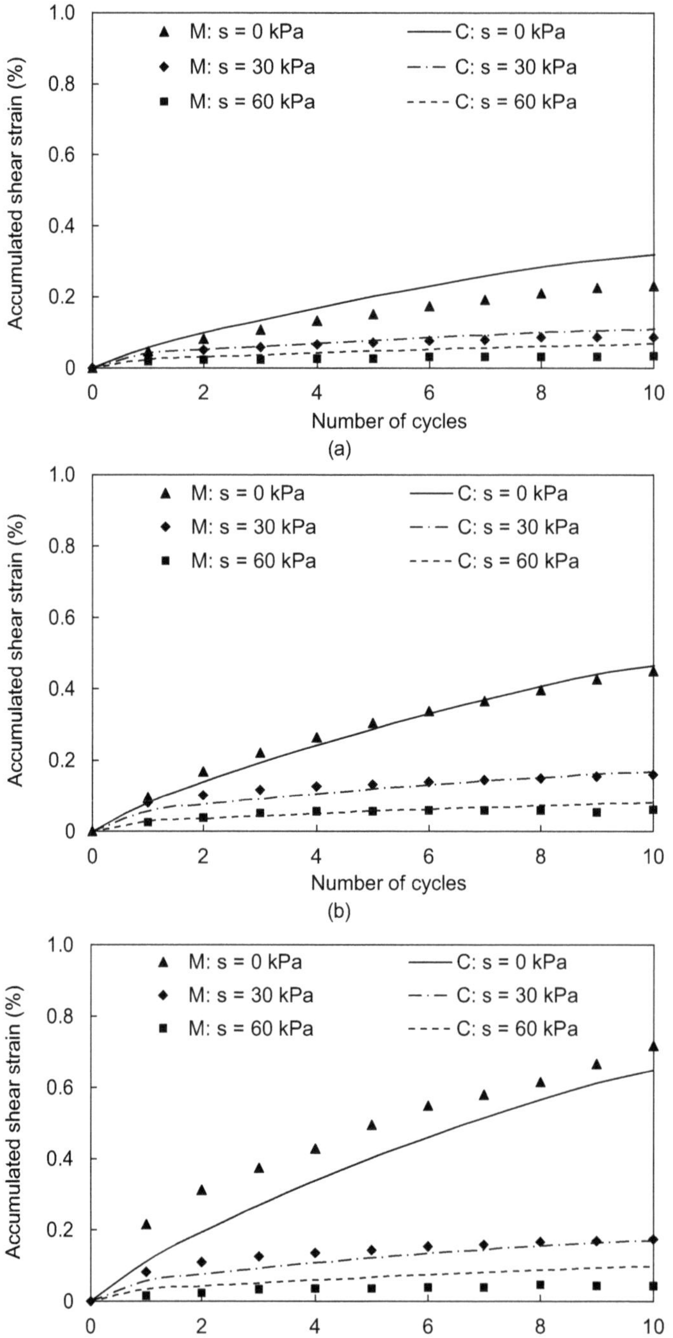

Figure 7.22 Comparisons between measured and computed accumulated shear strains at temperatures of (a) 20°C; (b) 40°C; (c) 60°C.

degree of saturation. Void ratio-dependent hydraulic hysteresis is coupled with stress-strain behaviour. To simulate elastoplastic soil behaviour during shearing, compression and suction change, three bounding surfaces are constructed. Unlike classic elastoplastic models, plastic strain is allowed within bounding surfaces to simulate nonlinear stress-strain behaviour at small strains. Procedures for calibrating the soil parameters have been described.

To evaluate the capability of the current model, experimental results from triaxial tests of a compacted unsaturated silt are used. The effects of suction, drying and wetting as well as stress on small-strain soil behaviour are verified and captured well. Consistent nonlinear stress-strain relationships between computed and measured data are obtained. The model can predict well the degradation of shear modulus with strain over a wide strain range, from 0.001% to 1%. Under cyclic shearing, the suction and thermal effects on the accumulation of plastic strain are also captured well.

SOURCES

Elastoplastic modelling of state- dependent soil behaviour (Ng et al., 2020a)
Simulating the cyclic behaviour at different temperatures using bounding surface plasticity approach (Zhou et al., 2015a; Zhou and Ng, 2016)

Chapter 8

Field monitoring and engineering applications

THE SOUTH-TO-NORTH WATER DIVERSION PROJECT, CHINA (NG ET AL., 2003)

Expansive soils can be found on almost all continents. The destructive effects of this type of soil have been reported in many countries, including the United States, Australia, South Africa, India, Canada, China and Israel (Nelson and Miller, 1992; Steinberg, 1998). According to Steinberg, the annual losses from damage caused by expansive soils is US$10 billion and up to ¥100 million in the United States and China, respectively. Clearly, there is an urgent need to improve both our understanding of the fundamental behaviour of expansive soils and design methodology for civil engineering structures constructed on these soils.

A major infrastructure project, the South-to-North Water Diversion Project (SNWDP) middle route, has been proposed to carry potable water from the Yangtze River region in the south to many arid and semi-arid areas in the northern regions of China, including Beijing. The proposed 1200 km "middle route" is likely to be an open channel with a trapezoidal cross-section formed by cut slopes and fills. At least 180 km of the proposed excavated canal will pass through areas of unsaturated expansive soils. One of the major geotechnical problems is to design safe and economical dimensions for the cut slopes in these unsaturated expansive soils, which have a high potential for swelling and shrinkage (Bao and Ng, 2000).

Although field studies of the effects of rainfall infiltration on slope stability have been carried out on residual soil slopes by some researchers (Lim et al., 1996), the fundamental mechanisms of rainfall infiltration into unsaturated expansive soils during wet seasons, and the complex interaction between changes in soil suction (or water content), the *in-situ* stress state and soil deformation leading to slope failure are not fully understood. To improve our understanding of the fundamental mechanisms of rain-induced retrogressive landslides in unsaturated expansive soils, an 11 m high cut slope in a typical medium-plastic expansive clay in the city of Zaoyang, close to the "middle route" of the SNWDP in Hubei, China, was selected for a comprehensive, well-instrumented field study of rainfall infiltration. The instrumentation includes tensiometers, thermal conductivity suction sensors, moisture probes, earth pressure cells, inclinometers, a tipping bucket rain gauge, a vee-notch flowmeter, and an evaporimeter.

The field study reported in this section was funded by a grant from the Research Grants Council of the HKSAR (HKUST6087/00E) to the first author of this book and Professor C.G. Bao, the former director of the Yangtze River Scientific Research

Institute (YRSRI), Wuhan, China. This work was the most comprehensive field study of the influence of two independent stress state variables, i.e., net stress and suction, on expansive soil slopes prior to the design of the canal. The findings of this field work laid the foundation for the design of the middle route across 180 km of unsaturated expansive soils. Subsequently, the YRSRI was appointed as the main designer for the middle route. The construction of the middle route was started in 2003 and successfully completed in 2014. Plates 8.1–8.3 show an aerial view of the South-to-North Water Diversion Project.

The test site

The test site is located on the intake canal of the Dagangpo second-level pumping station in Zaoyang, Hubei, China (Plate 8.4). It is about 230 km north-west of Wuhan and about 100 km south of the intake canal for the SNWDP (middle route) in Nanyang, Henan, China. The site is in a semi-arid area with an average annual rainfall of about 800 mm, and 70% of the annual rainfall occurs between May to September. The intake canal at the test site was excavated in 1970 with an average excavation depth of 13 m. The slope angle following excavation was 228°. Several years after construction, a number of slope failures took place in succession, and parts of a masonry retaining wall have been seriously deflected or destroyed. Most of the mass movement occurred during wet seasons, and the slip surfaces are on the order of 2 m deep.

The test site area was selected on a cut slope on the northern side of the canal (Plate 8.5). The area has a uniform slope angle of 22.8° and a uniform slope height of 11 m (measured from the top of the retaining wall). There was a 1 m wide berm at the mid-height of the

Plate 8.1 Aerial view of the ditch in the middle route of the South-to-North Water Diversion Project (https://m.thepaper.cn/newsDetail_forward_9480277).

378 Advanced Unsaturated Soil Mechanics

Plate 8.2 Aerial view of the Yellow River crossing ditch in the middle route of the South-to-North Water Diversion Project (www.163.com/dy/article/HLTS7FI10514R9NP.html).

Plate 8.3 Aerial view of the Hutuo River lock along the South-to-North Water Diversion Project (www.thepaper.cn/newsDetail_forward_7685637).

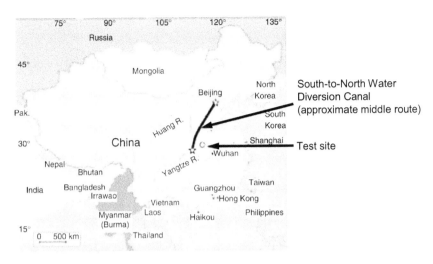

Plate 8.4 Site location for field trial.

Source: After Ng et al., 2003.

slope. The slope surface was well grassed, but no trees were present. The area has a significant depth of typical unsaturated expansive soil. The ground level at the toe of the slope is approximately +96 m OD (Ordnance Datum). About 5 m from the slope toe there is a 3-high masonry retaining wall. The depth of the canal below the slope toe is about 3–5 m. Just to the west of the selected testing area there are a number of typical shallow slips and retrogressive slope failures.

Soil profile and properties

Three groups of boreholes (BH1 and BH2, BH3 and BH4, BH5 and BH6) were drilled to a layer with hard and coarse calcareous concretions around the monitored area on the slope (Plate 8.6). Each group comprised two boreholes spaced 1 m apart. One borehole was used for sampling and standard penetration tests (SPT) and the other for dilatometer tests (DMT). Two boreholes (BH7 and BH8) were drilled in the monitored area and used to install inclinometers after sampling and SPTs. The soil profiles and geotechnical parameters obtained from the boreholes around mid-slope (BH5 and BH6) are shown in Figure 8.1.

As shown in Figure 8.1, the predominant stratum in the slope was a yellow-brown, stiff fissured clay. The clay layer was sometimes interlayered with thin layers of grey clay or iron concretions. The yellow-brown clay contained about 15% hard and coarse calcareous concretions (particle size generally from 30 to 50 mm). X-ray diffraction analyses indicated that the predominant clay minerals are illite (31–35%) and montmorillonite (16–22%), with a small percentage of kaolinite (8%) (Liu, 1997). The natural gravimetric water content was generally slightly above the plastic limit ($w_p = 19.5\%$, $I_p = 30$) with the exception of a relatively low water content within the top 1 m. The dry density profile down to 2 m indicated that a denser soil layer was present at a depth of about 1–5 m.

Plate 8.5 Overall view of the selected cut slope.

Source: After Ng et al., 2003.

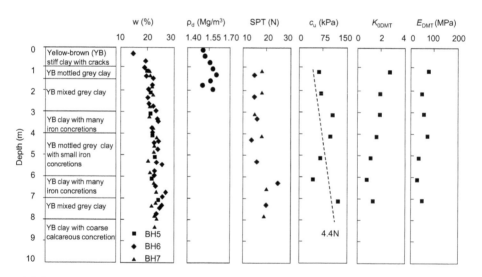

Figure 8.1 Soil profiles and geotechnical parameters from the boreholes located at mid-slope.

Source: After Ng et al., 2003.

The clay was typically over-consolidated because of desiccation, as indicated by the total stress value (K_o) obtained from dilatometer tests, and by its swelling pressure, which ranged from 30 to 200 kPa. The clay exhibited significant swelling and shrinkage characteristics upon wetting and drying, and an abundance of cracks and fissures were observed in the

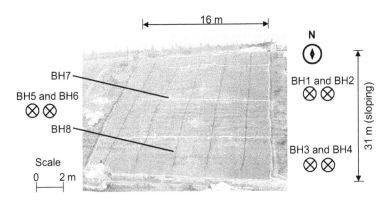

Plate 8.6 Instrumented slope and layout of boreholes.

Source: After Ng et al., 2003.

field. The width of some cracks was as large as 10 mm on the slope surface and were observed to extend as deep as 1 m.

The soil-water characteristics of the expansive clay near the test site were investigated by Wang (2000). Figure 8.2 shows some soil-water characteristic curves (SWCCs) measured on two undisturbed specimens with a dry density of 1.40 Mg/m^3, which is less than the measured average dry density for the soils in the monitoring area (1.56 Mg/m^3). Based on a study of the effects of soil density on the SWCCs of a sandy silt (Ng and Pang, 2000a), the SWCCs for the expansive soil with a density of 1.56 Mg/m^3 can be deduced approximately and constructed by shifting the measured SWCCs at 1.40 Mg/m^3, as shown in Figure 8.2. The shape of the SWCCs for the expansive soil is relatively flat, which indicates that the expansive soil has a high water-retention capacity. The air-entry value of the soil is about 30 kPa. The drying and wetting hysteresis between the desorption and adsorption curves is relatively insignificant.

Field instrumentation programme

An area 16 m wide by 31 m long with a cleared surface (Plate 8.6) was selected for instrumentation and artificial rainfall simulation tests. The instruments included jet-fill tensiometers, thermal conductivity suction sensors (Fredlund et al., 2000), ThetaProbes for determining water content (Delta-T Devices Ltd, Cambridge, UK), vibrating-wire earth pressure cells, inclinometers, a tipping-bucket rain gauge, a vee-notch flowmeter, and an evaporimeter. The layout and location of the instruments are shown in Figures 8.3 and 8.4, and the details for each instrument are summarised in Table 8.1.

Monitoring soil suction and water content

As shown in Figures 8.3 and 8.4, there were three rows of instrumentation for soil suction and water content monitoring: R1 in the upper part, R2 in the middle part, and R3 in the lower part of the slope. In each row, there were seven to nine suction sensors (i.e., jet-fill

382 Advanced Unsaturated Soil Mechanics

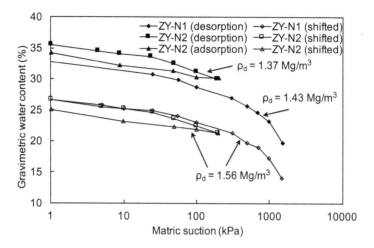

Figure 8.2 Soil-water characteristic curves for the soil from Zaoyang.

Source: After Ng et al., 2003.

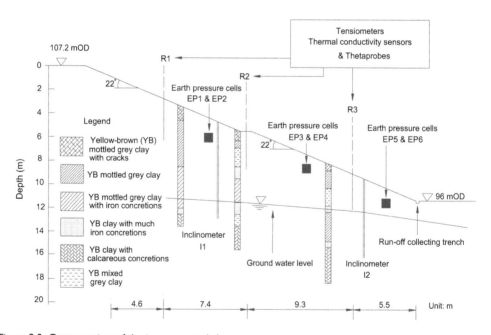

Figure 8.3 Cross-section of the instrumented slope.

Source: After Ng et al., 2003.

Field monitoring and engineering applications 383

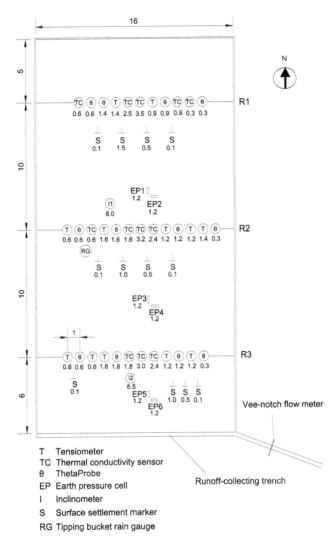

Figure 8.4 Layout of instruments. All dimensions are in metres; numbers denote depths.

Source: After Ng et al., 2003.

tensiometers or thermal conductivity sensors; Fredlund et al., 2000) and four ThetaProbes, which were used to measure the volumetric water content (VWC) indirectly. These probes use the standing-wave technique to measure the apparent dielectric constant of a soil, which is then correlated with the volumetric water content in the soil. The sensors in each row were spaced 1 m apart. Most of the sensors were embedded within a depth of 2 m. For each depth, there were generally two soil suction sensors and one ThetaProbe.

Disturbed samples were taken slightly below the three rows of instrumentation using a small-diameter auger. This was done every day to determine the gravimetric water content

Table 8.1 Summary of instruments in the bare area of the trial slope

Measurement	Type of instrument	Quantity	Measuring range	Source/references
Soil suction	Jet-fill tensiometer (2527 A)	12	Matric suction less than 90 kPa	Soilmoisture Equipment Corporation, Santa Barbara. USA
	Thermal conductivity sensor	12	Matric suction 20–1,500 kPa	Fredlund et al. (2000)
Water content	ThetaProbe (ML2x/w)	12	Volumetric water content, 0–50%	Delta-T Devices Ltd, Cambridge, UK
Horizontal stress	Vibrating-wire earth pressure cell (EPS-36-S)	6	0–1,000 kPa	Encardio-Rite Electronics Pvte Ltd, Lucknow, India
Rainfall intensity	Tipping-bucket rain gauge (ARC 100)	2	0.2 mm/tip	Environmental Measurements Ltd, UK
Run off	Vee-notch flowmeter	1	NA	–
Horizontal displacement	Inclinometer	2	±12°	Nil
Heave/settlement	Movement point	12	NA	Nil
Potential Evaporation rate	Evaporimeter	1	NA	Nil

Source: After Ng et al., 2003.

(GWC) profiles during the rainfall simulation period. All the auger holes were backfilled immediately after sampling.

Monitoring horizontal total stresses

Three pairs of earth pressure cells were installed to monitor the total horizontal stress in two orthogonal directions. As mentioned above, a mid-slope berm divided the slope into two parts. Two pairs of earth pressure cells were embedded 2.5 m above the toe in the upper half and the lower half of the slope, respectively. The other pair of earth pressure cells was located midway between the former two pairs. For each pair, one earth pressure cell was installed to measure the horizontal stress in the north-south direction (i.e., the inclination direction of the slope); the other was placed in the east-west direction (i.e., parallel to the longitudinal direction of the canal). All six earth pressure cells were installed vertically at a depth of 1.2 m. The installation procedure proposed by Brackley and Sanders (1992) was adopted to minimise soil disturbance caused by excavating a slot for a pressure cell.

Monitoring horizontal movements and surface heave

Two inclinometers were installed in two orthogonal directions: one (I1) near the toe of the upper portion of the slope, and the other (I2) near the toe of the lower portion of the slope. The inclinometers were bottomed at depths of 8.0 m and 6.5 m respectively (i.e., on the hard layer with coarse and hard calcareous concretions).

In order to measure the swelling of the unsaturated expansive soil resulting from rainwater infiltration, three rows of movement points were set up near the three main rows of instrumentation (R1, R2 and R3) respectively. The movement points were constructed with concrete blocks. Each row has four movement points found at depths of 0.1, 0.5 and 1.0 m respectively. Two levelling datum points were constructed 20 m outside the artificial rainfall area and found at a depth of 3 m. These two datum points were frequently monitored and checked using a city grid datum located over 100 m away from the site. Monitoring and checking confirmed that the two datum points were stable and were not affected by the artificial rainfall.

Monitoring rainfall intensity, runoff and evaporation

A tipping-bucket rain gauge was installed to record the intensity and duration of rainfall. Flowmeters were installed in the main water-supply line of a sprinkler system to record the total amount of water sprinkled onto the slope within a given time interval. A water collection channel was constructed along the toe of the slope to measure surface runoff using an automatic vee-notch flowmeter system installed at the end of the channel. An evaporimeter was installed at the middle of the slope outside the monitoring area to measure potential evaporation.

Other than the two inclinometers and the movement points, all the instruments were connected to a computerised data acquisition system housed at the top of the slope.

Artificial rainfall simulations

Rainfall was produced artificially using a specially designed sprinkler system. This was done to accelerate the field test programme. The sprinkler system comprised a pump, a main water supply pipe, five branches, and 35 sprinkler heads. The system could produce three levels of rainfall intensity (3, 6 and 9 mm/h).

The site was fairly dry from November 2000 to April 2001, with a total rainfall of only 60 mm. In May, when the wet season generally begins, there was only about 40 mm of rainfall. From June to 18 August, before starting the rainfall simulation tests, the monitored area was protected against rainfall infiltration with a plastic membrane. The rainfall simulation tests therefore started from a relatively dry soil condition.

Figure 8.5 shows the two simulated rainfall events during the one-month monitoring period, from 13 August to 12 September 2001. Two rainfall events were simulated. The first lasted for 7 days, from the morning of 18 August to the morning of 25 August 2001, with an average daily rainfall of 62 mm. The second simulated rainfall was from the morning of 8 September to the afternoon of 10 September 2001. During both rainfall periods, on the morning of each day, the artificial rainfall was stopped for 2–3 h to allow horizontal displacements and soil swelling to be measured, and also to auger disturbed specimens to determine gravimetric water content (GWC) profiles. Apart from this regular stoppage, the artificial rainfall intensity was maintained at a constant 3 mm/h.

The surface runoff from the artificial rainfall was measured by the vee-notch flowmeter. If the amount of infiltration during the two rainfall periods is assumed to be equal to the difference between the rain intensity and surface runoff, then the percentage of infiltration can be calculated by dividing the amount of infiltration by the rain intensity. During the first 1½ days after the beginning of the artificial rainfall, the percentage of infiltration was 100% (i.e., no runoff). Thereafter, the percentage of infiltration decreased with rainfall

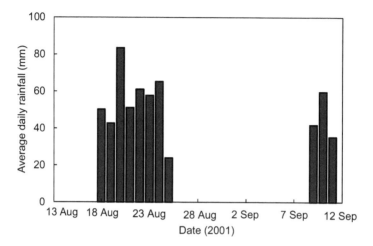

Figure 8.5 Intensity of rainfall events during the trial period.
Source: After Ng et al., 2003.

duration. After 4 days of rainfall, the percentage of infiltration tended towards a steady 30%. Note that the measured percentage of infiltration was corrected for evaporation. Based on the measurements made by the evaporimeter, it was found that the evaporation potential ranged from 2 to 8 mm/day, which is relatively small compared with the rainfall intensity of 60 mm/day.

Observed field performance

Effects of simulated rainfall on pore-water pressures (PWP) and soil suction

Responses of soil suction or pore-water pressure (PWP)

Since the changes in response to the simulated rainfalls of *in-situ* PWP or soil suction exhibited similar characteristics for the three different sections (R1, R2 and R3, located in the upper, middle and lower sections of the slope, respectively), typical monitoring results from R2 have been selected and are shown in Figure 8.6. The results recorded by four tensiometers and four thermal conductivity sensors are presented in terms of PWP and soil suction in Figures 8.6a and 8.6b respectively.

Immediately before the first artificial rainfall on 18 August 2001, negative PWPs ranging from 18 to 62 kPa were recorded by the four tensiometers (see Figure 8.6a). As expected, the higher the elevation of the tensiometer, the more negative the PWP. With the exception of the thermal conductivity sensor (R2-TC-0-6), which showed a high soil suction of about 250 kPa at 0.6 m below ground (see Figure 8.6b), soil suctions deduced from the remaining three sensors were generally consistent with the measurements obtained from the tensiometers.

After the first heavy artificial rainfall started on 18 August 2001, the negative PWP and soil suction measured by most sensors only began to decrease after about 2 days of rainfall

Field monitoring and engineering applications 387

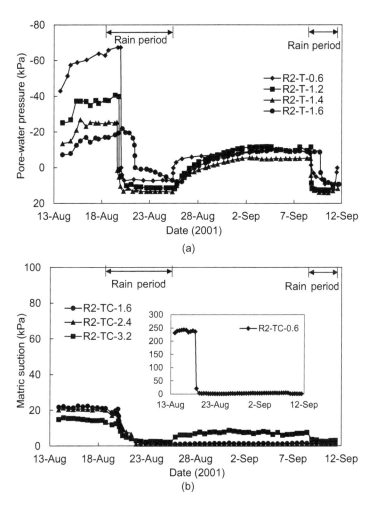

Figure 8.6 Response from suction sensors located at R2: (a) pore-water pressures measured by tensiometers; (b) soil suctions measured by thermal conductivity sensors.

Source: After Ng et al., 2003.

(about 90 mm of rain). As shown in Figure 8.6a, there was a clear delay in the response of pore-water pressure to rainfall infiltration, even at a depth of 0.6 m. Above 1.5 m, the duration of the delay appeared to decrease as depth increased. Based on field reconnaissance and observations in trial pits, it was found that many cracks and fissures occurred near the ground surface, and a layer of relatively impermeable material was identified at about 1.5 m below the ground surface. It is postulated that, as the intact expansive clay has a relatively low water permeability, water can enter the clay only through cracks and fissures. While rainwater flowed through the cracks and fissures initially, the tensiometers did not register any significant changes of soil suction around their tips, and this led to the initial delayed response. Subsequently, when the infiltrated rainwater started to rise from the bottom of the cracks, or from a perched water table formed caused by the presence of the impeding layer, and seep in all directions, the lower the tensiometer, the quicker the response (i.e., the shorter the delay). Obviously, a rapid response was shown by a sharp reduction in negative PWP when water reached the locations of the tensiometers above the impeding layer. The tensiometer located below the impeding layer showed the slowest and most gradual rate of response to rainfall and the lowest magnitude of reduction of negative PWP.

After the first rainfall, the lower three tensiometers showed a gradual increase (or recovery) in negative PWP and appeared to reach a steady state after 2 September, recording negative PWP from 3 kPa to 10 kPa. The rate of recovery was very similar for the lower three tensiometers. On the other hand, the highest one initially showed a much more rapid recovery. However, the final magnitudes of recovered negative PWP fell within a narrow range and did not appear to be governed strongly by the depths of the tensiometers.

During the second artificial rainfall, the lower three tensiometers showed almost no delay in response to the rainfall. There was a change in pore-water pressure from negative to positive, although the magnitude of the change (about 10 kPa) was not very significant. On the other hand, the top tensiometer showed a 1-day delay in response to the second rainfall. However, the "final equilibrium" PWP recorded during the second rainfall was similar to those during the first one.

The general responses to the two artificial rainfalls of indirect soil suction measured by thermal conductivity sensors (see Figure 8.6b) were similar to those recorded by the tensiometers, except that the former showed a slower rate of response than the latter. The magnitudes of pore-water pressures measured by the two types of sensor were generally consistent, particularly at the depth of 1–6 m below ground. However, the inconsistency between the two sensors located at the depth of 0–6 m before the first rainfall may result from the inherent limitation of tensiometers, caused by cavitation at high suction.

Responses of piezometric level

Figure 8.7 shows the variations of piezometric level (i.e., elevation head plus PWP head) with time for sections R1, R2 and R3. The piezometric levels were calculated from the monitored results of the suction sensors installed at three different depths (0.6, 1.2 and 2.4 m) in each section. The elevation head for each sensor was calculated according to the local datum (96 m OD) located at the slope toe. The responses of piezometric levels within and below the 1 m depth are shown in Figures 8.7a and 8.7b respectively.

Before the first rainfall event, the piezometric levels at a depth of 0.6 m decreased with increasing elevation of the three sections (i.e., from R3 to R1: see Figure 8.7a). This suggests that there was an up-slope water flow within 1 m below the ground surface, resembling the capillary rise of water in an inclined column of unsaturated soil. In contrast, at a given depth of 1 m or more below the ground surface, the piezometric levels increased with the elevation of the sections (see Figure 8.7b). This indicates a down-slope water flow below a depth of 1 m. The observed difference in the water flow directions above and below a depth of 1 m before the rainfall may be caused by the presence of large numbers of open cracks and fissures near the ground surface. These structural features enhance the process and rate of evaporation, and hence result in substantial high negative PWP or soil suction near the ground surface (see Figure 8.6). By contrast, as the number of open cracks and fissures decreases significantly with depth, the influence of cracks and fissures appears to be negligible at greater depths.

Field monitoring and engineering applications 389

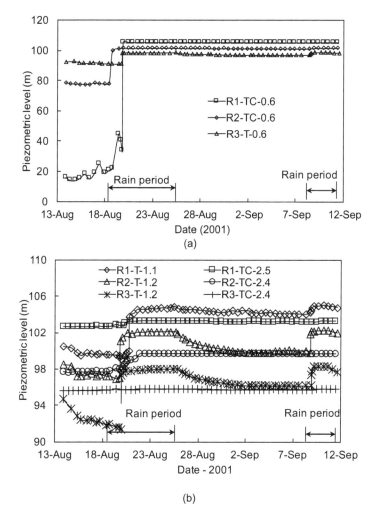

Figure 8.7 Variations of piezometric level at various depths: (a) by suction sensors within 1 m depth; (b) by suction sensors below 1 m depth.

Source: After Ng et al., 2003.

After the first rainfall started, the piezometric levels at each section increased owing to the decrease in negative PWP. At a given depth along the slope, the piezometric levels decreased with a reduction in the elevations. A down-slope water flow appeared as a result of rainfall infiltration. During the two-week non-rainfall period, the magnitude of the difference in the piezometric levels at sections R1, R2 and R3 remained the same, even though there was a slight recovery of negative PWPs in the soil, indicating that there was a water flow in the down-slope direction after the rainfall. These results were consistent with the observed exit of groundwater near the slope toe for several days after the first rainfall period.

Variations of in-situ PWP profiles

Figure 8.8 shows the PWP distributions with depth. Before the first rainfall period, the negative PWPs near the ground surface were substantially higher than those at greater depths, and hence the PWP profiles deviated significantly from theoretical hydrostatic conditions. The negative PWPs below a depth of 2 m were relatively low and decreased gently with an increase in depth.

After 3 days of about 180 mm of heavy rainfall, the PWPs increased significantly within the upper 2 m of soil. A positive PWP appeared at a depth of about 1.5 m below the ground at sections R2 and R3. The continued rainfall after 21 August resulted in a further increase in PWP but at a significantly reduced rate.

At the end of each of the two rainfall periods (i.e., 25 August and 10 September), the significant positive PWP observed by tensiometers within the upper 2 m of soil was greatest at a depth of about 1.5 m at each section. This seems to indicate the presence of a perched groundwater table about 1.5 m below the ground surface. The measured *in-situ* dry density profiles demonstrated that there was a denser ($\rho_d \geq 1.60$ Mg/m^3) soil layer, ranging from 0.3 to 0.5 m thick, located at about 1.5 m (see Figure 8.1). It is believed that the dense soil layer possesses a relatively low coefficient of water permeability, and hence the infiltrated rainwater is retained above this dense layer. Because of the impedance effect of the dense layer, the influence of rainfall on PWP below 1.5 m at this site was generally insignificant. The presence of the perched groundwater table at a depth of about 1.5 m caused the development of significant positive PWPs, which led to the expansion of the initially dry expansive soil upon wetting, resulting in a reduction in the shear strength of the soil layer. This may explain why most rain-induced landslides appear to be relatively shallow, generally found within 2 m depth (Bao and Ng, 2000).

Response of volumetric water content (VWC)

Figure 8.9 shows the VWC monitored by four ThetaProbes located at the R2 section during the two artificial rainfalls. The response of VWC recorded by the ThetaProbes was generally consistent with the corresponding pore-water pressure responses shown in Figure 8.6. With the exception of the ThetaProbe located 0.3 m below the ground surface (i.e., R2-θ-0.3), there was a delay of at least 2 days before VWC changed in response to the first artificial rainfall, which started on 18 August. The infiltration characteristics revealed by the lower three ThetaProbes were generally consistent with the pore-water pressure responses shown in Figure 8.6. The rapid response of R2-θ-0.3 might be attributed to the presence of a large number of cracks and fissures near the surface. For the lower three ThetaProbes, the one at l.2 m depth (R2-θ-1.2) responded first, followed by the one at 0.6 m depth, and finally the one at 1.6 m depth. The order of response may perhaps be explained by the presence of an impeding layer located at about 1.5 m depth, as discussed above.

After 3 days of rainfall, all VWCs measured at various depths appeared to reach a steady state. About 3 days after the cessation of rain, the VWCs at different depths began to decrease progressively to another steady-state condition. The shallow probes reached new equilibrium values first, followed by the deeper ones on 2 September. After the start of the second rainfall on 8 September, all the ThetaProbes responded quite rapidly, but the magnitude of the increase in VWC was generally smaller than that during the first rainfall.

Field monitoring and engineering applications 391

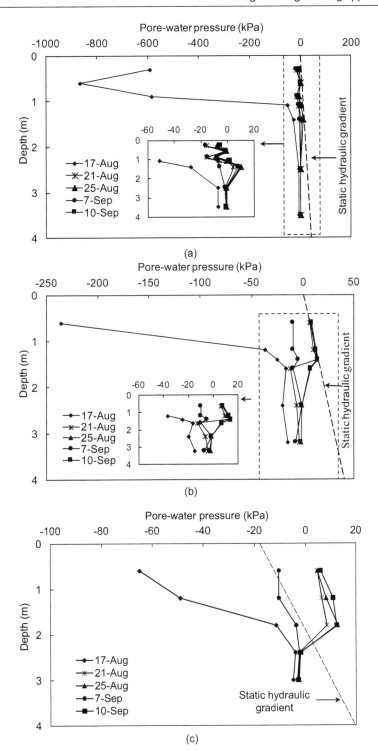

Figure 8.8 Variation of pore-water pressure: (a) at R1; (b) at R2; (c) at R3.

Source: After Ng et al., 2003.

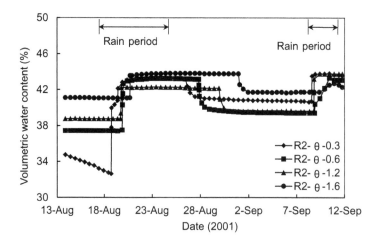

Figure 8.9 Volumetric water content changes in response to rainfall measured by ThetaProbes located at R2.

Source: After Ng et al., 2003.

Variations of in-situ water content profiles

Figures 8.10a, 8.10b and 8.10c illustrate the variations in water content profiles in response to rainfalls at sections R1, R2 and R3 respectively. In each figure, the GWC profiles were obtained by directly sampling the soil, and the two dashed lines in the middle indicate the calculated minimum and maximum VWC profiles based on the measured GWC profiles and the measured dry density profiles. The solid lines labelled with open symbols represent the VWC profiles measured by ThetaProbes.

Just before the first rainfall event started, the initial GWCs generally increased with depth within 1.5 m below ground, suggesting an upward flow of moisture via evaporation. The measured relationship of the initial GWC profiles and the measured initial negative PWP profiles (Figure 8.8) was approximately consistent with the wetting curve of the SWCCs shown in Figure 8.2. For example, the measured GWCs varied from 16.5% to 22.8% above 2 m depth at the R1 section and were reasonably consistent with the estimated range of GWC (17.5–25%) deduced from the wetting curve of the SWCCs and the negative PWP profiles measured by tensiometers, as shown in Figure 8.8.

After rainfall started, the measured GWC profiles were generally consistent with those of the PWP responses discussed previously (Figure 8.8). A significant increase in GWC could be found within the upper 1.5 m, and the influence of the rainfall on the GWCs below 1.5 m seemed to be essentially negligible, particularly at R2. This finding supports the previous supposition that there is a low-permeability layer at mid-slope. The reduction in GWC during the two-week no-rain period appeared to be small, even in the soil layer near the ground surface.

By comparing the measured VWC profiles attained using the ThetaProbes with the calculated bounds of VWC obtained using the measured GWC profiles, it can be seen that the measured VWC is, in general, significantly larger than the upper bound of VWC calculated from the GWC profiles at all three sections. Note that the lower and upper

Field monitoring and engineering applications 393

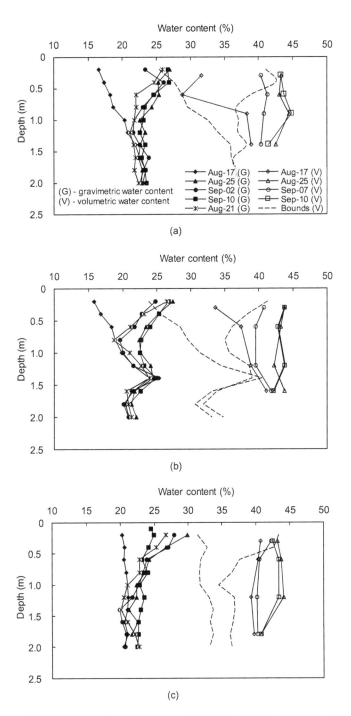

Figure 8.10 Changes of water content in response to rainfall: (a) at R1; (b) at R2; (c) at R3.

Source: After Ng et al., 2003.

bounds of the calculated VWC, represented by the dashed lines, were obtained using the envelopes of measured GWC profiles taken on 17 August and the maximum of the two data sets recorded on 25 August and 2 September respectively. The inconsistency between the measured and calculated VWCs may be attributed to the accuracy of the indirect measurements of VWC using the ThetaProbes. The measurement accuracy using ThetaProbes can be affected by many factors, such as variations in soil composition, dry density and cracks (Li et al., 2002). It is suggested that the measured VWC can only be interpreted as an indication of what is happening.

Response of horizontal total stresses to changes of pore-water pressure or suction

Figure 8.11 shows the monitored total stress ratio (σ_h/σ_v) against time from six vibrating-wire earth pressure cells (EPCs) installed in the slope. All EPCs were installed at a depth of 1.2 m, giving rise to an estimated total vertical stress (σ_v) of about 23.4 kPa, which corresponded to an average dry density of 1.56 Mg/m^3. Pressure cells EP1, EP3 and EP5 (see Figure 8.4) measured the stress changes acting in the east-west direction (i.e., perpendicular to the inclination of the slope), while EP2, EP4 and EP6 recorded pressures acting in the north-south direction (i.e., parallel to the inclination of the slope).

Before the first rainfall, the total stress ratios recorded by all the cells were below 0.3. An initial equilibrium stress ratio appeared to have been established for each cell shortly before the rainfall on 18 August. Two out of the six cells registered a small tensile stress, probably induced by soil drying. Please note that the vibrating-wire-type cells cannot record tensile stress. During installation, the clearance between the wall of the earth pressure cell and the soil was backfilled with an epoxy resin. This thin layer of epoxy resin attached the cell securely to the soil and allowed tensile force to be transmitted between the cell and the soil. This installation procedure, originally proposed by Brackley and Sanders (1992) and monitoring its results demonstrated that a tensile force can be detected by the vibrating-wire cell.

After the first rainfall started, none of the EPCs registered any significant change of stress for about 1.5 days. The delayed response of the pressure cells was consistent with the PWP and VWC measurements shown in Figures 8.6 and 8.9 respectively. Once the EPCs started to respond, the ratios (σ_h/σ_v) increased very rapidly and significantly within 1 day, and then approached a steady value during the first rainfall event. It appeared that the magnitude of increase in total horizontal stress was strongly related to the height of the EPCs and the initial negative PWP (see Figure 8.8). The higher the location of EPC, the larger the initial negative PWP present in the ground, and hence the greater the increase in σ_h/σ_v. This performance appeared to be consistent with the relationship between the swelling potential of expansive soils and initial soil suction: that is, the swelling potential of an expansive soil generally increases with increasing initial negative PWP or suction of the soil (Fredlund and Rahardjo, 1993; Alonso, 1998). For a given pair of pressure cells located at the same elevation, the measured stress ratio in the east-west direction was always higher than that in the north-south direction. This is probably related to a greater constraint imposed by ground sloping in the east-west direction as opposed to that in the north-south direction.

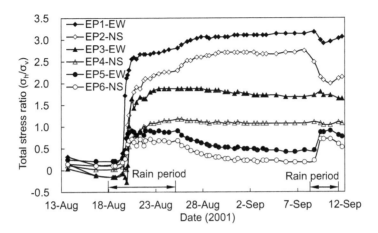

Figure 8.11 Changes of in-situ total stress ratio measured by earth pressure cells.
Source: After Ng et al., 2003.

During the two-week no-rain period, a further increase in σ_h/σ_v was observed at EP1 and EP2. On the other hand, the EP3 and EP4 pressure cells showed a slight decrease in stress ratio throughout the no-rain period, and EP5 and EP6 recorded a larger reduction in stress ratio than EP3 and EP4. The reduction in σ_h/σ_v appeared to be primarily caused by a decrease in the positive PWPs at a depth of 1–2 m during the no-rain period (see Figure 8.6). However, the continuous and gradual increase in σ_h/σ_v at EP1 and EP2 (but at a reduced rate) may be because of an ongoing "soaking" of the soil near the location of the EPCs at R2, even after the first rainfall event.

After the start of the second rainfall event, the responses at the three pairs of EPCs were distinctly different. At EP1 and EP2, σ_h/σ_v decreased rather than increased. This may be attributed to softening of the soil after prolonged swelling during the no-rain period. For the pressure cells (EP5 and EP6) near the toe of the slope, an increase in σ_h/σ_v was recorded resulting from the recovery of positive PWP during the second rainfall. The performance of EP3 and EP4 fell between the former two cases.

To make a comparison between the measured total stress ratios after the simulated rainfalls with the corresponding theoretical limiting conditions, the total stress ratios at the passive failure conditions were calculated using the *total stress* and *effective stress* approaches respectively (Fredlund and Rahardjo, 1993). For the total stress approach, the undrained shear strength was assumed to be 4.4 times the SPT N value obtained in the monitored area after the rainfalls, and the calculated total stress ratio for passive earth failure conditions was 4.4: these values were greater than the measured earth pressures except at cells EP1 and EP2, which appeared to be close to passive failure conditions.

For the effective stress approach, the calculated total passive stress ratio ranged from 2.0 to 3.1 if the saturated shear strength parameters, $c' = 5.15$ kPa and $\varphi' = 17°$ obtained from testing specimens with fissures and cracks (Liu, 1997), were used in the calculations.

The calculated passive stress ratios appeared to be close to the values measured by EP1 and EP2 after the simulated rainfalls. This seemed to suggest that the expansive soil, after the simulated rainfalls, may reach passive failure along existing cracks and fissures. This finding is consistent with that observed by Brackley and Sanders (1992). This further supports the possible softening behaviour of the soil upon prolonged wetting. The high *in-situ* stress ratios caused by the swelling of expansive soils upon wetting might be one of the main reasons for the retrogressive shallow failures found near the monitored slope.

Response of ground deformations in response to the simulated rainfalls

Horizontal displacements due to changes in soil suction

Figure 8.12 shows the horizontal displacements of the ground in response to the simulated rainfalls. All horizontal displacements shown in the figure were calculated by taking the rotations measured just before the rainfalls started as the reference datum. The calculated displacements of the inclinometers from the south to the north direction (i.e., the up-slope direction) and from the west to the east direction are defined as positive.

Figures 8.12a and 8.12b show the monitored horizontal displacements from inclinometer I1 (located just above the mid-slope) in the north-south and east-west directions respectively. The results indicate that the ground moves in the down-slope and east directions. The measured horizontal displacements in both directions illustrate similar characteristics. The horizontal displacements in the upper 1.5 m were understandably greater than those below this depth, which looks like a "cantilever" mode of deformation. The variations in horizontal displacement profiles were consistent with changes in PWP, showing that the most significant changes took place at shallow depth (i.e., less than 2 m), as discussed previously in Figure 8.8.

By comparing the displacements on 19 August and 21 August, it can be seen that there was a significant increase in horizontal displacements, particularly near the ground surface, in both directions. The observed increase in displacement after 3 days of rainfall appeared to be consistent with the 2-day delayed response in pore-water pressures, as discussed previously in Figure 8.6. As the rainfall continued, further changes in horizontal displacement were relatively small. After the first rainfall event, a recovery in horizontal displacement (i.e., shrinkage response) was observed with respect to both directions during the two-week no-rainfall period (i.e., from 25 August to 7 September 2001), owing to the increase in soil suction or decrease in positive pore-water pressures. The recovery of 2 mm at the ground surface from the east to the west direction was far more significant than that in the up-slope direction (i.e., only about 0.2 mm). The effects on the up-slope movements of an increase in soil suction were counteracted by the influence of gravity. At the end of the second rainfall event, the observed horizontal displacements were similar to those measured at the end of the first rainfall in both directions.

Figures 8.12c and 8.12d show the monitored horizontal displacements from inclinometer I2 in the down-slope and the east direction respectively. It can be seen that the magnitudes of displacement and the deformed shapes observed were consistent in both directions, but the magnitudes were significantly larger, and the depth of influence was substantially deeper (deep-seated) than those observed at the mid-slope (i.e., at I1). It is believed that the observed larger displacements near the toe are attributable to the lower initial negative

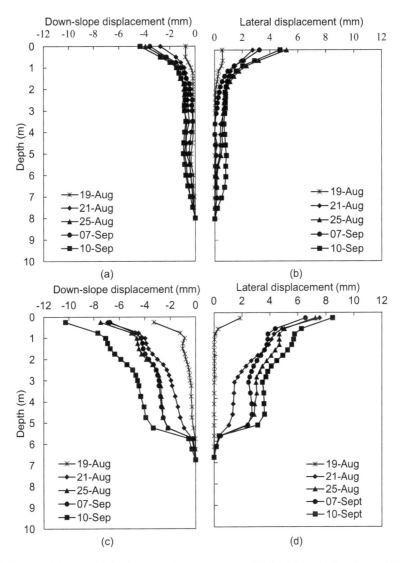

Figure 8.12 Observed horizontal displacement in response to rainfall: (a) from I1 (north-south); (b) from I1 (east-west); (c) from I2 (north-south); (d) from I2 (east-west).

Source: After Ng et al., 2003.

PWP (or soil suction) present at I2 than those at I1 (see Figure 8.8), resulting in a lower soil stiffness near the toe. The greater depth of influence near the toe of the slope was consistent with the deeper influence of the simulated rainfalls on the GWC measured at section R3, as opposed to section R2 (see Figure 8.10).

At both I1 and I2, the consistently observed eastern movements of the ground resulting from rainfall infiltration might be attributed to the direction of subsurface water flow from west to east caused by the presence of slightly dipping geological planes. On 23 August, a seepage exit point was observed about 15 m to the east along the lower part of the masonry wall, outside the monitoring site area. This observation seemed to support the hypothesis of the direction of water flow and ground movement.

Soil swelling at shallow depths upon wetting

Figures 8.13a, 8.13b and 8.13c show the measured vertical swelling in response to the simulated rainfalls near the three sections, R1, R2 and R3, respectively. It can be seen from Figure 8.13a that the top two movement points registered an upward soil movement of about 3 mm 1 day after the start of the first rainfall event. The rate of soil swelling was almost constant during the first 5 days of rainfall, and the soil continued to swell, but at a reduced rate, throughout the remainder of the monitoring period. On the other hand, there was no recorded swelling at the movement points embedded at both 0.5 m and 1.0 m depth during the first 4 days of rainfall. Thereafter, the measured rates of swelling at both depths were like those recorded at a depth of 0.1 m. As anticipated, the shallower the embedded movement point, the greater the magnitude of the measured swelling, owing to the larger changes in soil suction at the shallower depths associated with rainfall infiltration (see Figure 8.8) and the cumulative swelling of the underlying soil. During the two-week no-rain period, the ground continued to swell but at a reduced rate. The continuous soil swelling at all three depths may be attributed to the slow seepage of infiltrated water from open cracks and fissures into the surrounding soil, leading to the secondary swelling of the expansive soil. Marked secondary swelling behaviour has been observed and reported by Chu and Mou (1973), and Sivapullaiah et al. (1996). Alonso (1998) has postulated that the marked secondary swelling behaviour of expansive clays is caused by the slow and progressive hydration of expansive soil microstructures. Certainly, the observed characteristics of soil swelling can assist in explaining the measured increase in earth pressures at EP1 and EP2 during the no-rain period, as reported in Figure 8.12. During the second rainfall event, the rates of soil swelling increased, particularly at a depth of 0.1 m.

As shown in Figures 8.13b and 8.13c, the observed soil swelling patterns at R2 and R3 were similar to those observed at R1, except that the duration of the delayed response to rainfall was longer at a lower elevation. At a given depth, the magnitudes of soil swelling at R2 and R3 were less than that observed at R1. This is probably because of the smaller initial soil suction at R2 and R3 than at R1 (see Figure 8.8). Based on the measurements at the three sections, it can be generalised that the higher the initial suction, the greater the soil swelling at a given depth. Within the same section, the shallower the embedded movement point, the larger the measured soil swelling.

Summary

Based on the field observations, the following key findings should be noted:

- Before the rainfall events, high soil suctions (or negative pore-water pressures, PWPs) were measured within the top 1 m of soil. The high initial soil suction induced an upward flux of water and moisture both vertically and along the inclined slope. On

Field monitoring and engineering applications 399

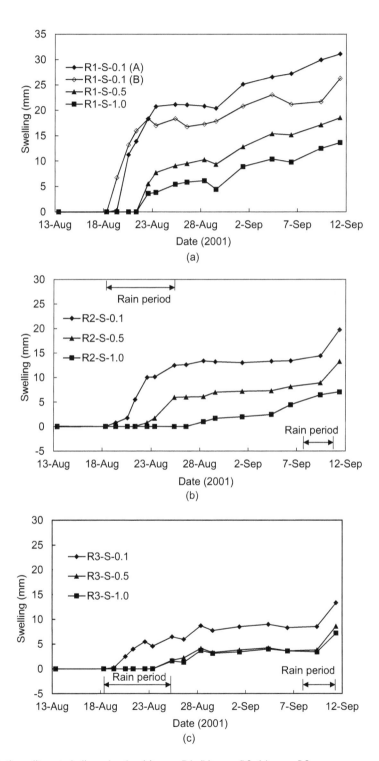

Figure 8.13 Soil swelling at shallow depths: (a) near R1; (b) near R2; (c) near R3.

Source: After Ng et al., 2003.

the other hand, a down-slope flux of water and moisture occurred at depths below 1.5 m, which may be attributed to the presence of a relatively impermeable soil layer at about 1.5 m below ground level.
- The observed responses in PWP, water content, horizontal stress and soil deformation generally showed a 1 to 2-day delay after the start of the rainfall event. The combined effects of the first 3 days of rainfall (i.e., about 180 mm) on the observed responses were much greater than the effects of the ongoing rainfall or the second rainfall event, except for the down-slope deformations at the toe of the slope.
- The effect of the simulated rainfalls on the variations of PWP, water content, horizontal stress and soil deformation was generally much more significant above 2 m below the ground surface than below a depth of 2 m.
- A significant perched water table was deduced at a depth of about 1.5 m below the slope surface. The perched water table is believed to be related to the presence of a dense soil layer without open cracks or fissures at that depth. The presence of the perched groundwater table caused the development of positive PWP and an expansion of the initially dry expansive soil upon wetting. This led to a reduction in the shear strength of the soil layer. This may help to explain why most rain-induced landslides occurring in similar unsaturated expansive soil slopes appear to be relatively shallow, with slip surfaces generally found within a 2 m depth.
- A significant increase in the total stress ratio (σ_h/σ_v) was observed after the simulated rainfalls, particularly in the soil layer with a high initial negative PWP or soil suction (at a high elevation on the slope). The maximum *in-situ* total horizontal stress after the simulated rainfalls was more than three times the total vertical stress. This indicated the possibility of passive pressure failures in the softened clay after the simulated rainfalls.
- Two distinct modes of down-slope deformation were observed: a cantilever deformation within the top metre at the mid-slope, and a deep-seated down-slope displacement near the toe of the slope. The observed modes of soil deformation appeared to be consistent with PWP responses and water content measurements. The deep-seated displacement was probably caused by the delayed subsurface water flows. An unusual but significant asymmetric horizontal soil displacement was observed towards the east, along the longitudinal axis of the slope, at both the mid-slope and slope toe.
- A substantial soil swelling was measured in the vertical direction after the simulated rainfalls, particularly in the soil layer with a high initial negative PWP (e.g., at shallow depths and at a high elevation of the slope). The higher the initial soil suction, the greater the soil swelling. The observed soil vertical movement also revealed a marked secondary swelling characteristic in the expansive clay.
- The abundant cracks and fissures in the expansive soils played an important role in the soil-water interaction in the expansive soil slope, and greatly affected the groundwater flows and soil suction.
- This work reported here was the most comprehensive field study of the influence of two independent stress state variables, i.e., net stress and suction, on expansive soil slopes before the design of the canal. The findings from this field work lay down the foundation for the design of the middle route across 180 km of unsaturated expansive soils.

FIELD MONITORING OF AN UNSATURATED SAPROLITIC HILLSLOPE ON LANTAU ISLAND IN HONG KONG (LEUNG ET AL., 2011)

Introduction

Rainfall-induced landslides are natural geological phenomena, which have caused untold numbers of catastrophes in terms of casualties and economic loss worldwide. These problems are especially serious in tropical and subtropical regions such as Malaysia, Japan, Brazil, Singapore, HKSAR, and part of mainland China. The intrinsic geological conditions in these regions, such as deep weathering of rocks, probably result in deep groundwater tables, and the soils above are frequently in an unsaturated condition. When subjected to seasonal climatic variations, the infiltration and deformation characteristics of heterogeneous unsaturated soil in the vadose zone may be complex. Reliable geological and hydrogeological models, which can be deduced from field inspections, ground investigation fieldwork, and field monitoring, are therefore always prerequisites to capturing the ground behaviour and understanding the landslide triggering mechanism of a hillslope subjected to severe rainfall events. The field monitoring reported in this section considered the influence of two independent stress state variables, i.e., net stress and suction, on unsaturated saprolitic soil slopes in Hong Kong.

Description of the study area

The selected hillslope in this study is located on Lantau Island, Hong Kong. Interpretation of aerial photographs from the years 1963 to 2004 and engineering, geological and geomorphological field mapping were carried out. The geomorphological setting and drainage characteristics of the study area are shown in Figure 8.14. A detailed description of the study area was reported by GEO (2007) and is briefly summarised in the following paragraphs. The natural terrain is moderately to densely vegetated. The terrain itself forms a blunt ridgeline located between a major stream channel on its northeastern side and a shallow topographic valley to the south and west. As a result, runoff from the northeast- and northwest-facing slopes may enter the stream channels at the toe of the hillslope quickly. Although limited mapping of the ground was carried out in the lower portion of the study area because of its extremely densely vegetated thick ground cover, the shallow gradient of the slope indicated that it was a depositional area with colluvial material. In the mid-portion of the study area, the topography itself formed a very slight bowl-shaped depression, flanked by shallow ridges on either flank and a concave break of slope at the crest and a convex break of slope at the toe. This shallow topographic depression may hence constrain any runoff in the central portion of the study area. Some prominent features of the active landslide body were identified in the mid-portion of the study area, and the details are given below.

Features of active landslide mass

To characterise the geological profile and groundwater conditions on the hillslope, fieldwork including boreholes, trial trenches and a geophysical survey were carried out. The locations of the boreholes (BH 1, 3, 5–8) and trial trenches (TT 1–5) are shown

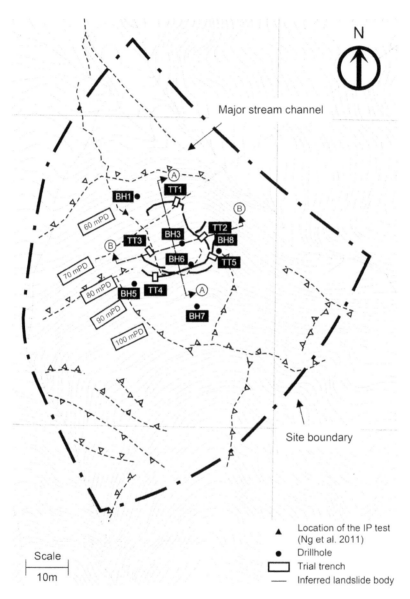

Figure 8.14 Overview of geomorphological setting of natural terrain and location of fieldworks. IP, instantaneous profile.

in Figure 8.14. In particular, the trial trenches were intentionally excavated across the prominent features of the active landslide body in continuity with the aerial photograph interpretation and field mapping. As shown in Figure 8.15, a series of subparallel tension cracks approximately 45 m long contouring the hillside were identified at an elevation between +84 and +86 mPD (mPD stands for "metres above Principle Datum" and refers to the height above mean sea level, approximately 1.23 m above the PD (SMO, 1995)) which forms the main scarp features. Trial trenches TT 4 and TT 5 excavated across the subparallel main scarps identified some poorly defined rupture surfaces at shallow depths of about 0.5 m. A number of lateral tension cracks extend north from the main scarps running down the slope from the eastern and the western side of the mid-portion of the study area. Trial trench TT 2 excavated across the eastern flank revealed that some cracks continue down to link with a relict joint within highly decomposed tuff (HDT) at about 2 m depth. Moreover, foliations identified in the decomposed tuff at 2 m below ground level (bgl) were slightly oblique to the dip of the hillslope, indicating significant past ground movement. In addition, a major thrust feature was found at +64 mPD, while some smaller thrust features were identified on both landslide flanks at +76 mPD.

An active landslide body approximately 45 m wide is inferred in the mid-portion of the study area, with a slope inclination of about 288°. The subparallel main scarp features identified, the lateral tension cracks, and the thrust features characterise the head, the flank, and the toe of the landslide body, respectively. Field mapping revealed that the past movements of the landslide body were oblique inwards from the flanks and towards the centre and the toe of the hillslope, resulting in ground destabilisation around the landslide flanks and causing distress in the area beneath the lowest of the subparallel main scarps.

Geological–hydrogeological model

Fieldwork including borehole exploration and geophysical surveys was conducted to identify the geological profiles, *in-situ* soil properties, and groundwater conditions of the natural hillslope. Some key features of the superficial and solid geology are summarised as follows. Colluvial materials, which comprised mainly sand and silt with some angular to subangular gravel- and cobble-sized rock fragments, were encountered in the superficial region. A relatively thick layer of colluvial deposits up to 3 m thick was found within the topographic depression in the mid-portion of the study area. The underlying saprolite is typically described as extremely to moderately weak, light grey, dappled light brown, completely decomposed coarse ash tuff (CDT) or HDT with occasional angular and subangular fine gravel.

For the solid geology, moderately strong, grey and dappled light brown, slightly or moderately decomposed coarse ash tuff (SDT, slightly decomposed tuff; MDT, moderately decomposed tuff) was encountered beneath the saprolites. Closely- to medium-spaced rock joints were infilled with kaolin. Evidence of hydrothermal alteration (silicification and honeycomb weathering of feldspar minerals) of MDT, which is considerably more resistant to weathering than the surrounding rock, was identified mainly in borehole BH 5 (near the western landslide flank) below depths of 1.7 m and in borehole BH 8 (near the eastern landslide flank) between depths of 8 and 12 m (see Figure 8.14). The drop of the rock head level from the western flank of the landslide to the eastern one possibly caused the formation of a bowl-shaped localised depression in the central portion of the landslide body and may, as a result, have some impact on the hydrogeological regime.

Based on the limited field records, a preliminary and simplified geological profile was established and is shown in Figure 8.16 for sections A – A and B – B. The groundwater profile immediately after instrumentation generally followed that of the MDT and was about 1–2 m above the decomposed rock head surface. The inferred rupture surfaces are also shown for reference. According to the classification system of types of slope failure by

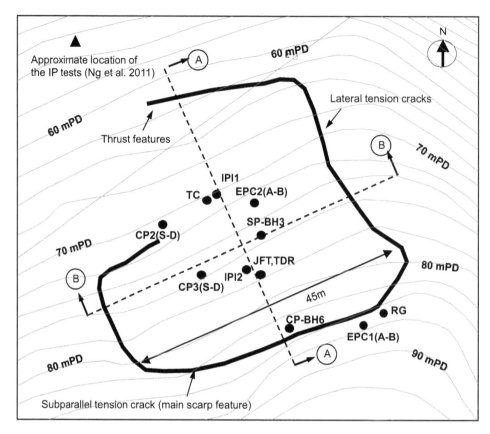

Figure 8.15 Inferred active landslide body and location of instruments. CP, Casagrande-type piezometer; EPC, earth pressure cell; IPI, in-place inclinometer; JFT, jet-filled tensiometer; SP, standpipe; TC, heat dissipation matric water potential sensor; TDR, water content reflectometer; RG, rain gauge.

Varnes (1978), the active landslide body in this study area appears to undergo a retrogressive translational-slide failure of limited mobility along poorly defined rupture surfaces at 3–5 m bgl, which are just beneath the colluvium-CDT interface.

Soil properties

A series of conventional laboratory tests were conducted to determine the soil properties of both the colluvium and the CDT. They included the determination of specific gravity, *in-situ* void ratio, *in-situ* dry density and water content, and the Atterberg limits. The measured index properties are summarised in Table 8.2.

Instrumentation programme and monitoring results

A comprehensive instrumentation scheme was implemented in the field monitoring programme. The instruments consisted of jet-filled tensiometers (JFTs), water content

Field monitoring and engineering applications 405

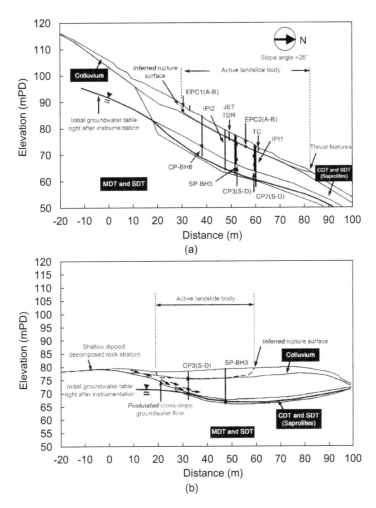

Figure 8.16 Preliminary geological model for (a) section A – A and (b) section B – B.

reflectometers (TDRs), heat dissipation matric water potential sensors (TCs), vibrating-wire earth pressure cells (EPCs), in-place inclinometers (IPIs), Casagrande-type piezometers (CPs), standpipes (SPs), and a tipping-bucket rain gauge (RG). These instruments were typically installed around the active landslide body. The diagrammatic instrument arrangement is shown in Figures 8.15 and 8.16.

Detailed information for each instrument is summarised in Tables 8.3 and 8.4. The sampling frequency of each instrument was 15 min. Since all instruments recorded significant responses during the two heavy rainstorms from 18 to 22 April and from 5 to 9 June 2008, the monitoring results between these two periods were selected to investigate the hillslope behaviour in this section.

Table 8.2 Summary of the measured index properties of the colluvium and the CDT

Measured index properties	Colluvium	CDT
In situ dry density: kg/m^3	1504	1600
In situ water content by mass: %	20.1	17.3
In situ void ratio	0.82	0.68
Specific gravity	2.73	2.68
Liquid limit: %	41	34
Plastic limit: %	17	20
Plasticity index: %	23	14
Gravel content: %	0–35	0
Sand content: %	20–25	35
Silt content: %	30–50	40
Clay content: %	15–25	25
Unified Soil Classification System (ASTM, 2011)	CL	CL

Note:
CL. inorganic silty clay of low to medium plasticity.

Rainfall characteristics

An RG was installed to record rain depth and to estimate rainfall intensity. Although the area sampled is small when compared to that of the terrain, it is assumed that the measured rain depth is representative and uniform over the entire study area. In comparison with the rainfall isohyet and the monthly rainfall amount reported by the Hong Kong Observatory (HKO), the measurements made by the RG are generally consistent and reliable. The recorded annual rainfall in 2008 was about 3084 mm. The wet season started in mid-April. It brought Hong Kong heavy rain on 19 April, with a maximum 24 h rolling rainfall of 187.7 mm (i.e., peak rainfall intensity of 62 mm/h). This required the first Black Rainstorm Warning (i.e., hourly rainfall exceeds 70 mm) issued by the HKO in 2008. In June, the weather was marked by heavy rain and squally thunderstorms. The RG recorded a rain depth of 1391 mm (45% of the annual rainfall in 2008) over the month. The peak 1 h rolling rainfall of 133.5 mm was recorded on 7 June and was the highest since the record began. The return period may be estimated using the procedures described by Evans and Yu (2001) based on the Gumbel method, though the estimation may not necessarily be applicable to other locations in Hong Kong (Peterson and Kwong, 1981; Lam and Leung, 1994; Evans and Yu, 2001). The return period of the peak 4 h rolling rainfall of 323.6 mm on 7 June is approximately 245 years. The passage of several tropical cyclones from July to October brought continuous storm surges, and the 24-hour rolling rainfall was typically about 200 mm.

Response of PWP and VWC

Three pairs of JFTs and TDRs were installed to measure PWPs and VWCs, respectively. Each pair of the instruments was installed at depths of 0.5, 1.5, and 2.5 m bgl near the sub-parallel main scarps (see Figures 8.15 and 8.16). Each TDR measures the travel time of an electrical pulse applied along the rods. The travel time primarily depends on the dielectric

Table 8.3 Some detailed information about each instrument

Instruments	Measurement	Installation depth: m bgl	Measurement range	Source
Tipping-bucket rain gauge	Rain depth	N/A	0–500 mm/h	Campbell Scientific
Jet-filled tensiometer	PWP	0.5, 1.5, 2.5	0 to –90 kPa	Soilmoisture Equipment Corp.
Heat dissipation matric water potential sensor	Measure temperature changes and deduce matric suction through calibration curve	0.2, 0.4, 0.6	10–2500 kPa	Campbell Scientific
Water content reflectometer	Measure period and deduce VWC through calibration curve	0.5, 1.5, 2.5	0–50%	Campbell Scientific
Casagrande-type piezometer	Measure positive PWP	Refer to Table 8.4	0–350 kPa	Dataqua Elektronikai Ltd.
Vibrating-wire earth pressure cell	Measure wire frequency and deduce total horizontal stress	2	1–350 kPa	Geokon
In-place inclinometer	Measure inclined angle and deduce horizontal displacement	0, 1, 3, 5, 7	15°	Geokon

Table 8.4 Installation records of standpipes and Casagrande-type piezometers

Instruments		Response zone			
Type	Symbol	From m bgl	To m bgl	Ground level: mPD	Material around the response zone
Standpipe	SP – BH3[a]	3.00	15.10	77.43	CDT and HDT
	SP – BH5	3.00	12.53	84.33	MDT to SDT
	SP – BH7	3.00	10.20	97.77	MDT to SDT and thin bands of HDT
	SP – BH9	3.00	12.72	114.25	MDT to SDT
Casagrande-type piezometer	CP – BH1[b]	14.02	15.32	58.90	HDT with a zone of fractured MDT
	CP – BH6	10.70	12.00	84.55	Interface of HDT and MDT
	CP1S	3.70	5.00	60.74	HDT to MDT
	CP1D	8.00	9.30		Competent bedrock strata
	CP2S	3.70	5.00	73.35	Interface of CDT to HDT
	CP2D	7.05	8.35		Interface of CDT to HDT
	CP3S	3.70	5.00	77.22	Colluvium
	CP3D	8.70	10.00		CDT to HDT

Notes:

[a] SP-BH3 means that a standpipe was installed in the drillhole BH 3.
[b] CP-BH1 means that a Casagrande-type piezometer was installed in the drillhole BH 1.

constant of the water in the surrounding soil, and VWC can thus be deduced indirectly using a soil-specific calibration curve.

Each JFT and TDR was inserted into 50 and 100-mm diameter predrilled holes, respectively. In particular, the rods of each TDR were carefully inserted into the ground to avoid any damage to the sensors. After installation, the lower part of each predrilled hole was backfilled with a 100 mm-thick layer of compacted *in-situ* soil, followed by bentonite cement grout. According to the borehole records, the instruments embedded at 0.5 and 1.5 m bgl were situated within the colluvium while the instruments embedded at 2.5 m bgl were situated within the CDT. The measurement range of JFTs is limited to about 80 kPa for negative PWPs because of the possibility of cavitation. Three calibrated TCs were installed in the colluvial deposits at 0.2, 0.4, and 0.6 m bgl. According to the manufacturer (CSI, 2009), each TC can measure matric suction indirectly between 10 and 2500 kPa, and its resolution is claimed to be up to 1 kPa at suctions greater than 100 kPa. Depending on the thermal conductivity of the ceramic water complex surrounding the sensor, temperature changes with constant power dissipation from the line heat source are measured at a specified sampling frequency.

Matric suction can then be deduced indirectly for each TC using a soil-specific calibration curve. To make reasonable comparisons with the JFT measurements, matric suctions deduced by the TCs are expressed in terms of PWP, assuming that the pore-air pressure equals the atmospheric pressure. Figure 8.17 shows the variation of PWP and VWC in response to the two rainstorms from 18 to 22 April and from 5 to 9 June 2008. The PWPs measured by the JFTs have been corrected to account for the difference in elevation head between the ceramic tip and the pressure transducer. Hence, maximum positive PWPs of 5, 15, and 25 kPa can be deduced for the JFT installed at 0.5, 1.5, and 2.5 m bgl, respectively. The matric suction deduced from each TC is temperature corrected. As shown in Figures 8.17a and 8.17c, the PWPs initially decreased from -10 kPa to -45 kPa with depth, while the VWCs increased from 21% to 32% with depth. For the rainstorms of 19 April, the PWPs at all depths increased rapidly to about 0 kPa within 30 min. This rapid advance of the wetting front seemed physically impossible for the colluvium, with an average saturated permeability of approximately 10^{-7} m/s, to percolate to 2.5 m depth within a short period of time. Similar rapid increases of PWP during rainfalls can be observed from other *in-situ* PWP measurements carried out by Gasmo et al. (1999), Ng et al. (2003), and Ng et al. (2011) when using JFTs.

At the peak rain depth on 19 April, positive PWPs were deduced at all depths, approaching the full hydrostatic condition at the ground surface (see Figure 8.17a). Further, the PWPs deduced from the three TCs increased from about -28 kPa to a maximum value of −15 kPa consistently. It is noted that when the soil moisture is too high, the change in the thermal conductivity of the water in the ceramic of each TC becomes indistinguishable. According to the calibration carried out in the laboratory, it is found that when the PWP is less than −60 kPa, the maximum deviation from the calibration curve and the precision of each TC is ±20% and 13 kPa, respectively. Therefore, any measurement made by each TC is expected to be less accurate for PWPs above -60 kPa. On the other hand, substantial but gradual increases in VWC are observed at depths of 0.5 and 1.5 m (see Figure 8.17c). The fairly close agreement with the laboratory-measured saturated VWC (34%) implies that ground above 1.5 m bgl may approach its saturation limit. By contrast, the observed rapid increase in VWC at 2.5 m bgl (from 21% to 36% within 30 min) seems to indicate

Field monitoring and engineering applications 409

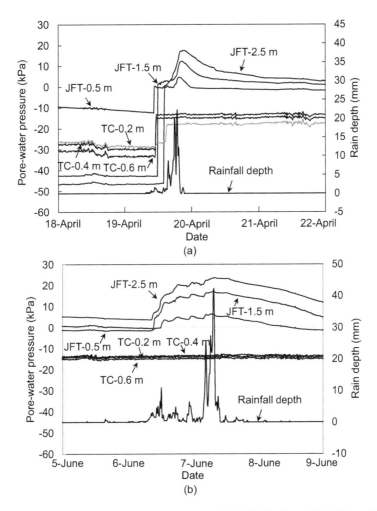

Figure 8.17 Variation of PWP with time during rainstorms (a) 18–22 April and (b) 5–9 June 2008; variation of VWC with time during rainstorms (c) 18–22 April and (d) 5–9 June 2008.

a problematic installation of the TDRs. Air pockets may possibly form if the rods of TDRs are not inserted fully into ground that is stiff or contains frequent gravel, cobbles, and boulders.

During the heavy rainstorms from 5 to 9 June, the responses of PWPs and VWCs generally exhibited similar features to those observed in the previous rainstorms. Significant positive PWPs were deduced at all depths. This full hydrostatic condition at the ground surface is rare but sometimes observed, which again suggests a problematic installation technique for the JFTs. A constant minimum PWP of -15 kPa was deduced by all three TCs during the entire rainstorms. In addition, the increase of VWC at 0.5 and 1.5 m bgl was gradual and limited, attaining a maximum value of about 36.5%.

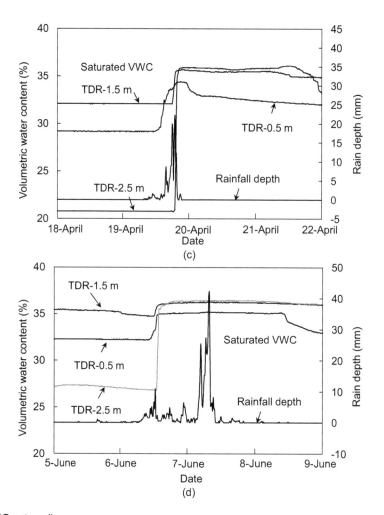

Figure 8.17 (Continued)

Piezometric-level variation

The SPs and CPs were installed at various elevations around the active landslide body to monitor variations in the piezometric level. They aimed to record any response of the groundwater table and any formation of transient perched groundwater tables during rainstorms. Each SP consisted of an open-ended, perforated PVC tube, and the gap between the tube and the borehole wall was backfilled with a gravel filter. The response zone was typically from about 3 to 12 m bgl. Each CP was equipped with a pressure transducer and was protected by a perforated rigid sheath, which was wrapped in a sand filter for backfilling. The response zone of a CP was narrower when compared to SP and commonly ranged from 1.5 to 2 m in height. Of the CPs, CP1, CP2, and CP3 consisted of a shallow device (S) and a deep device (D), which were installed in a single borehole.

Table 8.4 summarises the elevation and the range of the response zone of each SP and CP. The material surrounding each response zone is also shown for reference.

Focusing on the groundwater responses within the active landslide body, SP-BH3, CP-BH6, CP2S, CP2D, CP3S, and CP3D are selected for detailed assessment in this section. The variation in piezometric level, i.e., elevation head plus pressure head, during the rainstorms from 18 to 22 April and from 5 to 9 June 2008 are shown in Figure 8.18. It should be noted that any change of piezometric level indicates a positive change of the PWP head. As shown in Figure 8.18a, at the peak rain depth on 19 April, the shallower device CP2S recorded a substantial increase of piezometric head of 2 m while the deeper device CP2D did not respond. On the other hand, the piezometric head in SP-BH3 increased by 4 m and gradually dropped back to its tip level within 1 day. CP3D, which was installed at nearly the same elevation as SP-BH3 (see Table 8.4 and Figure 8.16b), recorded consistent and similar variations but with less increase in piezometric head (about 2.5 m). However, surprisingly, the shallower device, CP3S, showed a negligible response, indicating a complex hydrogeological regime in the hillslope.

The groundwater flow mechanism might be affected primarily by the geological setting of this particular hillslope and is discussed in detail later. During the heavy rainstorms from 5 to 9 June, all of the devices typically showed a significant increase of piezometric head which dropped gradually after the peak rain depth (see Figure 8.18b). In particular, both CP2S and CP3S recorded a peak increase of piezometrical head of 3 m while CPBH6 exhibited a gradual and limited increase of 2 m. SP-BH3 and CP3D again showed comparable piezometric head increases of 6 m in the deep regions because of their similar installation depth. These dramatic increases in water pressure may also be affected by the complex geological setting of the natural hillslope.

Subsurface total horizontal stress

Two pairs of EPCs were installed near the main scarps (EPC1) and in the central portion of the landslide body (EPC2) to monitor the total horizontal stress at 2 m depth (see Figures 8.2 and 8.3a). For each pair, the variations in total horizontal stress in both the down-slope (A) and the cross-slope (B) direction were recorded. Each EPC was inserted in a narrow slot, which was slightly oversized at the base of a trial pit at 2.5 m bgl. The resulting void between the CDT and the sensor was then backfilled with cement-bentonite grout.

Figure 8.19 shows the variation of *in-situ* total horizontal stress with time in response to the rainstorms from 18 to 22 April and from 5 to 9 June 2008. The deduced positive PWPs recorded by JFT-1.5 m are also shown for comparison. Initially, a few negative values (i.e., tensile stresses) of 0.5 kPa were recorded, which was probably because of stress alteration during installation and (or) shrinkage of the grout. Nevertheless, these negative readings are apparent, because the accuracy of EPCs is ±0.35 kPa (Geokon, 2007a). As shown in Figures 8.19a and 8.19b, the measured total horizontal stress of EPC1A and EPC1B recorded very similar variations during the two rainstorms. Peak total horizontal stresses of 12 and 18 kPa were attained at the peak rain depths on 19 April and 7 June, respectively. The stress then dropped gradually and stabilised at a slightly higher pressure as both the rainstorms ceased. Moreover, it can be seen that the measurements made by EPC1A and (or) EPC1B were reasonably close to the positive PWPs deduced from JFT-1.5 m.

412 Advanced Unsaturated Soil Mechanics

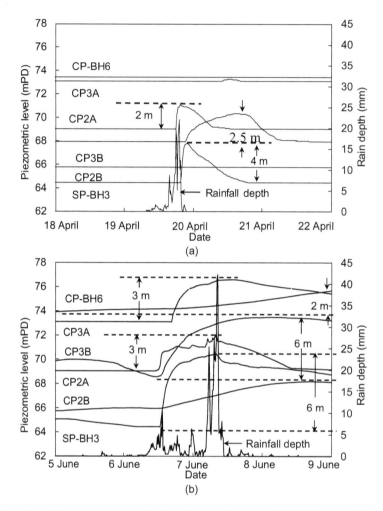

Figure 8.18 Variation of piezometric level with time during rainstorms (a) 18–22 April and (b) 5–9 June 2008.

Theoretically, it is anticipated that a decrease in matric suction would result in consequent elastic soil swelling. However, the above observations seem to suggest that the increases in stress near the main scarps were probably caused by the increase in positive PWPs instead of the earth pressure itself. In other words, the backfilled cement-bentonite grout may possibly strengthen the ground and thus reduce its compressibility upon matric suction changes.

In the central portion of the active landslide body, the trend of the measured total horizontal stress at both orientations (EPC2A and EPC2B) was similar, and the magnitude in the down-slope direction was consistently larger than that in the cross-slope direction after rainstorms (see Figures 8.19c and 8.19d). At peak rain depths on 19 April and 7 June, the maximum total horizontal stresses in the down-slope direction were 18 and 28 kPa, respectively. The stress difference between orientations increased from

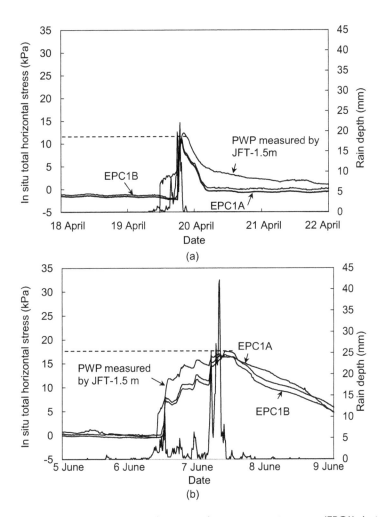

Figure 8.19 Variation of *in-situ* total horizontal stress with time near main scarps (EPC1) during rainstorms (a) 18–22 April and (b) 5–9 June 2008; variation of *in-situ* total horizontal stress with time at central portion of landslide body (EPC2) during rainstorms (c) 18–22 April and (d) 5–9 June 2008.

about 0 to 7 kPa on 19 April and from 5 to 17 kPa on 7 June. The stresses recorded by both EPC2A and EPC2B dropped steadily as the rainstorms ceased. Moreover, it can be seen that the variations in the total horizontal stress recorded by both EPC1B and EPC2B are close to the positive PWPs deduced from JFT-1.5 m (see Figure 8.19b and 8.19d). This suggests that the stress changes in the cross-slope direction in the active landslide body probably originated in the increase of positive PWPs. On the other hand, the larger stress changes recorded by EPC2A in the down-slope direction indicate either significant ground movement or the presence of preferential groundwater flow in the down-slope direction.

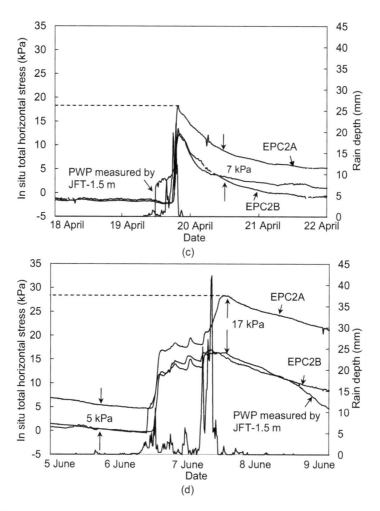

Figure 8.19 (Continued)

Subsurface horizontal deformation characteristic

Two IPIs (IPI1 and IPI2) were installed around the central portion of the active landslide body to measure subsurface horizontal displacement. Four tilt sensors were arranged suspended "in-place" along each inclinometer casing at the ground surface (i.e., 0 m bgl) and at 1, 3, and 5 m bgl. The horizontal displacement at 7 m bgl was assumed to be zero. The tilt sensors of each IPI recorded the inclination angle resulting from ground deformation continuously, with an accuracy of 0.0158° (Geokon, 2007b). As with the EPCs, horizontal displacements in both the down-slope and the cross-slope direction were monitored.

Figure 8.20 shows the down-slope and cross-slope horizontal displacement profiles of IPI1 and IPI2 on key days. The numbers shown in parentheses indicate the peak rain depth during the day. The estimated colluvium-CDT interface and the location of the inferred rupture surfaces as revealed by the ground investigations (refer to Figure 8.16a) are shown on the displacement profiles for comparison. During the rainstorms from 18 to 22 April, the ground generally exhibited "cantilever" mode deformation in the downslope direction. The peak displacement change at the ground surface was approximately 3 mm (see solid triangular symbol in Figure 8.20a and 8.20b). This deformation characteristic is consistent with the inferred shallow, translational-slide type of failure, where the rupture surfaces are just below the interface between the colluvium and the CDT.

In response to the heavy rainstorms from 5 to 9 June, significant down-slope ground movements were recorded at both IPI1 and IPI2 (see open triangular symbol in Figures 8.20a and 8.20b). In particular, IPI2 measured a nearly irrecoverable displacement of 40 mm, and the displacement profile was fairly uniform with depth. There is hence a possibility

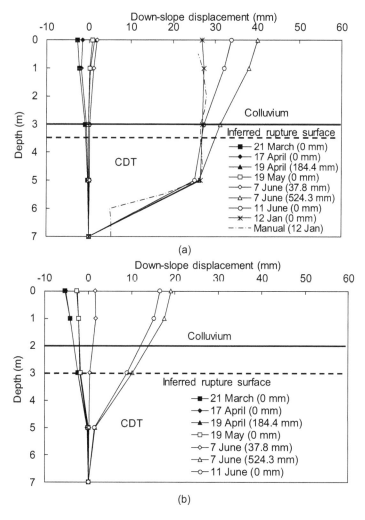

Figure 8.20 Down-slope displacement profiles of (a) IPI2 and (b) IPI1; cross-slope displacement profiles of (c) IPI2 and (d) IPI1 at central portion of active landslide body.

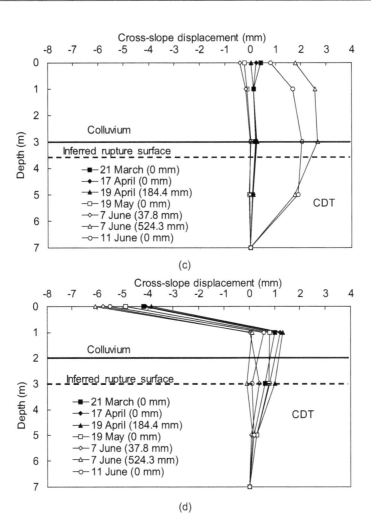

Figure 8.20 (Continued)

that movement in the upper soil profile is dragging the lower portion of the casing. The assumption of zero horizontal displacement at 7 m bgl may be questionable and needs to be examined in these circumstances. A conventional manual inclinometer survey was therefore undertaken to record the "real" deformation profiles accurately. The manual measurement was made on 9 January 2009 and was compared with the displacement recorded on the same day (see dotted line and cross symbol in Figure 8.20a). The comparison provides evidence that well-defined rupture surfaces have possibly developed at 5 m bgl or deeper because of the relatively large displacement between 5 and 7 m. In contrast, the ground displacement recorded by IPI1 exhibited a cantilever mode of deformation when the peak rain depth occurred on 7 June (see Figure 8.20b). The peak displacement increment at the ground surface was 25 mm. In contrast with the ground movements in the down-slope direction, the displacement change in the cross-slope direction appears to be limited and less sensitive to rainstorms (see Figures 8.20c and 8.20d). The ground generally tended to deform in the easterly direction, where a maximum displacement of 3 mm took place near the colluvium-CDT interface (see open circle and open triangle symbols in Figure 8.20c). Consistently with the observations recorded by the groundwater monitoring devices (see the variations in SP-BH3, CP3S, and CP3D in Figure 8.20), this unusual deformation characteristic might be affected by the complex groundwater flow and geology of the active landslide body. The detail is discussed in the next section.

Investigation of hillslope behaviour

In an attempt to capture the hillslope behaviour and the landslide-triggering mechanism in response to precipitation, a site-specific geological and hydrogeological model are established in this section. The field observations during the heavy rainstorms from 18 to 22 April and from 5 to 9 June 2008 are interpreted and explained. The site-specific infiltration and deformation characteristics of the hillslope are then identified.

A preliminary geological and hydrogeological model based on previous ground investigations, field mapping, and aerial photograph interpretation has been set up and was reported and described above. The simplified soil profile, initial groundwater table, and the inferred rupture surfaces are shown in Figure 8.16. Before the rainstorms, the ground above 1.5 m bgl was wet (see Figure 8.17c). This probably resulted from the considerable amount of preceding rainfall from January to March 2008. For ease of discussion, the hillslope responses and some observed key features at the peak rain depth on 7 June are illustrated schematically in Figure 8.21.

Infiltration characteristic and groundwater condition

Following rainstorms, significant increases in piezometric head and total horizontal stress at 2 m bgl were typically and consistently observed (see the variations in CP2S and CP3S in Figure 8.18 and variations of EPC1 and EPC2 in Figure 8.19). Positive PWPs as high as 20 kPa were frequently recorded. The VWCs also showed a gradual but limited increase to reach the *in-situ* saturation limits (about 36%) in shallow regions (see Figures 8.17c and 8.17d). As revealed by the geophysical survey, a lens of stiff material was found to overlay the ground surface of the hillslope. This indicates the presence of corestones or fractured decomposed rock material as a result of past mass wasting processes. Based on these field observations and measurements, shallow (i.e., top 3 m) transient perched groundwater tables are believed to have developed within the bouldery colluvial deposit in the landslide body (see Figure 8.19). This is consistent with the observation from the IP test that positive PWPs were recorded near the colluvium-CDT interface, probably because of the significant difference in unsaturated water permeability between the two strata (Ng et al., 2011).

During the rainstorms between 18 and 22 April, the significant increase of about 4 m (40 kPa of positive PWP) in the piezometric head at 10 m bgl (see the variations in SP-BH3 in Figure 8.18a) implies a possible rise in the main groundwater table, resulting from the continuous advance of rainwater in the subsurface. As shown in Figure 8.16b, a shallow decomposed rock stratum dipping in the easterly direction was identified across the hillslope at +75 mPD. The unusually complex geological setting of this particular hillslope may affect its hydrogeological regime and hence the groundwater flow mechanism upon

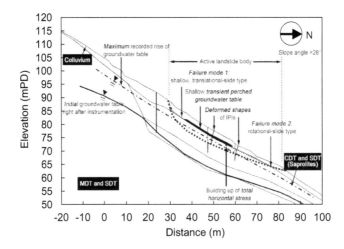

Figure 8.21 Hillslope responses at peak rain depth on 7 June 2008.

infiltration. It may be reasonable to postulate that a portion of the rainwater infiltering from the western flank may flow towards the central portion of the landslide body on top of this dipping rock head profile, resulting in possible cross-slope groundwater flow (see the arrows in Figure 8.16b).

As indicated by the close field- and laboratory-measured VWC, the ground is believed to have been significantly wetted after the rainstorms in April (see Figure 8.17d). The increase in PWP (or decrease of matric suction) (see Figure 8.17b) increased hydraulic flow paths in the ground, and its water permeability would hence increase. This " improvement" of ground permeability would thus allow more rainwater to seep into the subsurface during the subsequent rainstorms from 5 to 9 June. As a result, at the peak rain depth on 7 June, the main groundwater table probably rose by 6 m from its initial position (see the variations of the SP-BH3 and CP3D in Figure 8.18b), approaching the colluvium-CDT interface (see Figure 8.21).

Deformation characteristics and failure modes

The increases in total horizontal stress (see Figures 8.19c and 8.19d) resulting from the build-up of positive PWPs in shallow regions (see the variations in the three JFTs in Figures 8.17a and 8.17b and variations of CP2S and CP3S in Figures 8.18a and 8.18b) during rainstorms would probably result in soil deformation. According to the measured horizontal displacement, two distinct modes of deformation were identified in this hillslope (see Figures 8.20a and 8.20b). At the peak rain depth on 19 April, a cantilever mode of deformation with limited displacement (about 3 mm at the ground surface) was observed. This deformation characteristic suggests that the inferred shallow translational-slide type of movement along poorly defined rupture surfaces at 3–5 m bgl, which is near the colluvium-CDT interface, is probably reactivated (failure mode 1 in Figure 8.21). In response to the heaviest rainfall on 7 June, well-defined rupture surfaces have possibly developed at 5 m bgl or deeper in the central portion of the landslide body. A nearly irrecoverable lateral displacement of 40 mm was recorded at the ground surface (see Figure 8.21a). This observed "deep-seated" mode of deformation probably results from the significant increases in total horizontal stress (maximum 28 kPa) in the down-slope direction (see the variation in EPC2A in Figure 8.19d and Figure 8.21).

By comparing the measured data on 7 June 2008, 11 June 2008, and 12 Jan 2009 as shown in Figure 8.20a, it is interesting to note that some recoverable displacements were recorded after the heavy rainstorm on 7 June 2008. Similarly, some partially recoverable displacements are also demonstrated in Figure 8.12 showing the field monitoring of an unsaturated expansive soil slope. The deep-seated mode of slope movement implies that the landslide mass underwent another distinct failure mode. Considering field-observed discontinuities such as relict joints and foliations in the subsurface, a rotational-slide type of failure may have been triggered (failure mode 2 in Figure 8.21). According to the classification system for types of slope failure by Varnes (1978), the combined failure mode (failure modes 1 and 2 in Figure 8.21) of the landslide body may be described as multiple, translational-slide, and rotational-slide. Apart from the effects of stress changes following precipitation, the existence of the unusually complex geological setting may also influence the deformation characteristics of the landslide body. The possible cross-slope groundwater flow on the shallow, dipping decomposed rock head regime at +75 mPD (see Figure 8.16b) might result in notable easterly horizontal ground movements. The maximum change in displacement of about 3 mm took place at 3 m bgl, which is close to the colluvium-CDT interface (see Figure 8.20c). The measurement suggests that the landslide body tended to slide in the easterly direction upon failure.

Summary

Based on the field monitoring, the following observations may be summarised:

- Except for the measurements made by the JFTs and the TDR installed at a depth of 2.5 m, all the instruments provided generally reliable and good-quality data. The measurements show a strong correlation between instruments. During the rainstorm of 19 April (peak rainfall intensity of 62 mm/h), the VWC of the overlying colluvium attained an *in-situ* saturation limit of 36%. Shallow pressure transducers installed at 2 m below ground level frequently recorded positive PWPs up to 20 kPa, suggesting that transient perched groundwater tables may have developed within the bouldery colluvial deposit in the top 3 m. The increases of total horizontal stress resulting from the increasing positive PWPs resulted in slope movements. A cantilever mode of ground deformation was observed. This kind of deformation characteristic implies that the shallow translational-slide type of failure (failure mode 1) of the landslide body was probably reactivated. It slid along poorly defined rupture surfaces at 3–5 m in depth, near the colluvium-CDT interface.
- When the rainfall reached its peak intensity on 7 June (133.5 mm/h), the main groundwater table probably rose by 6 m, approaching the colluvium-CDT interface. The groundwater flow mechanism of this particular hillslope may have been affected by its complex geological setting, where a shallow decomposed rock stratum dipping towards the easterly direction was identified across the hillslope. Cross-slope groundwater flow in the central portion of the landslide body may have been possible.

- Because of the significant increase in total horizontal stress (maximum value of 28 kPa on 7 June) in the down-slope direction, a deep-seated mode of slope movement resulted. A nearly irrecoverable displacement of 40 mm was recorded at the ground surface, and the displacement profile was fairly uniform with depth. Well-defined rupture surfaces possibly developed at about 5 m below ground or deeper. Considering the discontinuities like relict joints and foliations in the subsurface identified *in-situ* this kind of slope movement indicates that a rotational-slide type of failure may have been triggered (failure mode 2). The combined failure mode (failure modes 1 and 2) may be described as multiple, translational-slide, and rotational-slide types.
- Based on the interpreted field monitoring results, groundwater flow and failure mechanisms are postulated. It will be useful to carry out a comprehensive three-dimensional coupled seepage-stress-deformation analysis to confirm the findings in future.

FIELD PERFORMANCE OF NON-VEGETATED AND VEGETATED THREE-LAYER LANDFILL COVER SYSTEMS USING CONSTRUCTION WASTE WITHOUT GEOMEMBRANE (NG ET AL., 2016B, 2019D, 2022B)

Introduction

As population and urbanisation increase, the production of municipal solid waste (MSW) also increases and is a global concern. In many countries, developing countries in particular, landfill is considered to be the simplest, cheapest and most cost-effective way to dispose of MSW. However, even in developed countries, MSW is landfilled. For instance, in the European Union, more than half of the member states still dispose of in excess of 50% of their waste in landfill (EEA, 2013). In the United States, 50% of total waste generated is also disposed of in landfills (USEPA, 2015). The closure standards for MSW landfills require owners/operators to install a final cover system to minimise downward migration of water into the waste, known as percolation, so as to prevent substantial leachate generation and groundwater contamination. All landfill covers are unsaturated. To satisfy this standard, most modern landfill cover systems use geotextile composites and geomembranes, which are supposed to be "impermeable" when they are in perfect condition. Albright et al. (2013) presented field data from seven large-scale test sections simulating landfill covers with composite hydraulic barriers (a geomembrane over a soil barrier or geosynthetic clay liner) in climates ranging from cool and humid to warm and arid. The annual percolation through the cover at the wettest site (Cedar Rapids, Iowa; average precipitation was 915 mm/yr) ranged between 0.1 and 6.2 mm/yr. The recommended equivalent percolation rate for covers with composite barriers is 3 mm/yr for humid climates. However, geomembranes are highly susceptible to the stability of the interface and defects or holes which compromise their water and gas permeability and reliability (Daniel, 1994; Koerner and Daniel, 1997; Amaya et al., 2006).

It is also common to rely on naturally occurring low-permeability materials such as clays. Typically, regulations (US EPA, 1993) require that a prescribed cover must have a saturated permeability of less than 10^{-9} m/s; that equates to 30 mm/yr of percolation if the barrier is continuously wetted with a hydraulic gradient of 1.0. However, unprotected clay barriers are prone to desiccation, induced cracking, which can compromise their integrity (Melchior, 1997; Albright et al., 2006).

Melchior (1997) reported that clay barriers in a cool and wet climate leaked 8 to 9% of the precipitation; he noted that at the end of an 8-year experiment, leakage rates were increasing. Albright et al. (2006) evaluated the performance of compacted clay barrier covers at three sites over 2 to 4 years. The climate at the sites was arid in California, humid in Iowa, and subtropical in Georgia. The as-built permeability of the clay barrier layers varied between 1.6×10^{-10} and 4.0×10^{-10} m/s. During the test period, the water permeability of the barriers increased by up to 800 times the as-built value. They concluded that large increases in the water permeability of clay barriers with time caused by desiccation cracks are not uncommon. Therefore, alternative covers are considered and used.

Alternative covers are defined as any cover used in place of the prescribed covers (US EPA, 1993). Current regulations require that alternative covers should be equivalent to prescribed covers in terms of their effectiveness in minimising water percolation. For instance, Benson et al. (2001) proposed the equivalency criterion of 30 mm/yr that is currently used to assess alternative covers. Most alternative covers rely on water storage principles (i.e., controlling percolation by storing water during periods of high precipitation and evapotranspiration during periods of low precipitation) and are often referred to as evapotranspirative (ET) covers and are found to be suitable for arid, semi-arid climates (Hauser et al., 2001; Albright et al., 2004; Bohnhoff et al., 2009) and sub-humid climates (Barnswell and Dwyer, 2011; Mijares and Khire, 2012). A cover with capillary barrier effect (CCBE), e.g., a layer of fine-grained soil (silt, clay) over a coarse geomaterial (sand, gravel, nonwoven geotextile), is sometimes added to increase the water storage capacity of the cover (Ross, 1990; Khire et al., 2000; Iryo and Rowe, 2005; Bouazza et al., 2006; McCartney and Zornberg, 2010; Siemens and Bathurst, 2010; Zornberg et al., 2010; Rahardjo et al., 2012). Several water infiltration column tests have been conducted to study the behaviour of CCBE under controlled laboratory conditions (Yang et al., 2006; Bathurst et al., 2007; McCartney and Zornberg, 2010; Rahardjo et al., 2012). These experimental studies have clearly shown the development of a capillary break which minimises the amount of water that can flow through the interface from the fine-grained soil into the coarse geomaterial until the overlying soil is nearly saturated. Field studies have also shown that employing capillary barriers can minimise percolation into underlying waste or contaminated soil effectively in arid and semi-arid regions (Benson and Khire, 1995; Khire et al., 1999; Khire et al., 2000; Zornberg et al., 2010). Although more attention has been paid to CCBEs as an alternative cover system in semi-arid and arid regions, the performance of CCBEs in humid climates has so far been unsatisfactory (Morris and Stormont, 1999; Khire et al., 2000; Albright et al., 2004; Rahardjo et al., 2006). These experimental studies report that there is a significant increase in the moisture content of the CCBE during periods of high precipitation and low ET, leading to water breaking through the barrier and the subsequent production of a large amount of leachate.

In humid climates such as Hong Kong, Malaysia, Philippines, Singapore, Cardiff city in the United Kingdom and some states in the U.S., that is, Hawaii, Louisiana, Alabama and Florida where an annual rainfall of over 1200 mm is not uncommon (World Bank, 2014), water is expected to break through the conventional two-layer CCBE cover.

In order to address this issue, a new three-layer landfill cover system based on the theory of unsaturated soil mechanics is reported to improve a capillary barrier for humid climatic conditions.

Theoretical considerations of newly proposed landfill cover

This new system consists of a fine-grained soil layer, such as clay, added beneath a two-layer CCBE cover (see Figure 8.22). It is intended and anticipated that the bottom clay layer will be protected from desiccation during dry seasons by the upper two soil layers. The feasibility and effectiveness of this proposed three-layer cover system are investigated by theoretical examination and by conducting a one-dimensional water infiltration test. The experiment is back-analysed and the computed results are compared with measured data. Moreover, a numerical parametric study is carried out to investigate the effectiveness of the proposed three-layer system if desiccation cracks form in the clay layer beneath the CCBE cover (Ng et al., 2016b). As shown in Figure 8.22a, the CCBE cover contains two soil layers which are a fine-grained soil layer overlying a coarse-grained soil layer. It relies on the capillary barrier effects between these two layers to prevent water infiltration. The infiltration mechanism is discussed further in the next section. By comparison, the proposed landfill soil cover is a three-layer cover system, which consists, of a compacted clay layer at the bottom, a coarse-grained layer and a fine-grained layer at the top, compacted successively from the bottom to the top of the system, as shown in Figure 8.22b. According to the water permeability functions illustrated in Figure 8.23, by introducing a compacted clay layer beneath a CCBE, water that infiltrates through the upper two layers can be intercepted and reduced by the bottom clay layer which has lower water permeability at high degrees of saturation (i.e., low suction) in a humid climate. On the other hand, the bottom clay layer is protected by the upper two soil layers from desiccation during dry seasons because the upper two soil layers have low water

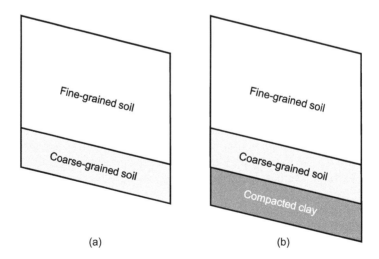

Figure 8.22 Conceptual diagrams of landfill covers: (a) Conventional capillary barrier landfill cover; (b) Newly proposed landfill cover.

Field monitoring and engineering applications 423

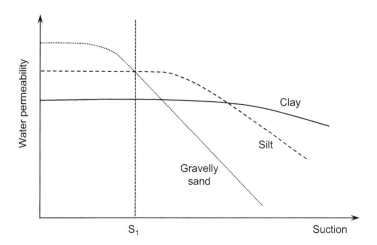

Figure 8.23 Schematic diagram showing water permeability functions of silt, gravelly sand and clay.

permeability at high suctions (i.e., low relative humidity). However, it should be noted that the clay layer of the proposed design may not be completely immune from desiccation because of vapour flow, which is favoured by the overlaying dry coarse layer and root intrusion.

In contrast to the regulatory requirements (US EPA, 1993) for a conventional three-layer compacted clay cover, the functionalities of the upper two layers of this proposed landfill cover are fundamentally different. As stated previously, the proposed upper two layers function as a CCBE cover, while the upper and lower layers of a conventional compacted clay system act as a vegetation support layer and a drainage layer, respectively, which may enhance desiccation by roots and water drainage. The upper layer of a conventional three-layer compacted clay cover is not necessarily designed as a CCBE. On the contrary, the bottom clay layer of the newly proposed landfill cover system is protected by the upper two layers to minimise desiccation. A comparison between these two types of cover is summarised in Table 8.5. Further, this new three-layer landfill cover system has been granted a U.S. patent (Ng et al., 2015d) and a Chinese invention patent (Ng et al., 2016f).

Fundamental principle of reducing water infiltration and percolation

Percolation is the process by which water migrates through a whole soil profile such as, in this case, a landfill cover. Percolation is closely related to but not equivalent to infiltration, which relates to the breakthrough of water into the soil through pores and/or cracks in the surface. Figure 8.23 shows a schematic diagram illustrating the relationship between the water permeability and matric suction of each soil layer. When a soil becomes drier and the water content decreases, the suction and water permeability increase and decrease respectively.

Table 8.5 Comparison between the traditional and new three-layer landfill cover for preventing water infiltration

	Working principle for preventing water infiltration	
	Traditional three-layer compacted clay cover	*Novel three-layer landfill cover*
Upper layer	**Vegetation support layer** – to retain water for growth of vegetation. The purposes of vegetation are to decrease runoff, increase evapotranspiration and minimise erosion.	**Vegetated Capillary barrier** – a fine-grained soil layer overlying a coarse-grained soil layer to minimise the downward movement of water through a capillary barrier effect. The principle is based on the contrast of unsaturated hydraulic properties (soil-water retention curves and permeability functions) of each layer. Infiltrated water is stored in the upper layer and ultimately removed by evaporation, evapotranspiration, and/or lateral drainage at arid and semi-arid regions. The presence of this two-layer capillary layer also protects the bottom clay from shrinkage cracks under arid and semi-arid climates. After water breakthrough, the coarse-grained soil acts as a drainage layer.
Intermediate layer	**Drainage layer** – to prevent ponding of water on the lower layer. Drains by gravity to toe drains.	
Lower layer	**Clay and/or geomembrane** – to prevent infiltration of water into the waste by utilising low-permeability materials. Since the clay layer is not protected by a capillary barrier, shrinkage cracks will occur under arid and semi-arid climates. This means that for heavy rainfall, the lower layer will not be effective in preventing infiltration. Typically, a geomembrane is installed overlying the compacted clay (Qian et al. 2001). However, geomembranes are highly susceptible to interface instability and defects/holes.	**Clay layer** – to prevent infiltration of water into the MSW in an event that the upper capillary barrier of the three-layer system fails in humid regions such as Singapore, Hong Kong and Brazil where annual rainfall is over 1200 mm. Clay has an inherently low saturated permeability. **No geomembrane is needed even under humid climates**

Source: Ng et al. 2016b.

When soil suction in the three-layer landfill cover is above point S_1 (shown in the figure), i.e., in semi-arid or arid climates, the cover soils are relatively dry. The water permeability of the silt layer is much greater than that of the gravelly sand layer. Infiltrated water is stored in and flows away through the silt layer, but no water infiltrates the gravelly sand layer. In other words, the two-layer CCBE cover works.

When soil suction in the landfill soil cover is below point S_1 (shown in the figure) under heavy or prolonged rainfall, i.e., in humid climates, cover soils are nearly saturated or saturated. The water permeability of the gravelly sand layer is the highest while that of the clay layer is the lowest. The capillary barrier effect formed by the upper silt layer and underlying gravelly sand layer will lose its function and water infiltrates the gravelly sand layer, since the water permeability of the gravelly sand layer is greater than that of the silt layer. At this point, the infiltrated water is stopped by the clay layer by its lowest water permeability and may drain away through the gravelly sand layer because of its relatively high saturated permeability. In this way, the head of water on the underlying clay layer is reduced and the amount of water percolation will be minimised thereby.

The addition of the compacted clay layer beneath the CCBE cover makes the proposed landfill cover applicable in any weather conditions. It should be noted that the focus of this study is to gain a fundamental understanding of the proposed landfill cover system in terms of water infiltration when water breaks through the CCBE cover.

Long-term field test of the new vegetated cover system using recycled concretes without geomembrane

Introduction

To promote environmental protection and sustainability, the use of plants and recycled waste in geotechnical constructions such as landfill covers is recommended. Without the use of geomembrane, Ng et al. (2016b) and Ng et al. (2015e) evaluated the performance of this three-layer landfill cover system, silt overlying gravelly sand overlying clay, by carrying out water infiltration tests in a one-dimensional soil column and two-dimensional flume model, respectively. Their results showed that this three-layer landfill cover system can certainly perform satisfactorily in humid regions under extreme rainfall conditions. However, the soils used for the previous studies (Ng et al., 2015a, 2016b) were artificial and idealised soils which would not typically be used for landfill covers. The use of pure kaolin clay as the low-permeability soil layer is impractical as it may not be readily available in the large amounts required to construct a landfill cover, aside from it being expensive and difficult to construct. Thus, a locally available soil with sufficiently low permeability is recommended for the lowest layer. Further, to promote sustainability, the use of recycled construction waste in geotechnical constructions is highly recommended (Basu et al., 2015). We thus propose that recycled construction waste materials should be used for at least one of the components of the novel three-layer landfill cover system. The use of recycled waste materials can help to reduce the amount of waste deposited in landfills, preserve our natural resources and help to alleviate the burden of landfill volume storage.

To further promote an environmentally friendly and sustainable landfill cover system, the use of grass is required after landfill cover placement. Grass extracts moisture from the landfill cover soil by evapotranspiration and hence influences soil suction and water infiltration (Sinnathamby et al., 2014; Ng et al., 2014b). Studies have been conducted to quantify the suction induced by plants in the field (Pollen-Bankhead and Simon, 2010; Smethurst et al., 2012) and in the laboratory (Ng et al., 2014b; Garg et al., 2015) and their effects on water infiltration. Some field studies have found that as compared to bare ground, the infiltration rates in vegetated ground were lower because of the presence of plant roots (Meek et al., 1992; Mitchell et al., 1995; Leung et al., 2015). However, most previous studies were carried out in a single-layer uniform soil. Hence, the effects of grass on the distribution of suction during drying and wetting in a layered soil such as the three-layer landfill cover system should be investigated.

A full-scale three-layer landfill cover system was constructed in Shenzhen, China. The main objective of this section is to evaluate the field performance of a novel vegetated three-layer landfill cover system using recycled crushed concrete without a geomembrane

Plate 8.7 The aerial views of the field test plot at the Xiaping landfill, Shenzhen, China.

at a humid site. Unsieved completely decomposed granite (CDG) and coarsely crushed concrete (CC) were used for the top layer and intermediate layer while sieved completely decomposed granite (CDG) was used as the low-permeability soil layer. One section of the field test plot was planted with *Cynodon dactylon* (Bermuda grass) while the other section was left bare for comparison. Field monitoring over a duration of 54 months (from 1 June 2016 to 1 January 2021) was carried out at a full-scale field site in Shenzhen Xiaping landfill, China. One-half of the sloping field cover was vegetated with *Cynodon dactylon*, while the other half remained without vegetation for comparison. Pore-water pressure (PWP), volumetric water content (VWC) and percolation were monitored continuously. In addition, numerical analysis was carried out to evaluate the significance of plants on the water balance in the cover system (Ng et al., 2020e). The life cycle analysis of this type of landfill cover system focusing on carbon neutrality was carried out by Ng et al. (2024b).

Site description

The field test plot evaluated in this study was constructed at the Xiaping landfill, which is located in Shenzhen City, China. The landfill began operating in October 1997. Currently, the Xiaping landfill, having a total area of 149 ha, is Shenzhen's biggest landfill and handles about 30% of the city's total waste (approximately 5000 t/day). The maximum height of the waste body is about 40 m. So far, the landfill has already reached its third phase. A 20 m long × 12 m wide embankment was selected in the west part of the Xiaping landfill to construct a three-layer landfill cover test plot (Plate 8.7). The slope inclination is 1.7H/1 V (H and V refer to horizontal and vertical, respectively). The age of the waste fill layer under the test plot was between 1 and 2 years at the time of construction.

The field test plot is located in a humid subtropical climate region, with approximately 80% of rainfall occurring between May and September. The annual cumulative rainfall and potential evapotranspiration (PET) are approximately 1900 mm and 1050 mm. The annual PET was estimated using the Penman-Monteith equation (Allen et al., 1998). The site is considered to have a humid climate since the ratio of annual cumulative rainfall to PET is greater than 0.75. The lowest monthly mean relative humidity during the driest months (i.e., October to February) is 64%. This implies that care must be taken when selecting the top layer of soil. Soils that are resistant to desiccation cracking should be selected. Silty sands, sandy silts and clayey sand are probably suitable for the top layer of soil (Khire et al., 2000).

Figures 8.24 and 8.25 show the plan and cross section view of the field test plot, respectively. The landfill cover consisted of three layers, 0.8 m thick sieved CDG, 0.4 m thick CC and 0.6 m thick unsieved CDG from the bottom to the top. The soils were compacted at their optimum moisture content as given in Table 8.6. As suggested earlier, coarsely crushed concrete (CC) was used as the intermediate layer of the novel three-layer landfill cover system to promote sustainability and alleviate the burden of landfill volume storage. A geotextile was placed between the fine- and coarse-grained layers to minimise fines migration into the CC layer. The slope had a total width of 12 m, a length of 20 m and an inclination of 30°. Haeri et al. (2000) studied the shear strength parameters of a sand-geotextile interface using direct shear tests. The measured friction angle of the interface was about 36 degrees. Since the soil used in this study is coarser, slope instability is unlikely

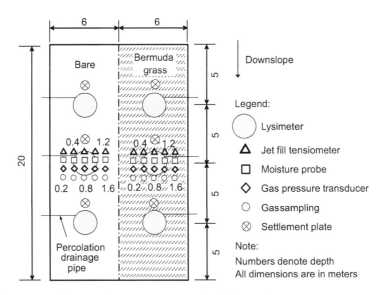

Figure 8.24 Plan view and layout of instrumentation in the field test plot.

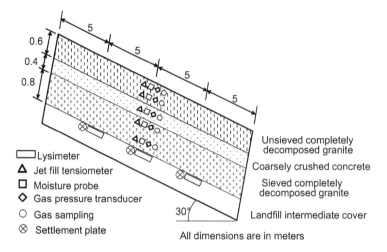

Figure 8.25 Typical cross section view and layout of instrumentation in the field test plot.

Table 8.6 Basic properties of CDG and construction waste

Property	Unsieved CDG	Sieved CDG	CC
Unified soil classification system	SC	SC	GP
Specific gravity, G_s	2.63	2.61	2.45
Atterberg limits			
Liquid limit, LL	37	37	-
Plastic limit, PL	20	20	-
Plasticity Index, PI	17	17	-
Standard compaction curve			
Maximum dry density, ρ_d: kg/m³	1860	1820	1890
Optimum moisture content: %	12.6	14.4	-
Saturated water permeability, k_s: m/s	5.7×10^{-5}	8.1×10^{-8}	7.5×10^{-2}

Notes:
CDG, completely decomposed granite; CC, coarsely crushed concrete.

for the 30-degree slope. Half of the field test plot (6 m width) was planted with Bermuda grass turf while the other half was left bare. The bare slope was then covered with a white nonwoven geotextile to prevent excessive surface erosion by reducing surface runoff (Bhattacharyya et al., 2010). Since the saturated permeability of geotextile (i.e., 7×10^{-2} m/s) was significantly higher than that of the uppermost soil layer (i.e., 5.7×10^{-5} m/s), water infiltration was not affected. The grass species, *Cynodon dactylon* (also known as Bermuda grass) is a warm-season grass species widely cultivated in many parts of Asia. This grass species has high drought tolerance and is commonly used for slope greening and ecological restoration (Hau and Corlett, 2003; Hu et al., 2010).

Although no geomembrane is used in the novel three-layer landfill cover system, the thickness of the lowest layer (i.e., 0.8 m thick) falls within the typical range (0.45 m to

0.9 m) of compacted fine-grained soil found in composite barriers (i.e., geomembrane overlying compacted fine-grained soil) (Albright et al., 2004; Loehr and Haikola, 2003).

The test plot was constructed at the Shenzhen Xiaping landfill in October 2015. The soil layers were placed in 200 mm thick lifts using a mini roller compactor having a compaction energy of 85 N m. Six specimens were randomly sampled after each two lifts were compacted in order to inspect the *in-situ* dry density. The dry density was determined using core cutters of known volume. The *in-situ* dry density of the sieved CDG, CC and unsieved CDG ranged from 92% to 98% of the standard Proctor maximum dry density (see Table 8.6).

Material properties

The CDG soil used to construct the three-layer landfill cover system was excavated from a slope near the test plot. For the low-permeability layer, the CDG soil was sieved to recover only the soil fraction less than 10 mm. The recycled CC was donated by a recycling plant in Shenzhen and delivered to the Xiaping landfill. The recycling plant in Shenzhen initially passes large concrete rubble through a primary jaw crusher which reduces the size to less than 250 mm. Impurities such as wood, paper and plastics are then removed manually while on the conveyor belt. The concrete rubble is then crushed using two secondary cone crushers, and sorted into 40 mm, 20 mm, and fine fractions (<10 mm) through a set of sieves. The market price of crushed concrete is about RMB 20 per ton, irrespective of the particle sizes.

Figure 8.26 shows the particle size analyses which were obtained from the sieve analysis described in the American Society for Testing and Materials D422 (ASTM, 2007). Specific gravity tests were performed in accordance with ASTM D854 (ASTM, 2010a). Atterberg limit tests were performed using ASTM D4318 (ASTM, 2010b). Based on their basic properties, the soils were classified in accordance with the Unified Soil Classification System (USCS) using ASTM D2487 (ASTM, 2011). Compaction curves for the sieved and unsieved CDG were determined in accordance with ASTM D698 (ASTM, 2012). The maximum and minimum dry densities of CC were determined by ASTM D4253 (ASTM, 2014b).

The saturated permeability (k_s) of the sieved and unsieved CDG was measured using the flexible wall permeameter as described in ASTM D5084 (ASTM, 2010b) while the constant-head method as described in ASTM D2434 (ASTM, 2006b) was used for CC. The measured saturated permeability of the sieved and unsieved CDG was 8.1×10^{-8} m/s and 5.7×10^{-5} m/s, respectively. The measured value for the saturated permeability of CC was 7.5×10^{-2} m/s. The basic properties of the cover materials are summarised in Table 8.6.

The WRC, which is also known as water retention curve, describes the relationship between the volumetric water content and matric suction of the soil. The drying and wetting WRCs of the sieved and unsieved CDG were obtained using the modified pressure plate apparatus (Ng and Pang, 2000b) by adopting the axis translation technique. A hanging column apparatus was used to obtain the drying and wetting path WRCs of CC. Figure 8.27 shows the measured WRCs of the soil and recycled concrete and the best-fit WRCs by using the van Genuchten (1980) WRC equation (see the fitted equation parameters in Table 8.7). As expected, the measurements show that there is hysteresis

430 Advanced Unsaturated Soil Mechanics

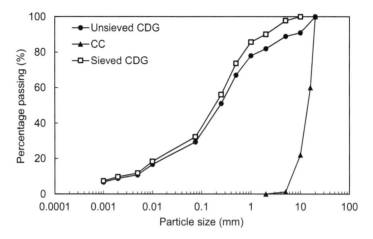

Figure 8.26 Particle size distributions of the unsieved completely decomposed granite (CDG), coarsely crushed concrete (CC) and sieved completely decomposed granite (CDG).

Table 8.7 van Genuchten (1980) fitting parameters of CDG and construction waste

	Symbol	Unsieved CDG	Sieved CDG	CC
Drying curve				
Saturated volumetric water content	θ_s	0.31	0.33	0.46
Residual volumetric water content	θ_r	0.01	0.03	0.02
van Genuchten (1980) fitting parameters	α	0.24	0.23	3.23
	n	1.51	1.38	4.87
	m	0.34	0.27	0.79
Wetting curve				
Saturated volumetric water content	θ_s	0.28	0.28	0.46
Residual volumetric water content	θ_r	10.01	0.01	0.02
van Genuchten (1980) fitting parameters	α	0.32	0.24	6.47
	n	1.51	1.35	3.89
	m	0.34	0.26	0.74

Note:
CDG, completely decomposed granite; CC, coarsely crushed concrete.

between the drying WRC and wetting WRC for sieved CDG, unsieved CDG and CC. Along the drying WRCs, the air-entry value of the sieved and unsieved CDG is estimated to be around 4 kPa and 2 kPa, respectively. The water-entry value of the CC is 0.4 kPa, which falls within the typical range (0.3 kPa to 1 kPa; Stormont and Anderson, 1999; Yang et al., 2006; Abdolahzadeh et al., 2011) that was utilised as the coarse-grained layer for CCBE. The ratio of the water-entry value between the fine-grained material and the coarse-grained material for the tested CCBE in this study (i.e., 200) is greater than the recommended minimum values (i.e., 10) suggested by Rahardjo et al. (2006). Figure 8.28 shows the predicted unsaturated permeability function from both the drying and wetting WRCs by using them in conjunction with the van Genuchten-Mualem equation (Mualem,

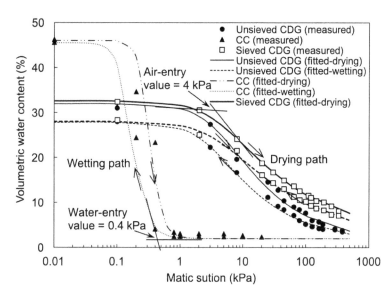

Figure 8.27 Measured and fitted WRCs of the unsieved completely decomposed granite (CDG), coarsely crushed concrete (CC) and sieved completely decomposed granite (CDG).

1976; van Genuchten, 1980). At 0.1 kPa, the permeability of the CC is approximately 4 and 7 orders of magnitude greater than that of the unsieved and sieved CDG, respectively. The low permeability of the sieved CDG when it is close to or at full saturation minimises water infiltration through the cover. The permeability of the CC approached the permeability of both the unsieved and sieved CDG as suction values increased. At 0.5 kPa suction, based on the wetting permeability function, the CC and unsieved CDG have the same value of permeability. At suction values greater than 1 kPa, the CC has a lower permeability than both the unsieved and sieved CDG. In this state, minimal water infiltrates into the CC layer during water infiltration and this is the fundamental working principle of CCBE.

Field instrumentation and monitoring

During the four-year monitoring, the field performance of the bare and grass-covered three-layer landfill cover systems was studied by monitoring percolation, PWP and VWC under humid climatic conditions. Percolation through the bare and grass covered landfill cover was monitored from 1 June 2016 to 1 January 2021. Six 1 m diameter lysimeters at a spacing of 5 m were installed at the bottom of the cover system (i.e., 1.8 m depth) to monitor percolation (see Figure 8.25). Each lysimeter was connected to an independent reservoir by using the drainage pipes. PWP and VWC within the landfill cover systems were measured from 1 May 2017 to 1 January 2021. At the mid-cross-section of each sloped cover, jet-fill tensiometers (JFTs, Soilmoisture Equipment Corporation) and SM300 moisture probes (Delta-T Devices Ltd) were installed at depths of 0.2 m, 0.4 m, 0.8 m, 1.2 m and 1.6 m to assess the variations of PWP and VWC (see Figure 8.25). To verify the measurements

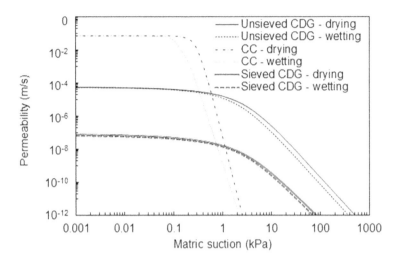

Figure 8.28 Predicted permeability functions of the unsieved completely decomposed granite (CDG), coarsely crushed concrete (CC) and sieved completely decomposed granite (CDG).

between them, each pair of JFTs and SM300 were installed at the same depth (Ng et al., 2019d). The JFTs were connected with pressure transducers and hence can measure PWP from –90 kPa to 100 kPa. Before installation, each JFT was fully saturated with de-aired water. Any air bubbles inside the JFTs were removed using a vacuum hand pump. To install the JFTs at the target depth, vertical soil holes with a diameter larger than that of a ceramic cup were drilled. Then, JFTs were installed in the predrilled holes. To ensure good contact between gravel and ceramic, the ceramic cup of the Jet-Fill tensiometer was covered by a slurry of kaolin clay (Zhan, 2003). SM300 moisture probes determine the VWC of soils by measuring their dielectric permittivity. Following the method used to install JFTs, vertical holes were also drilled to the target depth in each soil layer immediately after the construction of the layer was complete. The SM300 probes were inserted in the holes. To ensure good contact between the cover material and probe, each hole was backfilled with fine-grained CDG, CC and coarse-grained CDG at the bottom, in the middle and at the top layers, respectively, and recompacted to their desired bulk density. Before field measurement, laboratory and *in-situ* calibration were carried out for all the JFTs and SM300. The accuracy was ±1 kPa and ±3% for measuring PWP and VWC, respectively. Please note that the details of gas and settlement monitoring of the three-layer landfill cover system are beyond the scope of this paper.

To assist in the interpretation of monitored results, soil cores (76 mm in diameter and 100 mm in length) were sampled from the grass-covered cover system each year within the top 100 mm to determine k_s in the laboratory. Three replicates were collected from the grass covered cover system each time. Each of the soil samples was first fully saturated with de-aired water and then their k_s was measured in a flexible wall permeameter using the falling-head method (ASTM, 2010b). In the meantime, the LAI of the grass from the field site was measured each year by dividing the projected canopy area into total leaf area (Ng et al., 2016g). Three replicates were also obtained from the grass covered cover system each time. Climate parameters such as daily rainfall intensity, solar radiation, temperature, wind speed and humidity were monitored by an automatic weather station during the field monitoring from 1 June 2016 to 1 January 2021.

Numerical analysis

Hydrological model for vegetated three-layer landfill cover system

To interpret the field data from the bare and grass covered three-layer landfill cover systems, numerical transient seepage analyses were conducted using COMSOL Multiphysics (COMSOL 5.6, COMSOL Inc., Stockholm, Sweden). The soil-plant-water-atmosphere interaction models developed by Ni et al. (2018a) and Ng et al. (2021b) were input into the COMSOL Multiphysics software. This model can satisfy energy balance and soil-water balance using the radiant energy partitioning equation and Darcy-Richards equation, respectively.

Considering the two-dimensional (2D) water balance in grass-covered unsaturated soil, the water flow can be expressed using the modified Darcy-Richards equation:

$$\frac{d\theta(h)}{dt} = \frac{d}{dx}\left(K(h)\frac{dh}{dx}\right) + \frac{d}{dz}(k(h)\frac{dh}{dz} + k(h) - S_g(h,z) \tag{8.1}$$

where t is the elapsed time; h is the total hydraulic head; θ is the VWC; $k(h)$ is the water permeability as a function of h; $S_g(h,z)$ is the sink term, which is equal to the water volume transpired by grass from the entire root zone during a given time interval (Feddes et al., 1976).

$$S_g(h,z) = \alpha(h,z) \cdot G(\eta) \cdot \text{PET}\left(1 - \exp(-k_L \cdot LAI)\right) \tag{8.2}$$

where $\alpha(h,z)$ is the transpiration reduction function and ranges from 0 to 1 (Feddes et al., 2001); $G(\eta)$ is related to root distribution, $\eta(z)$; PET is potential ET; k_L is the parameter related to the radiation interception by plant leaves (0.5 for *Cynodon dactylon* (Bermuda grass), Ni et al., 2018b); LAI is the leaf area index of grass (see Table 8.8).

$$\alpha(h,z) = \begin{cases} 0 & h < h_{wp} \\ \dfrac{h - h_{wp}}{h_{fc} - h_{wp}} & h_{wp} \leq h \leq h_{fc} \\ 1 & h > h_{fc} \end{cases} \tag{8.3}$$

Table 8.8 Summary of the measured leaf area index (LAI) of grass and saturated water permeability (k_s) of grassed CDG during field monitoring

Parameter	Jun. 2016	Nov. 2017	Oct. 2018	Mar. 2019	May 2020
			Date		
LAI	1.91–2.07	2.10–2.20	2.28–2.60	2.40–3.01	2.80–3.24
	Mean: 1.99	Mean: 2.13	Mean: 2.45	Mean: 2.67	Mean: 2.95
	S.D.: ±0.08	S.D.: ±0.05	S.D.: ±0.13	S.D.: ±0.25	S.D.: ±0.21
k_s (10^{-5} m/s)	1.9 (bare CDG)	1.45–1.92	0.98–1.20	1.02–1.52	2.69–3.29
	-	Mean: 1.69	Mean: 1.06	Mean: 1.24	Mean: 2.99
	-	S.D.: ±0.19	S.D.: ±0.09	S.D.: ±0.21	S.D.: ±0.25

where h_{wp} and h_{fc} are the total hydraulic heads that correspond to the permanent wilting point and field capacity, respectively (Feddes et al., 2001).

According to numerical analyses conducted by Ni et al. (2018b), the root-induced changes in SWRC have minor effects on the PWP response in soil slope under heavy rainfall. Therefore, this study only considers root-induced changes in k_s while root effects on SWRC were ignored. The van Genuchten (1980) parameters of SWRC and water permeability function used in the numerical model are summarised in Table 8.7.

Model implementation and boundary conditions

Water infiltration into a three-layer landfill cover system was simulated two-dimensionally. In order to simulate the effects of grass on the landfill cover system, two domains were generated, one for rooted soil and the other for bare soil under the root zone. The measured k_s (refer to Table 8.8) at a depth of 0.1 m was assigned to the entire layer of coarse-grained CDG (i.e., root zone) in the numerical simulation.

Figure 8.29 shows the geometry and mesh size of the numerical model. For finite element discretisation, the minimum mesh size of the model was 5.46×10^{-4} m and the maximum size was 0.27 m. A time-dependent mixed flux boundary of evaporation and rainfall was applied to the top of the model (ABC in Figure 8.29) in the simulation. Based on the measured atmospheric parameters, the potential evapotranspiration (PET) was estimated by the Penman-Monteith equation as this equation can take the change in surface resistance due to plants into consideration (Monteith, 1965). The calculated PET was then used to determine the sink term in Equation (8.1). The flux boundary changes to a pressure boundary at a water pressure head of 1 mm (Chui and Freyberg, 2007, 2009), when rainfall intensity is greater than the k_s of the surface soil. By using this technique, water ponding on the surface of the cover system can be avoided (Ng et al., 2022b). A potential seepage surface was specified on the left and right boundaries (CD and AF in Figure 8.29). In order to simulate the boundary between the lysimeter and the bottom soil, the bottom of the cover system was set as a free drainage boundary.

The numerical analyses focused on the field performance of the cover systems from 1 June 2016 to 1 January 2021. Soil samples at different depths were taken on 1 June 2016 to measure the initial VWC distribution along the cover system. Based on the SWRCs of the cover materials in Figure 8.27 and the measured VWCs, the initial PWP distribution along the soil depth can be obtained.

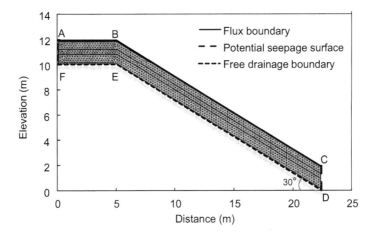

Figure 8.29 Finite element mesh and boundary conditions of three-layer landfill cover system.

Field monitoring results

Seasonal variations of climate conditions

Figure 8.30 shows the recorded atmospheric parameters during the 54-month monitoring period from 1 June 2016 to 1 January 2021. As shown in Figure 8.30a, the total amount of rainfall during the monitoring period was 9800 mm. The annual precipitation for the first, second, third and fourth years was 2166, 2120, 2616 and 1442 mm, respectively. The measured cumulative precipitation during the last 6 months of monitoring (from 1 June 2020 to 1 January 2021) was 1102 mm. 80% of the annual rainfall occurred in the wet season between April and October each year. Several rainfall patterns were observed at the site during the monitoring period. Large rainfalls (i.e., more than 100 mm) with long duration (i.e., more than 10 h) frequently appeared in the wet seasons from April to October. The highest daily rainfall intensity was nearly 230 mm/d (i.e., 30 August 2018), which is equivalent to rainfall with a return period of 15 years in Shenzhen (Shenzhen Meteorological Observatory, 2015).

A long-term dry period (with a daily rainfall depth of less than 10 mm) was observed from November to March each year in this study. Despite that, the leaf area index (LAI) of Bermuda grass still increased by 50% during the 54-month field monitoring (see Table 8.8). It was evident that this grass species can grow well despite a long-term dry period. Similar results were also found in a field study on a CDG cover with a high degree of compaction (i.e., 95%) by Ng et al. (2020e). In their study, the LAI and root length density of the Bermuda grass linearly increased by 5 and 7 times, respectively, over a one-year monitoring period which included a 4-month dry period. Those results demonstrate the long-term drought tolerance of Bermuda grass in the highly compacted CDG soil.

The rainfall depth measured in the wet seasons after December 2018 was below 100 mm, which may be because of climate change-induced long-term drying. Hence, the measured annual rainfall depth dropped by up to 2600 mm during the first three years to around 1400 mm during the fourth year of the monitoring.

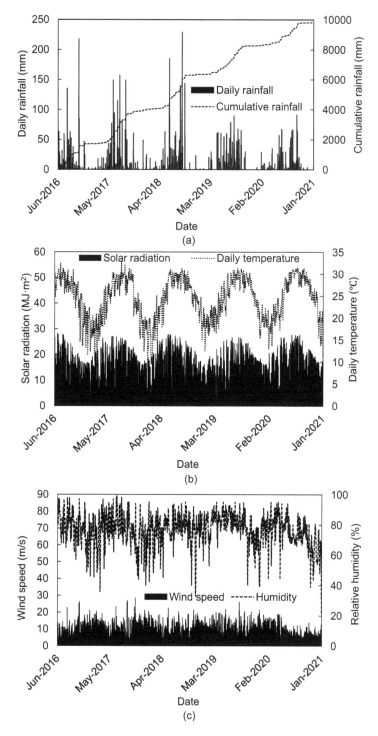

Figure 8.30 Recorded atmospheric parameters during the monitoring period (a) daily rainfall intensity and cumulative rainfall; (b) solar radiation and temperature; (c) relative humidity and wind speed.

As shown in Figure 8.30b, solar radiation increased during the wet seasons from March to September and decreased during the dry seasons from October to February. The highest solar radiation was approximately 28 MJ/m^2 in the summer, whereas it was only 17 MJ/m^2 during the winter period. Atmospheric temperature measured by the weather station throughout the monitoring period is also presented in Figure 8.30b. The seasonal temperature followed a sinusoidal pattern and ranged from 9°C to 32°C. Lower temperatures were measured during the dry seasons and higher temperatures during the wet seasons. The daily temperature fluctuation was about 10°C. Figure 8.30c shows the measured wind speed and daily relative humidity (RH). Wind speed 2 m above the cover system over the four-year monitoring period ranged from 5 to 30 m/s, with an average value of 11 m/s. Higher wind speeds tend to enhance the ET process, hence reducing the amount of water storage in the cover (Alam, 2017). That means that rainfall water infiltrated and stored in the cover system during wet seasons can be removed by the improved ET during the dry seasons. Therefore, the storage-release capacity of the three-layer landfill cover system can be enhanced, thereby leading to a potential reduction in percolation through the cover system during the dry seasons. Daily RH during the monitoring period did not have a regular trend as the site is in a subtropical humid climate region. In general, the daily RH was between 33% and 98%. The minimum daily RH was commonly found during December of each year, while its maximum value appeared during the wet seasons from April to October. However, higher RH is one of the factors that can decrease ET and tend to increase the percolation of the landfill cover system.

Seasonal responses of pore-water pressure

Figures 8.31a and 8.31b shows the measured variation in PWP at different depths in the bare and grass covered three-layer landfill cover systems from 1 May 2017 to 1 January 2021, respectively. Generally, negative PWP reached peak values in both of the landfill cover systems during dry periods. The magnitude of PWP induced by grass was commonly higher than that in the bare cover in dry seasons. When rainfall occurred during the wet season in the first 24 months, the PWP preserved after rainfall in the grass-covered cover system was lower than that in the bare cover. For instance, a PWP of -23 kPa was recorded at 0.2 m depth in both the bare and grass covered landfill covers on 27 August 2017 (see Figure 8.32a), which was immediately before a 10-year return period of rainfall with a total depth of 150 mm (Shenzhen Meteorological Observatory, 2015). After this heavy rainfall, the PWP preserved in the bare cover system was around 3 kPa higher than that of grass covered cover system (i.e., -9 kPa), despite the same initial PWPs. Similar results were also observed during the rainfall from 29 to 30 August 2018 (see Figure 8.32b). The lower retained PWP in the grass-covered cover system is attributed to the lower water permeability induced by soil pores being occupied by fresh roots (Chen et al., 2019; Ng et al., 2020e). This lower retained PWP can potentially reduce water percolation through the cover system.

Under the rainfalls in March 2019 (see Figure 8.32c), the PWPs retained in the grass-covered cover system at a depth of 0.2 m were higher since k_s of grass-covered soil increased to be greater than that of bare soil (see Table 8.8). For instance, the PWP measured at a depth of 0.2 m in the grass-covered cover system was observed to be around 20 kPa less than that in the bare cover system before the rainfall event on 26 January 2020 because of

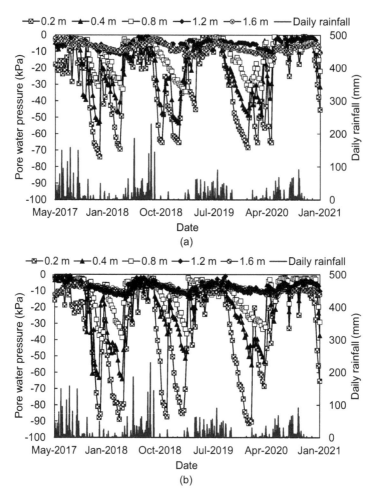

Figure 8.31 Variations of measured pore-water pressure at different depths in (a) non-grassed and (b) grassed three-layer landfill cover systems.

ET. However, after rainfall with a total depth of 29 mm, the PWP measured at this depth in the bare cover system was 4 kPa less than that in the grass-covered cover system (see Figure 8.32d). This is probably caused by the formation of root channels and their induced macropores during root growth (Ghestem et al., 2011; Leung et al., 2017), potentially increasing the k_s of the soil.

Moreover, similarities were also observed for the PWP responses under rainfall between the bare and grass covered cover systems. PWPs near the surface of the cover systems (i.e., 0.2 m) were easily influenced by rainfall with the greatest changes. With increasing soil depth, the response of PWP to the onset of rainfall events was progressively slower for both the bare and grass covered cover systems. As shown in Figure 8.31, rainfall events with small depths (i.e., less than 10 mm) seem to have negligible effects on the PWP variations in the bottom layers (i.e., 1.2 and 1.6 m). This is because CCBE forms at the

Field monitoring and engineering applications 439

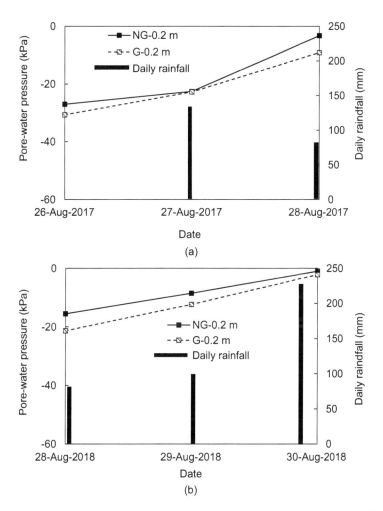

Figure 8.32 Variations of measured pore-water pressure at 0.2 m depth in non-grassed (NG) and grassed (G) three-layer landfill cover systems on (a) 27 August 2017; (b) 29–30 August 2017; (c) 1–11 March 2019 and (d) 26 January 2020.

interface of the upper two soil layers. The infiltrated water after a small rainfall can be stored in the top fine-grained layer, which has been demonstrated by previous field studies (Bossé et al., 2015; Knidiri et al., 2017; Ng et al., 2019).

Responses of pore-water pressure along depth under heavy rainfall

Figure 8.33 shows the variations in measured and computed PWPs along depth in both the bare (NG) and grass covered (G) landfill covers before and after the rainfall from 29 to 30 August 2018. The total rainfall depth during those two days was around 350 mm. During the entire monitoring period, the highest intensity daily rainfall of 230 mm over 17 hours

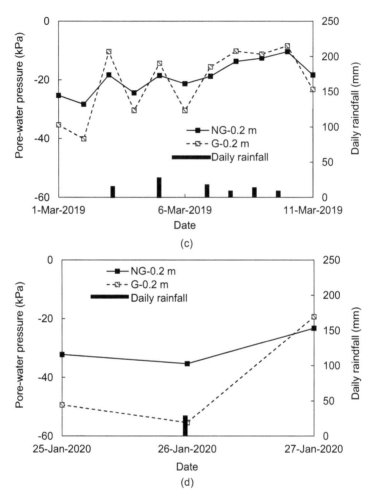

Figure 8.32 (Continued)

occurred on 30 August 2018, which is equivalent to a 15-year return period for rainfall in Shenzhen (Shenzhen Meteorological Observatory, 2015). Before this heavy rainfall, the test site experienced with a long preceding rainy season with a total rainfall of about 1900 mm. According to previous studies, most alternative cover systems (e.g., ET cover, Albright et al., 2004; three-layer inclined capillary barrier, Zhan et al., 2014b; dual capillary barrier, Rahardjo et al., 2016) cannot perform satisfactorily in such a rainy season. However, in this study, relatively low PWPs were still retained in both the grass-covered and bare three-layer landfill cover systems (i.e., nearly −10 kPa and −15 kPa, respectively) after the long rainy season. The measured PWPs in the grass-covered site were substantially lower than those measured in the bare site, especially at a shallow depth (i.e., 0.2 m and 0.4 m). Beyond the plant root zone (i.e., 0.6 m), the root system can still help to preserve low PWP through root water uptake (Ni et al., 2018a). Consequently, a lower PWP

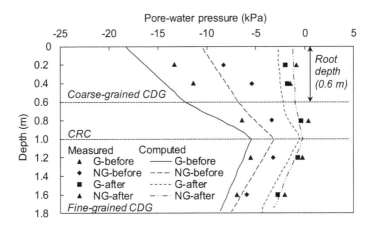

Figure 8.33 Variations of measured and computed pore-water pressure profiles in the non-grassed (NG) and grassed (G) three-layer landfill cover systems before and after the rainfall from 29 to 30 August 2018.

was measured in the middle CC layer of the grass-covered site than that in the bare site. The lower PWP in the middle layer potentially enhances the performance of the upper two-layer CCBE in reducing water infiltration into the bottom layer. This has been identified as one of the main reasons why less percolation was measured through the vegetated cover system (Ng et al., 2019d). The difference in the measured PWP between bare and grass-covered areas was up to 5 kPa, which was largely due to plant root blockage of soil pore throats (Ng et al., 2016 g) and ET.

Similar variations in PWP of bare and grass covered areas were also observed in numerical results. However, the computed PWPs were lower than the measured values, especially in the grass-covered area. The differences between the measured and computed PWPs are mainly attributable to water flow through tension cracks induced by differential settlement of the solid waste (Crawford and Smith, 1985). The numerical model does not consider water infiltration through cracks. In addition, the Bermuda grass roots may not be distributed uniformly in the root zone (Leung et al., 2015), while the k_s of the root zone was assumed to be the same along the depth. This would overestimate root water uptake and underestimate water infiltration, leading to a lower PWP after drying and rainfall, respectively.

At the end of the rainfall, both measured and computed PWPs at each depth increased substantially in both cover systems. Under this heavy rainfall, water broke through the top CCBE and PWP in the CC layer and the fine-grained CDG layer increased. PWPs in the middle layer in grass covered and bare covers reached about 0 kPa. After water broke through, the middle CC layer functioned as a drainage layer. This also demonstrates the feasibility of using CC as the middle layer of a three-layer landfill cover system. PWPs increased in the bottom fine-grained CDG layer for both cover systems, indicating water infiltration into the bottom soil layer. Even though increased percolation was observed after prolonged and heavy rainfall, the bottom layer remained unsaturated because of its low water permeability. The PWP at the bottom of the lowest layer increased to −3 kPa, which corresponded to a water permeability close to 1×10^{-9} m/s, meeting the value required by Benson et al. (2001) (see Figure 8.28b). The measured and computed results demonstrate that even after heavy rainfall, both the bare and grass covered three-layer landfill cover systems using construction waste without a geomembrane are still effective in minimising water infiltration.

Seasonal responses of volumetric water content

Figures 8.34a and 8.34b show the variation in VWC at different depths in both bare and grass covered three-layer landfill cover systems from 1 May 2017 to 1 January 2021, respectively. During monitoring, the measured VWC was correlated closely with the PWP measurement. Pronounced fluctuations in VWC were recorded at shallow depths, i.e., 0.2, 0.4, and 0.8 m. However, the changes in VWC at the bottom layer were very small (i.e., less than 5%), even under heavy rainfall or drought. VWCs near the cover surface (i.e., 0.2 and 0.4 m) were first influenced by the rainfall events, which was similar to the PWPs response (as shown in Figure 8.31). Because of the CCBE provided by the upper two soil layers, small rainfalls (i.e., rainfall depth less than 10 mm) could not influence the VWCs at depths of 0.8, 1.2, and 1.6 m. This demonstrates the effectiveness of the upper CCBE in minimising water infiltration during such rainfall events. However, an increase of VWC in the CC layer was measured during heavy rainfall events, which indicates that the capillary barrier effect can only last for a short period. Consistent findings were also reported by Knidiri et al. (2017) and Zhan et al. (2017) for a two-layer CCBE. Water infiltration through the three-layer landfill cover system without geomembrane was significantly reduced by the fine-grained CDG layer with low permeability. For both grass-covered and bare covers, the variations in VWC in response to rainfall were much faster after drying than after wetting. Post-drying rainfall events normally resulted in a sudden increase in VWC at all the measured depths (i.e., 6 June 2018), while pre-drying changes in VWC were gentler (i.e., from July to August 2018). On the one hand, the hydraulic gradient near the soil surface after drying can be much greater than that after wetting. Hence, the rainfall infiltration rate and VWC response to rainfall was faster after drying. On the other hand, the hydraulic hysteresis in SWRC can also play a role in this VWC response (as illustrated in Figure 8.28a). After drying, the soil state is located on the drying path of the SWRC. During the following rainfall, it changes from the drying path to the wetting path along the wetting scanning curve. This induces a quick response in the variation of VWC (Yang et al., 2004b).

The upper two-layer CCBE has a relatively low water permeability at a lower water content during drought. Therefore, the existence of the upper CCBE should prevent the bottom layer from water loss and desiccation during drying periods (Ng et al., 2016b). The monitored results also suggest that even during a long drought (i.e., September 2018 to March 2019), the decrease in VWC in the bottom fine-grained CDG layer was very small (i.e., less than 5%). Moreover, the drought also leads to a decrease in VWC in the coarse-grained CDG and CC layers, which helps to recover the CCBE after water breaks through. This implies that the proposed three-layer landfill cover system using construction waste without a geomembrane can work satisfactorily during the rainy and drought periods of a humid climate by using the upper CCBE as well as the bottom low-permeability layer.

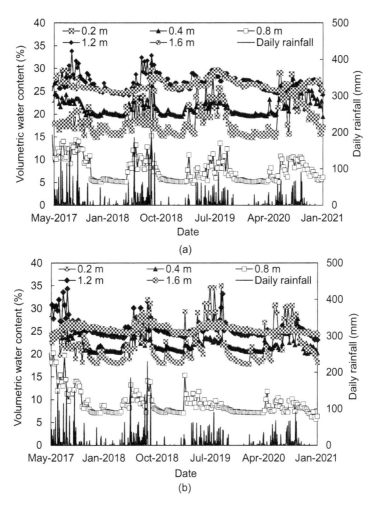

Figure 8.34 Variations of measured volumetric water content at different depths in (a) non-grassed and (b) grassed three-layer landfill cover systems.

Responses of volumetric water content along depth under heavy rainfall

Figure 8.35 shows the variations in measured and computed VWCs along depth in both the bare (NG) and grass covered (G) landfill covers before and after the persistent rainfall from 29 to 30 August 2018. The total rainfall depth was around 350 mm. Before the rain, the measured VWCs along the soil depth in the bare cover system were up to 8% higher than those in the grass-covered cover system. Lower VWCs along the soil depth in the grass-covered cover system were also computed, which was consistent with the lower PWPs observed in the cover system (see Figure 8.33). However, the computed VWCs were markedly lower than their measured values, especially within the top 1.2 m. On the one hand, the water flow through tension cracks was not considered in the numerical model.

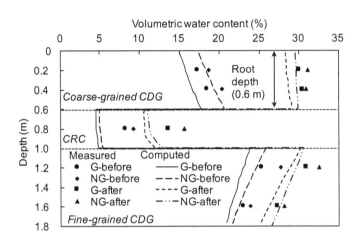

Figure 8.35 Variations of measured and computed volumetric water content profiles in the non-grassed (NG) and grassed (G) three-layer landfill cover systems before and after the rainfall 29 to 30 August 2018.

On the other hand, root water uptake and the effect of grass on the reduction of water infiltration were overestimated in the simulation. The VWCs in the bottom fine-grained CDG layer for bare and grass covered cover systems were significantly higher than those in the top unsieved CDG layer. The rainfall water stored in the top layer during past rainfalls can be removed by ET and lateral drainage under drying. However, the loss of water content in the bottom layer can be reduced by the protection of the upper two-layer CCBE during drying (Ng et al., 2016a). The high VWC in the bottom layer can help to prevent landfill gas emission and avoid desiccation cracks. These are two of the key functions of the three-layer landfill cover systems as described in Ng et al. (2016a) and (2019).

At the end of the two-day-long persistent rainfall, the measured VWCs along the soil depth in both bare and grass covered cover systems increased. The trend of the computed results was consistent with the measured one. However, the computed VWCs were slightly lower than their measured values, because of water infiltration through tension cracks. As with the PWPs (see Figure 8.33), the difference in the observed VWCs between the bare and grass-covered cover systems was not significant (i.e., 2%), which is within the measurement accuracy (±3%) of the SM 300 moisture probe. The VWCs in both bare and grass covered covers after rainfall were close to the saturated value. The difference in VWC between bare and grass covered covers is attributable to the effect of plants on k_s. The most significant increase in VWC (up to 70%) was observed at the shallow depths (i.e., 0.2 m and 0.4 m). The increase of VWC in the middle CC layer for both cover systems indicates a breakthrough of the upper two-layer CCBE. Since the measured VWC in the middle layer was far less than its saturated value, it prevented the water from ponding on the bottom fine-grained layer. Most of the infiltrated water was transferred to lateral drainage in the CC layer. This demonstrates the effectiveness and suitability of recycled concrete that was used as an alternative material for the three-layer landfill cover systems. Moreover, the increase in VWCs in the bottom layer for both the bare and grass covered layer implies that rainfall had infiltrated into the bottom layer of the three-layer landfill cover system.

Field monitoring and engineering applications   445

The initial VWC (or PWP), rainfall intensity and antecedent rainy season have a substantial influence on the performance of the cover system in reducing water infiltration.

Effects of grass on measured percolation

Figure 8.36 shows the measured cumulative percolation by lysimeters at three locations (i.e., crest, middle and toe) in both the bare (NG) and grass covered (G) three-layer landfill cover systems from 1 June 2016 to 1 January 2021. For ease of comparison, the measured cumulative rainfall over the whole four years is also included in the figure. During the four-year monitoring period, the total amount of rainfall was 9800 mm, with 80% occurring between April and October each year. As expected, the measured percolation increased steadily during the wet seasons and agreed well with the cumulative rainfall during the entire four-year monitoring period. Despite that there were no measured percolation data from December 2016 to May 2017, the measured percolation in May 2017 was the cumulative value during this period.

During the four-year field monitoring period, the maximum recorded annual percolation for all three locations for the bare landfill cover were 21 mm, 25 mm, 26 mm and 20 mm, respectively. The measured cumulative percolation during the last 6 months of monitoring (from 1 June 2020 to 1 January 2021) was about 13 mm. The measured average annual percolation of the bare cover was about 23 mm. However, the maximum annual percolation through the grass covered area was 18 mm in the first year and 20 mm in the second year, caused by the effect of grass. In the first two years of monitoring, the grass covered cover had up to 22% less annual percolation than the bare one. This was caused by the lower PWP induced by grass roots as illustrated in Figure 8.31. The measured results were also consistent with the one-dimensional column tests conducted by Ng et al. (2019a). The lower PWP led to a reduction in water permeability (Buczko et al., 2007; Scanlan and Hinz, 2010) and hence less percolation. However, root growth may form root channels and macropores (Ghestem et al., 2011; Leung et al., 2017). As shown

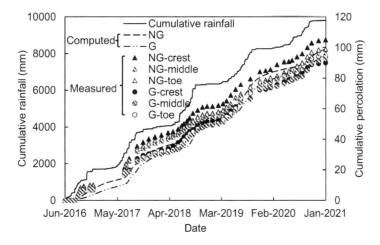

Figure 8.36 Measured and computed cumulative percolations in the non-grassed (NG) and grassed (G) three-layer landfill cover systems.

in Table 8.8, k_s in the grass covered landfill cover system increased progressively after March 2019, which led to an increase in infiltrated water and hence percolation through the grass covered cover system. As a result, the maximum annual percolation through the grass covered cover system was 25 mm in the third year and 22 mm in the fourth year of monitoring, while the measured cumulative percolation during the last 6 months of monitoring was about 13 mm. These were close to or greater than the measured percolation through a bare landfill cover system. At the end of the field monitoring, the measured average annual percolation of the grass-covered cover was about 21 mm. Nonetheless, the measured maximum annual percolation of both cover systems meets the design criterion (30 mm/year) recommended by the USEPA (1993). This demonstrates again the effectiveness of the three-layer landfill cover system using construction waste without a geomembrane in preventing excessive percolation.

Figure 8.36 also shows the computed variation in cumulative percolation against time for bare and grass covered three-layer landfill cover systems. The influence of root-induced changes in k_s on cumulative percolation was included in the numerical simulation. The computed percolation increased substantially during the wet seasons of each year (from April to October), while little percolation was predicted during drought periods. For the bare cover system, the computed cumulative percolation did not match the measured values very well during the initial two-year monitoring period. The computed percolation was up to 30% less than the measured results, especially from October 2017 to June 2018. The differences between the measured and computed results during this period of time are largely due to water flow through tension cracks induced by differential settlement of the solid waste. The numerical model does not account for preferential flow through cracks. The fill age of the solid waste under the cover system was about 1 year before the start of monitoring. As pointed out by Ivanova et al. (2008), settlement during this period is dominated by the delayed compression of the waste with high settlement rates. Therefore, tension cracks in the cover system were observed during the first two years of monitoring. Between June 2018 and January 2021, the computed percolation was close to the measured values through the bare cover system. During this period, the cracks were observed to close because the differential settlement at the measured points decreased, leading to a reduction of water passing through preferential flow paths.

For the grass covered cover system, the computed percolation was up to 40% less than the measured results between June 2016 to October 2017. The input k_s of grass-covered soil and LAI may overpredict surface runoff and plant ET-induced water loss. Therefore, a significantly low percolation was computed for the grass covered cover system. During the following 38 months (October 2017 to January 2021), the simulation results captured the measurements very well when taking the effects of grass root on k_s into consideration. The hydrological responses of the cover system and their interaction with plants were highly dynamic with soil depth (Ng et al., 2020e). During the early stage of grass growth (November 2017 to June 2019), very few roots (less than 5% of the total number of roots) penetrate deeply (i.e., 1.2 m and 1.6 m). Hence, k_s in the deep soil depth was not as low as that near the soil surface. The computed percolation during this period was slightly less than the measured value. After June 2019, root channels and macropores may have been formed by root growth, leading to an increase in k_s. The computed results were close to the measured percolation through the grass covered cover system. According to both the field monitoring and numerical simulation, it is obvious that the hydrological responses

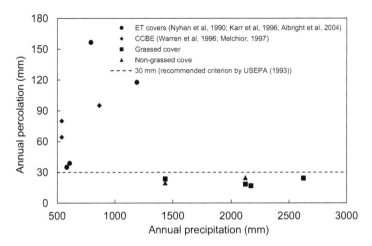

Figure 8.37 Comparison of measured annual percolation through the three-layer landfill cover systems with ET and CCBE cover systems.

induced in soil by plant growth have significant effects on water infiltration through the three-layer landfill cover system. To promote the performance of landfill cover, we suggest the use of plants with shorter root depth to reduce water percolation and limit the increase in k_s induced by plant growth.

Figure 8.37 shows a comparison of the measured annual percolation through three alternative cover systems, namely the three-layer landfill cover system, ET cover and CCBE. The performance of ET and CCBE cover systems reported in previous studies in humid regions of the USA (i.e., Nyhan et al., 1990; Warren et al, 1996; Karr et al., 1999; Albright et al, 2004) and Germany (i.e., Melchior, 1997) were considered for comparison. All the ET covers reported in the literature are from the USA. The annual rainfall for ET covers ranged from 579 mm to 1191 mm, with the corresponding minimum and maximum annual percolations of 35 mm and 157 mm, respectively. The ratio of annual percolation to annual rainfall was up to 19%. The CCBEs located in three field sites in the USA and Germany were also compared. The measured annual rainfall for CCBE was between 539 mm and 865 mm. The ratio of annual percolation to annual rainfall for CCBE ranged from 11% to 15%.

The measured annual percolation through ET cover and CCBE was greater than 30 mm, even though the annual precipitation was only around 500 mm. For ET cover, it is likely that the water permeability of the cover materials was not low enough even with plant ET. For CCBE, the fine- and coarse-grained soils cannot provide sufficient contrast in water permeability. With increasing annual precipitation, the annual percolation in those two types of alternative cover increased significantly. However, the annual percolation through the three-layer landfill cover system is far less than those of ET cover and CCBE even under an annual precipitation exceeding 2600 mm. The measured annual percolation through both the grass covered and bare covers was less than 30 mm and meets the design criterion recommended by the USEPA (1993). The measured percolation during the four-year monitoring period obviously demonstrates the effectiveness of the three-layer landfill

Effects of grass on soil-water balance

Figure 8.38a shows the computed water balance components of the bare three-layer landfill cover system during the 54-month monitoring period, including cumulative rainfall, actual evaporation, surface runoff, lateral drainage, soil-water storage and percolation. The cumulative rainfall during the four-year monitoring period was 9800 mm, nearly half of which evaporated through the cover surface. The field test site is in a subtropical humid climate area, with annual evaporation close to 1100 mm (Wei et al., 2008). The computed evaporation is consistent with its measured value. The total amount of surface runoff during the entire field monitoring period was only about 0.5% of the cumulative rainfall. Surface runoff increased substantially during the wet seasons when there was a large amount of rainfall (e.g., August 2018). The top flux boundary in the numerical model becomes a pressure boundary under this condition. Surface runoff from landfill covers was greatly governed by the conditions at the soil-atmosphere interface, especially the water permeability of the soil near the cover surface (Albright et al., 2004). During the monitoring period, water storage in the bare three-layer landfill covers ranged from 230 to 410 mm. Water storage increased with fluctuations during the wet seasons (from April to October), while it decreased during dry seasons (from November to March). This implies that the infiltrated rainfall can be stored in the cover system during wet periods and released by evaporation during dry seasons. If the rainfall exceeded the water storage capacity of the cover system, the infiltrated water can be also removed by lateral drainage in the middle CC layer. The computed cumulative lateral diversion for the bare cover system was 4988 mm, which accounted for about 51% of the total rainfall. As mentioned previously, the rainfall during wet periods exceeded the water storage capacity and was diverted laterally in the middle layer. Light rainfall during the dry seasons did not increase lateral drainage. Water can be stored in the upper coarse-grained CDG layer and then removed by evaporation.

As shown in Figure 8.38a, the deep percolation of the cover system appeared after water was diverted through the middle layer for a few days. That means that CCBE functioned well in the early stages of the prolonged rainfall. However, breakthroughs of the upper two soil layers of the CCBE occurred during wet seasons. This is because the contrasting permeability of the two upper soil layers cannot persist for long during prolonged and heavy rainfall events. The contribution of the middle CC layer in the three-layer landfill cover system is that even after the CCBE has been broken through, the middle layer can still function as a drainage layer, because of the contrasting k_s of CC and fine-grained CDG. This differs from the principle of the three-layer inclined capillary barrier proposed by Zhan et al. (2014b). The bottom layer can work as a hydraulic barrier to prevent water infiltration. This is the main reason that only about 1% of the cumulative rainfall percolated through the bare three-layer landfill cover system, indicating that a three-layer landfill cover system using construction waste without a geomembrane can perform satisfactorily to minimise water percolation into the solid waste.

Figure 8.38b shows the computed water balance components of the grass covered three-layer landfill cover system during the 54-month monitoring period. The potential

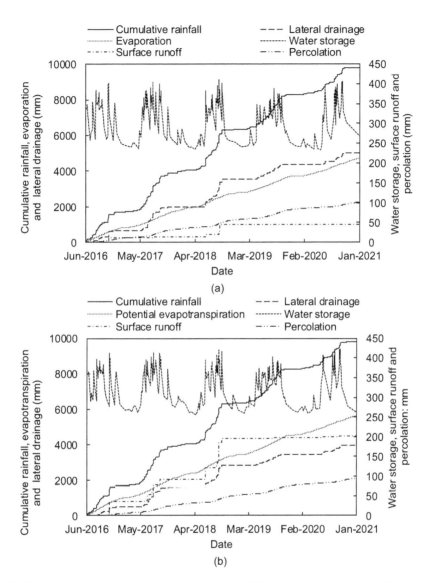

Figure 8.38 Computed water balance of (a) non-grassed and (b) grassed three-layer landfill cover system.

evapotranspiration (PET) was calculated using the Penman–Monteith equation (Monteith, 1965). The cumulative PET in the grass covered cover during the entire 54-month monitoring period was 5657 mm, which was about 58% of the cumulative rainfall and 20% higher than soil evaporation from bare cover. The cumulative surface runoff while monitoring was 202 mm, which was 2% of the cumulative rainfall. More surface runoff (i.e., 5 times) occurred in the grass covered landfill cover than in the bare one. The presence of grass reduced the infiltration rate in the early stages of plant growth because of the reduction in k_s of the CDG soil caused by plant roots and the lower initial PWP by plant

ET. However, k_s for the grass-covered soil increased beyond that of bare soil after August 2019. The computed results show that there was no surface runoff between August 2019 and January 2021. This is because the k_s of the surface grass covered CDG was higher than the rainfall intensity, so the rainfall infiltrated the deep soil. As pointed out by Albright et al. (2004), surface runoff from landfill covers is controlled by the water permeability of the surficial soil and the density of the vegetation. As shown in Figure 8.37b, water storage in the grass covered three-layer landfill covers ranged from 250 to 420 mm during the monitoring period, which was up to 9% higher than that of the bare landfill cover system. This is because plant roots increase the water-holding capacity of soil at any given PWP (Rahardjo et al., 2014; Yan and Zhang, 2015; Zhou and Chen, 2021). As more water was retained in the grass covered cover system, especially in the bottom layer, landfill gas emission was minimised. The computed cumulative lateral drainage for grass-covered cover was around 3965 mm, which accounted for more than 40% of the cumulative rainfall. The computed lateral drainage of the grass covered cover system was 20% less than that of the bare cover system because more rainfall water (more than 20%) was removed by ET, surface runoff and water storage. Because of the high k_s (10^{-2} m/s) of the CC layer, the water infiltrating the middle layer can be removed quickly as lateral drainage, leading to a reduction of ponding on the bottom soil layer. This indicates that the hydraulic properties and thickness of the CC layer are sufficient to minimise water infiltration into the bottom layer by providing lateral drainage. Through the enhanced water storage capacity, plant ET and sufficient lateral drainage of the CC layer, a lower cumulative percolation through the grass covered three-layer cover system resulted when compared with the bare cover system without a geomembrane.

Summary

Full-scale field monitoring was conducted in the Xiaping landfill site to study the use of construction waste and vegetation to form a three-layer landfill cover system without a geomembrane for 54 months (from 1 June 2016 to 1 January 2021). During the 54-month field monitoring period, plant characteristics, PWP, VWC, percolation and atmospheric parameters were monitored. In addition, numerical analyses were carried out to compare the water balance of the bare and grass covered cover systems. The following conclusions can be drawn:

- The surface runoff and water storage in the grass covered cover were about 5 times and 9% higher than those of the bare cover. The grass covered cover system can reduce percolation by employing ET, enhancing surface runoff and water storage. However, the long-term behaviour may be affected by an increase in saturated water permeability resulting from the growth of grass roots.
- During the field monitoring, a total of 9800 mm rainfall was recorded. More than 50% of the total rainfall was diverted at the middle CC layer. This indicates the effectiveness in facilitating lateral drainage of the hydraulic properties and thickness of the CC layer.
- The measured average annual percolation for the grass covered and bare cover was 23 mm and 21 mm during the monitoring period, respectively. The measured results

were consistent with numerical predictions. They all meet the US recommended criterion (USEPA, 1993). The presence of grass reduced the cumulative percolation during the entire 54 months by up to 14%.
- The field monitoring and numerical results consistently demonstrated that either the grass covered or bare three-layer cover system using construction waste but without a geomembrane are promising alternatives for humid climate regions with annual rainfall up to 2600 mm.

USE OF UNSATURATED SMALL-STRAIN SOIL STIFFNESS IN THE DESIGN OF WALL DEFLECTION AND GROUND MOVEMENT ADJACENT TO DEEP EXCAVATIONS (NG ET AL., 2020D)

Introduction

Excessive excavation-induced wall deflection and ground surface settlement can have serious consequences for surrounding buildings and services. Based on field data, many empirical and semi-empirical equations, and design charts have been proposed to predict excavation-induced wall deflection and ground surface settlement (Peck, 1969; Hsieh and Ou, 1998; Shi et al., 2019a, b; Wang et al., 2010; Tan and Wei, 2012). However, these equations and charts cannot consider explicitly the effects of small-strain soil stiffness and the degree of soil saturation on wall and ground movements. Very often, the initially saturated ground conditions can become unsaturated due to de-watering. Taking into account the effects of ground de-saturation during construction can make the economical design analysis of excavation-induced wall deflection and ground surface settlement possible.

It Is well-known that the shear stiffness of saturated soils decreases nonlinearly with increasing shear strain (Atkinson, 2000). For unsaturated soils, the small-strain stiffness increases significantly with increasing suction (Mancuso et al., 2002; Ng and Yung, 2008). Over small strains between 0.001% and 1%, the shear stiffness increases by up to 35% when the suction increases from 150 to 300 kPa (Ng and Xu, 2012). Further, Ng et al. (2009) found that small-strain stiffness is also affected by the drying and wetting paths (or hydraulic hysteresis). In terms of theoretical modelling, Sawangsuriya et al. (2009) and Khosravi and McCartney (2009, 2012) modified the general expression for the small-strain stiffness model proposed by Hardin and Black (1968). They replaced the original effective stress term for saturated soils with the mean effective stress defined using the approach of Bishop (1959). Based on experimental results, Ng and Yung (2008) proposed semi-empirical equations to predict the anisotropic small-strain stiffness of unsaturated soils as a power function of net stress and soil suction. Moreover, Biglari et al. (2011b) and Wong et al. (2014) incorporated the degree of saturation in their semi-empirical equations to reflect the effects of hydraulic hysteresis on soil stiffness. However, the use of these equations in design analysis is rarely reported.

In this section, the use of suction-dependent small-strain soil stiffness in the design analysis of a 15-m deep excavation in Tianjin, China is reported. In the design analyses, a Hardening Soil-Small Strain (HSS) model (Benz, 2007) was modified by incorporating the effects of suction on soil stiffness into Plaxis 2D (Brinkgreve et al., 2015). To ensure safe and economical construction progress, field measurements were compared with numerical predictions both with and without taking suction-dependent small-strain soil stiffness into account throughout the excavation.

The excavation site

The excavation site for high-rise buildings, approximately 181 m by 268 m, is situated in the downtown area of Tianjin, China (Plate 8.8). The northern side was retained by 29 m-long contiguous piles (each with a diameter of 0.9 m at 1.1 m spacing), while the other three sides were supported by diaphragm walls with a thickness of 1–1.2 m. On the northern side, an earth berm (19 m wide and 11.5 m high) was cut in front of the pile wall to provide extra support during excavation (Figure 8.39a). At the inner boundary of the earth berm, two rows of 21 m-long contiguous piles with row spacing of 3.2 m were installed (Figure 8.39b).

On the excavation site, there were three different soil types (i.e., fill, silt and silty clay) along the depth (Figure 8.39b). The top 5.5 m layer was fill material. The soil at depths of 9.5-11.5 m and 23.0-24.2 m was classified as silt. Soil at other depths was classified as silty clay. In order to determine the basic properties of the soils, intact soil samples were collected from the field for laboratory triaxial and oedometric tests (Zheng et al., 2018). The properties of these three soils are summarised in Table 8.9.

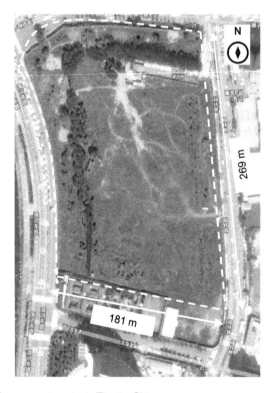

Plate 8.8 Overview of the excavation site in Tianjin, China.

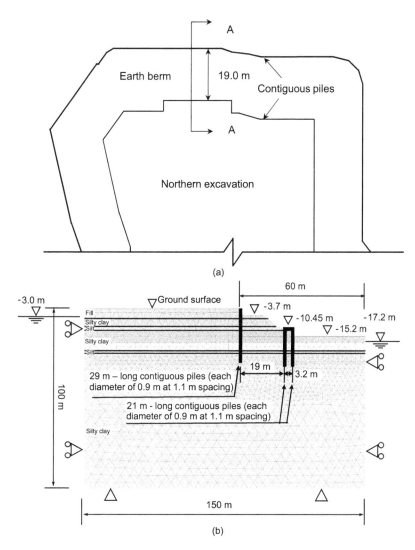

Figure 8.39 (a) Location of the selected cross-section A-A and (b) illustration of cross-section A-A of the excavation in the design analysis.

Theoretical basis

The HSS constitutive model developed by Benz (2007) was implemented in Plaxis. In the model, there are two key parameters controlling small-strain soil stiffness, that is, reference shear stiffness at a given stress level G_0^{ref} and a reference shear strain $\gamma_{0.7}$ at which the shear stiffness is 70% of G_0^{ref}. The stress-dependent small-strain soil stiffness G_0^{ref} for saturated soils can be expressed as follows

$$G_0 = G_0^{ref} \left(\frac{c' \cos \varphi' + \sigma_3' \sin \varphi'}{c' \cos \varphi' + p^{ref} \sin \varphi'} \right) \tag{8.4}$$

Table 8.9 Soil parameters used in the design analysis

Soil type	Depth: m	γ: kN/m³	e	c': kPa	φ': degrees	E_{50}^{ref}: MPa	E_{oed}^{ref}: MPa	E_{ur}^{ref}: MPa	G_0^{ref*}: MPa	0.7	n	k	m
Fill	0–5.5	18.5	0.94	12	16.1	4.4	4.4	26.3	71.0	0.0002	0.17	0.045	0.5
Silt	9.5–11.5 23.0–24.2	18.7	0.74	10	32.3	8.4	8.4	44.1	119.2	0.0002	0.17	0.045	0.5
Silty clay	5.5–9.5 11.5–23.0 >24.2	19.8	0.64	14	25.7	7.2	5.1	36.8	99.3	0.0002	0.17	0.045	0.5

Notes:

γ = unit weight of soil; e = void ratio; c' = effective cohesion; φ' = effective friction angle; E_{50}^{ref} = triaxial loading Young's modulus when shear stress is 50% of shear strength; E_{oed}^{ref} = oedometric loading modulus; E_{ur}^{ref} = unloading-reloading Young's modulus; G_0^{ref} = reference shear stiffness Y; $_{0.7}$ = reference shear strain at which shear stiffness is 70% of G_0^{ref}; m = the power function parameter in Equation (8.4).

* These values for G_0^{ref} do not consider the effects of net stress and soil suction.

where c' is the effective cohesion, φ' is the effective friction angle, σ'_3 is the minor effective principle stress, p^{ref} is the reference pressure and m is the power parameter for the stress dependence of stiffness.

With increasing shear strain γ, the corresponding small-strain shear stiffness G_s can be described by the following Equation (8.5).

$$\frac{G_z}{G_0} = \frac{1}{1+0.385\left(\dfrac{\gamma}{\gamma_{0.7}}\right)} \tag{8.5}$$

The HSS model can account for the reduction in shear stiffness with increasing strain at small strains. Finite element analyses using the HSS model have demonstrated the ability to predict the deformation of soils and retaining structures during excavation (Zheng et al., 2018; Kung et al., 2009; Zhang et al., 2015).

By using two stress variables, i.e., net mean stress and soil suction, Ng and Yung (2008) derived a small-strain stiffness model for unsaturated soils (see Equation (5.7)). Under isotropic conditions, their equation is as follows:

$$G_0 = C^2 f(e) \left(\frac{p}{p_{\text{ref}}}\right)^{2n} \left(1 + \frac{s}{p_{\text{ref}}}\right)^{2k_4} \tag{8.6}$$

Where G_0 is the small-strain stiffness, which takes into account the effects of net stress and soil suction; C is a constant reflecting the effects of inherent soil structure or stress-induced anisotropy; $f(e)$ is a void ratio function relating shear stiffness to void ratio, and this function adopts the formulation e^a for simplicity, where a is a regression parameter; p and s are mean net stress and matric suction $(u_a - u_w)$, respectively; p^{ref} is reference pressure for normalising p and is generally assumed to be 1 kPa for simplicity; n and k_4 are regression parameters. When the soil changes from the unsaturated to the saturated state, the transition of G_0 is smooth (Ng and Yung, 2008). This is because when the suction reaches zero, the pore-air pressure is equal to the pore-water pressure, and hence the net stress is equal to Terzaghi's effective stress. It should be noted that suction can exist in a saturated fine-grained soil in which Terzaghi's definition of effective stress is still valid. In addition, if a soil is wetted to zero suction, it may still be unsaturated because of trapped air. In this case, the net stress is not equal to Terzaghi's effective stress.

Equation (8.6) can model the effects of suction on G_0 below and above the air-entry value (AEV). When soil suction is below the AEV, the soil remains essentially saturated. When the soil suction is higher than the AEV, the soil desaturates, leading to a higher G_0. By substituting Equation (8.4) with Equation (8.6) in Plaxis, the effects of suction on small-strain soil stiffness can be considered in finite element simulations.

Design analysis

Plane-strain design analyses were carried out using the finite element software Plaxis 2D. In the analysis, a typical section through the northern side of the excavation (labelled

A-A, Figure 8.39a) was selected. The two-stress state variable approach (Fredlund and Rahardjo, 1993) is used to model unsaturated soils. Two analyses, with and without taking suction-dependent soil stiffness into consideration were conducted.

Finite element mesh and boundary conditions

Figure 8.39b shows the finite element mesh adopted in the analyses. According to Zheng et al. (2018), the soil was modelled using fifteen-node triangular elements, while the contiguous piles were simulated using plate elements. The thickness of the plates was estimated based on the equivalent values of the flexible stiffness (Chai et al., 2015). Both the horizontal and vertical displacements at the bottom boundary were fixed. At the two lateral boundaries, a vertical sliding boundary was set with rollers, but the horizontal displacement was constrained. The groundwater tables inside and outside the area before excavation were set at depths of −17.2 m and −3 m, respectively. During excavation, a fully automatic time-stepping procedure was adopted in Plaxis (Brinkgreve et al., 2015). No slip elements were used at the soil-wall interface. That means the soil elements adjacent to the pile wall were directly connected to the pile wall surface and hence the soil-wall interface displacement is continuous. This assumption of a perfect soil-structure interface has negligible effects on the simulated results (Shi, 2015). Numerical convergence was ensured by using a Newton-type iterative procedure (Brinkgreve et al., 2015).

Input parameters for soil and pile wall

The parameters C, a, n and k required by Equation (8.6) were determined from the experimental data by a least-squares method using a multiple linear regression model (Ng and Yung, 2008). They were calibrated as 65.5, −0.77, 0.17 and 0.045, respectively, for both silt and silty clay. For simplicity, the above calibrated parameters were also employed for the fill material. Poisson's ratio is a function of soil suction (Oh and Vanapalli, 2011) and ranges from 0.1 to 0.3 for unsaturated soils (Bowles, 1995). For simplicity, the Poisson's ratio in this study was assumed to be 0.2. Note that the effects of Poisson's ratio on ground settlement and basement heave can almost be neglected, as the difference was less than 6% when Poisson's ratio increased from 0.1 to 0.3 (Ng et al., 2020d). Young's modulus, Poisson's ratio and unit weight of the pile wall were 30 GPa, 0.2 and 25 kN/m³, respectively. A summary of other measured parameters used in the HSS model is given in Table 8.9.

Construction stages and simulation procedures

The simulation procedures were in accordance with the actual construction stages. The initial stress conditions of soils in the simulation were generated at 1 g (gravitational acceleration) by assuming that the coefficient of at-rest earth pressure of the soil (K_0) equals 1-sinφ' (Jaky, 1944). At construction Stage 1, the installation of the contiguous piles was modelled with a "wish-in-place" (WIP) wall for simplicity (Ng, 1993). Then, the plate elements of the contiguous piles were activated. At Stage 2, the water table inside the excavation was lowered to a depth of −17.2 m. From Stage 3 to Stage 6 (final stage), the ground was consecutively excavated to depths of −2 m, −3.7 m, −10.45 m and −15.2 m, respectively. The suction distribution above the groundwater table during excavation was assumed to follow the hydrostatic line. Excavation was simulated by removing nodes and elements in each stage.

Comparison of analyses with and without considering suction effects on soil stiffness

Deflection of pile wall

Figure 8.40a compares the measured and predicted wall deflection without taking suction effects on small-strain soil stiffness into account. It can be seen that a cantilever mode of wall deflection was measured and predicted after each excavation stage. From construction Stage 3 to the final stage, the magnitude of the wall deflection increased, especially near the ground surface. The measured maximum lateral wall deflection was around 0.3% of the excavation depth. This value is much less than Peck's data (2% of excavation depth; Peck, 1969), where there were lateral supporting systems. This implies that without using the lateral supporting systems in the current project, the presence of an unsaturated earth berm in front of the pile wall also reduced the wall deflection significantly.

The analysis without taking unsaturated soil stiffness into consideration shows that the predicted results were greater than the measured data, especially at Stage 5 and the final stage. At the end of excavation, the wall deflection near the ground surface was overestimated by 85%. However, the prediction used to control construction was improved significantly by including the effects of soil suction on soil stiffness in the model (Figure 8.40b). The analysis including soil suction effects predicted the wall deflection quite well at Stage 3. The prediction error was only 20% in the final stage. The comparison between Figures 8.40a and 8.40b reveals that the wall deflection was highly overestimated when the soil stiffness was determined from saturated soils. It also demonstrates the importance of modelling suction-dependent small-strain soil stiffness in the design analysis of deep excavations.

Ground surface settlement

Figure 8.41a shows a comparison of the ground surface settlement behind the pile wall using saturated soil stiffness. The ground settlement was only caused by excavation. The settlement caused by lowering the water table was not considered in this study. Based on the field measurement, the maximum surface settlement after the final stage was 46.5 mm (around 4 m away from the wall). The predicted results reveal that with increasing distance from the pile wall, the ground surface settlement first increased and then decreased gradually. The maximum ground surface settlement was located at a distance of 2.5 m away from the back of the wall. A concave settlement profile was observed. The analysis based on saturated soil stiffness shows that the maximum settlement in the final stage was greatly overestimated by 55%. However, the analysis predicted the ground surface settlement quite well when the unsaturated soil stiffness was included (see Figure 8.41b). A suction-induced increase in the soil stiffness of the earth berm restrained the lateral wall deflection (Figure 8.40) and hence reduced ground surface settlement behind the wall.

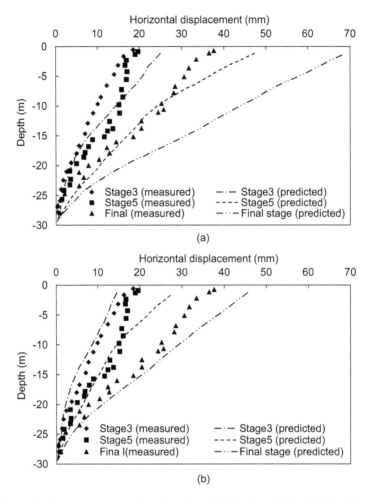

Figure 8.40 Comparison between measured and predicted deflection of pile wall: (a) without and (b) with considering suction-dependent soil stiffness.

By comparing Figures 8.41a and 8.41b, we see that the prediction of the ground surface settlement without taking unsaturated soil stiffness into account was too conservative and hence not economical in practical design.

Basement heave

Figure 8.42 shows the basement heave during excavation. The maximum measured heave was 43 mm, which was around 4 m away from the inner pile wall. The predicted results clearly show that the basement heave was convex in shape, with a maximum of about 3 m away from the pile wall. The heave amount became constant when the distance away from the pile wall was more than 20 m. Compared to the analysis based on saturated soil stiffness, the analysis including suction-dependent soil stiffness predicted the maximum

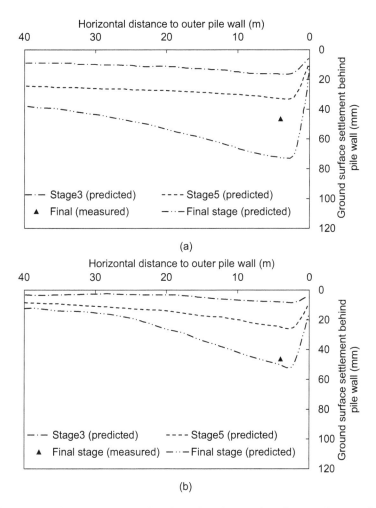

Figure 8.41 Comparison between measured and predicted ground surface settlement after outer pile wall: (a) without and (b) with considering suction-dependent soil stiffness.

basement heave better. The accuracy of the prediction improved by more than 40% when suction effects were considered. This improvement demonstrates that the unsaturated soil within the top 2 m of the basement can restrict ground heave due to the suction-induced increase in small-strain soil stiffness.

After including suction effects on the maximum shear modulus, the trends of wall deflection, ground settlement and basement heave remained similar (see Figures 8.40 to 8.42). In addition, the location of the maximum settlement and basement heave did not change (Figures 8.41 and 8.42). This observation is consistent with the measured results reported by Roy and Robinson (2009) and Zhang et al. (2018), who showed that the locations of peak ground movements were not affected by changes in effective stress due to dewatering. Based on the predicted and measured results in Figures 8.3–8.5, it is clear that the design analysis with suction-dependent small-strain soil stiffness correctly predicted

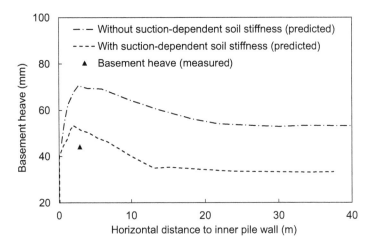

Figure 8.42 Comparison between measured and predicted basement heave with and without considering suction-dependent soil stiffness.

the field performance resulting from de-watering in a deep excavation. Hence, the analysis including unsaturated soil stiffness provided a safe and economical design during construction. It saved construction time and reduced construction costs.

Summary

By including suction-dependent small-strain stiffness to account for the effects of de-watering in design analyses of a 15 m-deep excavation in Tianjin, China, wall deflection and ground movements were predicted and compared with field measurements during the construction of the excavation. It was found that the analysis that did not include the unsaturated small-strain soil stiffness significantly overestimated the deflection of the pile wall by 85%, the ground surface settlement by 55% and basement heave by 40%. On the other hand, by including unsaturated soil stiffness, the analysis allowed for the safe and economical design and construction of the deep excavation. It is recommended that unsaturated small-strain soil stiffness resulting from de-watering should be considered during the construction of deep excavations in future.

SOURCES

The South- to- North Water Division Project, China (Ng et al., 2003)
Field monitoring of an unsaturated saprolitic hillslope on Lan Tau Island in Hong Kong (Leung et al, 2011)
Field performance of non- vegetated and vegetated three- layer landfill cover systems using construction waste without geomembrane (Ng et al., 2016b, 2019d, 2022b)
Use of unsaturated small- strain soil stiffness to the design of wall deflection and ground movement adjacent to deep excavation (Ng et al., 2020d)

References

AASHTO. (2008). *Mechanical empirical pavement design guide: a manual of practice*. American Association of State and Highway Transportation Officials, Washington DC, USA.

AASHTO. (2017). *Standard method of test for determining the resilient modulus of soils and aggregate materials*. American Association of State Highway and State Highway Officials, Washington DC, USA.

Abdolahzadeh, A. M., Lacroix Vachon, B. and Cabral, A. R. (2011). Evaluation of the effectiveness of a cover with capillary barrier effect to control percolation into a waste disposal facility. *Canadian Geotechnical Journal*, 48(7), 996–1009.

Abuel-Naga, H. M., Bergado, D. T., Bouazza, A. and Ramana, G. V. (2007a). Volume change behaviour of saturated clays under drained heating conditions: experimental results and constitutive modeling. *Canadian Geotechnical Journal*, 44(8), 942–956.

Abuel-Naga, H. M., Bergado, D. T. and Bouazza, A. (2007b). Thermally induced volume change and excess pore water pressure of soft Bangkok clay. *Engineering Geology*, 89(1–2), 144–154.

Agus, S. S. and Schanz, T. (2005). Comparison of four methods for measuring total suction. *Vadose Zone Journal*, 4(4), 1087.

Ahmad, M., Rajapaksha, A. U., Lim, J. E., Zhang, M., Bolan, N., Mohan, D., Vithanage, M., Lee, S. S. and Ok, Y. S. (2014). Biochar as a sorbent for contaminant management in soil and water: a review. *Chemosphere*, 99, 19–33.

Airey, D., Suchowerska, A. and Williams, D. (2012). Limonite – a weathered residual soil heterogeneous at all scales. *Géotechnique Letters*, 2, 119–122.

Aitchison, G. D. (1961). Relationship of moisture and effective stress functions in unsaturated soils. *Proceedings of Confeernce on Pore Pressure and Suction in Soils*, 47–52.

Aitchison, G. D. (1965). Soil properties, shear strength and consolidation. *Proceedings of the 6th International Conference on Soil Mechanics and Foundation Engineering*, 3, 318–321.

Al Khafaf, S. and Hanks, R. J. (1974). Evaluation of the filter paper method for estimating soil water potential. *Journal of Soil Science*, 117(4), 194–199.

Alam, M. J. B. (2017). Evaluation of plant root on the performance of Evapotranspiration (ET) cover system. PhD Thesis, The University of Texas, Arlington, USA.

Albright, W. H., Benson, C. H., Gee, G. W., Roesler, A. C., Abichou, T., Apiwantragoon, P., Lyles, B. F. and Rock, S. A. (2004). Field water balance of landfill final covers. *Journal of Environmental Quality*, 33(6), 2317–2332.

Albright, W. H., Benson, C. H., Gee, G. W., Abichou, T., Tyler, S. W. and Rock, S. A. (2006). Field performance of three compacted clay landfill covers. *Vadose Zone Journal*, 5(4), 1157–1171.

Albright, W. H., Benson, C. H. and Apiwantragoon, P. (2013). Field hydrology of landfill final covers with composite barrier layers. *Journal of Geotechnical and Geoenvironmental Engineering*, 139(1), 1–12.

Alexander, L.T. and Cady, J.G. (1962). *Genesis and hardening of laterite in soils*. US Department of Agriculture.

Allen, R. G., Pereira, L. S., Raes, D. and Smith, M. (1998). Crop evapotranspiration. Guidelines for computing crop water requirements. Irrigation and drainage paper 56, FAO, Rome, Italy.

Al-Mukhtar, M., Qi, Y., Alcover, J. F. and Bergaya, F. (1999). Oedometric and water-retention behavior of highly compacted unsaturated smectites. *Canadian Geotechnical Journal*, 36(4), 675–684.

Alonso, E. E. (1998). Modeling expansive soil behavior. *Proceedings of the 2nd International Conference on Unsaturated Soils*, 37–70.

Alonso, E. E., Gens, A., and Hight, D. (1987). General report: special problem soils. *Proceedings of the 9th European Conference of Soil Mechanics and Foundation Engineering*, 3, 1087–1146.

Alonso, E. E., Gens, A. and Josa, A. (1990). A constitutive model for partially saturated soils. *Géotechnique*, 40(3), 405–430.

Alonso, E. E., Iturralde, E. F. O. and Romero, E. E. (2007). Dilatancy of coarse granular aggregates. In *Experimental unsaturated soil mechanics*. Springer Berlin Heidelberg, Berlin, Heidelberg, 112, 119–135.

Alonso, E. E., Pereira, J. M., Vaunat, J., Olivella, S. and Nat, V. A. U. (2010). A microstructurally based effective stress for unsaturated soils. *Géotechnique*, 60(12), 913–925.

Alonso, E. E., Pinyol, N. M. and Gens, A. (2013). Compacted soil behaviour: initial state, structure and constitutive modelling. *Géotechnique*, 63(6), 463–478.

Alsherif, N. A. and McCartney, J. S. (2015). Thermal behaviour of unsaturated silt at high suction magnitudes. *Géotechnique*, 65(9), 703–716.

Amaya, P., Queen, B., Stark, T. D. and Choi, H. (2006). Case history of liner veneer instability. *Geosynthetics International*, 13(1), 36–46.

Apfel, R. E. (1970). The role of impurity in cavitation-threshold determination. *The Journal of the Acoustical Society of America*, 48(58), 1179–1186.

ASTM. (2006a). *Standard practice for classification of soils for engineering purposes (Unified soil classification system)*. American Society for Testing and Materials, West Conshohocken, PA.

ASTM. (2006b). *Standard test method permeability of granular soils (constant head)*. American Society for Testing and Materials, West Conshohocken, PA.

ASTM. (2007). *Standard test method for particle-size analysis of soils*. American Society for Testing and Materials, West Conshohocken, PA.

ASTM. (2010a). *Standard test methods for liquid limit, plastic limit, and plasticity index of soils*. American Society for Testing and Materials, West Conshohocken, PA.

ASTM. (2010b). *Standard test methods for measurement of hydraulic conductivity of saturated porous materials using a flexible wall permeameter*. American Society for Testing and Materials, West Conshohocken, PA.

ASTM. (2011). *Standard practice for classification of soils for engineering purposes (USCS)*. American Society for Testing and Materials, West Conshohocken, PA.

ASTM. (2012). *Standard test method for laboratory compaction characteristics of soil using standard effort*. American Society for Testing and Materials, West Conshohocken, PA.

ASTM. (2014a). *Standard test methods for determining the amount of material finer than 75-μm (No.200) sieve in soils by washing*. American Society for Testing and Materials, West Conshohocken, PA.

ASTM. (2014b). *Standard test methods for maximum index density and unit weight of soils using a vibratory table*. American Society for Testing and Materials, West Conshohocken, PA.

ASTM. (2016). *Standard test method for measurement of soil potential (Suction) using filter paper*. American Society for Testing and Materials, West Conshohocken, PA.

Atkinson, J. H. (2000). Non-linear soil stiffness in routine design. *Géotechnique*, 50(5), 487–508.

Atkinson, J. H. and Sallfors, G. (1991). Experimental determination of stress–strain–time characteristics in laboratory and in situ tests. General report to Session 1. *Proceedings of the 10th European Conference on Soil Mechanics and Foundation Engineering*, 3, 915–956.

Atkinson, J. H., Richardson, D. and Stallebrass, S. E. (1990). Effect of recent stress history on the stiffness of overconsolidated soil. *Géotechnique*, 40(4), 531–540.

Aubertin, M., Cifuentes, E., Apithy, S. A., Bussière, B., Molson, J. and Chapuis, R. P. (2009). Analyses of water diversion along inclined covers with capillary barrier effects. *Canadian Geotechnical Journal*, 46(10), 1146–1164.

Bai, B., Guo, L. and Han, S. (2014). Pore pressure and consolidation of saturated silty clay induced by progressively heating/cooling. *Mechanics of Materials*, 75, 84–94.

Baker, R. and Frydman, S. (2009). Unsaturated soil mechanics: critical review of physical foundations. *Engineering Geology*, 106(1–2), 26–39.

Bao, C. G. and Ng, C. W. W. (2000). Keynote lecture: some thoughts and studies on the prediction of slope stability in expansive soils. *Proceedings of the 1st Asian Conference on Unsaturated Soils*, 15–31.

Bardet, J. P. (1986). Bounding surface plasticity model for sands. *Journal of Engineering Mechanics*, 112(11), 1198–1217.

Barnswell, K. D. and Dwyer, D. F. (2011). Assessing the performance of evapotranspiration covers for municipal solid waste landfills in Northwestern Ohio. *Journal of Environmental Engineering*, 137(4), 301–305.

Barnswell, K. D., and Dwyer, D. F. (2012). Two-year performance by evapotranspiration covers for municipal solid waste landfills in northwest Ohio. *Waste Management*, 32(12), 2336–2341.

Barrón, V. and Torrent, J. (1996). Surface hydroxyl configuration of various crystal faces of hematite and goethite. *Journal of Colloid and Interface Science*, 177(2), 407–410.

Basile, A., Coppola, A., de Mascellis, R. and Randazzo, L. (2006). Scaling approach to deduce field hydraulic properties and behavior from laboratory measurements on small cores. *Vadose Zone Journal*, 5(3), 1005–1016.

Basu, D., Misra, A. and Puppala, A. J. (2015). Sustainability and geotechnical engineering: perspectives and review. *Canadian Geotechnical Journal*, 52(1), 96–113.

Bathurst, R. J., Ho, A. F. and Siemens, G. (2007). A column apparatus for investigation of 1-D unsaturated-saturated response of sand-geotextile systems. *Geotechnical Testing Journal*, 30(6), 100954.

Been, K. and Jefferies, M. G. (1985). A state parameter for sands. *Géotechnique*, 35(2), 99–112.

Benson, C. and Khire, M. (1995). Earthen final covers for landfills in semi-arid and arid climates. In *Landfill closures*. Edited by R. Dunn and U. Singh. ASCE, Reston, VA, 201–208.

Benson, C. H., Daniel, D. E. and Boutwell, G. P. (1999). Field performance of compacted clay liners. *Journal of Geotechnical and Geoenvironmental Engineering*, 125(5), 390–403.

Benson, C. H., Barlaz, M. A., Lane, D. T. and Rawe, J. M. (2007). Practice review of five bioreactor/recirculation landfills. *Waste Management*, 27(1), 13–29.

Benson, C., Abichou, T., Albright, W., Gee, G. and Roesler, A. (2001). Field evaluation of alternative earthen final covers. *International Journal of Phytoremediation*, 3(1), 105–127.

Benz, T. (2007). Small-strain stiffness of soil and its numerical consequences. PhD thesis, University of Stuttgart, Germany.

Bhattacharyya, R., Smets, T., Fullen, M. A., Poesen, J. and Booth, C. A. (2010). Effectiveness of geotextiles in reducing runoff and soil loss: a synthesis. *Catena*, 81(3), 184–195.

Biglari, M., Jafari, M.K., Shafiee, A., Mancuso, C. and D'Onofrio, A. (2011a). Shear Modulus and damping ratio of unsaturated kaolin measured by new suction-controlled cyclic triaxial device. *Geotechnical Testing Journal*, 34(5), 1–12.

Biglari, M., Mancuso, C., d'Onofrio, A., Jafari, M. K. and Shafiee, A. (2011b). Modelling the initial shear stiffness of unsaturated soils as a function of the coupled effects of the void ratio and the degree of saturation. *Computers and Geotechnics*, 38(5), 709–720.

Bishop, A. W. (1959). The principle of effective stress. *Tecknisk Ukeblad*, 106(39), 859–863.

Bishop, A. W. and Blight, G. E. (1963). Some aspects of effective stress in saturated and partly saturated soils. *Géotechnique*, 13(3), 177–197.

Blatz, J. A., Cui, Y. J. and Oldecop, L. (2008). Vapour equilibrium and osmotic technique for suction control. *Geotechnical and Geological Engineering*, 26(6), 661–673.

Blight, G. E. (1965). A study of effective stresses for volume change. In *Moisture Equilibria and Moisture Changes in Soils Beneath Covered Areas*, Proceedings of a Conference in Symposium, 259–269.

Bocking, K. A. and Fredlund, D. G. (1980). Limitations of the axis translation technique. *Proceedings of the 4th International Conference on Expansive Soils*, 117–135.

Bohnhoff, G. L., Ogorzalek, A. S., Benson, C. H., Shackelford, C. D. and Apiwantragoon, P. (2009). Field data and water-balance predictions for a monolithic cover in a semiarid climate. *Journal of Geotechnical and Geoenvironmental Engineering*, 135(3), 333–348.

Bolton, M. D. (1986). The strength and dilatancy of sands. *Géotechnique*, 36(1), 65–78.

Bossé, B., Bussière, B., Hakkou, R., Maqsoud, A. and Benzaazoua, M. (2015). Field experimental cells to assess hydrogeological behaviour of store-and-release covers made with phosphate mine waste. *Canadian Geotechnical Journal*, 52(9), 1255–1269.

Bouazza, A. and Rahman, F. (2007). Oxygen diffusion through partially hydrated geosynthetic clay liners. *Géotechnique*, 57(9), 767–772.

Bouazza, A. and Vangpaisal, T. (2003). An apparatus to measure gas permeability of geosynthetic clay liners. *Geotextiles and Geomembranes*, 21(2), 85–101.

Bouazza, A., Gates, W. P. and Ranjith, P. G. (2009). Hydraulic conductivity of biopolymer-treated silty sand. *Géotechnique*, 59(1), 71–72.

Bouazza, A., Zornberg, J. G., McCartney, J. S. and Nahlawi, H. (2006). Significance of unsaturated behaviour of geotextiles in earthen structures. *Australian Geomechanics*, 41(3), 133–142.

Bowles, J. E. (1995). *Foundation analysis and design* (5th edition). McGraw-Hill Companies, New York.

Brackley, I. J. A. and Sanders, P. J. (1992). In situ measurement of total natural horizontal stresses in an expansive clay. *Géotechnique*, 42(3), 443–451.

Brinkgreve, R. B. J., Kumarswamy, S. and Swolfs, W. M. (2015). *PLAXIS manual for version 2015*. The Netherlands.

Brown, R. W. and Bartos, D. L. (1982). A calibration model for screen-caged Peltier thermocouple psychrometers. Research paper INT-293, U.S. Department of Agriculture, Washington, DC.

Brown, R. W. and Collins, J. (1980). A screen-caged thermocouple psychrometer and calibration chamber for measurements of plant and soil water potential. *Agronomy Journal*, 72, 851–854.

Brown, S. F. (1996). Soil mechanics in pavement engineering. *Géotechnique*, 46(3), 383–426.

Brown, S. F. (1997). Achievements and challenges in Asphalt Pavement Engineering. *Proceedings of the 8th International Conference on Asphalt Pavements*, 1, 1–23.

Brown, S. F., Loach, S. C. and O'Reilly, M. P. (1987). Repeated loading of fine grained soils. Contractor report 72, Transportation Research Laboratory, Pavement Design and Maintenance Division.

Brown, S. F., Tam, W. S. and Brunton, J. M. (1986). Development of an analytical method for the structural evaluation of pavements. *Proceedings of the 2nd International Conference on Bearing Capacity of Roads and Airfields*, 267–276.

BSI. (1990a). *British standard method of test for soils for civil engineering purposes-part 2: classification tests.*, British Standard International, London.

BSI. (1990b). *British standard method of test for soils for civil engineering purposes-part 4: compaction related tests*. British Standard International, London.

Buckingham, E. (1907). Studies on the movement of soil moisture. Bureau of Soils No. 38, US Department of Agriculture, Washington, DC.

Buczko, U., Bens, O. and Hüttl, R. F. (2007). Changes in soil water repellency in a pinebeech forest transformation chronosequence: influence of antecedent rainfall and air temperatures. *Ecological Engineering*, 31(3), 154–164.

Budhu, M. (2008). *Mechanics and foundations* (2nd Edition). John Wiley & Sons, New York, USA.

Burdine, N. T. (1953). Relative permeability calculations from pore size distribution data. *Journal of Petroleum Technology*, 5(3), 71–78.

Burland, J. B. (1964). Effective stresses in partly saturated soils. *Géotechnique*, 14(1), 65–68.

Burland, J. B. (1965). Some aspects of the mechanical behaviour of partially saturated soils. In *Moisture equilibria and moisture changes beneath covered areas*, 270–278.

Burland, J. B. (1989). Ninth Laurits Bjerrum Memorial Lecture: "Small is beautiful" – the stiffness of soils at small strains. *Canadian Geotechnical Journal*, 26(4), 499–516.

Burland, J. B. (1990). On the compressibility and shear strength of natural clays. *Géotechnique*, 40(3), 329–378.

Burton, G. J., Pineda, J. A., Sheng, D. and Airey, D. (2015). Microstructural changes of an undisturbed, reconstituted and compacted high plasticity clay subjected to wetting and drying. *Engineering Geology*, 193, 363–373.

Bussière, B., Aubertin, M. and Chapuis, R. P. (2003). The behavior of inclined covers used as oxygen barriers. *Canadian Geotechnical Journal*, 40(3), 512–535.

Callisto, L. and Rampello, S. (2004). An interpretation of structural degradation for three natural clays. *Canadian Geotechnical Journal*, 41(3), 392–407.

Campanella, R. G. and Mitchell, J. K. (1968). Influence of temperature variations on soil behavior. *Journal of the Soil Mechanics and Foundations Division*, 94(3), 709–734.

Carbon Farming Initiative (CFI). (2013). *Guidelines for calculating regulatory baselines for legacy waste landfill methane projects*. Australian Government.

Cekerevac, C. and Laloui, L. (2004). Experimental study of thermal effects on the mechanical behaviour of a clay. *International Journal for Numerical and Analytical Methods in Geomechanics*, 28(3), 209–228.

Cekerevac, C. and Laloui, L. (2010). Experimental analysis of the cyclic behaviour of kaolin at high temperature. *Géotechnique*, 60(8), 651–655.

Chai, J. C., Shrestha, S., Hino, T., Ding, W. Q., Kamo, Y. and Carter, J. (2015). 2D and 3D analyses of an embankment on clay improved by soil–cement columns. *Computers and Geotechnics*, 68, 28–37.

Chandler, R. J. and Gutierrez, C. I. (1986). The filter-paper method of suction measurement. *Géotechnique*, 36(2), 265–268.

Chang, I., Im, J. and Cho, G. C. (2016a). Geotechnical engineering behaviors of gellan gum biopolymer treated sand. *Canadian Geotechnical Journal*, 53(10), 1658–1670.

Chang, I., Im, J. and Cho, G. C. (2016b). Introduction of microbial biopolymers in soil treatment for future environmentally-friendly and sustainable geotechnical engineering. *Sustainability*, 8(3), 251.

Chapman, P. J., Richards, B. E. and Trevena, D. H. (1975). Monitoring the growth of tension in a liquid contained in a Berthelot tube. *Journal of Physics E: Scientific Instruments*, 8(9), 731–735.

Chapuis, R. P., Masse, I., Madinier, B. and Aubertin, M. (2006). A drainage column test for determining unsaturated properties of coarse materials. *Geotechnical Testing Journal*, 30(2), 83–89.

Chávez, C. and Alonso, E.E. (2003). A constitutive model for crushed granular aggregates which includes suction effects. *Soils and Foundations*, 43(4), 215–227.

Chen, F. H. (1988). *Foundations on expansive soils* (2nd ed.). Elsevier, Amsterdam.

Chen, R. (2007). Experimental study and constitutive modelling of stress-dependent coupled hydraulic hysteresis and mechanical behaviour of an unsaturated soil. PhD thesis, The Hong Kong University of Science and Technology, China.

Chen, R., Huang, J., Chen, Z., Xu, Y., Liu, J. and Ge, Y. (2019). Effect of root density of wheat and okra on hydraulic properties of an unsaturated compacted loam. *European Journal of Soil Science*, 70(3), 493–506.

Chen, R., Lee, I. and Zhang, L. (2015). Biopolymer stabilization of mine tailings for dust control. *Journal of Geotechnical and Geoenvironmental Engineering*, 141(2), 04014100–1.

Chen, R., Xu, T., Lei, W., Zhao, Y. and Qiao, J. (2018). Impact of multiple drying–wetting cycles on shear behaviour of an unsaturated compacted clay. *Environmental Earth Sciences*, 77(19), 683.

Chen, Y. M., Zhan, L. T., Xu, X. B. and Liu, H. L. (2013). Geo-environmental problems in landfills of MSW with high organic content. *Proceedings of the 18th International Conference on Soil Mechanics and Geotechnical Engineering: Challenges and Innovations in Geotechnics*, 4, 3009–3012.

Chen, Z. K. (2016). Gas breakthrough and emission in unsaturated landfill final cover considering cracking effect. PhD Thesis, The Hong Kong University of Science and Technology, China.

Cheng, Q., Ng, C. W. W., Zhou, C. and Tang, C. S. (2019). A new water retention model that considers pore non-uniformity and evolution of pore size distribution. *Bulletin of Engineering Geology and the Environment*, 78(7), 5055–5065.

Cheng, Q., Zhou, C., Ng, C. W. W. and Tang, C. S. (2020). Effects of soil structure on thermal softening of yield stress. *Engineering Geology*, 269, 1–6.

Childs, E. (1969). *An Introduction to the Physical Basis of Soil Water Phenomena*. Wiley-Interscience, London.

Ching, R. K. H. and Fredlund, D. G. (1984). A small Saskatchewan copes with swelling clay problems. *Proceedings of the 5th Intenational Conference of Expansive Soils*, 306–610.

Chiu, C. F. (2001). Behaviour of unsaturated loosely compacted weathered materials. Phd Thesis, The Hong Kong University of Science and Technology, China.

Chiu, C. F. and Ng, C. W. W. (2003). A state-dependent elasto-plastic model for saturated and unsaturated soils. *Géotechnique*, 53(9), 809–829.

Chiu, C. F. and Ng, C. W. W. (2012). Coupled water retention and shrinkage properties of a compacted silt under isotropic and deviatoric stress paths. *Canadian Geotechnical Journal*, 49(8), 928–938.

Chiu, C. F., Ng, C. W. W. and Shen, C. K. (1998). Collapse behavior of loosely compacted virgin and non-virgin fills in Hong Kong. *Proceedings of the 2nd International Conference on Unsaturated Soils*, 1, 25–30.

Choo, L. P. and Yanful, E. K. (2000). Water flow through cover soils using modeling and experimental methods. *Journal of Geotechnical and Geoenvironmental Engineering*, 126(4), 324–334.

Chu, T. Y. and Mou, C. H. (1973). Volume change characteristic of expansive soils determined by controlled suction tests. *Proceedings of the 3rd Intenational Conference on Expansive Soils*, 177–185.

Chui, T. F. M. and Freyberg, D. L. (2007). The use of COMSOL for integrated hydrological modeling. *Proceedings of the COMSOL Conference*, 217–223.

Chui, T. F. M. and Freyberg, D.L. (2009). Implementing hydrologic boundary conditions in a multiphysics model. *Journal of Hydrologic Engineering*, 14(12), 1374–1377.

Clayton, C. R. I. (2011). Stiffness at small strain: research and practice. *Géotechnique*, 61(1), 5–37.

Clayton, C. R. I. and Heymann, G. (2001). Stiffness of geomaterials at very small strains. *Géotechnique*, 51(3), 245–255.

Clayton, C. R. I., Khatrush, S. A., Bica, A. V. D. and Siddique, A. (1989). The use of hall effect semiconductors in geotechnical instrumentation. *Geotechnical Testing Journal*, 12(1), 69–76.

Coleman, J. D. (1962). Stress-strain relations for partly saturated soil. *Géotechnique*, 12(4), 348–350.

Crawford, J. F. and Smith, P. G. (1985). *Landfill technology*. Butterwoths. London.

Cresswell, A. and Powrie, W. (2004). Triaxial tests on an unbonded locked sand. *Géotechnique*, 54(2), 107–115.

Crilly, M. S., Schreiner, H. D. and Gourley, C. (1991). A simple field suction probe. *Proceedings of the 10th Regional Conference for Africa on Soil Mechanics & Foundation Engineering*, 291–298.

Croney, D., Coleman, J. D. and Black, W. P. M. (1958). Movement and distribution of water in relation to highway design and performance. Highway Research Board Special Report No. 40, Highway Research Board, Washington.

CSI. (2009). Instruction manual for the 229 heat dissipation matric water potential sensor. Revision 5/09, Campbell Scientific, Inc. (CSI), Logan, Utah.

Cui, Y. J. and Delage, P. (1996). Yielding and plastic behaviour of an unsaturated compacted silt. *Géotechnique*, 46(2), 291–311.

Cui, Y. J., Sultan, N. and Delage, P. (2000). A thermomechanical model for saturated clays. *Canadian Geotechnical Journal*, 37(3), 607–620.

D'Onza, F., Gallipoli, D., Wheeler, S., Casini, F., Vaunat, J., Khalili, N., Laloui, L., Mancuso, C., Mašín, D., Nuth, M., Pereira, J. M. and Vassallo, R. (2011). Benchmark of constitutive models for unsaturated soils. *Géotechnique*, 61(4), 283–302.

Dafalias, Y. F. (1986). Bounding surface plasticity. I: mathematical foundation and hypoplasticity. *Journal of Engineering Mechanics*, 112(9), 966–987.

Dafalias, Y. F. and Manzari, M. T. (2004). Simple plasticity sand model accounting for fabric change effects. *Journal of Engineering Mechanics*, 130(6), 622–634.

Dalton, H. M. and March, P. E. (1998). Molecular genetics of bacterial attachment and biofouling. *Current Opinion in Biotechnology*, 9(3), 252–255.

Daniel, D. E. (1994). Surface barriers: problems, solutions, and future needs. *Proceedings of the Thirty-Third Hanford Symposium on Health and the Environment*, 441-87.

Dashko, R. and Shidlovskaya, A. (2016). Impact of microbial activity on soil properties. *Canadian Geotechnical Journal*, 53(9), 1386–1397.

Dehlen, G. L. (1969). The effect of non-linear material response on behavior of pavements subjected to traffic loads. Ph.D. Thesis, University of California, Berkeley, USA.

DeJong, J. T., Soga, K., Kavazanjian, E., Burns, S., Van Paasen, L. A., AL Qabany, A., Aydilek, A., Bang, S. S., Burbank, M., Caslake, L. F., … and Weaver, T. (2013). Biogeochemical processes and geotechnical applications: progress, opportunities and challenges. *Géotechnique*, 63(4), 287–301.

Delage, P. and Lefebvre, G. (1984). Study of the structure of a sensitive Champlain clay and of its evolution during consolidation. *Canadian Geotechnical Journal*, 21(1), 21–35.

Delage, P., Audiguier, M., Cui, Y. J. and Howat, M. D. (1996). Microstructure of a compacted silt. *Canadian Geotechnical Journal*, 33(1), 150–158.

Delage, P., Howat, M. D. and Cui, Y. J. (1998). The relationship between suction and swelling properties in a heavily compacted unsaturated clay. *Engineering Geology*, 50(1–2), 31–48.

Delage, P., Romero, E. and Tarantino, A. (2008). Recent developments in the techniques of controlling and measuring suction in unsaturated soils. *Proceedings of the 1st European Conference on Unsaturated Soils (E-UNSAT 2008)*, 33–52.

Di Donna, A. and Laloui, L. (2015). Response of soil subjected to thermal cyclic loading: experimental and constitutive study. *Engineering Geology*, 190, 65–76.

Dineen, K. and Burland, J. B. (1995). A new approach to osmotically controlled oedometer testing. *Proceedings of 1st International conference on Unsaturated Soils*, 2, 459–465.

Dong, Y., Lu, N. and McCartney, J. S. (2016). Unified model for small-strain shear modulus of variably saturated soil. *Journal of Geotechnical and Geoenvironmental Engineering*, 142(9), 04016039.

Dorsey, N. E. (1940). *Properties of the ordinary water substance in all its phases*. Amer. Chemical Society. Mono. Series, Reinhold, New York.

Duan, X., Zeng, L. and Sun, X. (2019). Generalized stress framework for unsaturated soil: demonstration and discussion. *Acta Geotechnica*, 14(5), 1459–1481.

Dudley, J. H. (1970). Review of collapsing soils. *Journal of Soil Mechanics and Foundations*, 96(SM3), 925–947.

Dyvik, R. and Madshus, C. (1985). Laboratory measurement of Gmax using bender elements. *Proceedings of ASCE Annual Convention: Advances in the Art of Testing Soils Under Cyclic Condition*, 186–196.

Edlefsen, N. E. and Anderson, A. B. C. (1943). Thermodynamics of soil moisture. *Hilgardia*, 15(2), 31–298.

EPA. (1999). U.S. Environmental Protection Agency. Municipal solid waste landfills, vol. 1. Summary of the requirements for the new source performance standards and emission guidelines for municipal solid waste landfills. EPA-453R/ 96-004. Office of Air Quality Planning and Standards. Research Triangle Park, NC.

Esteban, V. and Saez, J. (1988). A device to measure the swelling characteristics of rock samples with control of the suction up to very high values. *ISRM Symposium on Rock Mechanics & Power Plants*, Madrid, 2.

European Environment Agency (EEA). (2013). Managing municipal solid waste – a review of achievements in 32 European countries. Office for Official Publications of the European Communities, Environmental assessment report no. 2, Copenhagen.

Evans, N. C. and Yu, Y. F. (2001). Regional variation in extreme rainfall values – Geotechnical Engineering Office (GEO), Civil Engineering Department, report no. 115.

Fawcett, R. and Collis-George, N. (1967). A filter-paper method for determining the moisture characteristics of soil. *Australian Journal of Experimental Agriculture*, 7(25), 162.

Feddes, R. A., Kowalik, P., Kolinska-Malinka, K. and Zaradny, H. (1976). Simulation of field water uptake by plants using a soil water dependent root extraction function. *Journal of Hydrology*, 31(1–2), 13–26.

Feddes, R. A., Hoff, H., Bruen, M., Dawson, T., de Rosnay, P., Dirmeyer, P., Jackson, R. B., Kabat, P., Kleidon, A., Lilly, A. and Pitman, A. J. (2001). Modeling root water uptake in hydrological and climate models. *Bulletin of the American Meteorological Society*, 82(12), 2797–2809.

Feng, M. (1999). The effects of capillary hysteresis on the measurement of matric suction using thermal conductivity sensors. M.Sc. Thesis, University of Saskatchewan, Saskatoon, Canada.

Feng, M. and Fredlund, D. G. (2003). Calibration of thermal conductivity sensors with consideration of hysteresis. *Canadian Geotechnical Journal*, 40(5), 1048–1055.

Fisher, R. A. (1926). On the capillary forces in an ideal soil; correction of formulae given by WB Haines. *The Journal of Agricultural Science*, 16(3), 492–505.

Flemming, H.-C. and Wingender, J. (2010). The biofilm matrix. *Nature Reviews Microbiology*, 8(9), 623–633.

François, B. and Laloui, L. (2008). ACMEG-TS: a constitutive model for unsaturated soils under non-isothermal conditions. *International Journal for Numerical and Analytical Methods in Geomechanics*, 32(16), 1955–1988.

Fredlund, D. G. (1973). Volume change behaviour of unsaturated soils. PhD thesis, University of Alberta, Edmonton, AB.

Fredlund, D. G. (1987). The stress state for expansive soils. *Proceedings of the 6th International Conference on Expansive Soils*, 2, 1–9.

Fredlund, D. G. (1996). The emergence of unsaturated soil mechanics. In *The Fourth Spencer J. Buchanan Lecture,* Texas A & M, College Station, TX.

Fredlund, D. G. (2000). The 1999 R.M. Hardy Lecture: the implementation of unsaturated soil mechanics into geotechnical engineering. *Canadian Geotechnical Journal*, 37(5), 963–986.

Fredlund, D. G. (2002). Use of soil-water characteristic curve in the implementation of unsaturated soil mechanics. *Proceedings of the Third International Converence on Unsaturated Soils*, 57–80.

Fredlund, D. G. and Morgenstern, N. R. (1977). Stress state variables for unsaturated soils. *Journal of the Geotechnical Engineering Division*, 103(5), 447–466.

Fredlund, D. G. and Rahardjo, H. (1993). *Soil mechanics for unsaturated soils*. John Wiley & Sons, Inc., Hoboken, NJ.

Fredlund, D. G. and Xing, A. (1994). Equations for the soil-water characteristic curve. *Canadian Geotechnical Journal*, 31(6), 1026–1026.

Fredlund, D.G., Bergan, A.T. and Wong, P.K. (1977). Relationship between resilient modulus and stress conditions for cohesive subgrade soils. *Transportation Research Record*, 642, 73–81.

Fredlund, D.G., Morgenstern, N.R. and Widger, R.A. (1978). The shear strength of unsaturated soils. *Canadian Geotechnical Journal*, 15(3), 313–321.

Fredlund, D. G., Xing, A. and Huang, S. (1994). Predicting the permeability function for unsaturated soils using the soil-water characteristic curve. *Canadian Geotechnical Journal*, 31(4), 533–546.

Fredlund, D. G., Xing, A., Fredlund, M. D. and Barbour, S. L. (1996). The relationship of the unsaturated soil shear strength to the soil-water characteristic curve. *Canadian Geotechnical Journal*, 33(3), 440–448.

Fredlund, M. D., Fredlund, D. G. and Wilson, G. W. (1997). Prediction of the SWCC from grain size distribution and volume mass properties. *Proceedings of the 3rd Brazilian Symposium on Unsaturated Soils*, 1, 13–23.

Fredlund, D. G., Shuai, F. and Feng, M. (2000). Use of a new thermal conductivity sensor for laboratory suction measurement. *Proceedings of the 1st Asian Conference on Unsaturated Soils*, 275–280.

Fredlund, D. G., Fredlund, M. D. and Zakerzadeh, N. (2001a). Predicting the permeability function for unsaturated soils. *Proceedings of International Symposium on Suction, Swelling, Permeability and Structured Clays*, 215–221.

Fredlund, D. G., Rahardjo, H., Leong, E. C. and Ng, C. W. W. (2001b). Suggestions and recommendations for the interpretation of soil-water characteristic curves. *Proceedings of the 14th Southeast Asian Geotechnical Conference*, 1, 503–508.

Fredlund, D. G., Rahardjo, H. and Fredlund, M. D. (2012). *Unsaturated soil mechanics in engineering practice*. John Wiley & Sons, Inc., Hoboken, NJ.

Fung, Y. C. (1965). *Foundations of solid mechanics*. Prentice-Hall, Inc., Englewood Cliffs, NJ.

Futai, M. M., Almeida, M. S. S. and Lacerda, W.A. (2004). Yield, strength, and critical state behavior of a tropical saturated soil. *Journal of Geotechnical and Geoenvironmental Engineering*, 130(11), 1169–1179.

Gallé, C. (2000). Gas breakthrough pressure in compacted Fo–Ca clay and interfacial gas overpressure in waste disposal context. *Applied Clay Science*, 17(1–2), 85–97.

Gallen, P. M. (1985). The measurement of soil suction using the filter paper methods. M.Sc. Thesis, University of Saskatchewan, Saskatoon, Canada.

Gallipoli, D. (2012). A hysteretic soil-water retention model accounting for cyclic variations of suction and void ratio. *Géotechnique*, 62(7), 605–616.

Gallipoli, D., Wheeler, S. J. and Karstunen, M. (2003a). Modelling the variation of degree of saturation in a deformable unsaturated soil. *Géotechnique*, 53(1), 105–112.

Gallipoli, D., Gens, A., Sharma, R. and Vaunat, J. (2003b). An elasto-plastic model for unsaturated soil incorporating the effects of suction and degree of saturation on mechanical behaviour. *Géotechnique*, 53(1), 123–135.

Gallipoli, D., Bruno, A. W., D'onza, F. and Mancuso, C. (2015). A bounding surface hysteretic water retention model for deformable soils. *Géotechnique*, 65(10), 793–804.

Gan, J. K. M. and Fredlund, D. G. (1996). Shear strength characteristics of two saprolitic soils. *Canadian Geotechnical Journal*, 33(4), 595–609.

Gan, J. K. M., Fredlund, D. G. and Rahardjo, H. (1988). Determination of the shear strength parameters of an unsaturated soil using the direct shear test. *Canadian Geotechnical Journal*, 25(3), 500–510.

Gao, Y. and Sun, D. (2017). Soil-water retention behavior of compacted soil with different densities over a wide suction range and its prediction. *Computers and Geotechnics*, 91, 17–26.

Gardner, R. (1937). A method of measuring the capillary tension of soil moisture over a wide moisture range. *Soil Science*, 43(4), 277–284.

Gardner, W. R. (1956). Calculation of capillary conductivity from pressure plate outflow data. *Soil Science Society of America Journal*, 20(3), 317.

Gardner, W. R. (1958). Some steady state solutions of the unsaturated moisture flow equation with application to evaporation from a water-table. *Soil Science Society of America Journal*, 85(4), 228–232.

Garg, A., Leung, A. K. and Ng, C. W. W. (2015). Comparisons of soil suction induced by evapotranspiration and transpiration of S. heptaphylla. *Canadian Geotechnical Journal*, 52(12), 2149–2155.

Gasmo, J., Hritzuk, K., Rahardjo, H. and Leong, E. (1999). Instrumentation of an unsaturated residual soil slope. *Geotechnical Testing Journal*, 22(2), 134.

Gebert, J. and Gröngröft, A. (2006). Performance of a passively vented field-scale biofilter for the microbial oxidation of landfill methane. *Waste Management*, 26(4), 399–407.

Gee, G. W., Ward, A. L., Zhang, Z. F., Campbell, G. S. and Mathison, J. (2002). The influence of hydraulic nonequilibrium on pressure plate data. *Vadose Zone Journal*, 1(1), 172–178.

Gens, A. (2010). Soil–environment interactions in geotechnical engineering. *Géotechnique*, 60(1), 3–74.

Gens, A. and Potts, D. M. (1982). A theoretical model for describing the behavior of soil not obeying Rendulic's principle. In *International Symposium on Numerical Models in Geomechanics*, Balkema, Rotterdam.

Gens, A., Sánchez, M. and Sheng, D. (2006). On constitutive modelling of unsaturated soils. *Acta Geotechnica*, 1, 137–147.

GEO. (1994). *Report on the Kwun Lung Lau Landslide of 23 July 1994*, 2. GEO, N.R. Morgenstern and Geotechnical Engineering Department, Hong Kong Government.

GEO. (2007). Landslide assessment and monitoring work at four selected sites – feasibility study. Final geological assessment report for Tung Chung Foothills study area.

Geokon. (2007a). Instruction manual for MEMS in-place inclinometer – model 6150. Lebanon, NH 03766, USA.

Geokon. (2007b). Instruction manual for vibrating – wire earth pressure cells – for models 4800, 4810, 4815, 4820 and 4830. Lebanon, NH 03766, USA.

Germine, J. T. and Ladd, C. C. (1988). Triaxial testing of saturated cohesive soils. In *Advanced triaxial testing of soil and rock*. ASTM STP 977. American Society for Testing and Materials, Philadelphia, PA, 421–459.

Ghestem, M., Sidle, R. C. and Stokes, A. (2011). The influence of plant root systems on subsurface flow: implications for slope stability. *BioScience*, 61(11), 869–879.

Giavasis, I., Harvey, L. M. and McNeil, B. (2000). Gellan gum. *Critical Reviews in Biotechnology*, 20(3), 177–211.

Gidigasu, M. D. (1976). *Laterite soil engineering-pedogenesis and engineering principles-developments in geotechnical engineering*, Elsevier Scientific Publishing Company, Ghana.

Gittens, G. J. (1969). Variation of surface tension of water with temperature. *Journal of Colloid And Interface Science*, 30(3), 406–412.

Goh, S. G., Rahardjo, H. and Leong, E. C. (2014). Shear strength of unsaturated soils under multiple drying-wetting cycles. *Journal of Geotechnical and Geoenvironmental Engineering*, 140(2), 1–5.

Gourley, C. S. and Schreiner, H. (1995). Field measurement of soil suction. *Proceedings of the 1st International Conference on Unsaturated Soils*, 2, 601–607.

Graham, J. and Houlsby, G. T. (1983). Anisotropic elasticity of a natural clay. *Géotechnique*, 33(2), 165–180.

Graham, J., Halayko, K. G., Hume, H., Kirkham, T., Gray, M. and Oscarson, D. (2002). A capillarity-advective model for gas break-through in clays. *Engineering Geology*, 64(2–3), 273–286.

Guan, Y. and Fredlund, D. G. (1997). Use of the tensile strength of water for the direct measurement of high soil suction. *Canadian Geotechnical Journal*, 34(4), 604–614.

Gupta, S. C. and Larson, W. E. (1979). Estimating soil water retention characteristics from particle size distribution, organic matter percent, and bulk density. *Water Resources Research*, 15(6), 1633–1635.

Gupta, S. C., Sharma, P. P. and DeFranchi, S. A. (1989). Compaction Effects on Soil Structure. *Advances in Agronomy*, 42(C), 311–338.

Haeri, S. M., Noorzad, R. and Oskoorouchi, A. M. (2000). Effect of geotextile reinforcement on the mechanical behavior of sand. *Geotextiles and Geomembranes*, 18(6), 385–402.

Haeri, S. M., Khosravi, A., Garakani, A. A. and Ghazizadeh, S. (2016). Effect of soil structure and disturbance on hydromechanical behavior of collapsible loessial soils. *International Journal of Geomechanics*, 17(1), 04016021.

Han, Z. and Vanapalli, S. K. (2016). Stiffness and shear strength of unsaturated soils in relation to soil-water characteristic curve. *Géotechnique*, 66(8), 627–647.

Handy, R. L. (1995). A stress path model for collapsible loess. *Genesis and Properties of Collapsible Soils*, 468, 33–47.

Hardin, B. O. and Black, W. L. (1968). Vibration modulus of normally consolidated clay. *Journal of the Soil Mechanics and Foundations Division*, 94(2), 353–369.

Hardin, B. O. and Drnevich, V. P. (1972). Shear modulus and damping in soils: measurement and parameter effects (Terzaghi Leture). *Journal of the Soil Mechanics and Foundations Division*, 98(6), 603–624.

Harvey, E. N., Barnes, D. K., McElroy, A. H., Whiteley, A. H., Pease, D. C. and Cooper, K. W. (1944). Bubble formation in animals, I. Physical factors. *Journal of Cellular and Comparative Physiology*, 24(1), 1–22.

Harvey, E. N., McElroy, W. D. and Whiteley, A. H. (1947). On cavity formation in water. *Journal of Applied Physics*, 18(2), 162–172.

Hau, B. C. H. and Corlett, R. T. (2003). Factors affecting the early survival and growth of native tree seedlings planted on a degraded hillside grassland in Hong Kong, China. *Restoration Ecology*, 11(4), 483–488.

Hauser, V. L., Weand, B. L. and Gill, M. D. (2001). Natural covers for landfills and buried waste. *Journal of Environmental Engineering*, 127(9), 768–775.

Hayward, A. T. J. (1970). New law for liquids: don't snap, stretch. *New Scientific*, 196–199.

He, J., Chu, J. and Ivanov, V. (2013). Mitigation of liquefaction of saturated sand using biogas. *Géotechnique*, 63(4), 267–275.

Henderson, S. J. and Speedy, R. J. (1980). A Berthelot-Bourdon tube method for studying water under tension. *Journal of Physics E: Scientific Instruments*, 13(7), 778–782.

Highways Agency. (2006). *Design manual for roads and bridges* (vol. 7), National Highways. London, UK.

Hildenbrand, A., Schlömer, S. and Krooss, B. M. (2002). Gas breakthrough experiments on fine-grained sedimentary rocks. *Geofluids*, 2(1), 3–23.

Hilf, J. W. (1956). *An investigation of pore water pressures in compacted cohesive* soils. U.S. Bureau of Reclamation, Denver, CO.

Hillel, D. (1998). *Introduction to environmental soil physics*. Academic Press, San Diego, CA, USA.

Hillel, D., Krentos, V. D. and Stylianou, Y. (1972). Procedure and test of an internal drainage method for measuring soil hydraulic characteristic in-situ. *Soil Science*, 114(5), 395–400.

Ho, D. Y. F. and Fredlund, D. G. (1982). Increase in strength due to suction for two Hong Kong soils. *Proceedings of the ASCE Specialty Conference on Engineering and Construction in Tropical and Residual Soils*, 263–296.

Hoefler, A. C. (2004). *Hydrocolloids*. Egan Press Handbook, AACC, MA, USA.

Hong Kong Observatory. (2012). *Monthly meteorological normals for Hong Kong*. Hong Kong Observatory, Kowloon. www.hko.gov.hk/en/cis/normal/1981_2010/normals.htm#

Hong Kong Observatory. (2015). *Daily total rainfall (mm) at the Hong Kong International Airport*. Hong Kong Observatory, Kowloon. www.hko.gov.hk/en/cis/dailyElement.htm?stn=HKO&ele=RF

Hong, P. Y., Pereira, J. M., Tang, A. M. and Cui, Y. J. (2013). On some advanced thermo-mechanical models for saturated clays. *International Journal for Numerical and Analytical Methods in Geomechanics*, 37(17), 2952–2971.

Horseman, S. T. and McEwen, T. J. (1996). Thermal constraints on disposal of heat-emitting waste in argillaceous rocks. *Engineering Geology*, 41(1–4), 5–16.

Hossain, M. A. and Yin, J. H. (2010). Shear strength and dilative characteristics of an unsaturated compacted completely decomposed granite soil. *Canadian Geotechnical Journal*, 47(10), 1112–1126.

Hossain, M. A. and Yin, J. H. (2015). Dilatancy and strength of an unsaturated soil-cement interface in direct shear tests. *International Journal of Geomechanics*, 15(5), 04014081.

Houlsby, G. T. (1997). The work input to an unsaturated granular material. *Géotechnique*, 47(1), 193–196.

Houlsby, G. T. and Wroth, C. P. (1991). The variation of shear modulus of a clay with pressure and overconsolidation ratio. *Soils and Foundations*, 31(3), 138–143.

Hoyos, L. R., Velosa, C. L. and Puppala, A. J. (2014). Residual shear strength of unsaturated soils via suction-controlled ring shear testing. *Engineering Geology*, 172, 1–11.

Hoyos, L. R., Suescún-Florez, E. A. and Puppala, A. J. (2015). Stiffness of intermediate unsaturated soil from simultaneous suction-controlled resonant column and bender element testing. *Engineering Geology*, 188, 10–28.

Hsieh, P. G. and Ou, C. Y. (1998). Shape of ground surface settlement profiles caused by excavation. *Canadian Geotechnical Journal*, 35(6), 1004–1017.

Hsu, C. C. and Vucetic, M. (2004). Volumetric threshold shear strain for cyclic settlement. *Journal of Geotechnical and Geoenvironmental Engineering*, 130(1), 58–70.

Hsueh, Y. H., Ke, W. J., Hsieh, C. Te, Lin, K. S., Tzou, D. Y. and Chiang, C. L. (2015). ZnO nanoparticles affect bacillus subtilis cell growth and biofilm formation. *PLoS One*, 10(6), 1–23.

Hu, L., Wang, Z., Du, H. and Huang, B. (2010). Differential accumulation of dehydrins in response to water stress for hybrid and common bermudagrass genotypes differing in drought tolerance. *Journal of Plant Physiology*, 167(2): 103–109.

Hu, R., Chen, Y. F., Liu, H. H. and Zhou, C. B. (2013). A water retention curve and unsaturated hydraulic conductivity model for deformable soils: consideration of the change in pore-size distribution. *Géotechnique*, 63(16), 1389–1405.

Hueckel, T. and Borsetto, M. (1990). Thermoplasticity of saturated soils and shales: constitutive equations. *Journal of Geotechnical Engineering*, 116(12), 1765–1777.

Hueckel, T., Tutumluer, E. and Pellegrini, R. (1992). A note on non-linear elasticity of isotropic overconsolidated clays. *International Journal for Numerical and Analytical Methods in Geomechanics*, 16(8), 603–618.

Hueckel, T., François, B., Laloui, L., Francois, B. and Laloui, L. (2009). Explaining thermal failure in saturated clays. *Géotechnique*, 59(3), 197–212.

Hveem, F. N. (1955). Pavement deflections and fatigue failures. Highway Research Board Bulletin No. 114, Highway Research Board, Washington, DC, 43–87.

Iryo, T. and Rowe, R. K. (2005). Hydraulic behaviour of soil–geocomposite layers in slopes. *Geosynthetics International*, 12(3), 145–155.

Isobe, Y., Endo, K. and Kawai, H. (1992). Properties of a highly viscous polysaccharide produced by a Bacillus strain isolated from Soil. *Bioscience, Biotechnology, and Biochemistry*, 56(4), 636–639.

Israelachvili, J. (2011). Intermolecular and surface forces. In *Intermolecular and Surface Forces*. Academic Press, Elsevier.

Ivanova, L. K., Richards, D. J. and Smallman, D. J. (2008). The long-term settlement of landfill waste. *Proceedings of the Institution of Civil Engineers – Waste and Resource Management*, 161(3), 121–133.

Iwasaki, T., Tatsuoka, F. and Takagi, Y. (1978). Shear moduli of sands under cyclic torsional shear loading. *Soils and Foundations, Japanese Society of Soil Mechanics and Foundation Engineering*, 18(1), 39–56.

Jacobsz, S. W. (2018). Low cost tensiometers for geotechnical applications. In *Proceedings of the 9th International Conference on Physical Modelling in Geotechnics*, CRC Press, Boca Raton, FL, 305–310.

Jaky, J. (1944). The coefficient of earth pressure at rest. *Journal for Society of Hungarian Architects and Engineers*, 78(22), 355–358.

Jamiolkowski, M., Lancellotta, R. and Lo Presti, D. C. F. (1995). Remarks on the stiffness at small strains of six Italian clays. *Proceedings of the International Symposium on Pre-Failure Deformation of Geomaterials*, 2, 817–836.

Japanese Geotechnical Society (JGS). (1999). Standard of Japanese Geotechnical Society for laboratory shear test. JGS0550, Tokyo.

Jardine, R.J., Symes, M.J., Burland, J.B. (1984). The measurement of soil stiffness in the triaxial apparatus. *Géotechnique*, 34(3), 323–340.

Jardine, R. J., Kuwano, R., Zdravkovic, L. and Thornton, C. (1999). Some fundamental aspects of the pre-failure behaviour of granular soils. *Proceedings of the Second International Conference on Pre-failure Deformation Characteristics of Geomaterials*, 2, 1077–1111.

Jefferson, I., Tye, C. and Northmore, K. J. (2001). Behaviour of silt: the engineering characteristics of loess in the UK. *Proceedings of Symposium on Problematic Soils*, 37–52.

Jennings, J. E. (1960). A revised effective stress law for use in the prediction of the behaviour of unsaturated soils, pore pressure and suction in soils. *Proceedings of Conference on Pore Pressure and Suction in Soils*, 26–30.

Jennings, J. E. B. and Burland, J. B. (1962). Limitations to the use of effective stresses in partly saturated soils. *Géotechnique*, 12(2), 125–144.

Jin, M. S., Lee, K. W. and Kovacs, W. D. (1994). Seasonal variation of resilient modulus of subgrade soils. *Journal of Transportation Engineering – ASCE*, 120(4), 603–616.

Jones, W. M., Overton, G. D. N. and Trevena, D. H. (1981). Tensile strength experiments with water using a new type of Berthelot tube. *Journal of Physics D: Applied Physics*, 14(7), 1283–1291.

Junaideen, S. M., Tham, L. G., Law, K. T., Dai, F. C. and Lee, C. F. (2010). Behaviour of recompacted residual soils in a constant shear stress path. *Canadian Geotechnical Journal*, 47(6), 648–661.

Karimi, S. (1998). A study of geotechnical applications of biopolymer treated soils with an emphasis on silt. Ph.D. Thesis, University of Southern California, Los Angeles, USA.

Karr, L., Harre, B. and Hakonson, T.E. (1999). Infiltration control landfill cover demonstration at marine corps base, Hawaii. Naval Facilities Engineering Service Center Port Hueneme CA, Technical report TR-2108-ENV.

Kassif, G. and Ben Shalom, A. (1971). Experimental relationship between swell pressure and suction. *Géotechnique*, 21(3), 245–255.

Khalili, N. and Khabbaz, M. H. (1998). A unique relationship for χ for the determination of the shear strength of unsaturated soils. *Géotechnique*, 48(5), 681–687.

Khalili, N., Geiser, F. and Blight, G.E. (2004). Effective stress in unsaturated soils: review with new evidence. *International Journal of Geomechanics*, 4(2), 115–126.

Khalili, N., Habte, M. A. and Zargarbashi, S. (2008). A fully coupled flow deformation model for cyclic analysis of unsaturated soils including hydraulic and mechanical hystereses. *Computers and Geotechnics*, 35(6), 872–889.

Khalili, N., Uchaipichat, A. and Javadi, A. A. (2010). Skeletal thermal expansion coefficient and thermo-hydro-mechanical constitutive relations for saturated homogeneous porous media. *Mechanics of Materials*, 42(6), 593–598.

Khatami, H. R. and O'Kelly, B. C. (2013). Improving mechanical properties of sand using biopolymers. *Journal of Geotechnical and Geoenvironmental Engineering*, 139(8), 1402–1406.

Khire, M. V., Benson, C. H. and Bosscher, P. J. (1999). Field data from a capillary barrier and model predictions with UNSAT-H. *Journal of Geotechnical and Geoenvironmental Engineering*, 125(6), 518–527.

Khire, M. V., Benson, C. H. and Bosscher, P. J. (2000). Capillary barriers: design variables and water balance. *Journal of Geotechnical and Geoenvironmental Engineering*, 126(8), 695–708.

Kholghifard, M., Kholghifard, A., Ahmad, K. and Latifi, N. (2012). The influence of suction changes on collapsibility and volume change behavior of unsaturated clay soil. *Electronic Journal of Geotechnical Engineering*, 17, 2623–2631.

Khosravi, A. and McCartney, J. S. (2009). Impact of stress state on the dynamic shear moduli of unsaturated, compacted soils. *Proceedings of 6th Asia-Pacific Conference on Unsaturated Soils*, 1–6.

Khosravi, A. and McCartney, J. S. (2012). Impact of hydraulic hysteresis on the small-strain shear modulus of low plasticity soils. *Journal of Geotechnical and Geoenvironmental Engineering*, 138(11), 1326–1333.

Khoury, N. N. and Zaman, M. M. (2004). Correlation between resilient modulus, moisture variation, and soil suction for subgrade soils. *Transportation Research Record: Journal of the Transportation Research Board*, 1874, 99–107.

Kim, D. and Kim, J. R. (2007). Resilient behavior of compacted subgrade soils under the repeated triaxial test. *Construction and Building Materials*, 21(7), 1470–1479.

Knidiri, J., Bussière, B., Hakkou, R., Bossé, B., Maqsoud, A. and Benzaazoua, M. (2017). Hydrogeological behaviour of an inclined store-and-release cover experimental cell made with phosphate mine wastes. *Canadian Geotechnical Journal*, 54(1), 102–116.

Knox, D. P., Stokoe, K. H. I. and Kopperman, S. E. (1982). Effect of state of stress in velocity of low amplitude shear waves propa- gating along principal stress directions in dry sand. Master Thesis, University of Texas at Austin, Austin, TX.

Koerner, R. M. and Daniel, D. E. (1997). *Final covers for solid waste landfills and abandoned dumps*. American Society of Civil Engineers, Reston, VA.

Koerner, R. M., Hsuan, Y. G., Koerner, G. R. and Gryger, D. (2010). Ten year creep puncture study of HDPE geomembranes protected by needle-punched nonwoven geotextiles. *Geotextiles and Geomembranes*, 28(6), 503–513.

Komornik, A., Livneh, M. and Smucha, S. (1980). Shear strength and swelling of clays under suction. *Proceedings of the 4th International Confenernce on Expansive Soil*, 206–266.

Krahn, J. and Fredlund, D. G. (1972). On total, matric and osmotic suction. *Soil Science*, 114(5), 339–348.

Kruse, G. A. M., Dijkstra, T. A. and Schokking, F. (2007). Effects of soil structure on soil behaviour: Illustrated with loess, glacially loaded clay and simulated flaser bedding examples. *Engineering Geology*, 91(1), 34–45.

Kung, G. T. C., Ou, C. Y. and Juang, C. H. (2009). Modeling small-strain behavior of Taipei clays for finite element analysis of braced excavations. *Computers and Geotechnics*, 36(1–2), 304–319.

Kunhel, R. A. and van der Gaast, S. (1993). Humidity-controlled diffractometry and its applications. *Advances in X-Ray Analysis*, 36, 439–449.

Lacasse, S. and Berre, T. (1988). Triaxial testing methods for soils. *Advanced Triaxial Testing of Soil and Rock*, 977, 264–289.

Lagerwerff, J. V., Ogata, G. and Eagle, H. E. (1961). Control of osmotic pressure of culture solutions with polyethylene glycol. *Science*, 133, 1486–1487.

Lai, C. H. (2004). Experimental study of stress-dependent soil-water characteristics and their applications on numerical analysis of slope stability. PhD Thesis, The Hong Kong University of Science and Technology, Hong Kong, China.

Laloui, L. and Cekerevac, C. (2003). Thermo-plasticity of clays: an isotropic yield mechanism. *Computers and Geotechnics*, 30(8), 649–660.

Laloui, L. and Cekerevac, C. (2008). Non-isothermal plasticity model for cyclic behaviour of soils. *International Journal for Numerical and Analytical Methods in Geomechanics*, 32(5), 437–460.

Laloui, L. and François, B. (2009). ACMEG-T: soil thermoplasticity model. *Journal of Engineering Mechanics*, 135(9), 932–944.

Laloui, L., Salager, S. and Rizzi, M. (2013). Retention behaviour of natural clayey materials at different temperatures. *Acta Geotechnica*, 8(5), 537–546.

Lam, C. C. and Leung, Y.K. (1994). *Extreme rainfall statistics and design rainstorm profiles at selected locations in Hong Kong*. Hong Kong Royal Observatory, Hong Kong.

Lam, L., Fredlund, D. G. and Barbour, S. L. (1987). Transient seepage model for saturated-unsaturated soil systems: a geotechnical engineering approach. *Canadian Geotechnical Journal*, 24(4), 565–580.

Lambe, T. W. (1960). A mechanistic picture of shear strength in clay. *Proceedings of ASCE Research Conference on Shear Strength of Cohesive Soils*, 437, 555–580.

Larrahondo, J. M., Choo, H. and Burns, S. E. (2011). Laboratory-prepared iron oxide coatings on sands: submicron-scale small-strain stiffness. *Engineering Geology*, 121(1–2), 7–17.

Lee, I. M., Sung, S. G. and Cho, G. C. (2005a). Effect of stress state on the unsaturated shear strength of a weathered granite. *Canadian Geotechnical Journal*, 42(2), 624–631.

Lee, S. R., Kim, Y. K. and Lee, S. J. (2005b). A method to estimate soil-water characteristic curve for weathered granite soil. *Proceedings of the 16th International Conference on Soil Mechanics and Geotechnical Engineering*, 15(1), 543–546.

Lehmann, J. and Joseph, S. (2009). Biochar for environmental management: science, technology and implementation. In *Earthscan*. Routledge, London, UK.

Lekarp, F., Isacsson, U. and Dawson, A. (2000). State of the art. I: resilient response of unbound aggregates. *Journal of Transportation Engineering*, 126(1), 66–75.

Lemon, K. P., Earl, A. M., Vlamakis, H. C., Aguilar, C. and Kolter, R. (2008). Biofilm development with an emphasis on Bacillus subtilis. In *Current Topics in Microbiology and Immunology*. Edited by T. Romeo, Springer, Berlin, Germany, 1–16.

Lengeler, J. W., Drews, G. and Schlegel, H. G. (1999). *Biology of the prokaryote*. Blackwell Science, Oxford, UK.

Leong, E. C. and Rahardjo, H. (1997). Review of soil-water characteristic curve equations. *Journal of Geotechnical and Geoenvironmental Engineering*, 123(12), 1106–1117.

Leong, E. C., Cahyadi, J. and Rahardjo, H. (2006). Stiffness of a compacted residual soil. *Unsaturated Soils- UNSAT 2006*, 1169–1180.

Leung, A. K., Sun, H. W., Millis, S. W., Pappin, J. W., Ng, C. W. W. and Wong, H. N. (2011). Field monitoring of an unsaturated saprolitic hillslope. *Canadian Geotechnical Journal*, 48(3), 339–353.

Leung, A. K., Garg, A., Coo, J. L., Ng, C. W. W. and Hau, B. C. H. (2015). Effects of the roots of Cynodon dactylon and Schefflera heptaphylla on water infiltration rate and soil hydraulic conductivity. *Hydrological Processes*, 29(15), 3342–3354.

Leung, A. K., Boldrin, D., Liang, T., Wu, Z. Y., Kamchoom, V. and Bengough, A. G. (2017). Plant age effects on soil infiltration rate during early plant establishment. *Géotechnique*, 68(7), 646–652.

Li, D. and Selig, E. T. (1994). Resilient modulus for fine-grained subgrade soils. *Journal of Geotechnical Engineering*, 120(6), 939–957.

Li, J., Smith, D. W., Fityus, S. G. and Sheng, D. C. (2002). Quantitative analysis of moisture content determination in expansive soils using neutron probes. *Proceedings of the 3rd International Conference on Unsaturated Soils*, 363–368.

Li, X. and Zhang, L. M. (2009). Characterization of dual-structure pore-size distribution of soil. *Canadian Geotechnical Journal*, 46(2), 129–141.

Li, X. S. (2002). A sand model with state-dependent dilatancy. *Géotechnique*, 52(3), 173–186.

Li, X. S. (2005). Modelling of hysteresis response for arbitrary wetting/drying paths. *Computers and Geotechnics*, 32(2), 133–137.

Li, X. S. and Dafalias, Y. F. (2000). Dilatancy for cohesionless soils. *Géotechnique*, 50(4), 449–460.

Li, X. S., Yang, W. L., Shen, C. K. and Wang, W. C. (1998). Energy-injecting virtual mass resonant column system. *Journal of Geotechnical and Geoenvironmental Engineering*, 124(5), 428–438.

Li, X., Zhang, L. M. and Fredlund, D. G. (2009). Wetting front advancing column test for measuring unsaturated hydraulic conductivity. *Canadian Geotechnical Journal*, 46(12), 1431–1445.

Likos, W. J. (2014). Effective stress in unsaturated soil: accounting for surface tension and interfacial area. *Vadose Zone Journal*, 13(5), vzj2013.05.0095.

Likos, W. J. and Lu, N. (2003). Automated humidity system for measuring total suction characteristics of clay. *Geotechnical Testing Journal*, 26(2), 179–190.

Lim, T. T., Rahardjo, H., Chang, M. F. and Fredlund, D. G. (1996). Effect of rainfall on matric suctions in a residual soil slope. *Canadian Geotechnical Journal*, 33(4), 618–628.

Liu, M. D. and Carter, J. P. (2002). A structured Cam Clay model. *Canadian Geotechnical Journal*, 39(6), 1313–1332.

Liu, T. (1988). *Loess in China*. Springer-Verlag, Berlin.

Liu, T. H. (1997). *Problems of expansive soils in engineering construction*. Architecture and Building Press of China, Beijing, China.

Lloret, A. and Alonso, E. E. (1985). State surfaces for partially saturated soils. *Proceedings of the Eleventh International Conference on Soil Mechanics Foundation Engineering*, 2, 557–562.

Lloret, A, Villar, M. V, Sanchez, M., Gens, A, Pintado, X. and Alonso, E. (2003). Mechanical behaviour of heavily compacted bentonite under high suction changes. *Géotechnique*, 53(1), 27–40.

Lo Presti, D. C. F. (1989). Proprieta` dinamiche dei terreni. *XIV Conferenza Geotecnica di Torino, Department of Structural Engineering*. Politecnico di Torino, Turin, Italy.

Loach, S. C. (1987). Repeated loading of fine grained soils for pavement design. Ph.D. Thesis, University of Nottingham, Nottingham UK.

Loehr, R. C. and Haikola, B. M. (2003). Long term landfill primary and secondary leachate production. *Journal of Geotechnical and Geoenvironmental Engineering*, 129(11), 1063–1067.

Loughnan, F. C. (1969). *Chemical weathering of the silicate minerals*. Elsevier Publ. Co., Amsterdam and New York.

Lu, N. and Likos, W. J. (2004). *Unsaturated soil mechanics*. John Wiley and Sons, New York, USA.

Lu, N. and Likos, W. J. (2006). Suction stress characteristic curve for unsaturated soil. *Journal of Geotechnical and Geoenvironmental Engineering*, 132(2),131–142.

Lu, N., Godt, J.W. and Wu, D. T. (2010). A closed-form equation for effective stress in unsaturated soil. *Water Resources Research*, 46(5), 1–14.

Luo, T., Chen, D., Yao, Y. P. and Zhou, A. N. (2020). An advanced UH model for unsaturated soils. *Acta Geotechnica*, 15(1), 145–164.

Mair, R. J. (1993). UnWin Memorial Lecture 1992: developments in geotechnical engineering research: application to tunnels and deep excavations. *Proceedings of the Institution of Civil Engineers – London*, 93(1), 27–41.

Mancuso, C., Vassallo, R. and D'Onofrio, A. (2002). Small strain behavior of a silty sand in controlled-suction resonant column – torsional shear tests. *Canadian Geotechnical Journal*, 39(1), 22–31.

Mao, S. Z., Cui, Y. J. and Ng, C. W. W. (2002). Slope stability analysis for a water diversion canal in China. *Proceedings of the 3rd International Conference on Unsaturated Soils*, 2, 805–810.

Marinho, F. A. M. and Chandler, R. J. (1995). Cavitation and the direct measurement of soil suction. *Proceedings of Conference on Unsaturated Soils*, 623–630.

Marinho, F. A. M. and de Sousa Pinto, C. (1997). Soil suction measurement using a tensiometer. *Symposium on Recent Developments in Soil and Pavement Mechanics*, 249–254.

Marinho, F. A. M., Take, W. A. and Tarantino, A. (2008). Measurement of matric suction using tensiometric and axis translation techniques. *Geotechnical and Geological Engineering*, 26(6), 615–631.

Mašín, D. and Khalili, N. (2011). Modelling of thermal effects in hypoplasticity. *Computer Methods for Geomechanics: Frontiers and New Applications*, 1, 237–245.

McBean, E. A., Rovers, F. A. and Farquhar, G. J. (1995). *Solid waste landfill engineering and design*. Prentice Hall PTR, Englewood Cliffs, NJ.

McCarthy, E. L. (1934). Mariotte's bottle. *Science*, 80(2065), 100–100.

McCartney, J. S. and Khosravi, A. (2013). Field-monitoring system for suction and temperature profiles under pavements. *Journal of Performance of Constructed Facilities*, 27(6), 818–825.

McCartney, J. S. and Zornberg, J. G. (2010). Effects of infiltration and evaporation on geosynthetic capillary barrier performance. *Canadian Geotechnical Journal*, 47(11), 1201–1213.

McKeen, R. (1985). Validation of procedures for pavement design on expansive soils. Final report, US Department of Transportation, Washington, DC.

McQueen, I. S. and Miller, R. F. (1968). Calibration and evaluation of a wide-range gravimetric method for measuring moisture stress. *Soil Science*, 106(3), 225–231.

Meek, B. D., Rechel, E. R., Carter, L. M., DeTar, W. R. and Urie, A. L. (1992). Infiltration rate of a sandy loam soil: effects of traffic, tillage, and plant roots. *Soil Science Society of America Journal*, 56(3), 908–913.

Meerdink, J. S., Benson, C. H. and Khire, M. V. (1996). Unsaturated hydraulic conductivity of two compacted barrier soils. *Journal of Geotechnical Engineering*, 122(7), 565–576.

Melchior, S. (1997). In-situ studies of the performance of landfill caps (compacted soil liners, geomembranes, geosynthetic clay liners and capillary barriers). *Land Contamination & Reclamation*, 5(3), 209–216.

Merchán, V., Romero, E. and Vaunat, J. (2011). An adapted ring shear apparatus for testing partly saturated soils in the high suction range. *Geotechnical Testing Journal*, 34(5), 1–12.

Michels, W. C. (1961). *The international dictionary of physics and electronics* (2nd ed.). Van Nostrand Co., Princeton, NJ.

Mijares, R. G. and Khire, M. V. (2012). Field data and numerical modeling of water balance of Lysimeter versus actual Earthen cap. *Journal of Geotechnical and Geoenvironmental Engineering*, 138(8), 889–897.

Mitchell, A. R., Ellsworth, T. R. and Meek, B. D. (1995). Effect of root systems on preferential flow in swelling soil. *Communications in Soil Science and Plant Analysis*, 26(15–16), 2655-2666.

Mitchell, J. K. and Santamarina, J. C. (2005). Biological considerations in geotechnical engineering. *Journal of Geotechnical and Geoenvironmental Engineering*, 131(10), 1222–1233.

Mohan, D., Sarswat, A., Ok, Y. S. and Pittman, C. U. (2014). Organic and inorganic contaminants removal from water with biochar, a renewable, low cost and sustainable adsorbent – a critical review. *Bioresource Technology*, 160, 191–202.

Monteith, J. L. (1965). Evaporation and environment. *In Symposia of the Society for Experimental Biology*, Cambridge University Press (CUP), Cambridge, 19, 205–234.

Moon, S., Nam, K., Kim, J., Hwan, S. and Chung, M. (2008). Effectiveness of compacted soil liner as a gas barrier in the landfill cover system. *Waste Management*, 28(10), 1909–1914.

More, T. T., Yadav, J. S. S., Yan, S., Tyagi, R. D. and Surampalli, R. Y. (2014). Extracellular polymeric substances of bacteria and their potential environmental applications. *Journal of Environmental Management*, 144, 1–25.

Morgenstern, N. R. (1979). Properties of compacted soils. *Proceedings of the 6th Pan-Ameican Conference on Soil Mechanics & Foundation Engineering*, 3, 349–354.

Morikawa, M. (2006). Beneficial biofilm formation by industrial bacteria Bacillus subtilis and related species. *Journal of Bioscience and Bioengineering*, 101(1), 1–8.

Morikawa, M., Kagihiro, S., Haruki, M., Takano, K., Branda, S., Kolter, R. and Kanaya, S. (2006). Biofilm formation by a Bacillus subtilis strain that produces γ-polyglutamate. *Microbiology*, 152(9), 2801–2807.

Morris, C. E. and Stormont, J. C. (1999). Parametric study of unsaturated drainage layers in a capillary barrier. *Journal of Geotechnical and Geoenvironmental Engineering*, 125(12), 1057-1065.

Morvan, M., Wong, H. and Branque, D. (2010). An unsaturated soil model with minimal number of parameters based on bounding surface plasticity. *International Journal for Numerical and Analytical Methods in Geomechanics*, 34(14), 1512–1537.

Moslemy, P., Neufeld, R. J., Millette, D. and Guiot, S. R. (2003). Transport of gellan gum microbeads through sand: an experimental evaluation for encapsulated cell bioaugmentation. *Journal of Environmental Management*, 69(3), 249–259.

Mualem, Y. (1976). A new model for predicting the hydraulic conductivity of unsaturated porous media. *Water Resources Research*, 12(3), 513–522.

Mukherjee, S. (2012). *Applied mineralogy: applications in industry and environment*. Springer Science & Business Media, New Delhi, India.

Muñoz, E., Ochoa, A. and Cordão-Neto, M. (2018). Stochastic ecohydrological-geotechnical modeling of long-term slope stability. *Landslides*, 15(5), 913–924.

Munõz-Castelblanco, J., Delage, P., Pereira, J. M. and Cui, Y. J. (2011). Some aspects of the compression and collapse behaviour of an unsaturated natural loess. *Géotechnique Letters*, 1(2), 17–22.

Muñoz-Castelblanco, J. A., Pereira, J. M., Delage, P. and Cui, Y. J. (2012a). The influence of changes in water content on the electrical resistivity of a natural unsaturated loess. *Geotechnical Testing Journal*, 35(1), 11–17.

Muñoz-Castelblanco, J. A., Pereira, J. M., Delage, P. and Cui, Y. J. (2012b). The water retention properties of a natural unsaturated loess from northern France. *Géotechnique*, 62(2), 95–106.

Nealson, K. H. and Saffarini, D. (1994). Iron and manganese in anaerobic respiration: environmental significance, physiology, and regulation. *Annual Review of Microbiology*, 48, 311–343.

Nelson, J. D. and Miller, D. J. (1992). *Expansive soils: problems and practice in foundation and pavement engineering*. Wiley Interscience., New York.

Ng, C. W. W. (1993). Nonlinear modelling of wall installation effects. In *Retaining structures*. Thomas Telford, London, 160–163.

Ng, C. W. W. (2010). High suction double-cell extractor. US Patent No. US 7,793,552 B2; Granted on 14 September 2010.

Ng, C. W. W. (2014). *Humidity and osmotic suction-controlled box*. The Hong Kong University of Science and Technology, Kowloon.

Ng, C. W. W. and Chen, R. (2005). Keynote lecture: advanced suction control techniques for testing unsaturated soils (in Chinese). In *2nd National Conference on Unsaturated Soils*, Hangzhou, China.

Ng, C. W. W. and Chen, R. (2006). Advanced suction control techniques for testing unsaturated soils. *Yantu Gongcheng Xuebao/Chinese Journal of Geotechnical Engineering*, 28(2), 123–128.

Ng, C. W. W. and Chiu, A. C. F. (2001). Behavior of a loosely compacted unsaturated volcanic soil. *Journal of Geotechnical and Geoenvironmental Engineering*, 127(12), 1027–1036.

Ng, C. W. W. and Chiu, A. C. F. (2003). Laboratory study of loose saturated and unsaturated decomposed granitic soil. *Journal of Geotechnical and Geoenvironmental Engineering*, 129(6), 550–559.

Ng, C. W. W. and Coo, J. L. (2015). Hydraulic conductivity of clay mixed with nanomaterials. *Canadian Geotechnical Journal*, 52(6), 808–811.

Ng, C. W. W. and Leung, A. K. (2012a). In-situ and laboratory investigations of stress-dependent permeability function and SDSWCC from an unsaturated soil slope. *Geotechnical Engineering*, 43(1), 26–39.

Ng, C. W. W. and Leung, A. K. (2012b). Measurements of drying and wetting permeability functions using a new stress-controllable soil column. *Journal of Geotechnical and Geoenvironmental Engineering*, 138(1), 58–68.

Ng, C. W. W. and Lings, M. L. (1995). Effects of modeling soil nonlinearity and wall installation on back-analysis of deep excavation in stiff clay. *Journal of Geotechnical Engineering*, 121(10), 687–695.

Ng, C. W. W. and Menzies, B. (2007). *Advanced unsaturated soil mechanics and engineering*. Taylor and Francis, London and New York.

Ng, C. W. W. and Pang, Y. W. (2000a). Experimental investigations of the soil-water characteristics of a volcanic soil. *Canadian Geotechnical Journal*, 37(6), 1252–1264.

Ng, C. W. W. and Pang, Y. W. (2000b). Influence of stress state on soil-water characteristics and slope stability. *Journal of Geotechnical and Geoenvironmental Engineering*, 126(2), 157–166.

Ng, C. W. W. and Shi, Q. (1998). A numerical investigation of the stability of unsaturated soil slopes subjected to transient seepage. *Computers and Geotechnics*, 22(1), 1–28.

Ng, C. W. W. and Xu, J. (2012). Effects of current suction ratio and recent suction history on small-strain behaviour of an unsaturated soil. *Canadian Geotechnical Journal*, 49(2), 226–243.

Ng, C. W. W. and Yung, S.Y. (2008). Determination of the anisotropic shear stiffness of an unsaturated decomposed soil. *Géotechnique*, 58(1), 23–35.

Ng, C. W. W. and Zhou, C. (2014). Cyclic behaviour of an unsaturated silt at various suctions and temperatures. *Géotechnique*, 64(9), 709–720.

Ng, C. W. W. and Zhou, R. (2005). Effects of soil suction on dilatancy of an unsaturated soil. *Proceedings of 16th International Conference on Soil Mechanics and Geotechnical Engineering*, 559–562.

Ng, C. W. W., Pun, W. K. and Pang, R. P. L. (2000). Small strain stiffness of natural granitic saprolite in Hong Kong. *Journal of Geotechnical and Geoenvironmental Engineering*, 126(9), 819–833.

Ng, C. W. W., Zhan, L. T. and Cui, Y. J. (2002). A new simple system for measuring volume changes in unsaturated soils. *Canadian Geotechnical Journal*, 39(3), 757–764.

Ng, C. W. W., Zhan, L. T., Bao, C. G., Fredlund, D. G. and Gong, B. W. (2003). Performance of an unsaturated expansive soil slope subjected to artificial rainfall infiltration. *Géotechnique*, 53(2), 143–157.

Ng, C. W. W., Fung, W. T., Cheuk, C. Y. and Zhang, L.M. (2004a). Influence of stress ratio and stress path on behavior of loose decomposed granite. *Journal of Geotechnical and Geoenvironmental Engineering*, 130(1), 36–44.

Ng, C. W. W., Leung, E. H. and Lau, C. K. (2004b). Inherent anisotropic stiffness of weathered geomaterial and its influence on ground deformations around deep excavations. *Canadian Geotechnical Journal*, 41(1), 12–24.

Ng, C. W. W., Cui, Y., Chen, R. and Delage, P. (2007). The axis-translation and osmotic techniques in shear testing of unsaturated soils: a comparison. *Soils and Foundations*, 47(4), 675–684.

Ng, C. W. W., Xu, J. and Yung, S.Y. (2009). Effects of wetting–drying and stress ratio on anisotropic stiffness of an unsaturated soil at very small strains. *Canadian Geotechnical Journal*, 46(9), 1062–1076.

Ng, C. W. W., Wong, H. N., Tse, Y. M., Pappin, J. W., Sun, H. W., Millis, S. W. and Leung, A. K. (2011). A field study of stress-dependent soil-water characteristic curves and permeability of a saprolitic slope in Hong Kong. *Géotechnique*, 61(6), 511–521.

Ng, C. W. W., Lai, C. H., and Chiu, C. F. (2012). A modified triaxial apparatus for measuring the stress path-dependent water retention curve. *Geotechnical Testing Journal*, 35(3), 104203.

Ng, C.W.W., Zhou, C., Yuan, Q. and Xu, J. (2013). Resilient modulus of unsaturated subgrade soil: experimental and theoretical investigations. *Canadian Geotechnical Journal*, 50(2), 223–232.

Ng, C.W.W., Shi, C., Gunawan, A. and Laloui, L. (2014a). Centrifuge modelling of energy piles subjected to heating and cooling cycles in clay. *Géotechnique Letters*, 4(4), 310–316.

Ng, C. W. W., Leung, A. K. and Woon, K. X. (2014b). Effects of soil density on grass-induced suction distributions in compacted soil subjected to rainfall. *Canadian Geotechnical Journal*, 51(3), 311–321.

Ng, C. W. W., Zhou, C. and Leung, A. K. (2015a). Comparisons of different suction control techniques by water retention curves: theoretical and experimental studies. *Vadose Zone Journal*, 14(9), vzj2015–01.

Ng, C. W. W., Chen, Z. K., Coo, J. L., Chen, R. and Zhou, C. (2015b). Gas breakthrough and emission through unsaturated compacted clay in landfill final cover. *Waste Management*, 44, 155–163.

Ng, C. W. W., Liu, J. and Chen, R. (2015c). Numerical investigation on gas emission from three landfill soil covers under dry weather conditions. *Vadose Zone Journal*, 14(8), vzj2014.12.0180.

Ng, C. W. W., Xu, J. and Chen, R. (2015d). All-weather landfill soil cover system for preventing water infiltration and landfill gas emission. US Patent No. US 9,101,968 B2; Granted on 11 August 2015.

Ng, C. W. W., Liu, J., Chen, R. and Xu, J. (2015e). Physical and numerical modeling of an inclined three-layer (silt/gravelly sand/clay) capillary barrier cover system under extreme rainfall. *Waste Management*, 38, 210–221.

Ng, C. W. W., Sadeghi, H., Hossen, S. B., Chiu, C. F., Alonso, E. E. and Baghbanrezvan, S. (2016a). Water retention and volumetric characteristics of intact and re-compacted loess. *Canadian Geotechnical Journal*, 53(8), 1258–1269.

Ng, C. W. W., Coo, J. L., Chen, Z. K. and Chen, R. (2016b). Water infiltration into a new three-layer landfill cover system. *Journal of Environmental Engineering*, 142(5), 04016007.

Ng, C. W. W., Mu, Q. Y. and Zhou, C. (2016c). Effects of soil structure on the shear behaviour of an unsaturated loess at different suctions and temperatures. *Canadian Geotechnical Journal*, 54(2), 270–279.

Ng, C. W. W., Wang, S. H. and Zhou, C. (2016d). Volume change behaviour of saturated sand under thermal cycles. *Géotechnique Letters*, 6(2), 124–131.

Ng, C. W. W., Cheng, Q., Zhou, C. and Alonso, E. E. (2016e). Volume changes of an unsaturated clay during heating and cooling. *Géotechnique Letters*, 6(3), 192–198.

Ng, C. W. W., Xu, J. and Chen, R. (2016f). All-weather landfill soil cover system for preventing water infiltration and landfill gas emission: construction and application. Chinese Patent No. CN103572785B; Granted on 2 March 2016.

Ng, C. W. W., Ni, J. J., Leung, A. K., Zhou, C., and Wang, Z. J. (2016g). Effects of planting density on tree growth and induced soil suction. *Geotechnique*, 66(9), 711–724.

Ng, C. W. W., Sadeghi, H. and Jafarzadeh, F. (2017a). Compression and shear strength characteristics of compacted loess at high suctions. *Canadian Geotechnical Journal*, 54(5), 690–699.

Ng, C. W. W., Baghbanrezvan, S., Sadeghi, H., Zhou, C. and Jafarzadeh, F. (2017b). Effect of specimen preparation techniques on dynamic properties of unsaturated fine-grained soil at high suctions. *Canadian Geotechnical Journal*, 54(9), 1310–1319.

Ng, C. W. W., Kaewsong, R., Zhou, C. and Alonso, E. E. (2017c). Small strain shear moduli of unsaturated natural and compacted loess. *Géotechnique*, 67(7), 646–651.

Ng, C. W. W., Cheng, Q. and Zhou, C. (2018). Thermal effects on yielding and wetting-induced collapse of recompacted and intact loess. *Canadian Geotechnical Journal*, 55(8), 1095–1103.

Ng, C. W. W., Leung, A. K. and Ni, J. J. (2019a). *Plant-soil slope interaction*. Taylor & Francis.

Ng, C. W. W., So, P. S., Coo, J. L., Zhou, C. and Lau, S. Y. (2019b). Effects of biofilm on gas permeability of unsaturated sand. *Géotechnique*, 69(10), 917–923.

Ng, C. W. W., Akinniyi, D. B., Zhou, C. and Chiu, A. C. F. (2019c). Comparisons of weathered lateritic, granitic and volcanic soils: compressibility and shear strength. *Engineering Geology*, 249, 235–240.

Ng, C. W. W., Chen, R., Coo, J. L., Liu, J., Ni, J. J., Chen, Y. M., Zhan, L. T., Guo, H. W. and Lu, B. W. (2019d). A novel vegetated three-layer landfill cover system using recycled construction wastes without geomembrane. *Canadian Geotechnical Journal*, 56(12), 1863–1875.

Ng, C. W. W., Zhou, C. and Chiu, C. F. (2020a). Constitutive modelling of state-dependent behaviour of unsaturated soils: an overview. *Acta Geotechnica*, 15(10), 2705–2725.

Ng, C. W. W., Sadeghi, H., Jafarzadeh, F., Sadeghi, M., Zhou, C., and Baghbanrezvan, S. (2020b). Effect of microstructure on shear strength and dilatancy of unsaturated loess at high suctions. *Canadian Geotechnical Journal*, 57(2), 221–235.

Ng, C. W. W., So, P. S., Lau, S. Y., Zhou, C., Coo, J. L. and Ni, J. J. (2020c). Influence of biopolymer on gas permeability in compacted clay at different densities and water contents. *Engineering Geology*, 272, 105631.

Ng, C. W. W., Zheng, G., Ni, J. and Zhou, C. (2020d). Use of unsaturated small-strain soil stiffness to the design of wall deflection and ground movement adjacent to deep excavation. *Computers and Geotechnics*, 119, 103375.

Ng, C. W. W., Ni, J. J. and Leung, A. K. (2020e). Effects of plant growth and spacing on soil hydrological changes: a field study. *Géotechnique*, 70(10), 867–881.

Ng, C. W. W., Farivar, A., Gomaa, S. M. M. H. and Jafarzadeh. F. (2021a). Centrifuge modelling of energy pile groups in clay subjected to non-symmetrical thermal cycles. *Journal of Geotechnical and Geoenvironmental Engineering*, 147(12), 04021146.

Ng, C. W. W., Zhang, Q., Ni, J. and Li, Z. (2021b). A new three-dimensional theoretical model for analysing the stability of vegetated slopes with different root architectures and planting patterns. *Computers and Geotechnics*, 130, 103912.

Ng, C. W. W., Zhao, X. D., Zhang, S., Ni, J. J. and Zhou, C. (2022a). An elasto-plastic numerical analysis of THM responses of floating energy pile foundations subjected to asymmetrical thermal cycles. *Géotechnique*, www.icevirtuallibrary.com/doi/abs/10.1680/jgeot.22.00055.

Ng, C. W. W., Guo, H., Ni, J., Chen, R., Xue, Q., Zhang, Y., Feng, Y., Chen, Z., Feng, S. and Zhang, Q. (2022b). Long-term field performance of non-vegetated and vegetated three-layer landfill cover systems using construction waste without geomembrane. *Géotechnique*, www.icevirtuallibrary.com/doi/abs/10.1680/jgeot.21.00238.

Ng, C. W. W., Crous, P. A. and Jacobsz, S. W. (2023). Centrifuge and numerical modeling of liquefied flow and nonliquefied slide failures of tailings dams. *Journal of Geotechnical and Geoenvironmental Engineering*, 149(9), 04023075.

Ni, J. J. and Ng, C. W. W. (2019). Long-term effects of grass roots on gas permeability in unsaturated simulated landfill covers. *Science of the Total Environment*, 666, 680–684.

Ng, C. W. W., Zhang, Q., Zhang, S., Lau, S. Y., Guo, H., Li, Z. (2024a). A new state-dependent constitutive model for cyclic thermo-mechanical behaviour of unsaturated vegetated soil. *Canadian Geotechnical Journal*, https://doi.org/10.1139/cgj-2023-0268.

Ng, C.W.W., Chen, H., Guo, H., Chen, R., Xue, Q. (2024b). Life cycle analysis of common landfill final cover systems focusing on carbon neutrality. *Science of the Total Environment*, 912, 168863.

Ni, J. J., Leung, A. K. and Ng, C. W. W. (2018a). Modelling soil suction changes due to mixed species planting. *Ecological Engineering*, 117, 1–17.

Ni, J. J., Leung, A. K., Ng, C. W. W. and Shao, W. (2018b). Modelling hydro-mechanical reinforcements of plants to slope stability. *Computers and Geotechnics*, 95, 99–109.

Nikooee, E., Habibagahi, G., Hassanizadeh, S. M. and Ghahramani, A. (2013). Effective stress in unsaturated soils: a thermodynamic approach based on the interfacial energy and hydromechanical coupling. *Transport in Porous Media*, 96(2), 369–396.

Nishimura, T. (2000). Direct shear properties of a compacted soil with known stress history. *Proceedings of the Asia Conference on Unsaturated Soils*, 557–562.

Nishimura, T. and Fredlund, D. (2000). Relationship between shear strength and matric suction in an unsaturated silty soil. *Unsaturated Soils for Asia*, 563–568.

Nishimura, T. and Fredlund, D. (2001). Failure envelope of a desiccated, unsaturated silty soil. *Proceedings of the Fifteenth International Conference on Soil Mechanics and Geotechnical Engineering*, 1–3, 615–618.

Nowamooz, H. and Masrouri, F. (2008). Hydromechanical behaviour of an expansive bentonite/silt mixture in cyclic suction-controlled drying and wetting tests. *Engineering Geology*, 101(3–4), 154–164.

Nuth, M. and Laloui, L. (2008). Advances in modelling hysteretic water retention curve in deformable soils. *Computers and Geotechnics*, 35(6), 835–844.

Nyhan, J. W., Hakonson, T. E. and Drennon, B. J. (1990). A water balance study of two landfill cover designs for semiarid regions. *Journal of Environmental Quality*, 19(2), 281–288.

Nyunt, T. T., Leong, E. C. and Rahardjo, H. (2010). Effect of matric suction and loading rate on the stiffness-strain behaviour of kaolin. In *Experimental studies in unsaturated and expansive soils & theoretical and numerical advances in unsaturated soil mechanics*, 15–19, CRC Press.

O'Toole, G., Kaplan, H. B. and Kolter, R. (2000). Biofilm formation as microbial development. *Annual Review of Microbiology*, 54, 49–79.

Oh, S. and Lu, N. (2014). Uniqueness of the suction stress characteristic curve under different confining stress conditions. *Vadose Zone Journal*, 13(5), 1–10.

Oh, W. T. and Vanapalli, S. K. (2011). Relationship between Poisson's ratio and soil suction for unsaturated soils. *Proceedings of 5th Asia-Pacific Conference on Unsaturated Soils*, 239–245.

Olivella, S. and Alonso, E. E. (2008). Gas flow through clay barriers. *Géotechnique*, 58(3), 157–176.

Olivella, S., Carrera, J., Gens, A. and Alonso, E. E. (1994). Nonisothermal multiphase flow of brine and gas through saline media. *Transport in Porous Media*, 15(3), 271–293.

Or, D. and Tuller, M. (2002). Cavitation during desaturation of porous media under tension. *Water Resources Research*, 38(5), 19-1–19-14.

Orts, W. J., Sojka, R. E. and Glenn, G. M. (2000). Biopolymer additives to reduce erosion-induced soil losses during irrigation. *Industrial Crops and Products*, 11(1), 19–29.

Otálvaro, I. F., Neto, M. P. C. and Caicedo, B. (2015). Compressibility and microstructure of compacted laterites. *Transportation Geotechnics*, 5, 20–34.

Oteo-Mazo, C., Saez-Aunon, J. and Esteban, F. (1995). Laboratory tests and equipment with suction control. *Proceedings of the 1st Intenational Conference on Unsaturated Soils*, 3, 1509–1515.

Oztoprak, S. and Bolton, M. D. (2013). Stiffness of sands through a laboratory test database. *Géotechnique*, 63(1), 54–70.

Padday, J. F. (1969). *Theory of surface tension: surface and colloid science*, Wiley-Interscience, Toronto, Canada.

Pagano, A. G., Tarantino, A. and Magnanimo, V. (2018). A microscale-based model for small-strain stiffness in unsaturated granular geomaterials. *Géotechnique*, 69(8), 687–700.

Pasha, A. Y., Khoshghalb, A. and Khalili, N. (2017). Hysteretic model for the evolution of water retention curve with void ratio. *Journal of Engineering Mechanics*, 143(7), 1–16.

Pearsall, I. S. (1972). *Cavitation*. Mills and Boon Ltd, London.

Peck, A. J. and Rabbidge, R. (1969). Design and performance of an osmotic tensiometer for measuring capillary potential. *Proceedings of Soil Science Society of America*, 33(2), 196–202.

Peck, R. B. (1969). Deep excavation and tunneling in soft ground. *Proceedings of the 7th International Conference on Soil Mechanics and Foundation Engineering*, 225–290.

Peng, X. and Horn, R. (2007). Anisotropic shrinkage and swelling of some organic and inorganic soils. *European Journal of Soil Science*, 58(1), 98–107.

Pennington, D. S., Nash, D. F. T. and Lings, M. L. (1997). Anisotropy of G0 shear stiffness in Gault Clay. *Géotechnique*, 47(3), 391–398.

Pennington, D. S., Nash, D. F. T. and Lings, M. L. (2001). Horizontally mounted bender elements for measuring anisotropic shear moduli in triaxial clay specimens. *Geotechnical Testing Journal*, 24(2), 133–144.

Peterson, P. and Kwong, H. (1981). A design storm profile for Hong Kong. Technical note No. 58. Royal Observatory, Hong Kong.

Phene, C. J., Hoffman, G. J. and Rawlins, S. L. (1971). Measuring soil matric potential in situ by sensing heat dissipation within a porous body: theory and sensor construction. *Proceedings of Soil Science Society of America*, 35, 27–32.

Philip, J. R. (1977). Unitary approach to capillary condensation and adsorption. *Journal of Chemical Physics*, 66(11), 5069–5075.

Plesset, M. S. (1969). The tensile strength of liquids. *Proceedings of ASME Fluids Engineering and Applied Mechanics Conference*, 15–25.

Pollen-Bankhead, N. and Simon, A. (2010). Hydrologic and hydraulic effects of riparian root networks on streambank stability: is mechanical root-reinforcement the whole story? *Geomorphology*, 116(3–4), 353–362.

Potts, D. M. and Zdravkovic, L. (2001). *Finite element analysis in geotechnical engineering Application*. Imperial College of Science, Technology and Medicine, Thomas Telford Services Ltd.

Proto, C. J., DeJong, J. T. and Nelson, D. C. (2016). Biomediated permeability reduction of saturated sands. *Journal of Geotechnical and Geoenvironmental Engineering*, 142(12), 04016073.

Qian, X., Gray, D. H. and Woods, R.D. (1993). Voids and granulometry: effects on shear modulus of unsaturated sands. *Journal of Geotechnical Engineering*, 119(2), 295–314.

Qian, X., Koerner, R. M. and Gray, D. H. (2001). *Geotechnical aspects of landfill construction and design*. Prentice Hall, Inc., Englewood Cliffs, NJ.

Rahardjo, H., Tami, D. and Leong, E. C. (2006). Effectiveness of sloping capillary barriers under high precipitation rates. *Proceedings of the 2nd International Conference on Problematic Soil*, 39–54.

Rahardjo, H., Santoso, V. A., Leong, E. C., Ng, Y. S. and Hua, C. J. (2012). Performance of an instrumented slope covered by a capillary barrier system. *Journal of Geotechnical and Geoenvironmental Engineering*, 138(4), 481–490.

Rahardjo, H., Satyanaga, A., Leong, E. C., Santoso, V. A. and Ng, Y. S. (2014). Performance of an instrumented slope covered with shrubs and deep-rooted grass. *Soils and Foundations*, 54(3), 417–425.

Rahardjo, H., Satyanaga, A., Harnas, F. R. and Leong, E. C. (2016). Use of dual capillary barrier as cover system for a sanitary landfill in singapore. *Indian Geotechnical Journal*, 46(3), 228–238.

Rampello, S., Viggiani, G. M. B. and Amorosi, A. (1997). Small-strain stiffness of reconstituted clay compressed along constant triaxial effective stress ratio paths. *Géotechnique*, 47(3), 475–489.

Rao, B. H. and Singh, D. N. (2010). Establishing soil-water characteristic curve of a fine-grained soil from electrical measurements. *Journal of Geotechnical and Geoenvironmental Engineering*, 136(5), 751–754.

Raveendiraraj, A. (2009). Coupling of mechanical behaviour and water retention behaviour in unsaturated soils. PhD Thesis, University of Glasgow, Glasgow, UK.

Rawlins, S. L. and Dalton, F. N. (1967). Psychometric measurement of soil-water potential without precise temperature control. *Journal of Soil Science*, 31, 297–301.

Richards, B. G. (1965). Measurement of the free energy of soil moisture by the psychrometric technique using thermistors. *Moisture Equilibria and Moisture Changes in Soils Beneath Covered Areas*, 39–46.

Richards, B. E. and Trevena, D. H. (1976). The measurement of positive and negative pressures in a liquid contained in a Berthelot tube. *Journal of Physics D: Applied Physics*, 9(11), 1123–1126.

Richards, B. G. (1966). The significance of moisture flow and equilibria in unsaturated soils in relation to the design of engineering structures built on shallow foundations in Australia. In *Symposium on Permeability and Capillarity*, American Society for Testing and Materials, Atlantic City, NJ.

Richards, S. J. and Weeks, L. V. (1953). Capillary conductivity values from moisture yield and tension measurements on soil columns. *Soil Science Society of America Journal*, 17(3), 206–209.

Ridley, A. M. and Burland, J. B. (1993). A new instrument for the measurement of soil moisture suction. *Géotechnique*, 43(2), 321–324.

Ridley, A. M. and Burland, J. B. (1995). A pore pressure probe for the in situ measurement of soil suction. *Proceedings of Conference on Advances in Site Investigation Practice*, 510–520.

Ridley, A. M. and Wray, W. K. (1996). Suction measurement: A review of current theory and practices. *Proceedings of the 1st International Conference on Unsaturated Soils*, 3, 1293–1322.

Roberson, E. B. and Firestone, M. K. (1992). Relationship between desiccation and exopolysaccharide production in a soil pseudomonas sp. *Applied and Environmental Microbiology*, 58(4), 1284–1291.

Roesler, S. K. (1979). Anisotropic shear modulus due to stress anisotropy. *Journal of the Geotechnical Engineering Division*, 105(7), 871–880.

Rojas, E., Chávez, O., Arroyo, H., López-Lara, T., Hernández, J. B. and Horta, J. (2017). Modeling the dependency of soil-water retention curve on volumetric deformation. *International Journal of Geomechanics*, 17(1), 04016039.

Romero, E. (2013). A microstructural insight into compacted clayey soils and their hydraulic properties. *Engineering Geology*, 165, 3–19.

Romero, E., Gens, A. and Lloret, A. (1999). Water permeability, water retention and microstructure of unsaturated compacted Boom clay. *Engineering Geology*, 54(1–2), 117–127.

Romero, E., Gens, A. and Lloret, A. (2001). Temperature effects on the hydraulic behaviour of an unsaturated clay. *Geotechnical and Geological Engineering*, 19(3–4), 311–332.

Romero, E., Gens, A. and Lloret, A. (2003). Suction effects on a compacted clay under non-isothermal conditions. *Géotechnique*, 53(1), 65–81.

Romero, E., Della Vecchia, G. and Jommi, C. (2011). An insight into the water retention properties of compacted clayey soils. *Géotechnique*, 61(4), 313–328.

Ross, B. (1990). The diversion capacity of capillary barriers. *Water Resources Research*, 26(10), 2625–2629.

Rowe, P. W. (1969). The relationship between the shear strength of sands in triaxial compression, plane strain and direct shear. *Géotechnique*, 19(1), 75–86.

Rowe, R. K. (2005). Long-term performance of contaminant barrier systems. *Géotechnique*, 55(9), 631–678.

Roy, D. and Robinson, K. E. (2009). Surface settlements at a soft soil site due to bedrock dewatering. *Engineering Geology*, 107(3–4), 109–117.

Roy, S. and Rajesh, S. (2020). Simplified model to predict features of soil–water retention curve accounting for stress state conditions. *International Journal of Geomechanics*, 20(3), 1–14.

Russell, A. R. and Khalili, N. (2006). A unified bounding surface plasticity model for unsaturated soils. *International Journal for Numerical and Analytical Methods in Geomechanics*, 30(3), 181–212.

Sanchez-Salinero, I., Roesset, J.M. and Stokoe, K.H. (1986). Analytical studies of body wave propagation and attenuation. Report GR 86-15, University of Texas, Austin, TX.

Santamarina, J. C., Klein, K. A. and Fam, M. A. (2001). *Soils and waves – particulate materials behavior, characterization and process monitoring*. Wiley, New York, NY, USA.

Sawangsuriya, A., Edil, T. B. and Bosscher, P. J. (2009). Modulus-suction-moisture relationship for compacted soils in postcompaction state. *Journal of Geotechnical and Geoenvironmental Engineering*, 135(10), 1390–1403.

Scanlan, C. A. and Hinz, C. (2010). Insight into the processes and effects of root induced changes to soil hydraulic properties. *Proceedings of the 19th World Congress of Soil Science, Soil Solutions for a Changing World*, 2.1.2, 41–44.

Schofield, R. (1935). The pF of water in soil. *Transaction of the 3rd International Congress on Soil Science*, 3(2), 37–48.

Schwertmann, U. and Fitzpatrick, R.W. (1992). Iron minerals in surface environments. *Catena Supplement*, 21, 7–30.

Seed, H. B., Chan, C. K. and Lee, C. E. (1962). Resilience characteristics of subgrade soils and their relation to fatigue failures. *Proceedings of the International Conference on Structural Design of Asphalt Pavements*, 611–636.

Senetakis, K., Anastasiadis, A., Pitilakis, K. and Coop, M. R. (2013). The dynamics of a pumice granular soil in dry state under isotropic resonant column testing. *Soil Dynamics and Earthquake Engineering*, 45, 70–79.

Shemesh, M. and Chaia, Y. (2013). A combination of glycerol and manganese promotes biofilm formation in Bacillus subtilis via histidine kinase KinD signaling. *Journal of Bacteriology*, 195(12), 2747–2754.

Sheng, D. (2011). Review of fundamental principles in modelling unsaturated soil behaviour. *Computers and Geotechnics*, 38(6), 757–776.

Sheng, D. and Zhou, A. N. (2011). Coupling hydraulic with mechanical models for unsaturated soils. *Canadian Geotechnical Journal*, 48(5), 826–840.

Sheng, D., Fredlund, D. G. and Gens, A. (2008a). A new modelling approach for unsaturated soils using independent stress variables. *Canadian Geotechnical Journal*, 45(4), 511–534.

Sheng, D., Gens, A., Fredlund, D. G. and Sloan, S.W. (2008b). Unsaturated soils: from constitutive modelling to numerical algorithms. *Computers and Geotechnics*, 35(6), 810–824.

Sheng, D., Zhou, A. and Fredlund, D. G. (2011). Shear strength criteria for unsaturated soils. *Geotechnical and Geological Engineering*, 29(2), 145–159.

Shi, B., Jiang, H., Liu, Z. and Fang, H. Y. (2002). Engineering geological characteristics of expansive soils in China. *Engineering Geology*, 67(1–2), 63–71.

Shi, J. (2015). Investigation of three-dimensional tunnel responses due to basement excavation. PhD thesis, The Hong Kong University of Science and Technology, Hong Kong, China.

Shi, J., Fu, Z. and Guo, W. (2019a). Investigation of geometric effects on three-dimensional tunnel deformation mechanisms due to basement excavation. *Computers and Geotechnics*, 106, 108–116.

Shi, J., Wei, J., Ng, C. W. W. and Lu, H. (2019b). Stress transfer mechanisms and settlement of a floating pile due to adjacent multi-propped deep excavation in dry sand. *Computers and Geotechnics*, 116, 103216.

Shuai, F., Yazdani, J., Feng, M. and Fredlund, D. G. (1998). *Supplemental report on the thermal conductivity matric suction sensor development (Year II)*. University of Saskatchewan, Saskatoon, Canada.

Siemens, G. and Bathurst, R. J. (2010). Numerical parametric investigation of infiltration in one-dimensional sand–geotextile columns. *Geotextiles and Geomembranes*, 28(5), 460–474.

Sillers, W. S. (1997). Mathematical representation of the soilwater characteristic curve. M.Sc. Thesis, University of Saskatchewan, Saskatoon, Canada.

Simpson, B. (1992). Retaining structures: displacement and design. *Géotechnique*, 42(4), 541–576.

Sinnathamby, G., Phillips, D. H., Sivakumar, V. and Paksy, A. (2014). Landfill cap models under simulated climate change precipitation: impacts of cracks and root growth. *Géotechnique*, 64(2), 95–107.

Sivakumar, V. (1993). A critical state framework for unsaturated soils. Ph.D. thesis, University of Sheffield, Sheffield, UK.

Sivapullaiah, P. V., Sridharan, A. and Stalin, V. K. (1996). Swelling behaviour of soil-bentonite mixtures. *Canadian Geotechnical Journal*, 33(5), 808–814.

Slatter, E., Jungnickel, C., Smith, D. W. and Allman, M. A. (2000). Investigation of suction generation in apparatus employing osmotic methods. *Proceedings of Asia Confeernce on Unsaturated Soils*, 297–302.

Smethurst, J. A., Clarke, D. and Powrie, W. (2012). Factors controlling the seasonal variation in soil water content and pore water pressures within a lightly vegetated clay slope. *Géotechnique*, 62(5), 429–446.

SMO. (1995). *Explanatory notes on geodetic datums in Hong Kong*. Survey & Mapping Office Lands Department, Hong Kong Government.

Spanner, D. C. (1951). The Peltier effect and its use in the measurement of suction pressure. *Journal of Experimental Botany*, 11, 145–168.

Sposito, G. (1981). *The Thermodynamics of Soil Solutions*. Oxford Clarendon Press, London, UK.

Sridharan, A., Abraham, B. M. and Jose, B. T. (1991). Improved technique for estimation of preconsolidation pressure. *Géotechnique*, 41(2), 263–268.

Stannard, D. (1992). Tensiometers – theory, construction and use. *Geotechnical Testing Journal*, 1, 48–58.

Stark, T. D., Choi, H., Lee, C. and Queen, B. (2012). Compacted soil liner interface strength importance. *Journal of Geotechnical and Geoenvironmental Engineering*, 138(4), 544–550.

Steinberg, M. (1998). *Geomembranes and the control of expansive soils in construction*. McGraw-Hill, New York.

Stokoe, K. H., I., Hwang, S. K., Lee, J. N. -K. and Andrus, R. D. (1995). Effects of various parameters on the stiffness and damping of soils at small to medium strains. *Proceedings of the 1st*

International Conference on Pre-Failure Deformation Characteristics of Geo- materials: Pre-failure Deformation of Geomaterials, 2, 785–816.

Stormont, J. C., and Anderson, C. E. (1999). Capillary barrier effect from underlying coarser soil layer. *Journal of Geotechnical and Geoenvironmental Engineering*, 125(8), 641–648.

Sun, D., Sheng, D. and Xu, Y. (2007). Collapse behaviour of unsaturated compacted soil with different initial densities. *Canadian Geotechnical Journal*, 44(6), 673–686.

Sun, Z., Moldrup, P., Elsgaard, L., Arthur, E., Bruun, E. W., Hauggaard-Nielsen, H. and de Jonge, L. W. (2013). Direct and indirect short-term effects of biochar on physical characteristics of an Arable sandy loam. *Soil Science*, 178(9), 465–473.

Tadepalli, R. and Fredlund, D. G. (1991). The collapse behavior of a compacted soil during inundation. *Canadian Geotechnical Journal*, 28(4), 477–488.

Take, W. A. and Bolton, M. D. (2003). Tensiometer saturation and the reliable measurement of soil suction. *Géotechnique*, 53(2), 159–172.

Tami, D., Rahardjo, H., Leong, E. C. and Fredlund, D. G. (2004). Design and laboratory verification of a physical model of sloping capillary barrier. *Canadian Geotechnical Journal*, 41(5), 814–830.

Tan, Y. and Wei, B. (2012). Observed behaviors of a long and deep excavation constructed by cut-and-cover technique in Shanghai soft clay. *Journal of Geotechnical and Geoenvironmental Engineering*, 138(1), 69–88.

Tang, A. M., Cui, Y. J. and Barnel, N. (2008). Thermo-mechanical behaviour of a compacted swelling clay. *Géotechnique*, 58(1), 45–54.

Tarantino, A. (2009). A water retention model for deformable soils. *Géotechnique*, 59(9), 751–762.

Tarantino, A. and Mongiovi, L. (2000). A study of the efficiency of semi-permeable membranes in controlling soil matrix suction using the osmotic technique. *Proceedings on Unsaturated Soils for Asia*, 303–308.

Tarantino, A., Gallipoli, D., Augarde, C. E., de Gennaro, V., Gomez, R., Laloui, L., Mancuso, C., El Mountassir, G., Munoz, J. J., Pereira, J. M., … Wheeler, S. (2011). Benchmark of experimental techniques for measuring and controlling suction. *Géotechnique*, 61(4), 303–312.

Tatsuoka, F. and Kohata, Y. (1995). Stiffness of hard soils and soft rocks in engineering applications. *Pre-Failure Deformation of Geomaterials*, 2, 947–1063.

Taylor, D. W. (1948). *Fundamentals of soil mechanics*. John Wiley & Sons, New York.

Taylor, S. A. and Ashcroft, G. L. (1972). *Physical edaphology*. W.H. Freeman, San Francisco, USA.

Temperley, H. N. V. and L.L.G, C. (1946). The behaviour of water under hydrostatic tension: I. *Proceedings of the Physical Society*, 58(4), 420–436.

Terzaghi, K. (1936). The shear strength of saturated soils. *Proceedings of the First International Conference on Soil Mechanics and Foundation Engineering*, 1, 54–56.

Tessier, D. (1984). Etude expérimentale de l'organisation des matériaux argileux – Hydratation, gonflement et structuration au cours de la dessiccation et la réhumectation. *Unité des Sciences physiques de la Terre*, Institut national de la recherche agronomique.

Thauer, R. K. (1998). Biochemistry of methanogenesis: a tribute to Marjory Stephenson:1998 Marjory Stephenson Prize Lecture. *Microbiology*, 144(9), 2377–2406.

Toll, D. G. (1990). A framework for unsaturated soil behaviour. *Géotechnique*, 40(1), 31–44.

Toll, D. G., and Ong, B.H. (2003). Critical-state parameters for an unsaturated residual sandy clay. *Géotechnique*, 53(1), 93–103.

Tombolato, S. and Tarantino, A. (2005). Coupling of hydraulic and mechanical behaviour in unsaturated compacted clay. *Géotechnique*, 55(4), 307–317.

Towhata, I. (2008). *Geotechnical earthquake engineering*. Springer Science & Business Media, Berlin, Germany.

Trevena, D. H. (1987). *Cavitation and tension in liquids*. Adam Hilger, Bristol, UK.

Tuller, M., Or, D., Dudley, L. M., Dani, O. and Dudley, L. M. (1999). Adsorption and capillary condensation in porous media: liquid retention and interfacial configurations in angular pores. *Water Resources Research*, 35(7), 1949–1964.

Uchaipichat, A. and Khalili, N. (2009). Experimental investigation of thermo-hydro-mechanical behaviour of an unsaturated silt. *Géotechnique*, 59(4), 339–353.

USEPA. (1993). *Solid waste disposal facility criteria*. US Environmental Protection Agency. EPA-530-R-93-017.

USEPA. (2004). Technical guidance for RCRA/CERCLA final covers. EPA-540-R-04-007, USEPA, Washington, DC.

USEPA. (2015). *Advancing sustainable materials management: facts and figures*. US Environmental Protection Agency.

Uzan, J. (1985). Characterization of granular material. *Transportation Research Record*, 1022, 52–59.

Vallejo, L. E. and Zhou, Y. (1994). The mechanical properties of simulated soil-rock mixtures. *Poceedings of International Conference on Soil Mechanics and Foundation Engineering*, 365–368.

van Genuchten, M. T. (1980). A closed-form equation for predicting the hydraulic conductivity of unsaturated soils. *Soil Science Society of America Journal*, 44(5), 892–898.

Vanapalli, S. K., Fredlund, D. G. and Pufahl, D. E. (1999). The influence of soil structure and stress history on the soil–water characteristics of a compacted till. *Géotechnique*, 49(2), 143–159.

Vardanega, P. J. and Bolton, M. D. (2013). Stiffness of clays and silts: normalizing shear modulus and shear strain. *Journal of Geotechnical and Geoenvironmental Engineering*, 9, 1575–1589.

Varnes, D. J. (1978). Slope movement types and processes. *Landslides, analysis and control. Special Report* 176, 11–33.

Vassallo, R., Mancuso, C. and Vinale, F. (2007). Effects of net stress and suction history on the small strain stiffness of a compacted clayey silt. *Canadian Geotechnical Journal*, 44(4), 447–462.

Vepraskas, M. J. and Craft, C. B. (2016). *Wetland soils: genesis, hydrology, landscapes, and classification*. CRC Press, Boca Raton, USA.

Verma, R., Sharma, P.K. and Pandey, V. (2016). Study of shear strength behaviour of soil- rock mixture. *International Journal of Innovative Research in Science, Engineering and Technology*, 5(8), 15327–15333.

Viggiani, G. and Atkinson, J. H. (1995a). Interpretation of bender element tests. *Géotechnique*, 45(1), 149–154.

Viggiani, G. and Atkinson, J. H. (1995b). Stiffness of fine-grained soil at very small strains. *Géotechnique*, 45(2), 249–265.

Vlamakis, H., Chai, Y., Beauregard, P., Losick, R. and Kolter, R. (2013). Sticking together: building a biofilm the Bacillus subtilis way. *Nature Reviews Microbiology*, 11(3), 157–168.

Vucetic, M. (1994). Cyclic threshold shear strains in soils. *Journal of Geotechnical engineering*, 120(12), 2208–2228.

Vucetic, M. and Dobry, R. (1991). Effect of soil plasticity on cylic response. *Journal of Geotechnical Engineering*, 117(1), 89–107.

Walter, G. R. (2003). Fatal flaws in measuring landfill gas generation rates by empirical well testing. *Journal of the Air and Waste Management Association*, 53(4), 461–468.

Wan, R. G. and Guo, P. J. (1998). A simple constitutive model for granular soils- modified stress. *Computers and Geotechnics*, 22(2), 109–133.

Wan, R., Khosravani, S. and Pouragha, M. (2014). Micromechanical analysis of force transport in wet granular soils. *Vadose Zone Journal*, 13(5), vzj2013.06.0113.

Wang, B. (2000). Stress effects on soil-water characteristics of unsaturated expansive soils. MPhil Thesis, The Hong Kong University of Science and Technology, Hong Kong, China.

Wang, J. H., Xu, Z. H. and Wang, W. D. (2010). Wall and ground movements due to deep excavations in shanghai soft soils. *Journal of Geotechnical and Geoenvironmental Engineering*, 136(7), 985–994.

Wang, Y. H. and Leung, S. C. (2008). Characterization of cemented sand by experimental and numerical investigations. *Journal of Geotechnical and Geoenvironmental Engineering*, 134(7), 992–1004.

Wang, Y. H. and Yan, W. M. (2006). Laboratory studies of two common saprolitic soils in Hong Kong. *Journal of Geotechnical and Geoenvironmental Engineering*, 132(7), 923–930.

Warren, R. W., Hakonson, T. E. and Bostick, K. V. (1996). Choosing the most effective hazardous waste landfill cover. *Remediation Journal*, 6(2), 23–41.

Watson, K. (1966). An instantaneous profile method for determining the hydraulic conductivity of unsaturated porous materials. *Water Resources Research*, 2(4), 709–715.

Weeks, B. and Wilson, G. W. (2005). Variations in moisture content for a soil cover over a 10 year period. *Canadian Geotechnical Journal*, 42(6), 1615–1630.

Wei, G. H., Lian, S. and Cao, W. (2008). Gray relation analysis method of meteorological factors impacting evaporation in Shenzhen City (in Chinese). *Desert and Oasis Meteorlogy*, 2(3), 33–35.

Wei, H.Y. 2007. Experimental and numerical study on gas migration in landfill of municipal solid waste. Ph.D. Thesis, Zhejiang University, Hangzhou, China.

Wen, B. P. and Yan, Y. J. (2014). Influence of structure on shear characteristics of the unsaturated loess in Lanzhou, China. *Engineering Geology*, 168, 46–58.

Wendroth, O., Ehlers, W., Hopmans, J. W., Kage, H., Halbertsma, J. and Wösten, J. H. M. (1993). Reevaluation of the Evaporation Method for Determining Hydraulic Functions in Unsaturated Soils. *Soil Science Society of America Journal*, 57(6), 1436–1443.

Wheeler, S. J. and Karube, D. (1996). State of the art report-constitutive modelling. *Proceedings of the 1st Intenational Confeernce on Unsaturated Soils,* 3, 1323–1356.

Wheeler, S. J. and Sivakumar, V. (1995). An elasto-plastic critical state framework for unsaturated soil. *Géotechnique.* 45(1), 35–53.

Wheeler, S. J., Sharma, R. S. and Buisson, M. S. R. (2003). Coupling of hydraulic hysteresis and stress-strain behaviour in unsaturated soils. *Géotechnique*, 53(1), 41–54.

Wickramarachchi, P., Kawamoto, K., Hamamoto, S., Nagamori, M., Moldrup, P. and Komatsu, T. (2011). Effects of dry bulk density and particle size fraction on gas transport parameters in variably saturated landfill cover soil. *Waste Management*, 31(12), 2464–2472.

Williams, J. and Shaykewich, C. F. (1969). An evaluation of polyethylene glycol (PEG) 6000 and PEG 20000 in the osmotic control of soil–water matric potential. *Canadian Journal of Soil Science*, 102(6), 394–398.

Wilson-Fahmy, R. F., Narejo, D. and Koerner, R. M. (1996). Puncture protection of geomembranes part I: theory. *Geosynthetics International*, 3(5), 605–628.

Wong, J. T. F., Chen, Z., Ng, C. W. W. and Wong, M. H. (2016). Gas permeability of biochar-amended clay: potential alternative landfill final cover material. *Environmental Science and Pollution Research*, 23(8), 7126–7131.

Wong, K. S., and Mašín, D. (2014). Coupled hydro-mechanical model for partially saturated soils predicting small strain stiffness. *Computers and Geotechnics*, 61, 355–369.

Wong, K. S., Mašín, D. and Ng, C. W. W. (2014). Modelling of shear stiffness of unsaturated fine grained soils at very small strains. *Computers and Geotechnics*, 56, 28–39.

Wood, D. M. (1990). *Soil behaviour and critical state soil mechanics*. Press Syndicate of the University of Cambridge, Cambridge, UK.

World Bank. (2014). *World development indicators 2014*. World Development Indicators, Washington, DC.

Wu, S., Gray, D. H., Asce, A. M., Richart, F. E., Asce, F., Wu, B. S., Gray, D. H., Asce, A. M., Richart, F. E., Asce, F., Wu, S., Gray, D. H. and Richart, F. E. (1984). Capillary effects on dynamic modulus of sands and silts. *Journal of Geotechnical Engineering*, 110(9), 1188–1203.

Xu, J. (2011). Experimental study of effects of suction history and wetting-drying on small strain stiffness of an unsaturated soil. Mphil Thesis, The Hong Kong University of Science and Technology, Hong Kong, China.

Yan, W. M. and Zhang, G. (2015). Soil-water characteristics of compacted sandy and cemented soils with and without vegetation. *Canadian Geotechnical Journal*, 52(9), 1331–1344.

Yang, H., Rahardjo, H. and Leong, E. C. (2006). Behavior of unsaturated layered soil columns during infiltration. *Journal of Hydrologic Engineering*, 11(4), 329–337.

Yang, H., Rahardjo, H., Wibawa, B., & Leong, E. C. (2004a). A soil column apparatus for laboratory infiltration study. *Geotechnical Testing Journal*, 27(4), 347–355.

Yang, H., Rahardjo, H., Leong, E. C. and Fredlund, D. G. (2004b). Factors affecting drying and wetting soil-water characteristic curves of sandy soils. *Canadian Geotechnical Journal*, 41(5), 908–920.

Yang, S. R., Huang, W. H. and Liao, C. C. (2008). Correlation between resilient modulus and plastic deformation for cohesive subgrade soil under repeated loading. *Transportation Research Record: Journal of the Transportation Research Board*, 2053(1), 72–79.

Yao, Y. P. and Zhou, A. N. (2013). Non-isothermal unified hardening model: a thermo-elasto-plastic model for clays. *Géotechnique*, 63(15), 1328–1345.

Yao, Y. P., Hou, W. and Zhou, A. N. (2009). UH model: Three-dimensional unified hardening model for overconsolidated clays. *Geotechnique*, 59(5), 451–469.

Yimsiri, S. and Soga, K. (2002). Application of micromechanics model to study anisotropy of soils at small strains. *Soils and Foundations*, 42(5), 15–26.

Yoo, S. D. and Harcum, S. W. (1999). Xanthan gum production from waste sugar beet pulp. *Bioresource Technology*, 70(1), 105–109.

Young, F. R. (1989). *Cavitation*. McGraw-Hill Book Company, London, UK.

Zeng, Y., Su, Z., Wan, L. and Wen, J. (2011). Numerical analysis of air-water-heat flow in unsaturated soil: Is it necessary to consider airflow in land surface models? *Journal of Geophysical Research Atmospheres*, 116(20), 1–18.

Zhan, G., Keller, J., Milczarek, M. and Giraudo, J. (2014a). 11 years of evapotranspiration cover performance at the AA leach pad at Barrick Goldstrike Mines. *Mine Water and the Environment*, 33(3), 195–205.

Zhan, L. T. (2003). Field and laboratory study of an unsaturated expansive soil associated with rain-induced slope instability. PhD Thesis, The Hong Kong University of Science and Technology, Hong Kong, China.

Zhan, L. T. and Ng, C. W. W. (2006). Shear strength characteristics of an unsaturated expansive clay. *Canadian Geotechnical Journal*, 43(7), 751–763.

Zhan, L. T., Li, H., Jia, G. W., Chen, Y. M. and Fredlund, D. G. (2014b). Physical and numerical study of lateral diversion by three-layer inclined capillary barrier covers under humid climatic conditions. *Canadian Geotechnical Journal*, 51(12), 1438–1448.

Zhan, L. T., Qiu, Q., Xu, W. and Chen, Y. (2016). Field measurement of gas permeability of compacted loess used as an earthen final cover for a municipal solid waste landfill. *Journal of Zhejiang University-Science A*, 17(7), 541–552.

Zhan, L. T., Li, G., Jiao, W., Wu, T., Lan, J. and Chen, Y. (2017). Field measurements of water storage capacity in a loess–gravel capillary barrier cover using rainfall simulation tests. *Canadian Geotechnical Journal*, 54(11), 1523–1536.

Zhan, Z. Q., Zhou, C., Liu, C. Q., Ng, C. W. W. (2023). Modelling hydro-mechanical coupled behaviour of unsaturated soil with two-phase two-point material point method. *Computers and Geotechnics*, 155, 105224.

Zhang, J., Andrus, R. D. and Juang, C. H. (2005). Normalized shear modulus and material damping ratio relationships. *Journal of Geotechnical and Geoenvironmental Engineering*, 131(4), 453–464.

Zhang, L. and Chen, Q. (2005). Predicting bimodal soil–water characteristic curves. *Journal of Geotechnical and Geoenvironmental Engineering*, 131(5), 666–670.

Zhang, W., Goh, A. T. C. and Xuan, F. (2015). A simple prediction model for wall deflection caused by braced excavation in clays. *Computers and Geotechnics*, 63, 67–72.

Zhang, X. W., Kong, L. W. and Li, J. (2014). An investigation of alterations in Zhanjiang clay properties due to atmospheric oxidation. *Géotechnique*, 64(12), 1003–1009.

Zheng, G., Yang, X., Zhou, H., Du, Y., Sun, J. and Yu, X. (2018). A simplified prediction method for evaluating tunnel displacement induced by laterally adjacent excavations. *Computers and Geotechnics*, 95, 119–128.

Zhou, A. and Sheng, D. (2015). An advanced hydro-mechanical constitutive model for unsaturated soils with different initial densities. *Computers and Geotechnics*, 63, 46–66.

Zhou, A. N. (2013). A contact angle-dependent hysteresis model for soil-water retention behaviour. *Computers and Geotechnics*, 49, 36–42.

Zhou, A. N., Sheng, D., Sloan, S. W. and Gens, A. (2012). Interpretation of unsaturated soil behaviour in the stress-saturation space, I: Volume change and water retention behaviour. *Computers and Geotechnics*, 43, 178–187.

Zhou, C. and Chen, R. (2021). Modelling the water retention behaviour of anisotropic soils. *Journal of Hydrology*, 599, 126361.

Zhou, C. and Ng, C. W. W. (2014). A new and simple stress-dependent water retention model for unsaturated soil. *Computers and Geotechnics*, 62, 216–222.

Zhou, C. and Ng, C. W. W. (2015). A thermomechanical model for saturated soil at small and large strains. *Canadian Geotechnical Journal*, 52(8), 1101–1110.

Zhou, C. and Ng, C. W. W. (2016). Simulating the cyclic behaviour of unsaturated soil at various temperatures using a bounding surface model. *Géotechnique*, 66(4), 344–350.

Zhou, C. and Ng, C. W. W. (2018). A new thermo-mechanical model for structured soil. *Géotechnique*, 68(12), 1109–1115.

Zhou, C. and Ng, C.W.W. (2013). Experimental study of resilient modulus of unsaturated soil at different temperatures. *Proceedings of the 18th ICSMGE*, 2, 1055–1058.

Zhou, C., Ng, C. W. W. and Chen, R. (2015a). A bounding surface plasticity model for unsaturated soil at small strains. *International Journal for Numerical and Analytical Methods in Geomechanics*, 39(11), 1141–1164.

Zhou, C., Xu, J. and Ng, C. W. W. (2015b). Effects of temperature and suction on secant shear modulus of unsaturated soil. *Géotechnique Letters*, 5(3), 123–128.

Zhou, R. Z. B., Take, W. A. and Ng, C. W. W. (2006). A case study in tensiometer interpretation: centrifuge modelling of unsaturated slope behaviour. *Proceedings of the 4th International Confeernce on Unsaturated Soils*, 2, 2300–2311.

Zornberg, J. G., Bouazza, A. and McCartney, J. S. (2010). Geosynthetic capillary barriers: current state of knowledge. *Geosynthetics International*, 17(5), 273–300.

Zur, B. (1966). Osmotic control of the matric soil water potential: I soil-water system. *Soil science*, 102, 394–398.

Author index

AASHTO, 238, 240, 244, 247, 267, 269, 270
Abdolahzadeh, A. M., 430
Abuel-Naga, H. M., 296, 307, 312, 353, 354
Agus, S. S., 58
Ahmad, M., 130
Airey, D., 193, 198
Aitchison, G. D., 9, 60
Alam, M. J. B., 437
Albright, W. H., 109, 118, 420–421, 429, 440, 447–448, 450
Alexander, L. T., 191, 193
Al Khafaf, S., 22
Allen, R. G., 427
Al-Mukhtar, M., 52–54
Alonso, E. E., 1, 73, 76, 78, 122, 162, 167, 174, 178, 180, 182, 183, 187, 227, 264, 300, 301, 303, 313, 316, 322, 323, 324, 325, 344, 345, 367, 394, 398
Alsherif, N. A., 170, 272, 313
Amaya, P., 420
Anderson, A. B. C., 6, 59
Anderson, C. E., 430
Apfel, R. E., 15–16
Ashcroft, G. L., 4
ASTM, 23, 80, 112, 171, 191–192, 199, 406, 429, 433
Atkinson, J. H., 200, 202, 204, 215–216, 238, 246, 269, 287, 329, 451
Aubertin, M., 118, 123

Bai, B., 298, 313
Baker, R., 59, 61
Bao, C. G., 376, 390
Bardet, J. P., 324
Barnswell, K. D., 109, 421
Barron, V., 198
Bartos, D. L., 22
Basile, A., 107
Basu, D., 425
Bathurst, R. J., 421
Been, K., 323, 347
Ben Shalom, A., 47–48
Benson, C., 109, 112, 421, 442
Benz, T., 451, 453
Berre, T., 232
Bhattacharyya, R., 428

Biglari, M., 258, 331, 334, 451
Bishop, A. W., 3, 8, 9, 11, 148, 451
Blatz, J. A., 170, 250
Blight, G. E., 11
Bocking, K. A., 59
Bohnhoff, G. L., 421
Bolton, M. D., 33, 168, 181, 255, 260, 335–336, 353
Borsetto, M., 265
Bossé, B., 439
Bouazza, A., 131, 134, 139, 143, 421
Bowles, J. E., 456
Brackley, I. J. A., 384, 394, 396
Brinkgreve, R. B., 451, 456
Brown, R. W., 21, 22
Brown, S. F., 237, 247, 249, 265, 267
BSI, 55, 135, 163
Buckingham, E., 87
Buczko, U., 445
Budhu, M., 179
Burdine, N. T., 71
Burland, J. B., 9, 11, 16, 33–34, 47–48, 51, 59–60, 148, 200, 298, 304
Burton, G. J., 252, 255
Bussière, B., 118

Cady, J. G., 191, 193
Callisto, L., 272, 282
Campanella, R. G., 282, 311, 312, 348, 353
Carter, J. P., 304
Cekerevac, C., 265–267, 272, 282, 296, 305, 307, 348–349, 353
CFI, 116–117, 127–130
Chai, J. C., 143, 456
Chandler, R. J., 13, 15, 17, 18, 23, 33
Chang, I., 133–134
Chapman, P. J., 17
Chapuis, P. J., 88
Chávez, C., 322–324, 325
Chen, F. H., 5
Chen, R., 19, 36–41, 45, 47–48, 50–52, 56–58, 133–134, 172, 317, 325, 437, 450
Chen, Y. M., 124
Chen, Z. K., 140
Cheng, Q., 291, 315, 343
Childs, E., 65, 67

Ching, R. K. H., 22
Chiu, C. F., 13, 30, 37, 43–44, 162, 163, 165, 167, 168, 180, 192, 196, 208, 264, 309, 316, 320, 321, 322–328, 337, 347, 358, 365, 367, 370
Choo, L. P., 88
Chu, T. Y., 398
Chui, T. F., 434
Clayton, C. R. I., 200, 205, 221, 253, 329
Coleman, J. D., 11, 60, 79
Collins, J., 21
Collis-George, N., 22
Coo, J. L., 143
Corlett, R. T., 428
Craft, C. B., 145
Crawford, J. F., 441
Cresswell, A., 181
Crilly, M. S., 23
Croney, D., 9
Cui, Y. J., 47–49, 153, 155, 163–164, 168, 265, 307, 322–323, 328–329, 349, 354

Dafalias, Y. F., 323, 347, 351, 356–357
Dalton, F. N., 22
Dalton, H. M., 143
Daniel, D. E., 121, 124, 420
Dashko, R., 143
Dehlen, G. L., 241
DeJong, J. T., 143
Delage, P., 47–49, 51–52, 54, 59, 63, 74, 76–78, 153, 155, 163–164, 168, 276–277, 322–323, 328–329
de Sousa Pinto, C., 16–18
Di Donna, A., 282, 296, 311, 312
Dineen, K., 47–48, 51, 59
Dobry, R., 247, 257, 260
Dong, Y., 331–332, 334
D'onza, F., 316, 345
Dorsey, N. E., 4
Drnevich, V. P., 201, 215
Duan, X., 317
Dudley, J. H., 5
Dwyer, D. F., 109, 421
Dyvik, R., 202, 204, 221

Edlefsen, N. E., 6, 59
EEA, 420
EPA, 109, 112, 420, 421, 423
Esteban, V., 36, 52
Evans, N. C., 406

Fawcett, R., 22
Feddes, R. A., 433–434
Feng, M., 19, 24–29
Firestone, M. K., 144
Fisher, R. A., 347
Fitzpatrick, R. W., 192, 193
Flemming, H. C., 133, 143–144
François, B., 262, 266, 289, 291, 354
Fredlund, D. G., 1–9, 11–13, 18, 19, 21–31, 33–35, 37–38, 59, 65–67, 68, 69, 71–73, 95, 102, 104–107, 138–139, 146–150, 155, 157, 160, 163, 165, 170, 172–173, 243, 261, 338–340, 383–384, 394–395, 456
Freyberg, D. L., 434
Frydman, S., 59, 61
Fung, Y. C., 8
Futai, M. M., 197

Gallé, C., 109
Gallen, P. M., 22
Gallipoli, D., 11, 13, 75, 79, 265, 278, 288, 296, 320, 339–340, 343, 346–347, 349
Gan, J. K. M., 37, 43, 148, 160, 164
Gardner, R., 23, 71, 339–340
Garg, A., 425
Gasmo, J., 408
Gebert, J., 124
Gee, G. W., 62
Gens, A., 1, 38, 308, 316–317, 322, 345
GEO, 95, 401
Geokon, 407, 411, 414
Germine, J. T., 232
Ghestem, M., 438, 445
Giavasis, I., 133
Gidigasu, M. D., 192
Gittens, G. T., 297
Goh, S. G., 325
Gourley, C. S., 23
Graham, J., 110, 216
Gröngröft, A., 124
Guan, Y., 33
Guo, P. J., 323
Gupta, S. C., 104, 137
Gutierrez, C. I., 22

Haeri, S. M., 186, 188, 427
Haikola, B. M., 429
Handy, R. L., 5
Hanks, R. J., 22
Han, Z., 331–332, 334
Harcum, S. W., 134
Hardin, B. O., 201, 215, 451
Harvey, E. N., 14, 16–17, 33
Hau, B. C. H., 428
Hauser, V. L., 421
Hayward, A. T. J., 15
He, J., 143
Henderson, S. J., 17
Heymann, G., 221, 253
Hildenbrand, A., 115
Hilf, J. W., 29, 36–37, 58
Hillel, D., 92, 101
Hinz, C., 445
HKO, 143, 406
Ho, D. Y. F., 155, 157
Hoefler, A. C., 134
Hong, P. Y., 354
Horn, R., 172
Horseman, S. T., 297, 310
Hossain, M. A., 175, 176, 179, 277, 284
Houlsby, G. T., 1, 12–13, 60, 216, 238, 345

Hoyos, L. R., 178, 251, 257, 330
Hsieh, P. G., 451
Hsu, C. C., 258
Hsueh, Y. H., 143
Hu, I., 428
Hueckel, T., 265, 350, 353
Hveem, F. N., 237

Iryo, T., 421
Isobe, Y., 144
Israelachvili, J., 267
Ivanova, L. K., 446

Jacobsz, S. W., 33
Jaky, J., 365, 457
Jamiolkowski, M., 201
Jardine, R. J., 200, 232
Jefferies, M. G., 323, 347
Jefferson, I., 6
Jennings, J. E., 9, 60, 148
JGS, 90
Jin, M. S., 262
Jones, W. M., 17
Joseph, S., 130
Junaideen, S. M., 197

Karimi, S., 133
Karr, L., 447
Karube, D., 9–10, 12, 266, 288, 316
Kassif, G., 47–48
Khabbaz, M. H., 318, 331
Khalili, N., 269, 272, 282, 284, 317–318, 323–324, 334, 350, 353, 362
Khatami, H. R., 133
Khire, M. G., 122, 42, 427
Kholghifard, M., 304
Khosravi, A., 75, 330–331, 451
Khoury, N. N., 237, 243
Kim, D., 237, 243
Kim, J. R., 237, 243
Knidiri, J., 439, 442
Knox, D. P., 201
Koerner, R. M., 118, 121, 124, 420
Kohata, Y., 200
Krahn, J., 35
Kruse, G. A. M., 275, 278
Kung, G. T. C., 455
Kunhel, R. A., 52
Kwong, H., 406

Lacasse, S., 232
Ladd, C. C., 232
Lagerwerff, J. V., 47
Lai, C. H., 165
Laloui, L., 59, 262, 265–267, 272, 282, 289, 291, 296, 305, 307, 311, 312, 343, 348–349, 353–354
Lam, C. C., 406
Lam, L., 65, 66
Lambe, T. W., 9
Larrahondo, J. M., 192

Larson, W. E., 104
Lee, I. M., 80, 82, 84
Lee, S. R., 339
Lefebvre, G., 74, 76, 77, 78
Lehmann, J., 130
Lekarp, F., 237–238
Lemon, K. P., 144
Lengeler, J. W., 144
Leong, E. C., 72, 330–331, 335
Leung, A. K., 61, 145, 425, 438, 441, 445, 460
Leung, S. C., 181
Leung, Y. K., 406
Li, D., 237
Li, J., 347, 394
Li, X., 82, 88, 91
Li, X. S., 250, 323, 343, 347, 357–358
Likos, W. J., 53–55, 251, 257, 319
Lim, T. T., 376
Lings, M. L., 200
Liu, M. T., 304
Liu, T., 5
Liu, T. H., 379, 395
Lloret, A., 52, 73
Loach, S. C., 243
Loehr, R. C., 429
Lo Presti, D. C. F., 201
Loughnan, F. C., 196, 198
Lu, N., 53–55, 179, 251, 319, 334

Madshus, C., 202, 204, 221
Mair, R. J., 200–201, 329
Mancuso, C., 210–211, 233, 238, 250, 270, 330–331, 334, 451
Manzari, M. T., 347, 356
Mao, S. Z., 150, 157
March, P. E., 143
Marinho, F. A. M., 13, 15–18, 33, 62
Mašín, D., 343, 353
Masrouri, F., 300
McBean, E. A., 109, 114, 116
McCarthy, E. L., 88
Mccartney, J. S., 75, 170, 272, 313, 330–331, 421, 451
McEwen, T. J., 297, 310
McKeen, R., 23
McQueen, I. S., 22, 23
Meek, B. D., 425
Meerdink, J. S., 91
Melchior, S., 421, 447
Menzies, B., 1, 3, 6, 37–38, 61, 200, 250, 269, 289, 330
Merchán, V., 170, 183, 257
Michels, W. C., 7
Mijares, R. G., 421
Miller, D. J., 376
Miller, R. F., 22, 23
Mitchell, A. R., 425
Mitchell, J. K., 145, 282, 311, 312, 348, 353
Mohan, D., 130
Mongiovi, L., 51
Monteith, J. L., 434, 449

Moon, S., 118, 138
More, T. T., 133
Morgenstern, N. R., 1, 8, 12, 38
Morikawa, M., 143–144
Morris, C. E., 121, 127, 421
Morvan, M., 322, 431
Mukherjee, S., 193
Muñoz-Castelblanco, J., 272, 278

Nealson, K. H., 144
Nelson, J. D., 376
Ng, C. W. W., 1, 3–6, 13, 33, 36–52, 56–59, 61, 64, 69–70, 73–75, 80–83, 85–87, 92, 95, 98, 102, 104, 108, 109, 112, 116–118, 121, 124, 128, 133, 137, 140, 143, 145, 150–151, 153–156, 158–174, 180–181, 183–186, 187–188, 191–192, 196, 199–201, 205–210, 212, 214–215, 217–218, 220, 222, 224, 226, 228–230, 232–240, 242–246, 248–256, 259–262, 265, 268, 269, 272–273, 275, 276, 277, 278, 280–281, 284, 287, 288, 289–290, 293, 294, 297–298, 303, 305, 306–309, 312, 315–316, 319–337, 339–341, 345, 347–348, 350, 357–358, 360, 362–363, 365, 367, 370, 375–376, 379–384, 386–387, 389–393, 395, 397, 399, 404, 408, 417, 420, 422–426, 429, 432–435, 437–438, 440–446, 451, 455–456, 460
Ni, J. J., 140, 145, 433–434, 440
Nikooee, E., 257
Nishimura, T., 163, 168–170
Nowamooz, H., 300
Nuth, M., 343
Nyhan, J. W., 447
Nyunt, T. T., 245

Oh, W. T., 456
Oh, S., 179
Olivella, S., 120, 122
Ong, B. H., 195, 196
Or, D., 59, 61
Orts, W. J., 133
Otalvaro, I. F., 191
Oteo Mazo, C., 52, 55
O'Toole, G., 143
Ou, C. Y., 451
Oztoprak, S., 255, 335, 353

Padday, J. F., 1
Pagano, A. G., 330, 331
Pang, Y. W., 37–40, 61, 69, 74–75, 80–82, 92, 102, 104, 112, 246, 339, 381, 429
Pearsall, I. S., 14
Peck, A. J., 51
Peck, R. B., 451, 457
Peng, X., 172
Pennington, D. S., 200, 212, 232
Peterson, S., 406
Phene, C. J., 24
Philip, J. R., 59
Plesset, M. S., 14
Pollen-Bankhead, N., 425
Potts, D. M., 322, 367

Powrie, W., 181
Proto, C. J., 144

Qian, X., 329, 424

Rabbidge, R., 51
Rahardjo, H., 1, 3, 5, 7, 19, 21–23, 31, 34, 37, 64–65, 67, 68, 72, 146–147, 149–150, 165, 338, 394–395, 421, 430, 440, 450, 456
Rampello, S., 201, 272, 282
Rao, B. H., 123
Raveendiraraj, A., 360
Rawlins, S. L., 22
Richards, B. E., 17
Richards, B. G., 9, 20
Richards, S. J., 88
Ridley, A. M., 16, 19–24, 27, 30, 32–35
Roberson, E. B., 144
Robinson, K. E., 459
Roesler, S. K., 201
Romero, E., 76, 77, 91, 251, 257, 262, 268, 272, 274, 289, 313, 351, 362
Ross, B., 118, 421
Rowe, P. W., 181
Rowe, R. K., 143, 421
Roy, D., 459
Russell, A. R., 317, 323–324

Saez, J., 36, 52
Saffarini, D., 145
Sanchez-Salinero, I., 204
Sanders, P. J., 384, 394, 396
Santamarina, J. C., 145, 267
Sawangsuriya, A., 330–332, 334, 451
Scanlan, C. A., 445
Schanz, T., 59
Schofield, R., 28
Schreiner, H., 23
Schwertmann, U., 192, 193
Seed, H. B., 237, 243, 262
Selig, E. T., 237
Senetakis, K., 258
Shemesh, K., 143
Sheng, D. C., 272, 316, 320, 323, 324, 327, 340–341, 343, 345
Shi, B., 5
Shi, J., 451, 456
Shi, Q., 98, 145
Shidlovskaya, A., 143
Shuai, F., 24–25
Siemens, G., 421
Sillers, W. S., 72
Simon, A., 425
Simpson, B., 200, 221, 357
Singh, D. N., 123
Sinnathamby, G., 425
Sivakumar, V., 10, 162, 167, 274, 320
Sivapullaiah, P. V., 398
Slatter, E., 51
Smethurst, J. A., 425
Smith, P. G., 441

SMO, 402
Soga, K., 221
Spanner, D. C., 22
Speedy, R. J., 17
Sposito, G., 6, 59–60
Sridharan, A., 295
Stannard, D., 30
Stark, T. D., 118
Steinberg, M., 376
Stokoe, K. H. I., 221, 362
Stormont, J. C., 121, 127, 422, 430
Sun, D., 301, 303, 304
Sun, Z., 131

Tadepalli, R., 30
Take, W. A., 18, 33
Tami, D., 123
Tan, Y., 451
Tang, A. M., 262, 272, 282, 296, 313
Tarantino, A., 51, 59, 62–63, 274, 339–341
Tatsuoka, F., 200
Taylor, D. W., 181
Taylor, S. A., 4
Temperley, H. N. V., 36
Terzaghi, K., 1
Tessier, D., 58
Thauer, R. K., 144
Toll, D. G., 195, 197–198
Tombolato, S., 274
Torrent, J., 198
Towhata, I., 260
Trevena, D. H., 16–17
Tuller, M., 59, 61

Uchaipichat, A., 267, 272, 282, 284, 350, 362
USEPA, 121, 420, 446–447, 451
Uzan, J., 237

Vallejo, L. E., 197
Vanapalli, S. K., 73, 331–332, 334, 339, 456
van der Gaast, S., 52
van Genuchten, M. T., 72, 75, 112, 114, 339–340, 429–431, 434
Vangpaisal, T., 131, 139
Vardanega, P. J., 260, 335–336, 353
Varnes, D. J., 404, 419
Vassallo, R., 183, 251, 330
Vepraskas, M. J., 145
Verma, R., 197
Viggiani, G., 202, 204, 238, 246
Vlamakis, H., 143
Vucetic, M., 247, 253, 257–258, 260

Walter, G. R., 124
Wan, R. G., 296, 323
Wang, B., 155, 381
Wang, J. H., 451
Wang, Y. H., 181, 197
Warren, R. W., 447
Watson, K., 88
Weeks, B., 109, 115

Weeks, L. V., 88
Wei, B., 451
Wei, G. H., 448
Wei, H. Y., 124
Wen, B. J., 272
Wendroth, O., 88, 107
Wheeler, S. J., 9–13, 60, 162, 167, 213, 236, 244, 246, 265, 288, 296, 316, 318, 320, 331, 343
Wickramarachchi, P., 138
Williams, J., 47, 51, 63
Wilson, G. W., 109, 115
Wilson-Fahmy, R. F., 118
Wingender, J., 133, 143–144
Wong, J. T. F., 130
Wong, K. S., 331, 334, 343, 451
Wood, D. M., 130, 246
Wray, W. K., 19–24, 27, 30, 32, 34–35
Wroth, C. P., 238
Wu, S., 329, 331, 334

Xing, A., 25, 72–73, 163, 172–173, 339–340
Xu, J., 74, 215, 217–218, 220, 222, 224, 226, 228, 243, 253, 255, 260, 269, 303, 335–336, 357, 360, 362, 451

Yan, W. M., 197, 450
Yan, Y. J., 272
Yang, H., 88, 421, 430, 442
Yang, S. R., 237, 243
Yao, Y. P., 307, 324
Yimsiri, S., 221
Yin, J. H., 175, 176, 179, 277, 284
Yoo, S. D., 134
Young, F. R., 1, 13–14, 18, 36, 105
Yu, Y. F., 406
Yung, S. Y., 200, 205–206, 208–210, 212, 214, 233, 235–236, 238, 251, 319, 320, 330–332, 334–335, 362–363, 451, 455–456

Zaman, M. M., 237, 243
Zeng, Y., 120
Zhan, G., 109
Zhan, L. T., 41, 43, 45, 55, 58, 115, 138–139, 142, 164, 272, 281, 284, 432, 440, 442, 448
Zhang, G., 450
Zhang, J., 335
Zhang, L. M., 82, 172
Zhang, W., 455
Zhang, X. W., 193
Zheng, G., 452, 455–456
Zhou, A. N., 307, 320, 321, 324, 336, 340–341, 343
Zhou, C., 61, 73, 145, 261, 262, 264–265, 272, 284, 288, 296–297, 305, 307, 312, 315–316, 336, 341, 343, 345, 348, 350, 360, 363, 375, 450
Zhou, R. Z. B., 18, 37, 45–46, 164–167, 169, 181–182, 277, 281, 322
Zhou, Y., 198
Zornberg, J. G., 421
Zur, B., 36–37, 47–48, 58, 62–63

Subject index

Absorbed water, 278
Accumulated plastic strain, 262, 370
Adsorption/wetting curve, 72
Adsorption rate, 71–72, 102–103, 108, 270, 342
Aggregate, 76, 131, 137, 143, 183, 188–189, 191–193, 195, 197–199, 252, 257, 259, 261, 275–276, 278, 286, 298–299, 306–307, 313
Air-entry value, 17, 33, 37, 40, 42–43, 47, 69, 71, 72, 74–75, 77, 101–102, 105, 109, 112–113, 122, 124–126, 128–129, 151, 153, 155, 165–166, 168–169, 185, 207, 210, 213–214, 230, 245, 381, 430, 455
Air-water interface, 1, 65, 67, 244, 246, 248, 251, 255, 260, 270
Angle of dilation, 179, 182
Angle of internal friction, 146, 197
Anisotropy, 203–204, 212–214, 235–236, 334, 344, 455
Apparent cohesion, 51, 158–161, 281
Apparent over-consolidation ratio, 267
Atterberg limit, 112, 135, 206, 404, 428–429
Axis-translation technique, 36–37, 41, 47, 51, 58–59, 60–62, 64, 150–152, 155–162, 164, 263, 335

Bimodal pore-size distribution, 82
Biochar-amended soil, 131
Biofilm, 143–145
Bishop's stress, 364
Bounding surface plasticity, 322, 324, 345, 351, 357, 367, 370, 375
Breakthrough pressure, 109–110, 114–117
Bulk density, 100, 203, 432
Bulk water, 10, 12, 59–60, 236, 265, 288, 330

Cam-clay model, 246
Capillary barrier, 118–121, 125–130, 421–422, 424, 440, 442, 448
Capillary effects, 120
Capillary force, 172
Capillary fringe, 2–3
Capillary rise, 31, 389
Capillary tube, 49
Cavitation, 13–18, 33, 36–37, 47, 59, 388, 408
Ceramic disc, 151, 263

Coarse-grained soil, 62, 120–121, 178–179, 257, 259–261, 341, 422, 424, 447
Coefficient of air permeability, 118, 120, 127–129
Coefficient of surface tension, 266
Coefficient of water permeability, 120, 122, 125, 390
Compaction curve, 112, 135, 172, 428–429
Compressibility, 78, 169, 190, 191, 193, 195, 196, 198, 199, 261, 296, 298, 300, 305, 316, 319, 320, 321, 327, 412
Compression test, 74, 170, 196, 207, 208, 209, 213, 217, 219, 221, 231, 233, 234, 235, 236, 245, 278, 287, 291, 292, 293, 294, 295, 302, 319, 328, 335, 360, 362, 363, 364, 365, 367
Confining pressure, 4, 49, 59, 71, 155, 156, 157, 159, 160, 162, 238, 239, 240, 250, 287, 296, 297, 298, 299, 323, 324, 361, 365
Consolidation curve, 174
Consolidation pressure, 10, 78, 266, 267, 278, 284, 310, 312, 313, 348, 349, 354
Constant-p drying, 334–335
Constant-p shear, 337, 344
Constant-q compression, 344
Constant-s compression, 334
Constitutive model, 162, 168, 316, 317, 318, 319, 321, 322, 323, 324, 325, 327, 329, 331, 333, 335, 337, 339, 341, 343, 344, 345, 347, 349, 351, 353, 355, 357, 359, 361, 363, 365, 367, 369, 371, 373, 375, 453
Constitutive stress, 316, 317, 318, 334, 344
Constitutive variable, 301, 316, 330, 331, 345
Contact angle, 92, 101, 296, 343, 346
Contraction, 32, 43, 56, 175, 176, 177, 180, 181, 185, 197, 242, 273, 278, 281, 284, 286, 301, 302, 303, 304, 307, 309, 310, 312, 313, 314, 321
Cooling, 22, 130, 196, 292, 293, 294, 308, 309, 310, 311, 312, 313, 314, 315, 354, 363
Coupled seepage-stress-deformation analysis, 420
Coupling effect, 213, 215, 236, 272, 287, 328, 339
Cracking, 4, 237, 265, 421, 427
Creep, 43
Cross-anisotropic, 216, 232
Cyclic behaviour, 262, 269, 271, 345, 358, 370, 375

499

Cyclic shearing, 200, 237, 375
Cyclic wetting and drying, 246

Damping ratio, 200, 201, 203, 205, 207, 209, 211, 213, 215, 217, 219, 221, 223, 225, 227, 229, 231, 233, 235, 237, 239, 241, 243, 245, 247, 249, 250, 251, 253, 255, 257, 258, 259, 260, 261
Datum, 31, 87, 379, 385, 388, 396, 402
Deformation behaviour, 278, 281, 283, 285
Degradation of soil stiffness, 247, 345
Degree of anisotropy, 203, 213
Degree of compaction, 112, 113, 131, 134, 135, 136, 137, 138, 139
Degree of nonlinearity, 243
Degree of saturation, 9, 11, 13, 19, 38, 58, 60, 65, 67, 69, 75, 78, 79, 80, 81, 82, 84, 85, 86, 109, 111, 113, 114, 115, 116, 117, 118, 122, 124, 138, 143, 144, 145, 148, 150, 155, 165, 173, 175, 178, 179, 183, 185, 203, 221, 223, 225, 227, 229, 230, 251, 255, 257, 265, 268, 273, 296, 301, 304, 316, 317, 319, 329, 330, 331, 334, 341, 345, 351, 352, 353, 358, 375, 451
Degree of stiffness anisotropy, 203–204, 212, 235–236
Density-dependence of the SWRC, 339
Desorption/drying, 72
Desorption curve, 233
Desorption rate, 72, 101, 102, 105, 108, 207, 343
Differential pressure transducer, 41–42, 154, 294
Differential settlement, 237, 446
Diffusion, 14, 109
Dilatancy, 45, 46, 47, 162, 163, 164, 165, 166, 167, 168, 169, 170, 179, 180, 181, 182, 183, 188, 189, 190, 196, 199, 261, 281, 284, 286, 321, 322, 323, 324, 325, 326, 328, 329, 338, 344, 347, 352, 357, 358, 364, 365
Dilation, 43, 168, 175, 176, 177, 179, 180, 181, 182, 183, 197, 242, 278, 281, 282, 284, 316, 321, 329, 347
Dissolved air, 14, 17, 33
Double-cell total volume measurement system, 151–152, 154
Drainage condition, 90
Drainage layer, 423–424, 441, 448
Drying-induced shrinkage, 326–327

Earth pressure, 146, 376, 382, 384, 394, 395, 398, 404, 405, 412, 456
Effective cohesion, 146, 454–455
Effective degree of saturation, 178, 179, 334
Effective mean stress, 76, 77, 79, 195, 196, 198, 353, 366
Effective stress, 1, 2, 3, 8, 9, 10, 11, 13, 75, 79, 147, 175, 178, 180, 181, 183, 201, 202, 210, 213, 216, 261, 301, 317, 318, 319, 330, 331, 334, 345, 395, 451, 455, 459, 461
Effective stress parameter, 9, 179
Elastic behaviour, 255, 362, 367
Elastic moduli, 353
Elastic strain, 338, 352, 353, 354

Elastic threshold, 221, 227, 253, 255, 257, 258, 336, 356
Elastoplasticity, 345, 351

Fabric, 138, 181, 203, 212, 213, 214, 235, 298, 330
Failure criterion, 146, 150, 181, 199, 261
Failure envelope, 49, 51, 146, 147, 148, 149, 150, 158, 159
Fine-grained soil, 63, 112, 118–119, 122, 138, 178, 188–189, 251, 257, 259, 260–261, 267, 421–422, 424, 429, 455
Flexible wall, 112, 130–131, 133, 143, 429, 432
Flow path, 66, 67, 91, 92, 94, 108, 113, 115, 140, 418, 446
Flow rule, 162, 163, 170, 322, 358, 359, 361, 366, 369
Freeze-dried, 185

Gauge pressure, 3, 13, 36–37
Geological and hydrogeological model, 401, 417
Geomaterial, 170, 188, 330, 421
Geomembrane, 118, 119, 420, 424, 425, 428, 429, 442, 446, 448, 450, 451, 460
Groundwater flow, 96, 411, 413, 417, 418, 419, 420

Hall-effect transducers, 205, 209, 219, 223, 241, 287
Hardening effects, 282, 284, 286, 287
Hardening law, 352, 359, 365–366, 369
Hardening parameter, 361
Heating and cooling, 308, 310, 312, 313, 315, 354
Heating-induced plastic strain, 287
Heating-induced volumetric strain, 281
Heating-induced yielding, 310
High-capacity suction probe, 218
Hydraulic conductivity, 65
Hydraulic gradient, 86, 87, 125, 391, 420, 442
Hydraulic head, 65, 87, 88, 433, 434
Hydraulic hysteresis, 13, 342, 343, 350, 351, 354, 358, 375, 442, 451
Hydraulic properties, 124, 272, 424, 450
Hydrogeological model, 401, 403, 417
Hydromechanical behaviour, 317, 328
Hydromechanical coupling, 316
Hydrostatic condition, 3, 390, 408, 409
Hydrostatic line, 4, 98, 125, 457
Hydrostatic pressure, 98

Illite, 155, 312, 315, 381
Inclinometer, 376, 379, 381, 382, 384, 385, 396, 404, 405, 407, 414, 416
Infiltration, 4, 43, 88, 90, 106, 108, 124, 125, 376, 384, 385, 386, 387, 389, 390, 397, 401, 417, 418, 421, 422, 423, 424, 425, 428, 431, 434, 441, 442, 444, 445, 447, 448, 449, 450
Inner cell, 41–43, 70, 292–294

Intact loess, 171, 183, 185, 186, 187, 188, 189, 275, 276, 286, 291, 292, 297, 298, 299, 301, 302, 303, 304, 306, 307, 308, 315
Intact specimen, 186, 188, 250, 253, 273, 274, 276, 277, 278, 280, 281, 282, 283, 284, 286, 287, 293, 297, 298, 301, 302, 303, 304, 306, 307, 308, 310, 312, 313, 315
Inter-aggregate pores, 76, 193, 195, 197, 313
Intergranular stress, 330, 331
Interlocking, 193, 195, 197, 198, 199
Inter-particle bonding, 202
Inter-particle contact, 236, 244, 260, 298, 306–307
Inter-particle contact areas, 306
Inter-particle contact force, 244
Inter-particle force, 11, 267
Inter-particle normal force, 249, 291, 298, 307, 336, 346, 347
Intra-aggregate pore, 76, 189, 193, 276, 278

Jet fill tensiometer, 427–428

Macropore, 76, 185, 186, 188, 189, 445, 446
Macrovoid, 183, 185, 187, 188, 189, 190
Matric suction, 1, 2, 7, 9, 11, 12, 15, 19, 20, 23, 24, 25, 26, 27, 28, 29, 33, 35, 36, 37, 38, 40, 43, 44, 47, 48, 55, 58, 59, 60, 65, 67, 72, 73, 75, 79, 81, 82, 84, 85, 86, 90, 91, 94, 95, 101, 102, 103, 104, 105, 106, 107, 108, 113, 123, 139, 140, 142, 146, 147, 148, 149, 150, 155, 161, 162, 165, 169, 202, 203, 204, 207, 208, 209, 210, 211, 212, 213, 215, 216, 217, 230, 231, 232, 233, 234, 235, 236, 238, 239, 240, 242, 244, 247, 267, 271, 280, 281, 297, 300, 302, 317, 357, 360, 361, 382, 384, 407, 408, 412, 418, 423, 429, 455
Mechanical behaviour, 12, 59, 148, 190, 265, 269, 316, 317, 324, 339, 344, 345
Mechanical loading, 282
Mechanical model, 221, 266, 289, 307, 312, 314, 345, 353
Menisci, 10, 210, 211, 257, 288, 320
Meniscus water, 10, 59, 60, 221, 223, 227, 229, 233, 236, 245, 250, 251, 260, 265, 266, 269, 270, 288, 289, 296, 330, 336, 347, 349, 357
Mercury intrusion porosimetry, 185, 257
Metastability, 16
Microscopic analysis, 322
Microstructure, 77, 91, 103, 137, 183, 185, 188, 189, 191, 193, 199, 249, 252, 255, 256, 257, 258, 261, 272, 274, 306, 329, 330, 398
Microvoid, 183, 185, 186, 188, 189
Mineralogy, 191, 197
Modulus, 74, 75, 201, 202, 203, 212, 213, 215, 216, 217, 219, 220, 221, 222, 223, 224, 225, 226, 227, 228, 229, 230, 232, 233, 234, 236, 237, 238, 240, 241, 242, 243, 244, 245, 246, 248, 249, 250, 251, 252, 253, 254, 255, 256, 257, 258, 260, 261, 262, 267, 268, 269, 270, 271, 287, 288, 291, 315, 321, 324, 329, 334, 335, 336, 345, 352, 353, 360, 362, 363, 367, 368, 369, 370, 371, 372, 375, 454, 456, 459
Mohr-Coulomb failure criterion, 146, 150
Mohr-Coulomb failure envelope, 146, 147–149
Montmorillonite, 5, 153, 379

Negative pore-water pressure, 2, 12, 30, 36, 51, 56, 140, 398
Non-associated flow rule, 162, 322
Nonlinearity, 33, 79, 243, 247, 261, 367
Nucleation, 16
Nuclei, 14, 16–18, 143
Numerical analysis, 343, 426, 433

Odour, 140
Oedometer, 38, 48, 53, 54, 173, 174, 312
One-dimensional column, 121, 445
One-dimensional downward flow, 96
Optimum moisture content, 100, 113, 135, 428
Optimum moisture content, 100, 113, 135, 427, 428
Optimum water content, 113, 163, 165, 172, 192, 206, 270, 275, 276
Osmotic solution, 47, 48
Osmotic suction, 7, 19, 34, 35, 47, 48, 64, 151, 154
Osmotic technique, 36, 47, 48, 49, 50, 51, 54, 55, 58, 60, 61, 63, 64, 150, 151, 155, 157, 159, 160, 161, 162, 199, 261
Over-consolidated, 246–247, 266–267, 289, 322, 324, 327, 380
Over-consolidation ratio, 215, 266, 267, 278, 284

Passive earth failure, 395
Passive failure conditions, 395
Path-dependent deformation characteristics and stiffness, 200
Pavement, 201, 237, 262, 265, 269, 270, 271
Peak dilation, 179
Peak dilation rate, 179
Peak shear strength, 175, 176, 178, 179, 281, 282, 286
Permeability, 5, 33, 65, 66, 67, 68, 69, 71, 73, 75, 77, 79, 81, 83, 85, 86, 87, 88, 89, 90, 91, 92, 93, 94, 95, 97, 98, 99, 101, 103, 105, 106, 107, 108, 109, 111, 112, 113, 115, 117, 118, 119, 120, 121, 122, 123, 124, 125, 127, 128, 129, 130–131, 133, 134, 135, 136, 137, 138, 139, 140, 141, 142, 143, 144, 145, 387, 390, 392, 408, 417, 418, 420, 421, 422, 423, 424, 425, 426, 428, 429, 430, 431, 432, 433, 434, 437, 441, 442, 445, 447, 448, 450
Permeameter, 112, 130–131, 133, 143, 429, 432
Plant, 4, 22, 133, 140, 142, 143, 425, 426, 428, 429, 433, 434, 440, 441, 444, 446, 447, 449, 450
Plant characteristics, 450
Plant leaves, 22, 433
Plant roots, 142, 425, 449, 450
Plastic behaviour, 74, 258
Plastic contraction, 281, 284, 309, 312, 313

Plastic deformation, 75
Plastic limit, 100, 113, 135, 153, 163, 192, 206, 379, 406, 428
Plastic mechanism, 352, 354
Plastic modulus, 345, 360, 369, 370
Plastic potential, 322, 324
Plastic strain, 10, 246, 262, 265, 266, 267, 269, 270, 271, 282, 284, 287, 302, 312, 327, 351, 352, 359, 360, 369, 370, 375
Plaxis 2D, 451, 455
Poisson's ratio, 237, 456
Pore-air pressure, 66, 75, 111, 112, 114, 122, 124, 126–127, 146–147, 151, 157, 166, 178, 208, 217, 408, 455
Pore-size distribution, 24, 73–77, 82, 94, 137–138, 185, 340
Pore-water energy, 20
Pore-water pressure, 2–5, 9, 11–13, 27, 30, 36–37, 41, 44, 47, 51, 56, 59, 65–66, 68, 75, 86, 95, 98, 100, 122, 124–127, 140, 146–147, 151, 164, 166, 208, 217–218, 240, 262, 264–265, 274, 281, 386–388, 390–391, 394, 396, 398, 426, 437–439, 441, 455
Porous ceramic, 27, 28, 30, 33, 34
Porous stone, 17, 38, 40, 263, 294
Post-yield, 174, 193
Precipitation, 3, 5, 143, 145, 417, 419, 421, 435, 447
Pre-consolidation pressure, 267, 268, 284, 310, 312–313, 348–349, 354
Preferential flow, 108, 113, 130, 446
Psychrometer, 20–22, 35, 51

Radial deformation, 38
Radial strain, 90, 209, 230, 231, 232, 233, 236, 241
Rainfall-induced landslides, 401
Rainstorm, 405, 406, 408, 409, 410, 411, 412, 413, 415, 417, 418, 419
Receiver element, 204
Recompacted soil, 74, 252, 272, 298, 307, 309, 313
Reconstituted soil, 304
Relative compaction, 82, 312
Relative density, 181
Relative humidity, 7, 20, 22, 51, 52, 53, 54, 60, 170, 171, 423, 427, 436, 437
Relative movement, 43
Relative permeability, 122
Residual degree of saturation, 257, 319
Residual friction angle, 170
Residual shear strength, 178
Residual shrinkage zone, 173
Residual soils, 188
Residual suction, 71, 72, 163, 166, 168, 169, 170, 178, 185, 255, 260
Resilient modulus, 237, 238, 240, 241, 242, 243, 244, 245, 246, 248, 249, 262, 267, 268, 269, 270, 271
Resonant column, 201, 250, 258, 270, 329, 330
Retaining wall, 146, 200, 201, 377, 379

Retention behaviour, 40, 75, 78, 79, 80, 82, 83, 85, 172, 316, 339, 342, 343, 344, 351, 362
Retention characteristics, 65, 67, 69, 71, 73, 75, 77, 79, 81, 83, 85, 87, 89, 91, 93, 95, 97, 99, 101, 103, 105, 107, 109, 111, 113, 115, 117, 119, 121, 123, 125, 127, 129, 131, 133, 135, 137, 139, 141, 143, 145, 183
Retention curve, 13, 38, 61, 64, 65, 66, 68, 69, 72, 73, 123, 172, 179, 181, 183, 350, 353, 362, 424, 429
Root depth, 447
Root distribution, 433
Root length, 435
Root water uptake, 440, 441, 444
Runoff, 385, 401, 424, 428, 446, 448, 449, 450

Saline soils, 24
Sampling method, 246
Sampling quality, 253
Sand, 14, 55, 62, 63, 80, 82, 83, 84, 100, 110, 120, 121, 122, 123, 124, 125, 126, 127, 128, 130, 138, 141, 143, 145, 153, 163, 171, 178, 181, 191, 192, 195, 196, 198, 206, 221, 249, 255, 257, 260, 261, 312, 321, 322, 324, 329, 330, 335, 337, 381, 384, 394, 396, 403, 406, 410, 421, 423, 424, 425, 427
Saprolite, 403
Saturated condition, 12, 46, 66, 118, 167, 323, 401
Scanning curves, 26, 27, 28, 29, 343
Scanning electron microscope, 143, 257, 272, 308
SDSWCC, 37, 38, 40, 69, 73, 95, 97, 101, 102, 103, 104, 105, 108, 171, 207, 231, 233, 235, 236
SDWRC, 37, 38, 69, 73
Semi-permeable membrane, 48, 153, 157, 161
Serviceability limit state, 110
Sesquioxide, 191, 193, 197–198
Shallow failures, 396
Shear behaviour, 170, 188, 189, 191, 198, 219, 221, 223, 225, 227, 262, 272, 273, 277, 278, 279, 281, 283, 285, 286, 315
Shear box, 41, 43, 45, 151, 154, 163, 164, 165, 166, 167, 170, 171, 272, 273
Shear modulus, 74, 75, 201, 202, 203, 212, 213, 215, 216, 217, 219, 220, 221, 222, 223, 224, 225, 226, 227, 228, 229, 230, 232, 233, 234, 236, 237, 246, 249, 250, 251, 252, 253, 254, 255, 256, 257, 258, 260, 261, 287, 288, 291, 315, 321, 329, 334, 335, 336, 345, 353, 362, 363, 367, 368, 371, 372, 375, 459
Shear modulus degradation, 250, 251, 252, 253, 255, 256, 258, 261, 362, 367
Shear modulus reduction, 74, 220, 221, 222, 223, 224, 225, 226, 227, 228, 229, 230, 260
Shear stiffness, 200, 201, 203, 205, 207, 209, 211, 213, 214, 215, 217, 219, 221, 223, 225, 227, 229, 230, 231, 233, 235, 236, 237, 239, 241, 243, 245, 247, 249, 250, 251, 253, 255, 256, 257, 259, 261, 278, 280, 281, 282, 284, 286, 288, 451, 453, 454, 455

Subject index

Shear strain, 163, 201, 216, 219, 220, 221, 222, 223, 225, 226, 227, 228, 229, 250, 251, 252, 253, 254, 255, 256, 257, 258, 259, 260, 261, 287, 288, 289, 290, 291, 351, 352, 359, 362, 365, 366, 367, 368, 369, 370, 371, 372, 373, 374, 451, 453, 454, 455
Shear strain threshold, 256, 258, 260, 261
Shear strength, 3, 8, 37, 49, 64, 146, 147, 148, 149, 150, 151, 153, 155, 157, 158, 159, 160, 161, 162, 163, 165, 167, 169, 170, 171, 173, 174, 175, 176, 177, 178, 179, 180, 181, 182, 183, 185, 187, 188, 189, 190, 191, 193, 195, 196, 197, 198, 199, 239, 261, 280, 281, 282, 284, 286, 316, 347, 390, 395, 400, 427
Shear stress, 45, 56, 146, 147, 148, 149, 176, 177, 274, 277, 281, 284, 367, 368, 371, 372, 454
Shear tests, 36, 46, 55, 56, 58, 150, 151, 155, 170, 183, 191, 239, 264, 270, 272, 273, 274, 277, 286, 321, 341, 360, 361, 365, 427
Shear wave, 200, 201, 202, 203, 204, 207, 208, 209, 210, 211, 212, 213, 229, 230, 231, 232, 233
Shrinkage, 4, 5, 172, 173, 185, 244, 252, 286, 326, 327, 376, 380, 396, 411, 424
Sieving, 131, 163, 191, 197, 275
Silicon, 192, 195, 204, 205
Silt, 55, 74, 80, 81, 82, 83, 84, 85, 86, 91, 94, 95, 96, 100, 107, 110, 120, 122, 123, 124, 125, 126, 127, 128, 129, 138, 141, 142, 153, 163, 164, 166, 168, 169, 170, 171, 178, 181, 188, 189, 191, 192, 198, 206, 238, 249, 257, 259, 260, 261, 262, 270, 274, 275, 276, 277, 282, 284, 286, 298, 300, 304, 306, 311, 312, 314, 315, 319, 320, 321, 326, 327, 328, 329, 330, 334, 336, 339, 341, 375, 381, 403, 406, 421, 423, 424, 425, 427, 452, 453, 454, 456
Slope, 1, 40, 43, 53, 65, 68, 71, 72, 73, 88, 99, 121, 142, 145, 146, 148, 163, 182, 193, 219, 243, 245, 278, 288, 296, 316, 338, 348, 349, 350, 353, 362, 367, 376, 377, 379, 380, 381, 382, 384, 385, 386, 388, 389, 392, 394, 395, 396, 397, 398, 400, 401, 403, 405, 411, 412, 413, 414, 415, 416, 417, 418, 419, 420, 427, 428, 429, 431, 434, 460
Small-strain behaviour, 223, 225, 334, 335, 360, 369, 370, 372
Small-strain shear behaviour, 219, 221, 223, 225, 227
Small-strain shear moduli, 225, 232
Small-strain shear modulus, 215, 217, 227, 352, 367
Small-strain soil stiffness, 451–453, 455, 457, 459–460
Small-strain stiffness, 217–219, 316, 344, 451, 455, 460
Softening, 23, 46, 56, 167, 168, 183, 196, 265, 266, 267, 270, 277, 282, 284, 297, 298, 299, 300, 304, 305, 306, 307, 324, 395, 396
Soil-atmosphere interface, 448
Soil behaviour, 1, 10, 13, 60, 162, 168, 192, 193, 215, 245, 262, 263, 265, 267, 269, 270, 271, 272, 273, 275, 277, 279, 281, 283, 284, 285, 287, 289, 291, 293, 295, 297, 299, 301, 303, 305, 307, 309, 311, 312, 313, 315, 316, 319, 338, 345, 346, 356, 361, 375
Soil-biopolymer interaction, 137, 140
Soil classification, 96, 113, 153, 191, 192, 206, 406, 428, 429
Soil density, 74, 134, 139, 169, 203, 242, 246, 339, 381
Soil-plant-water-atmosphere, 433
Soil-structure interface, 456
Soil-water characteristic curve, 13, 38, 68–70, 73, 106, 114, 145, 168, 207, 381–382
Soil-water retention curve, 183, 424
South-to-North water diversion project, 376–378
Specific air-water interface, 251
Specific entropy, 60
Specific gravity, 100, 112, 113, 135, 192, 203, 206, 346, 404, 406, 428, 429
Specific volume, 6, 60, 160, 162, 305, 346, 370
Specimen preparation method, 275, 278
Squeezing technique, 34
Standard compaction curve, 428
Standard compaction tests, 206
State-dependent behaviour, 316–317, 319, 321, 323, 325, 327, 329, 331, 333, 335, 337, 339, 341, 343, 345, 347, 349, 351, 353, 355, 357, 359, 361, 363, 365, 367, 369, 371, 373, 375
State-dependent dilatancy, 163, 323, 328, 344
State-dependent hydro-mechanical behaviour, 344
State-dependent model, 375
State parameter, 188, 189, 323, 346, 347, 348, 364, 370
State variables, 7, 8, 12, 13, 38, 40, 60, 61, 94, 146, 148, 150, 202, 212, 237, 238, 334, 377, 400, 401
Steady state, 110, 388, 390
Stiffening effects, 261, 320
Stiffness, 78, 175, 183, 200, 201, 203, 204, 205, 207, 209, 210, 211, 212, 213, 214, 215, 217, 218, 219, 221, 223, 225, 227, 229, 230, 231, 232, 233, 235, 236, 237, 238, 239, 241, 243, 245, 246, 247, 249, 250, 251, 253, 255, 256, 257, 259, 261, 267, 269, 270, 278, 280, 281, 282, 284, 286, 288, 289, 290, 316, 330, 331, 334, 335, 336, 344, 345, 353, 367, 369, 370, 397, 451, 452, 453, 454, 455, 456, 457, 458, 459, 460
Stiffness-strain relationship, 245
Strain-hardening, 277
Strain-softening, 277
Strength, 3, 8, 9, 13, 33, 36, 37, 41, 49, 58, 64, 146, 147, 148, 149, 150, 151, 153, 155, 157, 158, 159, 160, 161, 162, 163, 165, 167, 168, 169, 170, 171, 173, 174, 175, 176, 177, 178, 179, 180, 181, 182, 183, 185, 187, 188, 189, 190, 191, 193, 195, 196, 197, 198, 199, 239, 261, 280, 281, 282, 284, 286, 316, 322, 347, 390, 395, 400, 412, 427, 454
Stress-dependent small-strain soil stiffness, 453
Stress-dependent soil-water characteristic curve, 69, 145
Stress-dependent SWCC, 171, 207

Subject index

Stress-dependent SWRC, 338–339
Stress-induced anisotropy, 203
Stress-state variable, 79, 94
Stress-strain behaviour, 221, 223, 225, 227, 238, 243, 249, 269, 278, 282, 321, 367
Stress-strain curve, 219, 221, 225, 227, 288
Stress-strain relationship, 157, 158, 197, 216, 219–223, 224, 226, 228, 243, 354, 367, 375
Subgrade soil, 237, 238, 240, 244, 246, 262, 269, 271
Sub-loading surface, 322, 324
Sub-stepping stress point algorithm, 367
Subsurface water flow, 397, 400
Suction, 1, 2, 6, 7, 9, 10, 11, 12, 13, 15, 16, 17, 18, 19, 20, 21, 22, 23, 24, 25, 26, 27, 28, 29, 30, 31, 32, 33, 34, 35, 36, 37, 38, 39, 40, 41, 43, 44, 45, 46, 47, 48, 49, 51, 52, 53, 54, 55, 56, 57, 58, 59, 60, 61, 62, 63, 64, 65, 67, 68, 69, 71, 72, 73, 74, 75, 76, 78, 79, 80, 81, 82, 83, 84, 85, 86, 90, 91, 92, 94, 95, 98, 101, 102, 103, 104, 105, 106, 107, 108, 112, 113, 120, 123, 124, 125, 128, 129, 139, 140, 142, 146, 147, 148, 149, 150, 151, 154, 155, 156, 157, 158, 159, 160, 161, 162, 163, 164, 165, 166, 167, 168, 169, 170, 171, 172, 173, 174, 175, 177, 178, 179, 180, 181, 182, 183, 184, 185, 186, 187, 188, 189, 190, 199, 202, 203, 204, 207, 208, 209, 210, 211, 212, 213, 214, 215, 216, 217, 218, 219, 220, 221, 222, 223, 224, 225, 226, 227, 228, 229, 230, 231, 232, 233, 234, 235, 236, 237, 238, 239, 240, 241, 242, 243, 244, 245, 246, 247, 248, 249, 250, 251, 252, 253, 254, 255, 256, 257, 258, 259, 260, 261, 262, 263, 264, 265, 266, 267, 268, 269, 270, 271, 272, 273, 274, 275, 277, 278, 279, 280, 281, 282, 284, 286, 287, 288, 289, 290, 291, 292, 293, 294, 295, 296, 297, 298, 299, 300, 301, 302, 303, 307, 308, 309, 310, 313, 315, 316, 317, 318, 319, 320, 321, 322, 323, 324, 325, 326, 327, 328, 329, 330, 331, 332, 333, 334, 335, 336, 337, 338, 339, 341, 342, 343, 344, 345, 346, 347, 349, 351, 352, 353, 354, 356, 357, 358, 360, 361, 362, 363, 364, 365, 367, 368, 369, 370, 373, 375, 376, 377, 381, 382, 383, 384, 386, 387, 388, 389, 394, 396, 397, 398, 400, 401, 407, 408, 412, 418, 422, 423, 424, 425, 429, 431, 451, 452, 454, 455, 456, 457, 458, 459, 460
Suction-controlled apparatus, 330
Suction-dependent small-strain soil stiffness, 451, 452, 457, 459
Suction-dependent soil stiffness, 456, 458–460
Suction-induced anisotropy, 344
Suction-induced desiccation, 251, 253, 255–256, 259, 261
Suction-induced dilatancy, 322
Suction-induced hardening, 270, 313, 315
Suction-induced volume changes, 327–328
SWCC, 13, 37, 38, 40, 58, 68, 69, 71, 72, 73, 95, 97, 101, 102, 103, 104, 105, 106, 108, 153, 155, 163, 164, 165, 168, 171, 207, 231, 233, 235, 236, 381, 392

Tamping, 214
Tangent shear modulus, 329
Temperature control, 51, 53, 293, 308, 361
Tensile strength, 13, 33, 36, 322
Tensiometer, 15, 16, 18, 30, 31, 32, 33, 34, 35, 36, 51, 89, 95, 96, 99, 101, 106, 140, 141, 172, 376, 381, 382, 383, 384, 386, 387, 388, 390, 392, 404, 407, 427, 428, 431, 432
Tension, 4, 9, 14, 15, 16, 18, 19, 27, 28, 36, 149, 266, 268, 269, 270, 291, 297, 402, 403, 404, 441, 443, 444, 446
Terzaghi's effective stress, 1, 3, 75, 317, 345, 455
Thermal conductivity, 24, 25, 26, 35, 376, 381, 382, 383, 384, 386, 387, 388, 408
Thermal contraction, 310
Thermal effects, 61, 262, 263, 265, 266, 267, 268, 269, 270, 271, 273, 275, 277, 279, 281, 282, 283, 285, 286, 287, 288, 289, 290, 291, 292, 293, 295, 296, 297, 299, 300, 301, 303, 305, 307, 309, 311, 313, 315, 345, 346, 348, 349, 350, 362, 370, 375
Thermal equalization, 262–264, 274, 287, 293–294
Thermal expansion, 32, 297, 313, 352, 353
Thermal hardening, 284
Thermal loading, 264, 269, 274, 287
Thermal softening, 265, 266, 267, 282, 284, 297, 304, 305, 306, 307
Thermocouple, 20, 21, 22, 35, 262, 263, 272, 273, 292, 293, 294
Thermodynamic, 6, 51, 52, 59, 60, 62, 64, 345
Thermo-hydro-mechanical, 264, 267, 274–275, 293–295
Thermostat, 262, 263, 272, 273, 292, 293, 294
Thetaprobe, 89, 381, 382, 383, 384, 390, 392, 394
Time-domain reflectometry, 95

Ultimate limit state, 110, 115, 272
Unconfined compression, 170
Undrained condition, 197
Undrained shear strength, 395
Undrained triaxial compression, 335, 364
Unsaturated behaviour, 9
Unsaturated soil, 1, 2, 3, 4, 5, 6, 8, 9, 10, 11, 12, 13, 14, 15, 16, 18, 19, 20, 22, 24, 26, 28, 30, 32, 34, 36, 37, 38, 40, 41, 42, 43, 44, 45, 46, 47, 48, 50, 51, 52, 54, 56, 58, 59, 60, 61, 62, 64, 65, 66, 67, 68, 69, 70, 71, 72, 73, 74, 75, 76, 78, 79, 80, 82, 84, 85, 86, 87, 88, 90, 92, 94, 96, 98, 100, 102, 104, 106, 108, 109, 110, 112, 114, 116, 118, 120, 122, 124, 126, 128, 130, 132–133, 134, 136, 138, 140, 142, 144, 145, 147, 148, 149, 150, 152, 154, 156, 158, 160, 162, 164, 166, 167, 168, 170, 172, 174, 175, 176, 178, 180, 182, 183, 184, 186, 188, 190, 192, 194, 196, 198, 200, 202, 203, 204, 206, 208, 209, 210, 212, 213, 214, 215, 216, 218, 220, 222, 224, 226, 228, 230, 232, 234, 236, 238, 240, 242, 244, 245, 246, 248, 249, 250, 252, 254, 256, 257, 258, 260, 262, 264, 265, 266, 268, 269, 270, 272, 274, 276, 278,

280, 281, 282, 284, 286, 288, 289, 290, 292, 294, 296, 298, 300, 301, 302, 304, 306, 308, 310, 312, 314, 316, 317, 318, 319, 320, 322, 323, 324, 326, 327, 328, 330, 331, 332, 334, 336, 338, 339, 340, 342, 343, 344, 345, 346, 347, 348, 349, 350, 352, 353, 354, 356, 358, 360, 362, 364, 366, 368, 369, 370, 372, 374, 378, 380, 382, 384, 386, 388, 390, 392, 394, 396, 398, 400, 401, 402, 404, 406, 408, 410, 412, 414, 416, 418, 420, 422, 424, 426, 428, 430, 432, 433, 434, 436, 438, 440, 442, 444, 446, 448, 450, 451, 452, 454, 455, 456, 457, 458, 459, 460
Upper bound, 88, 324, 392

Vacuum gauge, 31–32, 34
Vacuum hand pump, 432
Vadose zone, 2–3, 5–6, 401
Vapour cavity, 15
Vapour equilibrium technique, 183, 249
Vapour pressure, 6, 14–15, 19–20, 22
Vegetated capillary barrier, 424
Vegetated ground, 425
Vegetated soil, 140, 142
Vegetation, 142, 423–424, 426, 450
Volumetric behaviour, 9, 308, 309, 313, 322
Volumetric contraction, 185, 286, 301, 302, 303, 304, 307, 312, 313
Volumetric pressure plate, 38, 39, 69, 70, 102, 153
Volumetric strain, 10, 43, 44, 163, 216, 225, 227, 241, 242, 252, 267, 274, 280, 281, 282, 284, 287, 290, 291, 299, 300, 301, 302, 303, 304, 307, 309, 310, 311, 313, 315, 316, 317, 320, 321, 326, 338, 344, 346, 351, 359
Volumetric strain hardening, 284
Volumetric water content, 65, 66, 69, 70, 72, 75, 87, 92, 94, 95, 98, 101, 105, 106, 112, 113, 114, 123, 124, 128, 130, 383, 390, 392, 393, 426, 429, 430, 442, 443, 444

Water adsorption, 54, 58, 257
Water balance, 426, 433, 448–450, 461
Water content, 23, 24, 36, 38, 54, 55, 57, 58, 61, 63, 64, 65, 66, 67, 68, 69, 70, 71, 72, 73, 75, 76, 82, 83, 87, 92, 94, 95, 98, 101, 105, 106, 111, 112, 113, 114, 118, 123, 124, 128, 129, 130, 133, 134, 135, 136, 137, 138, 139, 140, 143, 144, 145, 150, 155, 156, 163, 165, 166, 171, 172, 192, 203, 206, 219, 223, 225, 227, 229, 230, 233, 236, 240, 245, 246, 263, 270, 273, 275, 276, 278, 283, 285, 293, 309, 321, 329, 330, 334, 335, 346, 376, 379, 381, 382, 383, 384, 385, 390, 392, 393, 400, 404, 406, 407, 423, 426, 429, 430, 442, 443, 444
Water diversion, 376, 377, 378
Water flow, 23, 30, 40, 59, 65, 66, 67, 68, 86, 87, 88, 91, 96, 98, 208, 293, 387, 388, 389, 397, 398, 400, 411, 413, 417, 418, 419, 420, 433, 441, 443, 446
Water infiltration, 384, 421, 422, 423, 424, 425, 428, 431, 434, 441, 442, 444, 445, 447, 448, 450
Water permeability, 65, 66, 68, 87, 88, 97, 106, 109, 122, 123, 125, 128, 134, 143, 145, 387, 390, 417, 418, 421, 422, 423, 424, 428, 434, 441, 442, 445, 447, 448, 450
Wetting-drying cycles, 95, 97, 101, 102, 104, 106, 108, 215
Wetting-induced collapse, 172, 174, 291–292, 299–302, 304, 307, 313, 315, 316, 359
Wetting-induced plastic contraction, 281
Wetting-induced softening, 298–300, 304, 307
Wetting-induced strain, 303
Wetting-induced swelling, 9
Wetting-induced yielding, 300, 302–304

Xanthan gum, 134, 136, 138–139
Xiaping landfill, 426–427, 429, 450
X-ray, 52, 191–192, 194–195, 379

Yield characteristics, 299
Yield curve, 310–312, 322
Yield function, 324
Yield stress, 173, 174, 193, 209, 264, 265, 266, 267, 270, 275, 288, 289, 290, 292, 295, 296, 297, 298, 300, 301, 302, 304, 305, 306, 307, 322, 352
Yield surface, 162, 227, 265, 266, 267, 310, 311, 312, 313, 314, 320, 322, 324, 327, 337, 345, 356, 357, 360, 362, 370
Young–Laplace equation, 104
Young's modulus, 237, 245–246, 454, 456

www.ingramcontent.com/pod-product-compliance
Ingram Content Group UK Ltd.
Pitfield, Milton Keynes, MK11 3LW, UK
UKHW052108190125
453847UK00007B/81